MOLECULAR BIOLOGY PROBLEM SOLVER

MOLECULAR BIOLOGY PROBLEM SOLVER

A LABORATORY GUIDE

Edited by

Alan S. Gerstein

A JOHN WILEY & SONS, INC., PUBLICATION

New York • Chichester • Weinheim • Brisbane • Singapore • Toronto

Published simultaneously in Canada.

For ordering and cutomer service information please call 1-800-CALL-WILEY.

Library of Congress Cataloging-in-Publication Data:

Molecular biology problem solver: a laboratory guide / edited by Alan S. Gerstein.
p. cm.
Includes bibliographical references.
ISBN 0-471-37972-7 (pbk.)
1. Molecular biology—Methodology. 2. Molecular biology—Laboratory manuals.
I. Gerstein, Alan S., 1957–
QH506.M6629 2001
572.8′078—dc21 2001023491

Printed in the United States of America.

10 9 8 7 6 5 4 3

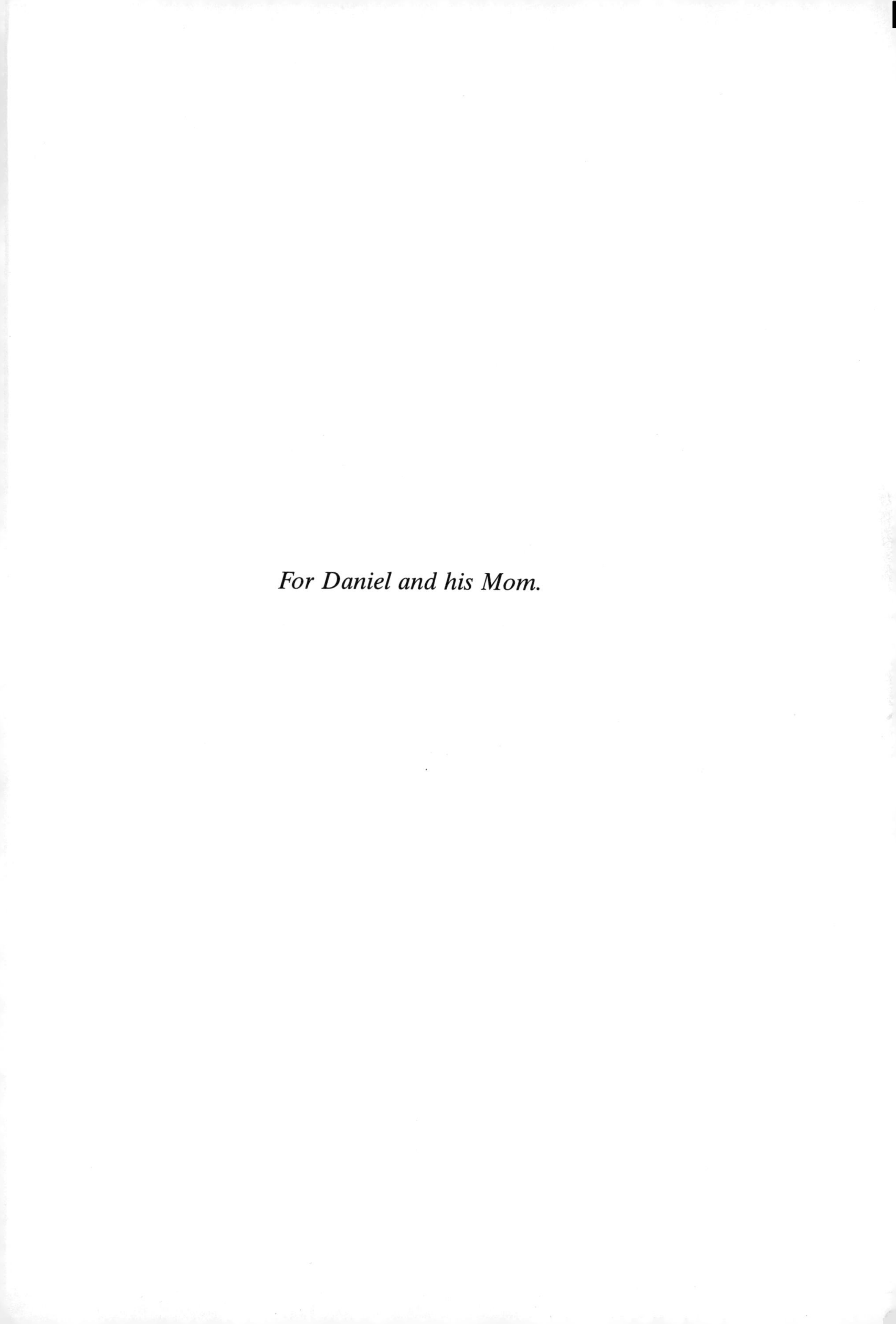

For Daniel and his Mom.

Tis better to ask some of the questions than to know all of the answers.

Unknown, Indiana

Contents

Preface ix

Contributors xi

Chapter 1. **Preparing for Success in the Laboratory**
Phillip P. Franciskovich 1

Chapter 2. **Getting What You Need from a Supplier**
Tom Tyre and Greg Krueger 11

Chapter 3. **The Preparation of Buffers and Other Solutions: A Chemist's Perspective**
Edward A. Pfannkoch 31

Chapter 4. **How to Properly Use and Maintain Laboratory Equipment**
Trevor Troutman, Kristin A. Prasauckas, Michele A. Kennedy, Jane Stevens, Michael G. Davies, and Andrew T. Dadd 49

Chapter 5. **Working Safely with Biological Samples**
Constantine G. Haidaris and Eartell J. Brownlow 113

Chapter 6. **Working Safely with Radioactive Materials**
William R. J. Volny Jr. 141

Chapter 7. **DNA Purification**
Sibylle Herzer 167

Chapter 8. **RNA Purification**
Lori A. Martin, Tiffany J. Smith, Dawn Obermoeller, Brian Bruner, Martin Kracklauer, and Subramanian Dharmaraj 197

Chapter 9. Restriction Endonucleases
Derek Robinson, Paul R. Walsh, and
Joseph A. Bonventre 225

Chapter 10. Nucleotides, Oligonucleotides, and Polynucleotides
Alan S. Gerstein 267

Chapter 11. PCR
Kazuko Aoyagi 291

Chapter 12. Electrophoresis
Martha L. Booz 331

Chapter 13. Western Blotting
Peter Riis 373

Chapter 14. Nucleic Acid Hybridization
Sibylle Herzer and David F. Englert 399

Chapter 15. *E. coli* Expression Systems
Peter A. Bell 461

Chapter 16. Eukaryotic Expression
John J. Trill, Robert Kirkpatrick, Allan R. Shatzman,
and Alice Marcy 491

Index 543

Preface

This book celebrates the importance of the question; it is not meant to be a collection of facts or procedures. The writing of this book was inspired by 16 years of queries from the research community. The contributors and I have tried to meet two primary objectives:

- Enhance the reader's ability to identify the critical elements of any technique, reagent, or procedure, in order to address questions for which documented answers might be unavailable.
- Clarify theory and practice that is taken for granted yet frequently misapplied.

Why is this book organized as a series of questions? For one, the researchers (and people in general) who greatly impress me are those who when faced with a seemingly impassible dilemma, can identify the question(s) that point the way to an eventual solution. Second, I'm fairly certain that I was most useful to others when all I did was to help them identify the questions that enabled them to solve their own problems.

Who should read this book? I can only say that the contributing authors, many of whom work within a technical support group or have previously done so, were asked to compose their chapters based on questions that they were chronically asked, or based on questions that they wish had been asked by those requesting assistance.

What are the strategies for working with this book? While I've been harping on the importance of the question, you might only have time to locate an answer. For readers in search of quick information, you might want to begin your search with a review of the index. A second approach would be to review the tables of content at the beginning of each chapter, which list the questions addressed within them.

I strongly recommend that at some point you read through a chapter of interest, focusing only on the questions being asked and the subheadings contained in the answer. The authors and I would like to think that this information and the questions they inspire will provide insight and perspective to help you solve problems that go beyond the content of this book.

So many friends, colleagues, and others who could have been classified as competitors gave selflessly to make this project a reality. It is all too likely that someone will be forgotten, and to all those individuals whose help I have not acknowledged, I sincerely apologize.

Among those that I remember are Peter Herzer, Billi Herzer, Martha Booz, Alice Marcy, Bob Dunst, Mary Ann Fink, George Donzella, Kathie Gorski, Lou Hosta, Claire Wheeler John Graziadei, Joseph Stencel, Phil Franciskovich, Tom Myers, Holly Hogrefe, Carl Baker, John SantaLucia, Patti Taranto, Phil Beckett (Cheers, me old mucker), Howard Coyer, Anita Gradowski, David Remeta, Cica Minetti, Peter Chiang, Martha Cole, Matt Szap, Barb Kaboard, Julie DeGregaro, and Paul Hoderlein for the invaluable service they provided by reviewing manuscripts. I am grateful to Terri Sunquist and colleagues at Promega Biotech for data on RNA polymerases, Bengt Bjellqvist for data on agarose, to Bjorn Lundgren for centrifugation data, to Bronwen Harvey and her research team for providing intriguing hybridization data, to Carl Fuller for sharing his contacts and enzyme expertise, and to Gene Stircak for access to search services. I am especially grateful to my colleagues at Amersham Pharmacia Biotech for their support, good wishes and collective sense of humor. With such a talented group of supporters and contributing authors, you can blame me for any inadequacies you note within these pages.

There are probably innumerable people at John Wiley and Sons that I should include, but I can only name a few. Ann Boyle and Virginia Benson Chanda have my sincere gratitude for their roles in converting an idea into a publication. Special thanks go to my editor, Luna Han, for her bottomless well of patience and professional guidance.

I wish I could thank my parents, Bernard and Florence, who urged me to focus on learning, not test scores. I'm glad I can thank my wife Sharon for her love and ability to ignore my mood swings during this project, and my son Daniel, whose uncanny knack for getting me out of bed before dawn hastened me to the finish line.

Alan Gerstein

P.S. Several authors (listed on page xi–xiii) provide their electronic mail address to receive your inquiries and comments. I would greatly appreciate your forwarding me (*mbproblemsolver@earthlink.net*) a copy of any correspondence you send to an author. Thank you in advance.

Contributors

Aoyagi, Kazuko, Millennium Pharmaceuticals, Inc., Cambridge, MA

Bell, Peter A., Orchid BioSciences, Inc., Princeton, NJ, pbell@orchid.com

Bonventre, Joseph A., New England Biolabs, Inc., Beverly, MA, info@neb.com

Booz, Martha L., Bio-Rad Laboratories, Hercules, CA, martha_booz@bio-rad.com

Brownlow, Eartell J., University of Cincinnati College of Medicine, Cincinnati, OH

Bruner, Brian, Ambion, Inc., Austin, TX

Dadd, Andrew T., Biochrom, LTD., Cambridge, UK

Davies, Michael G., Biochrom, LTD., Cambridge, UK, enquiries@biochrom.co.uk

Dharmaraj, Subramanian, Ambion, Inc., Austin, TX

Englert, David F., Packard Bioscience, Meriden, CT, support@packardinstrument.com

Franciskovich, Phillip P., Motorola Life Sciences, Tempe AZ, apf008@email.mot.com

Gerstein, Alan S., Amersham Pharmacia Biotech, Piscataway, NJ, Mbproblemsolver@earthlink.net, alan.gerstein@am.apbiotech.com

Haidaris, Constantine G., University of Rochester School of Medicine and Dentistry, Rochester, NY

Herzer, Sibylle, Amersham Pharmacia Biotech, Piscataway, NJ, sibylle.herzer@am.apbiotech.com

Kennedy, Michele A., Brinkmann Instruments, Inc., Westbury, NY, info@brinkmann.com

Kirkpatrick, Robert, GlaxoSmithKline, King of Prussia, PA

Kracklauer, Martin, Ambion, Inc., Austin, TX

Krueger, Gregory, Amersham Pharmacia Biotech, Piscataway, NJ

Obermoeller, Dawn, Ambion, Inc., Austin, TX

Marcy, Alice, Merck Research Labs, Rahway, NJ, alice_marcy@merck.com

Martin, Lori A., Ambion, Inc., Austin, TX, moinfo@ambion.com

Pfannkoch, Edward A., Gerstel Corporation, Baltimore, MD

Prasauckas, Kristin A., Packard Bioscience, Meriden, CT, kprasauckas@packardinst.com

Riis, Peter, Chicago, IL

Robinson, Derek, New England Biolabs, Beverly, MA

Shatzman, Alan R., GlaxoSmithKline, King of Prussia, PA

Smith, Tiffany J., Ambion, Inc., Austin, TX

Stevens, Jane, Thermo Orion, Beverly, MA, Domcsl@thermoorion.com

Trill, John J., GlaxoSmithKline, King of Prussia, PA

Troutman, Trevor, Sartorius Inc., Edgewood, NY

Tyre, Tom, Pierce Milwaukee, Milwaukee, WI,
Tom.tyre@piercenet.com

Volny, William R. J. Jr., Amersham Pharmacia Biotech, Piscataway, NJ

Walsh, Paul R., New England Biolabs, Beverly, MA

1

Preparing for Success in the Laboratory

Phillip P. Franciskovich

The Project 2
If You Don't Define the Project, the Project Will Define You 2
Which Research Style Best Fits Your Situation? 2
Do You Have the Essential Resources? 2
Expect the Unexpected 3
What If Things Go Better Than Expected? 4
When Has the Project Been Completed? 4
Was the Project a Success? 4
A Friendly Suggestion 4
The Research 5
Are Bad Data a Myth? 5
What Constitutes a Successful Outcome? 5
What Source of Data Would Be Most Compelling? 5
Do You Have the Expertise to Obtain These Types of Data? 5
What Can You Do to Maximize the Reliability of Your Data? 6
Are You on Schedule? 7
Which Variables Require Controls? 7
The Roles of Reporting 8
The Rewards 9
Bibliography 9

Molecular Biology Problem Solver, Edited by Alan S. Gerstein.
ISBN 0-471-37972-7

THE PROJECT

If You Don't Define the Project, the Project Will Define You

One of the first and toughest questions researchers must answer to foster success in the lab is: What do I have to accomplish? This requires you to understand your purpose to the larger task at hand. If your research is self-directed, the answer will most likely differ from that for someone working as part of a team effort or answering to an immediate supervisor or experimental designer. Ask them (or yourself) what the ultimate goals are and what constitutes a successful outcome. Establish what constitutes compelling evidence. By projecting ahead it becomes much easier to characterize the nature of the desired outcome.

This approach allows for problem reduction and reasonable task planning. The greatest mistake one can make is to react hastily to the pressures of the research by jumping in unprepared. By starting with the big picture, the stage is set for working backward and reducing what might otherwise appear to be a daunting undertaking into a series of reasonably achievable tasks. This exercise also establishes the criteria for making the many decisions that you will face during the course of your work.

Which Research Style Best Fits Your Situation?

Certain decisions will have a profound impact on the nature and quality of your efforts. Some scientists favor deliberate attention to detail, careful planning and execution of each experiment. Others emphasize taking risks, skipping ahead and plunging in for quick results. You might want to consider which approach would best satisfy your superior(s) and colleagues. Each of these "styles" has its benefits and risks, but a well-balanced approach takes advantage of each. Sometimes it is essential to obtain a quick answer to a question before committing a substantial amount of time to a more diligent data-collecting phase. Be sure everyone involved is in agreement and then plan your activities accordingly.

Do You Have the Essential Resources?

Evaluate your circumstances with a critical eye. Look at your schedule and that of your collaborators. Is everyone able to devote the time and energies this project will demand with a minimum of distractions? Check your facilities; do you have access to the materials and methods to do the job? Do you have the support of the decision-makers and budget managers for the duration of the work?

Whether or not problems were uncovered, share your findings with your director and collaborators; the objective of this phase is to build a consensus to proceed with no further changes.

Expect the Unexpected

How flexible is your research plan? Have you allowed yourself the freedom to adapt your strategy in light of unanticipated outcomes? This happens frequently and is not always bad news. Unexpected results might require slowing down the process or stopping altogether until a new path can be selected. Perhaps whole elements of the work might be skipped. In any case you should plan on midcourse corrections in your schedule. You can't always eliminate these redirections, but if you plan for them, you can avoid many unnecessary surprises. There are likely to be multiple paths to the desired outcome. If the unexpected occurs, consider categorizing problems as either technical or global. Technical problems are usually procedural in nature. The data obtained are either unreliable or untenable. In the former case the gathering of data may need to be repeated or the procedure optimized to the new conditions in order to increase data reliability. In the later case the procedure may prove to be inadequate and an alternative needs to be found. A global problem is one in which reliable data point you in a direction far removed from the original plan.

Technical problems are ultimately the responsibility of the principal investigators, so keep them informed. They might provide the solution, or refer you to another resource. Sometimes these problems can take forever to fix, so an upper limit should be agreed upon so that long delays will not be an unpleasant surprise to the other participants. Delays can be the source of much resentment among team members but should be considered an unavoidable consequence of research.

Global problems might require more drastic rethinking. The challenge for the investigator is to decide what constitutes a solvable technical glitch and what comprises a serious threat to the overall objectives. Experience is the best guide. If you have handled similar problems in the past, then you are the best judge. If you haven't, locate someone who has. In any case communicate your concerns to all involved parties as early as possible.

What If Things Go Better Than Expected?

How can you use good fortune to your best advantage? Most research triumphs are a blend of good times and bad. When good things happen during the course of your work, you may find yourself ahead of schedule or gaining confidence in the direction of your efforts. If you find yourself ahead of schedule, think ahead and use the extra time to stay ahead.

More often than not there will be subsequent phases of the work for which too little time has been allocated. Start the next step early or spend the time to address future problem areas of the plan. If the nature of the success you have achieved is to eliminate the necessity for some of the future work planned, you may be tempted to skip ahead. Such a change would constitute a significant departure from the original plan, so check with your superiors before proceeding on this altered course.

When Has the Project Been Completed?

A project will end when the basic objectives have been met. This view of the end is comforting in that you have specific objectives and a plan to achieve them, but disconcerting if the objectives change for reasons described above. If changes were controlled, discussed and documented throughout, endpoints should still be easy to identify. This is another reason why it is so important to establish a written consensus for each deviation in the plan.

Was the Project a Success?

If you stuck to your original plan and encountered no problems along the way, you were lucky. If problems required you to adapt your thinking, then real success was achieved. Remember, true failures are rare. The process of conducting research is one of constant evolution. If you have maintained an open mind and based your decisions on the facts uncovered by your work, your efforts were successful.

A Friendly Suggestion

If you are a new investigator or otherwise engaged in research that is new to you, take a lesson from the "old-timers." It's not that they have all the answers, it's just that they know how to ask better questions. They have had numerous opportunities to make their own mistakes, and if they have been successful, it is because they have learned from them.

THE RESEARCH

Are Bad Data a Myth?

Data are the medium of the scientific method, and can neither be good or bad. Data are the answers to the questions we pose, and it is the way we pose these questions that can be good or bad. Data could have intrinsic values: indeterminate, suggestive, or compelling in nature. Poorly posed questions often lead to indeterminate results, while exquisitely framed questions more often lead to compelling data. Therefore the secret to good research is in its design.

What Constitutes a Successful Outcome?

The answer to this question requires another: What are the specific objectives of your work? Must you produce a publication (basic research), a working model (industrial research), a reliable technique (applications research), or a prophetic example (intellectual property development)?

The specifications for success may vary significantly among these outcomes, so it might be worthwhile to verify your objectives with your supervisor or your collaborators.

What Source of Data Would Be Most Compelling?

If the answer isn't apparent, imagine yourself presenting data in front of a group of critical reviewers. What sort of questions or objections would you expect to hear? Answers to this question can be gleaned from seminars on topics similar to yours and from the scientific literature. The data published in peer-reviewed journals have stood up to the test of the review process and have been condensed to the most compelling evidence available to the author. You might also learn that the author applied an unexpected statistical analysis to support their conclusions.

Do You Have the Expertise to Obtain These Types of Data?

Do you have access to the specific equipment, materials, and methods necessary to perform your work? Finding access to one of these elements can provide access to the other, as can a network of friends and colleagues. Your desire for training might inspire someone to loan you the use of their equipment, along with their expertise.

What are your options if the equipment or expertise are unavailable to you? A review of the scientific literature might provide you with an alternative approach. For example, if technique *A* isn't available, the literature describing the development of that method will undoubtedly discuss techniques *B* and *C* and why they are inferior to technique *A*. Even if you have access to technique *A*, verifying your data via technique *B* or *C* might prove useful.

What Can You Do to Maximize the Reliability of Your Data?

Equipment and Reagents

Is your instrumentation working properly? When was it last checked for accuracy? An inaccurate spectrophotometer or pH meter could affect many aspects of your research. Do you possess all necessary reagents and have you proved their potency?

Have you considered your current and future sample needs? Will you employ statistical sampling in your experimental plans? You might save time, trouble, and money by analyzing your statistical sampling needs at the start of the project instead of returning to an earlier phase of the research to repeat a number of experiments. How will the data be collected, stored, and analyzed? How will statistics be applied, if at all?

Sample Issues

Replicates

A discussion about statistical analysis is beyond this book, but Motulsky (1995) provides practical guidance into the use of statistics in experimental design. Consider the use of statistics when determining the number of required replicates. Otherwise, you might find yourself returning to an earlier phase of your project just to repeat experiments for the purpose of statistical validation.

Quantity

How much material will you require over the short and long terms? Will the source of your material be available in the future, or is it rare and difficult to obtain? Will the physiological or chemical properties of the source change with time? What is the likelihood that the nature of your work will change, introducing new sample demands that require frequent sample preparations?

Should you prepare enough material in one episode to last the duration of your project? Sounds like a sure approach to minimize batch to batch variations, or is it? If the sample requirements make it practical to prepare an extraordinarily large amount of material, what do you know about the storage stability of the prepared material? Will chemical stabilizers interfere with the research now or in the future? Periodic control assays of material stored over a long term might prove helpful.

If the sample is subject to minimal batch-to-batch variation during preparation, then multiple small samplings may be the most convenient approach, for this provides an additional benefit of providing fresh sample.

If you can verify or control for the long-term stability of your sample, large-scale sample preparations are usually preferred, since most samples reflect the state of their source at the time that they are obtained.

Quality

Generally speaking, samples of high purity require much more starting material, so one approach to controlling demand on sample quantities is to establish the requisite levels of purity for your application. Many assays and experiments have some degree of tolerance for impurities and will work well with samples that are only moderately pure. If you test the usefulness of different sample purities in your research, you might uncover opportunities to reduce the required amount of sample.

Are You on Schedule?

You will likely be asked for precise estimates of when you plan to complete your work, or for time points of certain research milestones. The answers to the previous questions should provide you with the big picture of the research and how the individual parts could affect one another. An accurate sense of the overall timing of the research ahead should follow.

This is also a good point to search your memory, or that of a colleague who has done similar work, to identify potential pitfalls. The goal is to eliminate surprises that tend to get you off schedule.

Which Variables Require Controls?

Consider the converse question: Which variables don't require controls? You might have to switch sample origins, reagents, reagent manufacturers, or instrumentation. As discussed in Chapter 2, "Getting What You Need from a Supplier," suppliers don't always notify the research community of every modification to a commercial product. Even control materials require their own controls. As mentioned above, you'll want to have proof that your large quantity of frozen control material is not degrading with time. Considering the possible changes that can occur during the course of a research project, it's risky to conclude that there exists any variable that doesn't merit a carefully documented control.

The Roles of Reporting

When Should You Report Your Research Results?

In general, most project leaders and collaborators prefer to be kept informed, good news or bad. When your data are reproducible, discuss it with your research leader or senior colleague. These meetings also provide an opportunity to check that your colleagues' expectations for your research still coincide with your own.

If either party consistently appears surprised or misled, you might want to reevaluate the frequency and form of reporting. As discussed earlier, few research projects proceed exactly as planned, and these changes might require a change in the nature and scope of your reporting.

What Are Your Expectations When You Report Your Data?

Like most of life's endeavors, a research project begins with and is motivated by at least two very human desires. One desire is to uncover the truth no matter the outcome (the noblest case), while a second is to achieve our personal goals (the practical case). Research is not done in a vacuum and inherently contains biases. Consider these conscious and subconscious factors when you and your colleagues interpret data and offer conclusions.

Ideally there will be only one tenable interpretation of the data, but this is a rare outcome. Providing a fair treatment of the various interpretations in the report should lead to a dispassionate discussion that produces a consensus next step if not a conclusion.

What Are Your Options When Someone Attacks Your Data, Interpretations, or Conclusions?

The most common (and very human) initial response is to become defensive, to focus your energy on finding a weakness in your detractor's attack. A more productive route would be to welcome and embrace any contrary opinions. Pay attention to the details; make sure you thoroughly understand every aspect of their criticism. If you can objectively analyze your detractor's comments, the worst that can happen is for your research to be improved. One of the most productive phrases in the human vocabulary is the statement; "Maybe you're right, let's think about it some more." Unfortunately, it is also one of the most underutilized.

THE REWARDS

Money is usually not the sole motivator. The practice of science requires much patience, a willingness to take risks, and the ability to wait months or years for the rewards. This requires a special kind of personal and professional commitment. Why did you choose to practice science in the first place? Curiosity and awe in the workings of nature? "Science in the service of knowledge and society" might elicit chuckles from some within and outside the scientific community, and that's a shame.

Hopefully you will find ways to enjoy the scientific process on a daily level, working to achieve the big things while relishing small accomplishments. Consider the benefit of recognizing and rewarding the achievements of others and you, and by all means, have fun along the way.

BIBLIOGRAPHY

Motulsky, H. 1995. *Intuitive Biostatistics*. Oxford University Press, New York.

2

Getting What You Need from a Supplier

Tom Tyre and Greg Krueger

How Can You Work Most Efficiently With Your Supplier? . . . 12
All Companies Are the Same? . . . 12
Big Is Better, Small Is Better? . . . 12
Is the Product Manufactured by the Company That Sells It? . . . 13
Does a Company Test Every Application for a Product? . . . 13
How Well Will the Product Perform? . . . 13
Are Identical Products Manufactured Identically? . . . 14
Will a Company Inform You When They Change the Product? . . . 14
How Can You Work Most Efficiently—and Pleasantly—with a Sales Representative? . . . 14
What Can a Sales Rep Do for You? . . . 15
What Should You Expect from a Sales Rep? . . . 15
How Can You Get What You Want from a Sales Rep? . . . 16
Ordering a Custom Product . . . 18
Know Exactly What You Need . . . 18
Know Your Quantity Needs and Frequency of Delivery . . . 18
Know Your Spending Limits . . . 18
Document Your Needs . . . 18
Identify the Right Manufacturer . . . 19
Obtain a Document That Details the Order Acknowledgment . . . 19

Molecular Biology Problem Solver, Edited by Alan S. Gerstein.
ISBN 0-471-37972-7

Resolving Problems 19
Solving Problems by Yourself 20
Example of Using the Six Problem-Solving Steps: The DNA That Wasn't There 23
Solving Problems with the Help of the Supplier 25
Contacting the Supplier 26

HOW CAN YOU WORK MOST EFFICIENTLY WITH YOUR SUPPLIER?

Companies hire researchers, license ideas, generate much useful data that aren't always published, and fund scholarships. Familiarity with the corporate mindset, structure and resources can help you obtain what you need and avoid problems you don't want.

All Companies Are the Same?

All companies are not the same, and this fact is becoming truer everyday. Today a company selling research reagents may consist of a scientist turned entrepreneur working out of a home office. From a home in the midwest, the scientist might incorporate in Delaware. Once set up as a corporation, she may find someone else to make the wonder reagent in California and then arrange for some other company to package, label, and distribute the wonder reagent. No board rooms, no business lunches, and practically no one for a customer to complain to when things go wrong.

At the other end of the spectrum is a corporation doing business in 50 countries with sales in the hundreds of millions of dollars. Of course, with a well-known name on the tip of every scientist's tongue and a great reputation, super big company is much easier to find and much easier to reach for help you need. Don't count on it.

Each company has its own goals, dreams (i.e., visions) and personality. Within large companies, each division might have a distinct philosophy and operating strategy. Satisfaction with the products and services from an instrument division doesn't guarantee similar performance from a reagent division.

Big Is Better, Small Is Better?

Whether Big is better or Small is better depends on whether they fulfill your needs. Small will often have the greater desire, since even the smallest amount of business you send to them will be significant to Small's bottom line. But it will often lack the resources, knowledge, or external contacts to fulfill your needs that are out of the mainstream of its operation. This conflict may result in Small promising you something it can't deliver.

Big on the other hand will tend to have access to more internal and external resources. A special request may be easily within Big's knowledge and capacity to deliver. But how much are you willing to buy? If it isn't enough, Big won't have the incentive to do something unique for you. It just wouldn't make economic sense. If Big does its job right, you will quickly know it isn't willing to deliver, and you can go looking for another supplier.

Is the Product Manufactured by the Company That Sells It?

Some companies only sell products which they conceive, develop, and manufacture. Other suppliers only distribute products manufactured by other firms. Many, perhaps most companies, do some of both. The true manufacturer of a product may not be indicated on a package. If you are satisfied with the product's performance and support, its origin isn't an issue. But it may become an issue when problems arise, since the original manufacturer will generally have the most knowledge about the product.

Does a Company Test Every Application for a Product?

The research community regularly generates novel applications for commercial products. Combine this with limited application resources by suppliers, and the result is that a company tests only those applications it judges most important to the majority of the research community. If your application isn't mentioned by the manufacturer, odds are that application hasn't been tested or has been attempted an insignificant number of times.

It never hurts to contact the company. While the company may not have tested the product in your particular application, your call might persuade the company to do so. It is not uncommon for suppliers to provide product at little or no cost in exchange for application data generated by the customer. Manufacturers also might have a database of researchers who've attempted your application. The *Methods and Reagents* bulletin board located in the Biosci Web site (*http://www.bio.net/hypermail/methods/*) is a productive location to ask if a product has ever been tested in your particular application. This site can also help you locate and obtain hard-to-find reagents.

How Well Will the Product Perform?

As alluded to throughout this chapter, it is impossible for a supplier to guarantee the performance of every product with every sample source. But in today's competitive marketplace any reputable supplier will do its best to guarantee that advertising claims match actual performance. In addition third-party reviewers help ensure advertising claims aren't overly exaggerated. These third party (and hopefully objective) reviews of commercial products are provided at the following Web sites:

- The Scientist, *http://www.thescientist.com*
- BIOSCI Methods Group, *http://www.bio.net/hypermail/methods*
- Biowire, *http://www.biowire.com*
- Biocompare, *http://www.biocompare.com*

Are Identical Products Manufactured Identically?

When different companies seem to manufacture identical items, there may be differences in the production methods. For example, company *A* might quantitate the activity of *Taq* DNA polymerase after packaging because company *A*'s automated dispensing equipment might cause foaming of the protein and thus instability. Alternatively, company *B* may never test the activity of the *Taq* polymerase after packaging because it is manually dispensed, a procedure that doesn't harm the activity of the enzyme. The difficulty for you is that switching manufacturers may change performance more than you expect.

Will a Company Inform You When They Change the Product?

Manufacturers prefer not to change production strategies, but sometimes no choice exists: raw materials become unavailable, broken equipment can't be replaced, or people leave the company and take away the knowledge for synthesizing a product.

Changes are not always announced to the public. Responsible companies try to judge the impact of a change and determine its effect on the research community, but it is impossible to correctly predict the impact for everyone. If the change is thought to be significant, products might be labeled *New and Improved*, instructions might be changed, or packaging might be changed. If you're not sure if the changes will affect your research, contact the company and get the details of the modifications. The manufacturer might have experimental data that will help you evaluate their impact.

HOW CAN YOU WORK MOST EFFICIENTLY—AND PLEASANTLY—WITH A SALES REPRESENTATIVE?

The preceding section discussed the inner workings of equipment and reagent manufacturers. The next discussion focuses on strategies to manage your relationship with a company's sales representative (sales rep).

What Can a Sales Rep Do for You?

A good sales rep can help you determine what you need, what you don't need, and the most cost-effective way to get it. As a conduit to a company's administrative and scientific resources, a sales rep can help you resolve bureaucratic problems, receive technical information in a timely fashion, make sure you clearly understand all the nuances of a price quote, and help you obtain special order items.

What Should You Expect from a Sales Rep?

While you and your sales rep may think differently, you should be made to feel confident that advancing your research is important to your sales rep. Respect for you and your time, and the confidentiality of your research should also be maintained. As discussed below, good salespeople love to know "the inside scoop" and take personal pride in their customer's research, but you shouldn't have to worry that their exuberance for your work results in confidential details discussed with your competitors. The best way to determine a rep's trustworthiness is to discuss other work in the field. If you're suddenly learning details about the competition that you would never share with the outside world until papers are published, you have reason to wonder if your ideas are being similarly discussed. Discussions about what someone else is buying (unless the researcher has agreed to serve as a reference) also is cause for concern. You have every right to expect that even your most mundane dealings with the company are kept confidential.

Is it reasonable to expect your rep to be thoroughly familiar with the technical aspects of their products? If they represent a catalog of 13,000 items, probably not. If the product line is more limited and highly technical in nature, you should expect a high degree of technical competence. In either case a good rep employed by a company that truly cares about their customers should be able to deliver answers to any questions within two to three business days.

As is true with business in general, your sales representative is probably managing her territory by the Pareto principle. That is, 80% of her business comes from 20% of the customers. While the majority of reps want desperately to assist all customers and treat them equally, the reality is that the elite 20% are going to get the lion's share of her attention. This is simple survival, as losing all or part of the business at those key accounts is likely to cost her significant commissions, and quite possibly her job. This doesn't mean that you should ever feel like one of "the-less-than-elite" 80%. You should always feel like the only person in the world when working with a sales representative, and a cell phone ringing in a briefcase is not something you should have to deal with.

How Can You Get What You Want from a Sales Rep?

Understand Their Motivation

A sales rep has at his core a rational self-interest. That is, he must do the things that will benefit his performance and ensure survival. While some reps are self-centered, others recognize the interdependent relationship he has with his customers. Your success is his success, though the converse is not true.

Companies typically motivate their representatives through sales contests and commission structures. Top salespeople often receive paid vacations, and commissions are often structured to move certain product lines. It is true that sales positions are some of the best paid positions in a company, and most sales people are to some extent money-motivated. But you still have every right to expect that products are being offered to you because they will solve your problem and not because they will make your rep the most money.

The best salespeople truly enjoy helping others. They enjoy the bonds that are established, and revel in the feeling that they are "on the inside" regarding research. At their heart, many salespeople also have a "need to please," and they receive a real boost when they've done something for you and you've noticed. If you have criticism, also feel free to relate it, and express your expectation that something be done to improve the situation. While a poor rep may avoid you once you've complained, the good ones will recognize your comments as an opportunity to change your opinion of them, their company, and therefore create a satisfied customer that will likely buy more product. The need to please can be a great motivational tool to get what you need from your sales representative.

Manage the Relationship

Evaluating what you need from the company, and how you want those needs managed will maximize value from the relationship. Your sales rep doesn't know what to expect because every customer is different. He deals with multiple people at each account:

the researcher, purchasing, receiving, safety, and so on. The relationship with your sales representative is a lot like dating; it can be ruined by unexpressed expectations. For that reason it is imperative that you express exactly what you need from this person. Do you need to see her every week? Do you want to be on the top of the list for trying new products? Do you simply want to see them on your terms, that is, "don't call me, I'll call you?" There is nothing wrong with expressing your wants. Rather, you are giving direction to someone in desperate need of it.

A good sales representative keeps a profile on important customers. Items in that profile may include area of research, money available, general temperment, and if you tell him, exactly how you like to be handled. Your rep will appreciate this, since it provides him a chance to better manage *your* expectations. There may be ground rules he can't accept, such as a weekly visit. He may have distant accounts that will demand his time. Perhaps you can compromise on an email inquiry along with a bi-weekly visit. Don't wait to discuss your needs; tell him on his first visit, reach compromises if necessary, and start working together.

You'd never dream of running an experiment without proper controls and measurements, so why treat this vital relationship any differently? If you've laid out your expectations, you now have the means to evaluate your sales rep. Exceptional sales representatives will automatically measure themselves against your expressed wishes. Feel free to ask for evidence when you review the relationship. Good sales reps will have an answer ready.

You should expect to review the relationship at some regular interval. Perhaps your needs have changed, or you've noticed some slippage or improvement in the performance of your rep. Don't hesitate to ask for a quick meeting to reassess.

A sales rep can enhance the relationship if she helps you manage your expectations of the company. She may ask you to forecast repeat usage or estimate future needs as a way to give you current information on availability and delivery. If you need one liter of a reagent, and that volume represents three months' production for that company, your rep must help you manage your expectation for immediate delivery.

Leverage

Serving as a reference is a great way to gain influence with your sales rep. There is no sales tool more powerful than a satisfied customer. If you're happy with your representative, her product, and the company, offer to serve as a reference. Your sales rep will be delighted, and this could help get you preferred treatment. Don't hesitate to explain why you're making this offer, and what you expect in return. This is part of "negotiating the relationship," and you don't want to make such a generous offer without expecting something in return.

ORDERING A CUSTOM PRODUCT

A product whose composition or quantity differs from the catalog item may be considered custom by many manufacturers. Such specialized items tend to be expensive; the following suggestions are provided to help you obtain the desired item at minimal cost and aggravation.

Know Exactly What You Need

Vague specifications cause problems. If you call a company and ask for 100 liters of phosphate buffer at pH 7.5, will it matter how the pH is adjusted? Does it matter whether sodium or potassium phosphate is employed?

Complete and detailed communication with the manufacturer is crucial. You as the buyer must take charge to ensure that the company tells you what information must be provided, specifications, and all other details. Ideally a supplier will ask several detailed, and maybe obvious, questions in order to truly understand your needs. Be suspicious of companies that ask little and promise everything. Some custom products are simple to specify, but it might not be feasible to thoroughly describe complex, or novel, products. In these cases it may be helpful to describe to the manufacturer what you don't want as well as what you do.

Know Your Quantity Needs and Frequency of Delivery

Manufacturers can't determine cost, nor their ability to deliver the proposed product, without knowing accurate quantity requirements and the frequency of orders.

Know Your Spending Limits

Although you do not want to negotiate price immediately with a manufacturer, you should know what you are willing to pay for the custom product. This will shorten your list of prospective manufacturers.

Document Your Needs

A thorough, comprehensive record of your answers to the preceding questions will prove invaluable during your conversations with suppliers.

Identify the Right Manufacturer

Determining which suppliers truly want your business is not a perfect science. Some manufacturers will tell you when they can't fulfill your needs, while others will hint but won't say no outright. Hints can include requests to buy a very large minimum quantity, suggestions of alternate products, or the news that delivery is not possible for ages. The trick is that these comments aren't always hints to chase away your business, but legitimate technical or business concerns that can't be avoided. After dealing with two or three potential suppliers, you will be able to identify those suppliers who are serious about your business.

Obtain a Document That Details the Order Acknowledgment

Require the company to document, in excruciating detail, what they will produce for you. If the description isn't complete, detailed, and accurate, make the company do it again. If something goes wrong, it will be your only proof of what the supplier promised to do for you.

RESOLVING PROBLEMS

There is nothing inherently negative in the word "problem." Its origin lay in phrases meaning "anything thrown forward" or "to lay before." A problem provides opportunities to sharpen your research skills and ultimately improve the reliability of your data. Keeping an open mind and an inquisitive nature when problems arise will minimize your frustration and speed the problem's resolution.

Problem prevention is faster than problem solving. To prevent problems from occurring, read the information supplied with the product. Suppliers usually work hard to determine what information is required to successfully use a product. Although reading directions may be boring, it can prevent many problems.

If you wish to use a product in a way that is not clearly described by the directions, consider asking the supplier the following questions before proceeding with the experiment:

- Has the product been successfully applied to your intended application? Even when the answer is no, suppliers could help you determine the likelihood of success.
- Is it safe to modify the procedure in the manual? Changing the volumes of reagents, incubation times, sample preparation, temperature of reactions, or any number of other seemingly minor

changes to the procedure may have large effects on the results obtained. Before deviating from any of the directions supplied with a product, it is best to call the manufacturer and see if they have any information on the effects of making that change in the procedure.

- Can the storage conditions be modified? Storage temperatures other than what the supplier recommends may compromise the stability of the product. This is especially likely if the product undergoes a phase change when stored at an alternate temperature. In addition, and maybe more important, products may become hazardous when stored at the wrong temperature.

Solving Problems by Yourself

In a perfect world, six steps will solve any problem:

1. Define the problem. What do you see?

The first step in any problem-solving activity is to fully understand the nature of the problem without drawing conclusions. Understanding a problem consists of describing all the factual aspects about the problem. Do not try to determine what caused the problem. That step comes later in the process. If a PCR reaction failed to give a product, the problem description is simply that no product was obtained. You may have used a new brand of *Taq* DNA polymerase in the reaction, and this is an important fact to state. But it is not a good idea to immediately draw the conclusion that the correlation of the new brand of *Taq* polymerase and the lack of a product means that the enzyme is bad. Rather, a more thorough analysis of all of the parameters involved should be done. Did the reaction buffer change? Did the thermocycler function properly? Was the template DNA the same as previous reactions that worked? Was a different method of DNA preparation used? And so on. Once the problem is fully described with all reasonable parameters understood, then some simple, obvious causes can be ruled out.

2. List *all* the theoretical explanations that could cause the problem reported in step 1, including the obvious ones.

The majority of problems stem from the most likely causes. Before searching for the esoteric sources, rule out the most likely explanations. This step of troubleshooting is often aided by asking another person for some help. An outside party will have a different perspective on the situation and may think of an obvious cause that escapes you. Obvious causes of a problem are always hardest for the person closest to the problem to see. For example, is a piece of equipment plugged in? This is so obvious that it is an often overlooked source of equipment problems. In the *Taq* polymerase example above, the equivalent question asks if enzyme was added to the reaction. Both are very likely simple errors that can lead to endless hours of troubleshooting until they get accidentally stumbled over.

3. Gather all the data that you have regarding the problem.

Was the control tested?

Instruments are often supplied with a standard for verifying the operation of the equipment. Analogously, reagents kits are often supplied with a control sample. If you have never used an instrument or a kit previously, consider testing the standard or control supplied before proceeding with any experiments.

Standards and controls are also extremely valuable when things go wrong. If the standard or control was not used and a problem appears, the first experiment to do is to test the standard or control. Changing experimental variables will be a complete waste of time if an instrument is out of calibration or the kit has deteriorated in some way. In addition, if you ask the supplier for help, one of the first questions that the supplier of the instrument or the reagents will ask is whether the standard was tested or whether the control in the kit was used. If your answer is no, it is very likely that you will be asked to test the standard or control and then call back. The reason for the question is that the supplier is trying to determine whether their product is the cause of your headaches or whether some other experimental variable is the problem.

How long was the product stored and under what conditions?

Properly maintained, common laboratory instruments do not deteriorate over time during storage in dry conditions if protected from dust. Instruments need routine maintenance and regular calibration, but aging is not a typical problem.

On the other hand, many chemicals and biochemicals do deteriorate over time. This deterioration is often accelerated by improper storage conditions. Before using any chemicals or biochemicals, verify that the chemical has been stored under recommended conditions. If it has not, either do not use it or call the supplier to see if they have information on the effect of alternate storage conditions.

Even if the chemical/biochemical has been kept at proper storage conditions, it is a good idea to determine the approximate age of the chemical or biochemical. It is risky to use a reagent whose age can't be determined.

Manufacturers may or may not have expiration dates on their chemicals. If they don't, the manufacturer should still be able to tell you, when given the lot number, when the product was made, and some estimate for how long a chemical can be safely stored under recommended conditions.

When asking about expected shelf life, have a clear idea of what you really need to know and why you are asking. Many manufacturers have never performed formal stability tests on their products and therefore can only give you anecdotal information from their experience. In many cases this will be sufficient.

If there is a chance that your research may lead to a commercial product that will be regulated through cGMP (Federal Register 21 CFR parts 210, 211, and 820) regulations, determine if the information the supplier has will be sufficient for your needs. Also be aware that the manufacturers will only be able to give you information about their product, in their packaging, under their recommended storage conditions. If you take that chemical and prepare a buffer or any other type of formulation with the chemical, their information cannot be extrapolated to your use of the chemical, and you will need to be responsible for the stability data on your formulations.

For this very reason, if you ask a manufacturer for the storage stability of their reagent once it has been applied to a procedure, they are likely to respond to you that they don't know. This answer is not to be difficult but is to prevent giving misleading information. You are likely to get more useful information from the manufacturer if you explain why you need the information. The manufacturer will then be able to give you a more complete and useful answer.

Getting an expiration date from a manufacturer is only as helpful as knowing exactly how that date was derived and knowing what it means. Products that pass their expiration dates may very well be sufficiently active for your purposes, since the date may be very conservative. If you are performing noncritical work, it may be acceptable to use chemicals past their expiration dates, once you know how a manufacturer determined the expiration date applied to the package.

Was the procedure modified?

If you deviated from the manufacturer's instructions, be sure to be able to exactly describe all changes. Even the slightest deviation may lead to suboptimal results.

4. Eliminate explanations from step 2 based on the data described in step 3.

5. Design and execute experiments that address the remaining explanations.
6. Eliminate the remaining explanations from step 2 based on the new data generated from the experiments of step 5.

Example of Using the Six Problem-Solving Steps: The DNA That Wasn't There

Step 1. Define the problem. What do you see?

One hundred ng (as quantitated by spectrophotmetry) of a 500 bp DNA fragment were loaded onto a 1.0% agarose gel and electrophoresed under standard conditions; ethidium bromide staining revealed the marker bands, but not the 500 bp fragment of interest. No staining was observed in the wells. Also two different DNA markers ranging from 1000 to 100 bp were loaded on the gel and ran as expected.

Step 2. List all the theoretical explanations that could cause the observations in step 1, and don't forget the obvious ones. Don't worry about the feasibility of your explanations yet.

a. The DNA was destroyed by a nuclease contaminant.
b. The DNA never migrated away from the loading well.
c. The DNA ran off the gel.
d. DNA was never present in the loaded sample.

Step 3. What data do you have?

a. Two lanes with different DNA markers appeared as expected.
b. The same gel box, power supply, and ethidium bromide used in your work successfully visualized DNA before and after your experiment.
c. Your spectrophotometer correctly quantitated a series of DNA standards in a concentration range similar to your 500 bp sample.

Step 4. Eliminate explanations from step 2 based on the data described in step 3.

a. *The DNA was destroyed by a nuclease contaminant.*

Then why weren't the two different markers similarly digested?

b. *The DNA never migrated away from the loading well.*

Not likely. Ethidium bromide staining did not appear at the wells.

c. *The DNA ran off the gel.*

Some of the markers were smaller than 500 bp, and they didn't run off the gel.

d. *DNA was never present in the loaded sample.*

Spectrophotometer data suggest that DNA was present. The same spectrophotometer accurately calculated the concentration of other DNA samples.

Step 5. Design and execute experiments that address the remaining explanations.

At face value all the possible experimental explanations have been eliminated. Or have they? Perhaps we should take a closer look at the spectrophotometer data.

The spectrophotometer used in the experiment was programmed to report the concentration of the samples in micrograms per milliliter.

Concentration (μg/ml) 500 bp Fragment	
Reading 1	40
Reading 2	35
Reading 3	40

The data look reasonable and reproducible, but just to be thorough, let's look at the absorbance values at 260 nm from these readings.

Absorbance at 260 nm 500 bp Fragment	
Reading 1	0.008
Reading 2	0.007
Reading 3	0.008

Concentration calculation:

$$0.008A_{260} \times 50\,\mu\text{g/ml} \times 100 \text{ (dilution factor of sample)} = 40\,\mu\text{g/ml}$$

The samples were very dilute, outside the preferred range for correlating absorbance with concentration and possibly beyond the sensitivity of the spectrophotometer, as discussed in

Chapter 4, "How To Properly Use and Maintain Laboratory Equipment." Furthermore this sample was a 1:100 dilution of the stock material, increasing concern that the sample was too dilute for accurate quantitation.

Step 6. Eliminate the remaining explanations from step 2 based on data generated from the experiments of step 5.

Measure the absorbance at 260nm of a 1:10 and 1:100 dilution of the DNA sample.

Absorbance at 260nm

	1:10	1:100
Reading 1	0.006	0.008
Reading 2	0.008	0.007
Reading 3	0.009	0.008

The experiment generated nearly identical absorbance values for both dilutions, implying that the samples are below the sensitivity of the spectrophotometer. Repeat the absorbance measurements of the undiluted stock to determine an accurate concentration.

Solving Problems with the Help of the Supplier

Gather All Pertinent Product Information

Once you determine that the control or standard has failed, the product is not extremely old and you didn't modify procedures from those recommended by the manufacturer, it is time to start thinking about calling the supplier. But before picking up the phone, gather all the information that you will need. The supplier will want the product number and the batch or serial number.

If it is an instrument, the supplier will usually ask for the serial number. Ideally this number is best recorded when the equipment is first received. Once an instrument is installed, it may be practically impossible to get to the number because of the inaccessible place the manufacturer chose to put it.

Reagents do not typically have a serial number but will often have a lot or batch number. This number is key to the supplier because it will give them the information that they need to be able to determine when the product was made and to trace back to the original manufacturing records. These records will help the manufacturer determine whether anything unusual happened during the manufacture of the product that might be causing your problems.

Are Comparisons Truly Side-by-Side?

If you are planning to describe to the supplier comparison experiments you did to troubleshoot the problem, be prepared to describe the exact conditions of the experiment. The supplier will want to know whether any comparisons performed were truly side-by-side. A true side-by-side comparison is one in which all variables are identical except for one. For example, a problem might be that a first-strand cDNA synthesis reaction failed to yield first-strand cDNA after changing to a new vial of reverse transcriptase. On the surface it may seem that the two reactions are side-by-side. But, if the mRNA applied in the reaction is from a different preparation than the mRNA used in the successful reaction, then two variables are different—different vials of reverse transcriptase and different mRNA samples. If the mRNA was degraded in the second sample, this would cause the first-strand reaction to fail and make it appear as if the reverse transcriptase is at fault. Being able to accurately describe how similar comparisons truly are will speed the problem's resolution.

Contacting the Supplier

Who to Call?

When calling a company for help, don't have a preconceived idea on who you should be speaking with. Some companies may have you deal with research and development scientists, others with full time technical support people, and some may first have you deal with your salesperson before passing you on. Don't assume one way is better than another. Each method has its positives and negatives, and each when successfully implemented by the company should be able to get you the help you need. Asking a company to follow the method that you think is best may cause several problems for the company including lack of documentation of your call, inability to authorize credit if required, and general confusion by the person who initially handles your call.

Record All the Details of the Conversation

You will want to write down all the basic information about each person you deal with, including the person's full name, their department, and the date and time of your call. If the situation continues over several days or weeks, what seems like basic facts you can't possibly forget will start to blur. Keeping an accurate log of each contact will also increase your credibility with the company, a benefit if you ultimately need to pursue the issue with supervisors and managers.

Finally, recording the name of the department will be useful when trying to contact the same person in the future. Even small companies may have two people working who have very similar or identical names and the department name will help locate the correct "John Smith." In large companies, service calls may be routed to various parts of the country, and it will be impossible to contact the same person in the future without knowing the department or even the city where the representative works.

State the Problem, Not the Conclusions

Describing the facts of the problem and not stating your theories on the underlying cause has several benefits. First, it gets you an unbiased opinion from your supplier's representative. If you give your ideas on the underlying cause and the person agrees with you, you may not have gotten the person's best judgment of the situation. Second, calling a representative and stating that you know there is a problem with their product can make some people very defensive and uncooperative. This may result in both parties being angry. The company may lose a customer, but your problem won't be any closer to being resolved. Finally, by stating just the facts, it will help you keep an open mind to the information that the company representative is telling you.

Ask If Anything Has Changed with the Product

If you are experiencing a problem with a product with which you have a history of success, it is useful to investigate whether anything has changed with the product. If you ask the representative whether anything has changed with the product and the person quickly says no, follow up the question with a list of specific items. Ask whether raw materials, equipment used in manufacturing, product specifications, or employees making the product have changed. The point of being specific in asking what may have been modified is that the person on the phone may not consider the wide range of alterations that could affect product performance. By specifically listing various potential changes, you are more likely to get the person to fully investigate whether everything is identical about the product since the last time you bought it.

Let People Call You Back

Good answers to your questions often require further investigation by several parties. Your question is likely just one of 20 to 50 handled that day by the representative. The person might have to check records, speak with people having specific knowledge, or just quietly analyze what you have said and consider possible causes. Whatever the reason, it is to your advantage to let the person take the time and investigate further before calling you back.

Remember to Thank the Person

As obvious or silly as it may seem, thank the person who has been helping you on the phone. Even helpful people usually try a little harder to assist those who treat them well.

If the person is unhelpful or obnoxious, keeping a polite, professional approach will increase your credibility with company superiors who later get involved with the problem. Losing your cool will only make management feel that their employee was abused.

If You're Still Unhappy

Even after trying to get a problem resolved with the company, you may still be very unhappy with the results. You might not have been treated fairly, or perhaps your expectations about what the company could do for you were too high.

What Is Reasonable to Expect?

Generally, it is only reasonable to expect the company to reimburse you for the product purchased. A statement indicating this is typically included in catalogs and is often present in the invoice that arrived with your order. The statement will exclude liability for your time, other products you may have used, lost research time, or other real costs you incurred due to a product that failed. Expecting reimbursement for any of these items is very unlikely, even if the company finds it was at fault for causing your headaches. You may be able to negotiate more than the replacement cost of the product you bought, but it will definitively require negotiation.

Who to Complain To?

Often the representatives who work directly with customers have very little freedom in what they can do to satisfy a customer. If you request reimbursement or assistance beyond what is typical, you will need to work your way up the corporate hierarchy. (Also it never hurts to contact your sales representative in these situations; they might be anxious to serve as your advocate.) Ask to speak with a supervisor. If they don't directly solve your problem, they can usually help you find the appropriate people. In some cases, calling the president or the person responsible for the manufacturing site may get the best response. It will just take patience working up the corporate ladder until you find someone who has the authority and resources to give help beyond the ordinary.*

**Editor's note: Yelling rings most effectively in the ears of upper management, not low-level personnel.*

3

The Preparation of Buffers and Other Solutions: A Chemist's Perspective

Edward A. Pfannkoch

Buffers 32
Why Buffer? 32
Can You Substitute One Buffer for Another? 32
How Does a Buffer Control the pH of a Solution? 32
When Is a Buffer Not a Buffer? 33
What Are the Criteria to Consider When Selecting a Buffer? 33
What Can Generate an Incorrect or Unreliable Buffer? 35
What Is the Storage Lifetime of a Buffer? 37

Editor's note: Many, perhaps most, molecular biology procedures don't require perfection in the handling of reagents and solution preparation. When procedures fail and logical thinking produces a dead end, it might be worthwhile to carefully review your experimental reagents and their preparation. The author of this discussion is an extremely meticulous analytical chemist, not a molecular biologist. He describes the most frequent mistakes and misconceptions observed during two decades of experimentation that requires excruciating accuracy and reproducibility in reagent preparation.

Molecular Biology Problem Solver, Edited by Alan S. Gerstein.
ISBN 0-471-37972-7

Reagents 39
Which Grade of Reagent Does Your Experiment Require? 39
Should You Question the Purity of Your Reagents? 39
What Are Your Options for Storing Reagents? 40
Are All Refrigerators Created Equal? 41
Safe and Unsafe Storage in Refrigerators 41
What Grades of Water Are Commonly Available in the Lab? 42
When Is 18 MΩ Water Not 18 MΩ Water? 44
What Is the Initial pH of the Water? 44
What Organics Can Be Present in the Water? 45
What Other Problems Occur in Water Systems? 46
Bibliography 47

BUFFERS

Why Buffer?

The primary purpose of a buffer is to control the pH of the solution. Buffers can also play secondary roles in a system, such as controlling ionic strength or solvating species, perhaps even affecting protein or nucleic acid structure or activity. Buffers are used to stabilize nucleic acids, nucleic acid–protein complexes, proteins, and biochemical reactions (whose products might be used in subsequent biochemical reactions). Complex buffer systems are used in electrophoretic systems to control pH or establish pH gradients.

Can You Substitute One Buffer for Another?

It is rarely a good idea to change the buffer type—that is, an amine-type buffer (e.g., Tris) for an acid-type buffer (e.g., phosphate). Generally, this invites complications due to secondary effects of the buffer on the biomolecules in the system. If the purpose of the buffer is simply pH control, there is more latitude to substitute one buffer for another than if the buffer plays other important roles in the assay.

How Does a Buffer Control the pH of a Solution?

Buffers are solutions that contain mixtures of weak acids and bases that make them relatively resistant to pH change. Conceptually buffers provide a ready source of both acid and base to either provide additional H^+ if a reaction (process) consumes H^+, or combine with excess H^+ if a reaction generates acid.

The most common types of buffers are mixtures of weak acids and salts of their conjugate bases, for example, acetic acid/sodium acetate. In this system the dissociation of acetic acid can be written as

$$CH_3COOH \rightarrow CH_3COO^- + H^+$$

where the acid dissociation constant is defined as $K_a = [H^+][CH_3COO^-]/[H_3COOH]$.

Rearranging and taking the negative logarithm gives the more familiar form of the Henderson-Hasselbalch equation:

$$pH = pK + \log\frac{[CH_3COO^-]}{[CH_3COOH]}$$

Inspection of this equation provides several insights as to the functioning of a buffer.

When the concentrations of acid and conjugate base are equal, $\log(1) = 0$ and the pH of the resulting solution will be equal to the pK_a of the acid. The ratio of the concentrations of acid and conjugate base can differ by a factor of 10 in either direction, and the resulting pH will only change by 1 unit. This is how a buffer maintains pH stability in the solution.

To a first approximation, the pH of a buffer solution is independent of the absolute concentration of the buffer; the pH depends only on the ratio of the acid and conjugate base present. However, concentration of the buffer is important to buffer capacity, and is considered later in this chapter.

When Is a Buffer Not a Buffer?

Simply having a weak acid and the salt of its conjugate base present in a solution doesn't ensure that the buffer will act as a buffer. Buffers are most effective within ± 1 pH unit of their pK_a. Outside of that range the concentration of either the acid or its salt is so low as to provide little or no capacity for pH control. Common mistakes are to select buffers without regard to the pK_a of the buffer. Examples of this would be to try to use K_2HPO_4/KH_2PO_4 ($pK_a = 6.7$) to buffer a solution at pH 4, or to use acetic acid ($pK_a = 4.7$) to buffer near neutral pH.

What Are the Criteria to Consider When Selecting a Buffer?

Target pH

Of primary concern is the target pH of the solution. This narrows the possible choices to those buffers with pK_a values within 1 pH unit of the target pH.

Concentration or Buffer Capacity

Choosing the appropriate buffer concentration can be a little tricky depending on whether pH control is the only role of the buffer, or if ionic strength or other considerations also are important. When determining the appropriate concentration for pH control, the following rule of thumb can be used to estimate a reasonable starting concentration.

1. If the process or reaction in the system being buffered does not actively produce or consume protons (H^+), then choose a moderate buffer concentration of 50 to 100 mM.
2. If the process or reaction actively produces or consumes protons (H^+), then estimate the number of millimoles of H^+ that are involved in the process (if possible) and divide by the solution volume. Choose a buffer concentration at least 20× higher than the result of the estimation above.

The rationale behind these two steps is that a properly chosen buffer will have a 50:50 ratio of acid to base at the target pH, therefore you will have 10× the available capacity to consume or supply protons as needed. A 10% loss of acid (and corresponding increase in base species), and vice versa, results in a 20% change in the ratio ($[CH_3COO^-]/[CH_3COOH$ from the Henderson-Hasselbalch example above]) resulting in less than a 0.1 pH unit change, which is probably tolerable in the system. While most biomolecules can withstand the level of hydrolysis that might accompany such a change (especially near neutral pH), it is possible that the secondary and tertiary structures of bioactive molecules might be affected.

Chemical Compatibility

It is important to anticipate (or be able to diagnose) problems due to interaction of your buffer components with other solution components. Certain inorganic ions can form insoluble complexes with buffer components; for example, the presence of calcium will cause phosphate to precipitate as the insoluble calcium phosphate, and amines are known to strongly bind copper. The presence of significant levels of organic solvents can limit solubility of some inorganic buffers. Potassium phosphate, for example, is more readily soluble in some organic solutions than the corresponding sodium phosphate salt.

One classic example of a buffer precipitation problem occurred when a researcher was trying to prepare a sodium phosphate buffer for use with a tryptic digest, only to have the Ca^{2+} (a necessary enzyme cofactor) precipitate as $Ca_3(PO_4)_2$. Incompatibilities can also arise when a buffer component interacts with a surface. One example is the binding of amine-type buffers (i.e., Tris) to a silica-based chromatography packing.

Biochemical Compatibility

Is the buffer applied at an early stage of a research project compatible with a downstream step? A protein isolated in a buffer containing 10 mM Mg^{2+} appears innocuous, but this cation concentration could significantly affect the interaction between a regulatory protein and its target DNA as monitored by band-shift assay (Hennighausen and Lubon, 1987; *BandShift Kit Instruction Manual*, Amersham Pharmacia Biotech, 1994). Incompatible salts can be removed by dialysis or chromatography, but each manipulation adds time, cost, and usually reduces yield. Better to avoid a problem than to eliminate it downstream.

What Can Generate an Incorrect or Unreliable Buffer?

Buffer Salts

All buffer salts are not created equal. Care must be exercised when selecting a salt to prepare a buffer. If the protocol calls for an anhydrous salt, and the hydrated salt is used instead, the buffer concentration will be too low by the fraction of water present in the salt. This will reduce your buffer capacity, ionic strength, and can lead to unreliable results.

Most buffer salts are anhydrous, but many are hygroscopic—they will pick up water from the atmosphere from repeated opening of the container. Poorly stored anhydrous salts also will produce lower than expected buffer concentrations and reduced buffering capacity. It is always wise to record the lot number of the salts used to prepare a buffer, so the offending bottle can be tracked down if an error is suspected.

If a major pH adjustment is needed to obtain the correct pH of your buffer, check that the correct buffer salts were used, the ratios of the two salts weren't switched, and finally verify the calculations of the proper buffer salt ratios by applying the Henderson-Hasselbalch equation. If both the acid and base components of the buffer are solids, you can use the Henderson-Hasselbalch equation to determine the proper mass ratios to blend and give your target pH and concentration. When this ratio is actually prepared, your pH will usually need some minor adjustment, which should be very minor compared to the overall concentration of the buffer.

pH Adjustment

Ionic strength differences can arise from the buffer preparation procedure. For example, when preparing a 0.1 M acetate buffer of pH 4.2, was 0.1 mole of sodium acetate added to 900 ml of water, and then titrated to pH 4.2 with acetic acid before bringing to 1 L volume? If so, the acetate concentration will be significantly higher than 0.1 M. Or, was the pH overshot, necessitating the addition of dilute NaOH to bring the pH back to target, increasing the ionic strength due to excess sodium? The 0.1 M acetate buffer might have been prepared by dissolving 0.1 mole sodium acetate in 1 liter of water, and the pH adjusted to 4.2 with acetic acid. Under these circumstances the final acetate concentration is anyone's guess but it will be different from the first example above.

The best way to avoid altering the ionic concentration of a buffer is to prepare the buffer by blending the acid and conjugate base in molar proportions based on Henderson-Hasselbalch calculations such that the pH will be very near the target pH. This solution will then require only minimal pH adjustment. Dilute to within 5% to 10% of final volume, make any final pH adjustment, then bring to volume.

Generally, select a strong acid containing a counter-ion already present in the system (e.g., Cl^-, PO_4^{3+}, and OAc^-) to adjust a basic buffer. The strength (concentration) of the acid should be chosen so that a minimum (but easily and reproducibly delivered) volume is used to accomplish the pH adjustment. If overshooting the pH target is a problem, reduce the concentration of the acid being used. Likewise, choose a base that contains the cations already present or known to be innocuous in the assay (Na^+, K^+, etc.)

Solutions of strong acids and bases used for final pH adjustment usually are stable for long periods of time, but not forever. Was the NaOH used for pH adjustment prepared during the last ice age? Was it stored properly to exclude atmospheric CO_2, whose presence can slowly neutralize the base, producing sodium bicarbonate ($NaHCO_3$) which further alters the buffer properties and ionic strength of the solution?

Buffers from Stock Solutions

Stock solutions can be a quick and accurate way to store "buffer precursors." Preparing 10× to 100× concentrated buffer salts can simplify buffer preparation, and these concentrated solutions can also retard or prevent bacterial growth, extending almost indefinitely the shelf stability of the solutions.

The pH of the stock solutions should not be adjusted prior to dilution; the pH is the negative log of the H^+ ion concentration, so dilution by definition will result in a pH change. Always adjust the pH at the final buffer concentrations unless the procedure explicitly indicates that the diluted buffer is at an acceptable pH and ionic concentation, as in the case with some hybridization and electrophoresis buffers (Gallagher, 1999).

Filtration

In many applications a buffer salt solution is filtered prior to mixing with the other buffer components. An inappropriate filter can alter your solution if it binds with high affinity to one of the solution components. This is usually not as problematic with polar buffer salts as it can be with cofactors, vitamins, and the like. This effect is very clearly demonstrated when a solution is prepared with low levels of riboflavin. After filtering through a PTFE filter, the filter becomes bright yellow and the riboflavin disappears from the solution.

Incomplete Procedural Information

If you ask one hundred chemists to write down how to adjust the pH of a buffer, you'll probably receive one hundred answers, and only two that you can reproduce. It is simply tedious to describe in detail exactly how buffer solutions are prepared. When reading procedures, read them with an eye for detail: Are all details of the procedure spelled out, or are important aspects left out? The poor soul who tries to follow in the footsteps of those who have gone before too often finds the footsteps lead to a cliff. Recognizing the cliff before one plunges headlong over it is a learned art. A few prototypical signposts that can alert you of an impending large first step follow:

- Which salts were used to prepare the "pH 4 acetate buffer"? Sodium or potassium? What was the final concentration?
- Was pH adjustment done before or after the solution was brought to final volume?
- If the solution was filtered, what type of filter was used?
- What grade of water was used? What was the pH of the starting water source?

What Is the Storage Lifetime of a Buffer?

A stable buffer has the desired pH and buffer capacity intended when it was made. The most common causes of buffer failure are pH changes due to absorption of basic (or acidic) materials in the storage environment, and bacterial growth. Commercially prepared buffers should be stored in their original containers. The storage of individually prepared buffers is discussed below. The importance of adequate labeling, including preparation date, composition, pH, the preparer's name, and ideally a notebook number or other reference to the exact procedure used for the preparation, cannot be overemphasized.

Absorption of Bases

The most common base absorbed by acidic buffers is ammonia. Most acidic buffers should be stored in glass vessels. The common indicator of buffer being neutralized by base is failure to achieve the target pH. In acidic buffers the pH would end up too high.

Absorption of Acids

Basic buffers can readily absorb CO_2 from the atmosphere, forming bicarbonate, resulting in neutralization of the base. This is very common with strong bases (NaOH, KOH), but often the effect will be negligible unless the system is sensitive to the presence of bicarbonate (as are some ion chromatography systems) or the base is very old. If high concentrations of acids (e.g., acetic acid) are present in the local environment, basic buffers can be neutralized by these as well. A similar common problem is improper storage of a basic solution in glass. Since silicic materials are acidic and will be attacked and dissolved by bases, long-term storage of basic buffers in glass can lead to etching of the glass and neutralization of the base.

Microbial Contamination

Buffers in the near-neutral pH range can often readily support microbial growth. This is particularly true for phosphate-containing buffers. Common indicators of bacterial contamination are cloudiness of the solution and contamination of assays or plates.

Strategies for avoiding microbial contamination include sterilizing buffers, manipulating them using sterile technique, refrigerated storage, and maintaining stock solutions of sufficiently high ionic concentration. A concentration of 0.5 M works well for phosphate buffers. For analytical chemistry procedures, phosphate buffers in target concentration ranges (typically 0.1–0.5 M) should be refrigerated and kept no more than one week. Other buffers could often be stored longer, but usually not more than two weeks.

REAGENTS

Which Grade of Reagent Does Your Experiment Require?

Does your application require top-of-the-line quality, or will technical grade suffice? A good rule of thumb is that it is safer to substitute a higher grade of reagent for a lower grade, rather than vice versa. If you want to apply a lower grade reagent, test the substitution against the validated grade in parallel experiments.

Should You Question the Purity of Your Reagents?

A certain level of paranoia and skepticism is a good thing in a scientist. But where to draw the line?

New from the Manufacturer

The major chemical manufacturers can usually be trusted when providing reagents as labeled in new, unopened bottles. Mistakes do happen, so if a carefully controlled procedure fails, and you eliminate all other sources of error, then consider the reagents as a possible source of the problem.

Opened Container

Here's where the fun begins. Once the bottle is opened, the manufacturer is not responsible for the purity or integrity of the chemical. The user must store the reagent properly, and use it correctly to avoid contamination, oxidation, hydration, or a host of other ills that can befall a stored reagent. How many times have you been tempted to use that reagent in the bottle with the faded label that is somewhere over 40 years old? A good rule of thumb is if the experiment is critical, use a new or nearly new bottle for which the history is known. If an experiment is easily repeated should a reagent turn out to be contaminated, then use your judgment when considering the use of an older reagent.

How can you maintain a reagent in nearly new condition? Respect the manufacturer's instructions. Storage conditions (freezer, refrigerator, dessicator, inert atmosphere, etc.) are often provided on the label or in the catalog. Improper handling is more likely than poor storage to lead to contamination of the reagent. It is rarely a good idea to pipette a liquid reagent directly from the original bottle; this invites contamination. Instead, pour a portion into a second container from which the pipetting will be done. Solids are less likely to be contaminated by removing them directly from the bottle, but that is not always the case. It's usually satisfactory to transfer buffer salts from a bottle, for instance, but use greater care handling a critical enzyme cofactor.

Reagents Prepared by Others

Never blindly trust a reagent prepared by someone other than yourself, especially for critical assays. It's a lot like packing your own parachute—it's your responsibility to prepare your important solutions. If you want to trust the outcome of an important experiment to something someone else may have prepared while thinking about an upcoming vacation, it's up to you. Prepare critical solutions yourself until you have a solid working relationship with whomever you plan to share solutions with. Even then, don't get offended if they don't trust your solutions!

Reagents Previously Prepared by You

How reliable are your solutions? Your solutions are probably fine to use if:

- Your labeling and record-keeping are contemporary and accurate.
- You don't share solutions with anyone who could have mishandled and contaminated them.
- Your material is within it's expected shelf life.

What Are Your Options for Storing Reagents?

Storage is half the battle (handling is the other half) in keeping reagents fit for use. Follow the manufacturer's recommendations.

Shelf (Room Temperature)

Solids, like buffer salts, are usually stored on the shelf in sealed bottles. Sometimes it is appropriate (e.g., for hygroscopic materials) to store them in a dessicator on a shelf. Many nonflammable liquid reagents can be also stored on a shelf. Care should be taken to store incompatible chemicals separately. For example, store acids and bases separated; store strong oxidizers away from other organics.

Vented Flammables Cabinet

Flammables or reagents with harmful vapors (e.g., methylene chloride) should be stored in ventilated cabinets designed for chemical storage. These cabinets are designed to minimize the chance of fire from flammable vapors; they often are designed to contain minor leaks, preventing wider contamination and possible fire. It is a good practice to use secondary spill containers (e.g., polypropylene or Teflon™ trays) in the flammables cabinet if they are not already built into the design.

Refrigerators

Many reagents require refrigeration for storage stability. Working buffers, particularly phosphates, will usually last a little longer if refrigerated between uses. Refrigerators used for storing chemicals must not be used to store foodstuffs.

Freezer

Check the label; many standards require freezer temperatures for long-term stability. Check that the freezer is functioning properly.

Are All Refrigerators Created Equal?

Household Refrigerator

It is cheap, stays cold, and is often perfectly fine for storing aqueous samples. It can have serious problems storing flammable organics, however, since the thermostat controls are usually located inside the refrigerator, which can spark and ignite flammable vapors.

Flammable Storage Refrigerator

The thermostat controls have been moved outside the cooled compartment. Unless a refrigerator is specifically labeled "Flammable Storage" by the manufacturer, don't assume it is appropriate for storing flammables.

Explosion-Proof Refrigerator

These units meet specific requirements regarding potential spark sources and can be used in hazardous environments. They are usually extremely expensive.

Safe and Unsafe Storage in Refrigerators

Volumetric Flasks and Graduated Cylinders

How tempting to prepare a fresh solution in a volumetric flask and store it in the refrigerator. Then, an hour later, you reach into the refrigerator to grab a sample prepared the previous week, and accidentally knock over the flask. Tall narrow vessels like volumetric flasks and graduated cylinders are unstable, especially if they sit on wire refrigerator shelves. Solutions should be transfered to a more stable bottle or flask before storing in the refrigerator.

The Shelf in the Door

A long time ago in a basement laboratory, reagents were stored within a shelf in a refrigerator door. The refrigerator was opened, the shelf broke, and bottles spilled onto the floor, breaking two of them. One was dimethyl sulfate, a strong alkylating reagent, and the other was hydrazine, which is pyrophoric. Upon exposure to the air, the hydrazine burst into flame, vaporizing the dimethyl sulfate. It was several days before it was clear that the people exposed to the vapors wouldn't die from pulmonary edema. It may be 20 years before they know whether they have been compromised in terms of lung cancer potential.

Hazardous reagents should not be stored on shelves in refrigerator doors.

Poorly Labeled Bottles

A heavily used, shared refrigerator quickly begins to resemble a dinosaur graveyard. Rummage around in back, and you find a jumble of old, poorly or unlabeled bottles for which nobody assumes responsibility. Ultimately someone gets assigned the task of sorting out and discarding the chemicals. It is much simpler to put strict refrigerator policies in place to avoid this situation, and conduct regular refrigerator purges, so no ancient chemicals accumulate.

What Grades of Water Are Commonly Available in the Lab?

Tap Water

Tap water is usually of uncontrolled quality, may have seasonal variations such as level of suspended sediment depending on the source (municipal reservoir, river, well), may contain other chemicals purposely added to drinking water (chlorine, fluoride), and is generally unsuitable for use in important experiments. Tap water is fine for washing glassware but should always be followed by a rinse with a higher-grade water (distilled, deionized, etc.).

Distilled Water

Distillation generally eliminates much of the inorganic contamination and particularly sediments present in tap water feedstock. It will also help reduce the level of some organic contaminants in the water. Double distilling simply gives a slightly higher grade distilled water, but cannot eliminate either inorganic or organic contaminants.

Distilled water is often produced in large stills that serve an entire department, or building. The quality of the water is dependent on how well the equipment is maintained. A significant stir occurred within a large university's biochemistry department when the first mention of a problem with the house distilled water was a memo that came out from the maintenance department that stated: "We would like to inform you that the repairs have been made to the still serving the department. There is no longer any radium in the water." The next day, a follow-up memo was issued that stated: "Correction—there is no longer any sodium in the distilled water."

Deionized Water

Deionized water can vary greatly in quality depending on the type and efficiency of the deionizing cartridges used. Ion exchange beds used in home systems, for instance, are used primarily to reduce the "hardness" of the water usually due to high levels of divalent cations such as magnesium and calcium. The resin bed consists of a cation exchanger, usually in the sodium form, which releases sodium into the water in exchange for removing the divalent ions. (Remember that when you attempt to reduce your sodium intake!) These beds therefore do not reduce the ionic content of the water but rather exchange one type of ion for another.

Laboratory deionizing cartridges are usually mixed-bed cartridges designed to eliminate both anions and cations from the water. This is accomplished by preparing the anion-exchange bed in the hydroxide (OH^-) form and the cation-exchange resin in the acid (H^+) form. Anions or cations in the water (including monovalent) are exchanged for OH^- or H^+, respectively, which combine to form neutral water. Any imbalance in the removal of the ions can result in a pH change of the water. Typically water from deionizing beds is slightly acidic, often between pH 5.5 to 6.5.

The deionizing resins can themselves increase the organic contaminant level in the water by leaching of resin contaminants, monomer, and so on, and should always be followed by a bed of activated carbon to eliminate the organics so introduced.

18 MΩ Water (Reverse Osmosis/MilliQ™)

The highest grade of water available is generally referred to as 18 MΩ water. This is because when the inorganic ions are completely removed, the ability of the water to conduct electric current decreases dramatically, giving a resistance of 18 MΩ. Commercial systems that produce this grade of water usually apply a multiple-step cleanup process including reverse osmosis, mixed-bed ion exchangers, carbon beds, and filter disks for particulates. Some may include filters that exclude microorganisms, resulting in a sterile water stream. High-grade 18 MΩ water tends to be fairly acidic—near pH 5. Necessary pH adjustments of dilute buffer solutions prepared using 18 MΩ water could cause discrepancies in the final ionic concentration of the buffer salts relative to buffers prepared using other water sources.

When Is 18 MΩ Water Not 18 MΩ Water?

Suppose that your research requires 18 MΩ water, and you purchased the system that produces 500 ml/min instead of the 2 L/min version. If your research doesn't require a constant flow of water, you can connect a 20 L carboy to your system to store your pristine water. Bad Move.

18 MΩ is not the most inert solvent; in practice, it is very aggressive. Water prefers the presence of some ions so as your 18 mΩ water enters the plastic carboy, it starts leaching anything it can out of the plastic, contaminating the quality of the water. The same thing happens if you try to store the water in glass. 18 mΩ water loves to attack glass, leaching silicates and other ions from the container. If you need the highest purity water, it's best not to store large quantities, but rather prepare it fresh.

For the same reason, the tubing used to transfer your high-grade water should always be the most inert available, typically Teflon™ or similar materials. Never use highly plasticized flexible plastic tubing. Absolutely avoid metals such as copper or stainless steel, as these almost always guarantee some level of contaminants in your water.

What Is the Initial pH of the Water?

As mentioned above, the initial pH of typical laboratory-grade distilled and deionized water is often between 5.5 and 6.5. Check your water supply from time to time, particularly when deionizing beds are changed to ensure that no major change in pH has occurred because of seasonal variation or improperly conditioned resin beds.

Although the initial pH of laboratory water may be slightly acidic, the good news is deionized water should have little or no buffer capacity, so your normal pH adjustment procedures should not be affected much. Pay particular attention if your buffer concentrations are very low (<10 mM) resulting in low buffer capacity.

What Organics Can Be Present in the Water?

The answer to this important question depends on the upstream processing of the water and the initial water source. Municipal water drawn from lakes or streams can have a whole host of organics in them to start with, ranging from petroleum products to pesticides to humic substances from decaying plant material to chlorinated species like chloroform resulting from the chlorination process. Well water may have lower levels of these contaminants (since the water has been filtered through lots of soil and rock, but even groundwater may contain pesticides and chlorinated species like trichloroethylene depending on land use near the aquifer.

Municipal processing will remove many organic contaminants from the tap water, but your in-lab water purifier is responsible for polishing the water to a grade fit for experimental use. Most commercial systems do a good job of that, but as mentioned previously, care must be taken to not introduce contaminants after the water has been polished. Plasticizers from tubing or plastic storage tanks, monomer or resin components from deionizer beds, and surfactants or lubricants on filters or other system components are the most common type of organic to be found in a newly installed system.

Another common, yet often overlooked source, is microbial contamination. In one case, a high-grade water purifier mounted on a wall near a window suddenly started showing evidence of organic background. Changing the carbon cartridge did not help the situation. Close inspection of the system showed the translucent plastic tubing connecting the reverse osmosis holding tank to the deionizer beds, and ultimately the lines that delivered the polished water to the spigot, had been contaminated by microbial growth. It was surmised that the intense sunlight during part of the day was providing a more hospitable environment for microorganisms to gain a foothold in the system. The clear tubing was replaced with opaque tubing and the problem disappeared.

In a second instance, a facility changed its water source from wells to a river draw-off. This drastically changed the stability of the incoming water quality. During periods of heavy rain, silt levels in the incoming water increased dramatically, quickly destroying expensive reverse osmosis cartridges in the water purifier system. The solution was to install two pre-filters of decreasing porosity in line ahead of the reverse osmosis unit. The first filter needed replacing monthly, but the second filter was good for three to six months. The system functioned properly for a while, but then problems reappeared in the reverse osmosis unit. Inspection showed heavy microbial contamination in the second prefilter which had a clear housing, admitting sunlight. After cleaning and sterilizing the filter unit, the outside of the housing was covered with black electrical tape, and the microbial contamination problem never returned.

As discussed in Chapter 12, dispensing hoses from water reservoirs resting in sinks can also lead to microbial contamination.

What Other Problems Occur in Water Systems?

Leaks

Leaks are sometimes one of the most serious problems that can occur with in-lab water purification systems. Leaks come in three kinds, typically. Leaks of the first kind start as slow drips, and can be spotted and corrected before developing into big unfriendly leaks.

Leaks of the second kind are generally caused by a catastrophic failure of a system component (tubing, valve, automatic shutoff switch, or backflush drain). Although highly uncommon, they usually occur around midnight on Fridays so as to maximize the amount of water that can escape from the system, therefore maximizing the resulting flooding in the lab. The likelihood of a leak of the second kind seems to increase exponentially with the cost of instrumentation in laboratories on floors directly below the lab with the water purifier system.

Leaks of the third kind result when a person places a relatively large vessel beneath the water system, begins filling, and walks away to tend to a few minor tasks or is otherwise distracted. The vessel overflows, flooding the lab with the extent of the flood depending on the duration of the distraction.

Leaks of the third kind are by far the most common type of leak, and are also the most preventable. Locating the water purification system immediately above a sink, so that any vessel being filled can be placed in the sink, usually prevents this type of catastrophe. If placement above a sink is not possible, locating the water purification system in a (relatively) high-traffic or well-used location in the lab can also minimize or eliminate the possibility of major spills, since someone is likely to notice a spill or leak.

Leaks of the first or second type are highly uncommon, but do occur. The best prevention is to have the system periodically inspected and maintained by qualified personnel, and never have major servicing done on a Friday. Problems seem to be most likely after the system has been poked and prodded, so best to do that early in the week. Then the system can be closly watched for a few days afterward before leaving it unattended.

BIBLIOGRAPHY

BandShift Kit Instruction Manual, Revision 2. Amersham Pharmacia Biotech, 1994.

Hennighausen, L., and Lubon, H. 1987, Interaction of protein with DNA in vitro. *Meth. Enzymol.* 152:721–735.

Gallagher, S. 1999. One-dimensional SDS gel electrohoresis of proteins. In Ausubel, F. M., Brent, R., Kingston, R. E., Moore, D. D., Seidman, J. G., Smith, J. A., and Struhl, K., eds., *Current Protocols in Molecular Biology*. Wiley, New York, pp 10.2A.4–10.2A.34.

4

How to Properly Use and Maintain Laboratory Equipment

Trevor Troutman, Kristin A. Prasauckas,
Michele A. Kennedy, Jane Stevens, Michael G. Davies,
and Andrew T. Dadd

Balances and Scales 51
How Are Balances and Scales Characterized? 51
How Can the Characteristics of a Sample and the Immediate Environment Affect Weighing Reproducibility? 51
By What Criteria Could You Select a Weighing Instrument? 54
How Can You Generate the Most Reliable and Reproducible Measurements? 54
How Can You Minimize Service Calls? 55
Centrifugation 55
Theory and Strategy 55
Practice 58
Centrifugation of DNA and RNA 63
Troubleshooting 64
Pipettors 67

Data on the performance characteristics of different protein concentration assays were generously provided by Bio Rad Inc.

Molecular Biology Problem Solver, Edited by Alan S. Gerstein.
ISBN 0-471-37972-7

Which Pipette Is Most Appropriate for Your Application? 67
What Are the Elements of Proper Pipetting Technique? . . . 68
Preventing and Solving Problems 68
Troubleshooting 77
pH Meters 77
What Are the Components of a pH Meter? 77
How Does a pH Meter Function? 80
How Does the Meter Measure the Sample pH? 81
What Is the Purpose of Autobuffer Recognition? 82
Which Buffers Are Appropriate for Your Calibration Step? 83
What Is Temperature Compensation and How Does One Choose the Best Method for an Analysis? 84
How Does Resolution Affect pH Measurement? 85
Why Does the Meter Indicate "Ready" Even as the pH Value Changes? 85
Which pH Electrode Is Most Appropriate for Your Analysis? 85
How Can You Maximize the Accuracy and Reproducibility of a pH Measurement? 87
How Do Lab Measurements Differ from Plant or Field Measurements? 90
Does Sample Volume Affect the Accuracy of the pH Measurement? 90
How Do You Measure the pH of Viscous, Semisolid, Low Ionic Strength, or Other Atypical Samples? 90
How Can You Maximize the Lifetime of Your pH Meter? 91
Troubleshooting 92
Is the Instrument the Problem? 92
Service Engineer, Technical Support, or Sales Rep: Who Can Best Help You and at the Least Expense? 94
Spectrophotometers 94
What Are the Criteria for Selecting a Spectrophotometer? 94
Beyond the Self-Tests Automatically Performed by Spectrophotomters, What Is the Best Indicator That an Instrument Is Operating Properly? 98
Which Cuvette Best Fits Your Needs? 100
What Are the Options for Cleaning Cuvettes? 101
How Can You Maximize the Reproducibility and Accuracy of Your Data? 101
What Can Contribute to Inaccurate A_{260} and A_{280} Data? 103

Does Absorbance Always Correlate with Concentration? 104
Why Does Popular Convention Recommend Working Between an Absorbance Range of 0.1 to 0.8 at 260 nm When Quantitating Nucleic Acids and When Quantitating Proteins at 280 nm 105
Is the Ratio, $A_{260}:A_{280}$, a Reliable Method to Evaluate Protein Contamination within Nucleic Acid Preparations? 106
What Can You Do to Minimize Service Calls? 106
How Can You Achieve the Maximum Lifetime from Your Lamps? 107
The Deuterium Lamp on Your UV-Visible Instrument Burned Out. Can You Perform Measurements in the Visible Range? 107
What Are the Strategies to Determine the Extinction Coefficient of a Compound? 108
What Is the Extinction Coefficient of an Oligonucleotide? 108
Is There a Single Conversion Factor to Convert Protein Absorbance Data into Concentration? 108
What Are the Strengths and Limitations of the Various Protein Quantitation Assays? 109
Bibliography 110

BALANCES AND SCALES (Trevor Troutman)

How Are Balances and Scales Characterized?

Balances are classified into several categories. Top-loaders are balances with 0.001 g or 1 mg readability and above, where readability is the lowest possible digit that is seen on the display. Analytical balances are instruments that read 0.1 mg. Semimicro balances are those that are 0.01 mg. Microbalances are 1 μg. Finally, the ultramicrobalances are 0.1 μg.

How Can the Characteristics of a Sample and the Immediate Environment Affect Weighing Reproducibility?

Moisture

Condensation forms on reagents that are not kept airtight while they equilibrate to room temperature. Similarly moisture is generated if the sample is not allowed to reach the temperature of the weighing instrument. This is especially problematic when weighing very small samples. You the researcher are also a source of moisture that can be transmitted quite easily to the sample in the form of fingerprints and body oils.

Air Buoyancy

Akin to water keeping something afloat, samples can be "lifted" by air, artificially decreasing their apparent weight. This air buoyancy can have a significant effect on smaller samples.

Electrostatic Forces

Electrostatic charges are almost always present in any environment, particularly in areas with very low humidity. If there are considerable charges present in a sample to be weighed on a high-precision instrument, it will manifest itself in the form of drifting, constant increase or decrease of weight readings, or nonreproducible results. Variability occurs when these electrical forces build up on the sample and the fixed parts of the balance that are not connected to the weighing pan. Substances with low electrical conductivity (e.g., glass, plastics, filter materials, and certain powders and liquids) lose these charges slowly, prolonging the drift during weighing.

The charges most likely originate when the sample is being transported or processed. Examples include friction with air in a convection oven, friction between filters and the surface they contact, internal friction between powders and liquids during transportation, and direct transfer of charged particles by persons. This charge accumulation is best prevented by use of a Faraday cage, which entails shielding a space in metallic walls. This frees the inside area from electrostatic fields. A metallic item can serve the same purpose. Surrounding the container that houses the reagent in foil can also reduce charge accumulation. For nonhygroscopic samples, adding water to increase the humidity inside the draft chamber can reduce static electricity. Accomplish this by placing a beaker with as much water as possible into the draft chamber. An alternative is to bombard the sample with ions of the opposite charge, as generated by expensive ionizing blowers and polonium radiators. A simple and effective solution is to place an inverted beaker onto the weigh pan, and then place the sample to be weighed onto the inverted beaker. This strategy increases the distance between the sample and the weigh pan, thus weakening any charge effects.

Temperature

Airing out a laboratory or turning the heat on for the first time with the change of seasons has a profound effect on an analytical balance. The components of a weighing system are of different size and material composition, and adapt to temperature changes at different rates. When weighing a sample, this variable response to temperature produces unreliable data. It is recommended to keep a constant temperature at all times in an environment where weighing instruments are kept. When room temperature changes, allow the instrument to equilibrate for 24 hours.

Air Currents or Drafts

The flow rate of ambient air should be minimized to get quick and stable results with weighing equipment. For balances with a readability of 1 mg, an open draft shield (glass cylinder) will suffice. Below 0.1 mg, a closed draft chamber is needed.

These shields or chambers should be as small as possible to eliminate convection currents within the chamber to minimize temperature variation and internal draft problems.

Magnetic Forces and Magnetic Samples

Magnetic forces are produced when a sample is magnetized or magnetizable, which means it contains a percentage of iron, cobalt, or nickel. Magnetic effects manifest themselves in the sample's loss of reproducibility. But unlike electrostatic forces, magnetic forces can yield a stable measurement. Changing the orientation of the magnetic field (moving the reagent sample) relative to the weigh system causes the irreproducible results.

Magnetic effects are thus difficult to detect unless the same sample is weighed more than once. Placing an inverted beaker or a piece of wood between the sample and the pan can counteract the magnetic force. Some instruments allow for *below balance weighing*, in which a hook used to attach magnetic samples lies underneath the weigh pan at a safe distance in order to eliminate magnetic effects.

Gravitational Tilt

A balance must be level when performing measurements on the weighing pan. Gravity operates in a direction that points straight to the center of the earth. Thus, if the weigh cell in not directly in this path, the weight will end up somewhat less. For example, say we weigh a 200 g sample that is 0.2865° (angle = α) out of parallel. We have

$$\text{Apparent weight} = \text{weight} * \cos\alpha$$

$$\text{Apparent weight} = 200 * \cos 0.2865 = 199.9975\,\text{g}$$

This result represents a 2.5 mg deviation. This is a significant quantity when working with analytical samples.

By What Criteria Could You Select a Weighing Instrument?

Capacity and Readability

Focus on determining your true readability needs. Cost increases significantly with greater readability. Also bear in mind that a quality balance usually has internal resolutions that are better than the displayed resolution. Stability is a more desired trait in analytical balances due to the small sample size.

Calibration

Most high-end lab balances will calibrate themselves or will have some internal weight that the user can activate.

Applications

Most lab applications require straight weighing (or "weigh only"). Analytical balances may have an air buoyancy correction application, or determination of filament diameters. Other functions include:

- *Checkweighing*. Sets a target weight or desired weight; then weighs samples to see if they hit the target weight.
- *Accumulation*. Calculates how much of a pre-set formula has been filled by the material being weighed.
- *Counting*. Calculates the number of samples present based on the reference weight of one sample.
- *Factor calculations*. Applies a weighed sample into a formula to calculate a final result.
- *Percent weighing*. A measured sample is represented as a percentage of a pre-set desired amount.
- *Printout of sample information and weights*.

Computer Interface

Some instruments can share data with a computer.

How Can You Generate the Most Reliable and Reproducible Measurements?

Vessel Size

Use the smallest container appropriate for the weighing task to reduce surface and buoyancy effects.

Sample Conditioning

The sample temperature should be in equilibrium with the ambient temperature and that of the balance. This will prevent convection currents at the surface. Cold samples will appear heavier, and hot samples will be lighter.

Humidity

If there is low humidity in the weighing environment, plastic vessels should not be used, mainly for their propensity to gain electrostatic charges. Vessels comprised of 100% glass are preferred because they are nonconductive.

Sample Handling

If high resolution is required (1 mg or less), the sample should not make contact with the user's skin. Traces of sweat add weight and attract moisture (up to 400 micrograms [μg]). The body will also transfer heat to the sample, creating problems addressed above.

Sample Location

The sample should be centered as much as possible on the weighing pan. Off-center loading creates torque that cannot be completely counterbalanced by the instrument. This problem is called the off-center load error.

Hygroscopic Samples

Weigh these samples in a closed container.

How Can You Minimize Service Calls?

Ideally weighing equipment should be calibrated daily, and a certified technician should occasionally clean the balance and the internal calibration weights so that they keep their accuracy. More involved problems can be averted by minimizing the handling of the instruments.

If instruments must be handled or moved, apply great care. A drop of inches can cause thousands of dollars of damage to an analytical balance. For moves, use the original shipping packaging, which was specially designed for your particular instrument. Avoid any jarring movements, which can ruin an instrument's calibration.

CENTRIFUGATION (Kristin A. Prasauckas)

Theory and Strategy

Do All Centrifugation Strategies Purify via One Mechanism?

Zonal, or rate-zonal, centrifugation separates particles based on mass, which reflects the particle's sedimentation coefficient. The greater the migration distance of the sample, the better is the resolution of separation. The average size of synthetic nucleic acid polymers are frequently determined by zonal centrifugation.

Isopycnic or equilibrium density centrifugation separates based on particle density, not size. Particles migrate to a location where the density of the particle matches the density of the centrifugation medium. Purification of plasmid DNA on cesium chloride is an example of isopycnic centrifugation. Voet and Voet (1995) and Rickwood (1984) discuss the techniques and applications of isopycnic separations.

Pelleting exploits differences in solubility, size, or density in order to concentrate material at the bottom of a centrifuge tube (Figure 4.1). The rotors recommended for these procedures are described in Table 4.1.

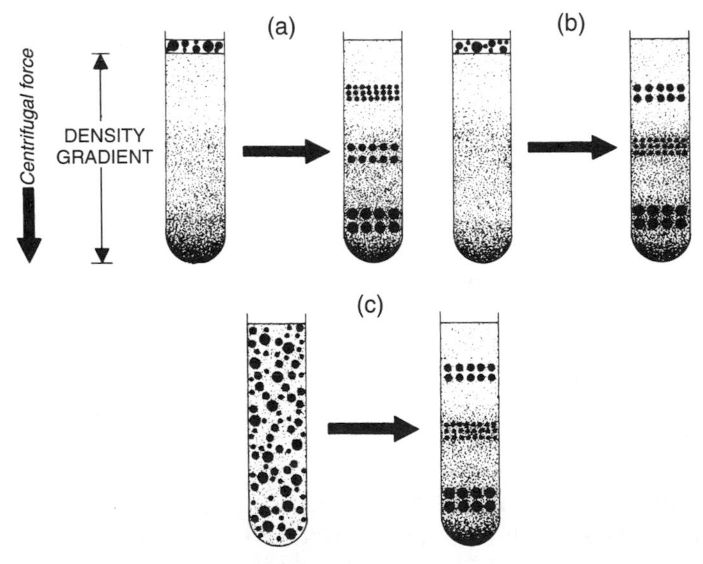

Figure 4.1 Types of density gradient centrifugation. (*a*) Rate-zonal centrifugation. The sample is loaded onto the top of a preformed density gradient (*left*), and centrifugation results in a series of zones of particles sedimenting at different rates depending upon the particle sizes (*right*). (*b*) Isopycnic centrifugation using a preformed density gradient. The sample is loaded on top of the gradient (*left*), and each sample particle sediments until it reaches a density in the gradient equal to its own density (*right*). Therefore the final position of each type of particle in the gradient (the isopycnic position) is determined by the particle density. (*c*) Isopycnic centrifugation using a self-forming gradient. The sample is mixed with the gradient medium to give a mixture of uniform density (*left*). During subsequent centrifugation, the gradient medium re-distributes to form a density gradient and the sample particles then band at their isopycnic positions (*right*). From Centifugation: A Practical Approach (2nd Ed.). 1984. Rickwood, D., ed. Reprinted by permission of Oxford University Press.

What Are the Strategies for Selecting a Purification Strategy?

Procedures for nucleic acid purification are abundant and reproducible (Ausubel et al., 1998). Methods to isolate cells and subcellular components are provided in the research literature and by manufacturers of centrifugation media. But frequently they require optimization, especially when the origin of your sample differs from that cited in your protocol.

If you can't locate a protocol in the research literature and texts, contact manufacturers of centrifugation media. If a reference isn't found for your exact sample, consider a protocol for a related sample. If all else fails, search for the published density of your sample. Manufacturers of centrifuge equipment and media can provide guidance in the construction of gradients based on your sample's density. Rickwood (1984) also provides excellent instructions on gradient construction.

Which Centrifuge Is Most Appropriate for Your Work?

Your purification strategy will dictate the choice of centrifuge, but these general guidelines provide a starting point. An example is provided in Table 4.2.

Table 4.1 Rotor Application Guide

Rotor Type	Swinging-Bucket	Fixed-Angle	Vertical
Pelleting	Good	Best	Not recommended
Rate-zonal	Best	Not recommended	Good
Isopycnic	Good	Better	Best

Table 4.2 Centrifuge Application Guide

Application for Pelleting	Low Speed (7000 rpm/ 7000 × g)	High Speed (30,000 rpm/ 100,000 × g)	Ultra (100,000 rpm/ 1,000,000 × g)
Cells	Yes	Yes	Feasible, but not recommended
Nuclei	Yes	Yes	Feasible, but not recommended
Membranous organelles	Some	Yes	Yes
Membrane fractions	Some	Some	Yes
Ribosomes	No	No	Yes
Macromolecules	No	No	Yes

Source: Data from Rickwood (1984).

Practice

Can You Use Your Existing Rotor Inventory?

Most rotors are compatible only with centrifuges produced by the same manufacturer. Confirm rotor compatibility with the manufacturer of your centrifuge.

What Are Your Options If You Don't Have Access to the Same Rotor Cited in a Procedure?

Ideally you should use a rotor with the angle and radius identical to that cited in your protocol. If you must work with alternative equipment, consider the *g* force effect of the rotor format when adapting your centrifugation strategy. The *g* force is a universal unit of measure, so selecting a rotor based on a similar *g* force, or RCF, will yield more reproducible results than selecting a rotor based on rpm characteristics.

Conversion between RCF and rpm

$$g \text{ force} = 11.18 \times R\,(\text{rpm}/1000)^2$$

Rotor Format

Protocols designed for a swinging-bucket rotor cannot be easily converted for use in a fixed-angle rotor. The converse is also true.

Rotor Angle

The shallower the rotor angle (the closer to vertical), the shorter the distance traveled by the sample, and the faster centrifugation proceeds. This parameter also alters the shape of a gradient generated during centrifugation, and the location of pelleted materials. The closer the rotor is to horizontal, the closer the pellet will form to the bottom of the tube (Figure 4.2).

Radius

The radius exerts several influences on fixed-angle and horizontal rotors. The *g* force is calculated as follows, and holds true for standard and microcentrifuges:

$$g \text{ force} = 1.12 \times 10^{-5} \times r \times \text{rpm}$$

$$r = \text{radius in cm}$$

Centrifugation force can be described as a maximum (*g*-max), a miminum (*g*-min), or an average *g* force (*g*-average). If no suffix is given, the convention is to assume that the procedure refers to *g*-max. These various *g* forces are defined for each rotor by the manufacturer.

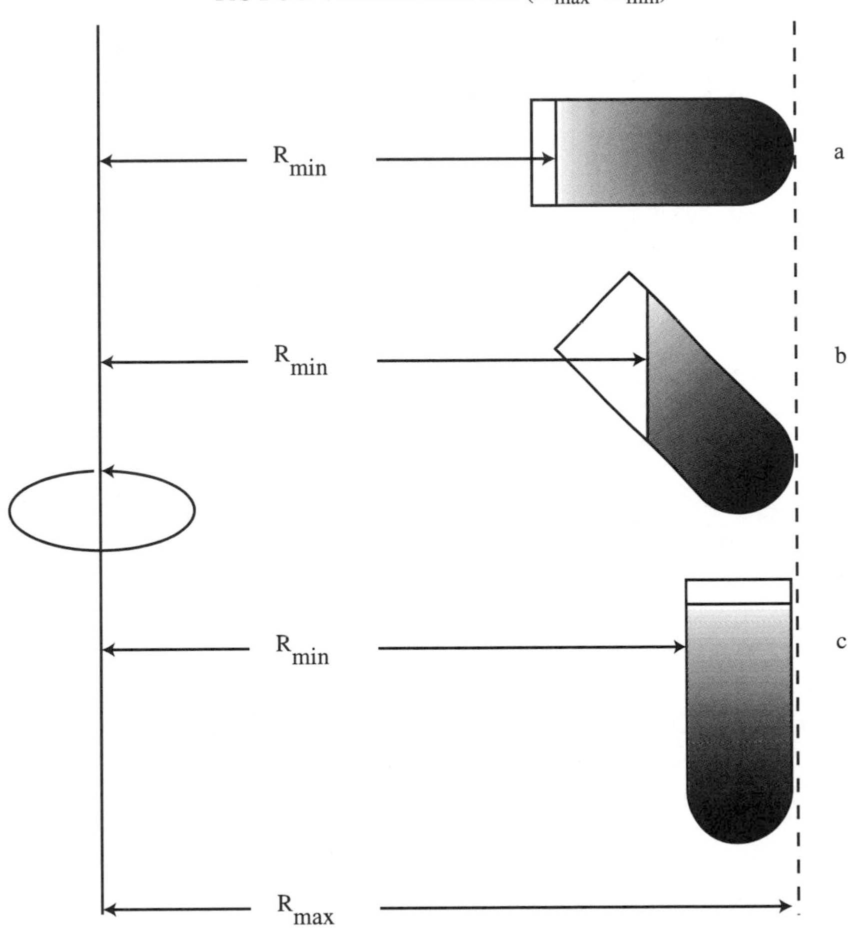

Figure 4.2 Effect of rotor angle on centrifugation experiments. Reproduced with permission of Kendro Laboratory Products. Artwork by Murray Levine.

The radius of a swinging bucket rotor is the distance between the center of the rotor and the bottom of the bucket when it is fully horizontal (Figure 4.2*a*). The greater the rotor radius, the greater is the *g* force.

The distance between the center of a fixed angle rotor and the bottom of the tube cavity determines the radius of a fixed-angle rotor (Figure 4.2*b*). Again the *g* force increases directly as the radius increases.

The greater the rotor angle, the greater is the distance the sample must travel before it pellets (Figure 4.2*b*). This travel distance also affects the shape of a density gradient (Figure 4.3).

The k-factor of a fixed-angle rotor provides a method to predict the required time of centrifugation for different fixed-angle rotors.

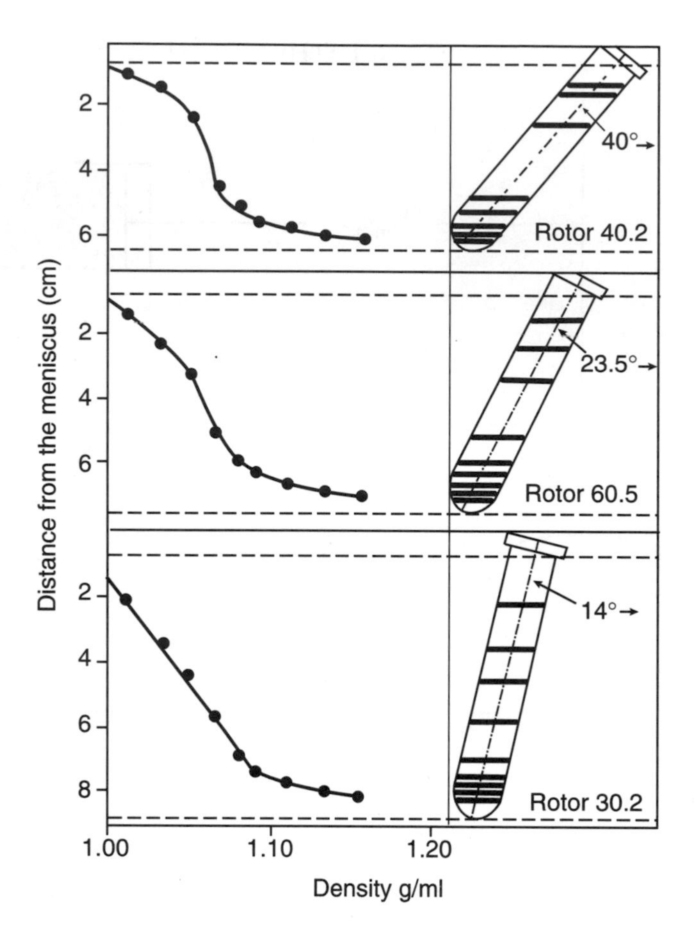

Figure 4.3 Effect of rotor angle on gradient formation. Reproduced with permission of Amersham Pharmacia Biotech.

The k-factor is a measure of pelleting efficiency; rotors with smaller k-factors (smaller fixed angles or vertical angles) pellet more efficiently, requiring shorter run times. The k-factor can also calculate the time required to generate a gradient when switching between different rotors. The k-factor can be determined by

$$k = \left(\frac{2.53 \times 10^{11}}{(\text{rpm})^2} \right) \ln(r_{max}/r_{min}) \tag{1}$$

Equation (2) uses the k-factor to predict the time required for centrifugation for different fixed-angle rotors:

$$\frac{T_1}{k_1} = \frac{T_2}{k_2} \tag{2}$$

T_1 is the run time in minutes for the established protocol. First, calculate the k_1 factor at the appropriate speed for the rotor that is referenced. Next, calculate the k_2 factor at the chosen speed for your rotor. Finally, solve for T_2. This strategy is not appropriate to convert protocols between rotor types (fixed-angle, horizontal, and vertical).

Should Flamables, Explosive, or Biohazardous Materials Be Centrifuged in Standard Centrifugation Equipment?

Centrifuge manufacturers strongly recommend that standard laboratory centrifuges should not be exposed to any materials capable of producing flammable or explosive vapors, or extreme exothermic reactions. Specialized equipment exists for centrifuging dangerous substances.

Which Centrifuge Tube Is Appropriate for Your Application?

A broken or leaking sample container can seriously damage a centrifuge by knocking the rotor out of balance or exposing mechanical and electrical components to harsh chemicals. Damage can occur at any speed. Use only the tubes that are recommended for centrifugation use. If unsure, contact the tube manufacturer to assess compatability.

With the trend toward smaller sample size and greater throughput, microplates have become very popular. Other protocols call for vials and slides. Never attempt to create your own adapter for these containers; ask the rotor manufacturer about the availability of specialized equipment.

Fit

Correct tube fit is critical, especially at higher *g* forces. Tubes or containers that are too large can get trapped in rotors, while tubes that fit loosely can leak or break. Never use homemade adapters. While a broken tube doesn't sound costly, poorly fitted containers can lead to costly repairs.

g Force

Many tubes are not suitable for high stress centrifugation. When in doubt about *g* force limitations, contact the tube's manufacturer. If this isn't feasible, you can test the tube by filling it with water, centrifuging at low rpm's, and inspecting the tube for damage or indications of stress while slowly increasing the speed.

Chemical Compatibility

Confirm the tube's resistance with the manufacturer. Containers that are not resistant to the sample might survive one or more centrifugations but will surely be weakened. Chemically resistant containers should always be inspected for signs of stress before using them. Repeated centrifugation can damage any container.

A Checklist for Centrifuge Use

Inspect the centrifuge for frost on the inside chamber. Accumulated frost must be removed because it can prevent proper temperature control. Previous spills should also be cleaned before starting the centrifuge.

If your instrument uses rotor identification codes, does your instrument have the appropriate software to recognize and operate your rotor? Don't apply the identification code of one rotor for a second rotor that does not possess it's own code. Manually confirm the speed limitations of your rotor if identification codes aren't relevant.

Inspect the rotor for signs of corrosion and wear-and-tear. If you see any pitting or stress marks in the rotor cavities, do not use the rotor. If it is difficult to lock the rotor lid down or lock the rotor to the centrifuge, don't use the rotor. Check that all O-rings on the rotor and sample holders are present, clean, in good physical condition, and well lubricated. Many fixed-angle rotors have a cover O-ring, while many rotors that get locked to the drive have a drive spindle O-ring. If you have concerns about the rotor's condition, don't use it. Request an inspection from the manufacturer.

All the buckets and/or carriers within a swinging bucket rotor must be in place, even when these positions are empty. Utilize the proper adaptors and tubes, as described above. Balance your tubes or bottles. Refer to the manufacturer's instructions for balance tolerance, which vary with different rotors. Place the rotor onto the drive shaft and check that it is seated properly. Many rotors must be secured to the drive. Gently try to lift the rotor off the drive as a final check that the rotor is properly installed.

Begin centrifugation. Even though most imbalances occur at lower speed, monitor the centrifuge until it approaches final speed. If an imbalance occurs, reinspect the balancing of the tubes and the placement of the rotor.

Should the Brake Be Applied, and If So, to What Degree?

If a brake is turned completely off, it could take hours for the rotor to stop, depending on the top speed and instrument conditions (i.e., vacuum run). The stiffer the brake setting, the greater the jolt to the sample, so take note that the default setting of most instruments is the hardest, quickest brake rate. A reduced brake rate is recommended when separating samples of similar densities, when high resolution gradients and layers are required, and when fluffy, noncompacted pellets are produced. The degree of jolting, the braking technique, and the terminology varies among manufacturers, so consult your operating manual or contact the manufacturer to discuss the most appropriate brake setting for your application. [The reverse is true when looking at the deceleration.) If you have an option as to where to control to (or from), 1000 to 1500 rpm is recommended.

Centrifugation of DNA and RNA

How Does a Deration Curve Affect Your Purification Strategy?

Deration describes the situation where a rotor should not attain its maximum speed because a high-density solution is used in the separation. For example, centrifuged at high *g* force, dense solutions of CsCl will precipitate at a density of about 1.9 g/ml, a situation that can blow out the bottom of the rotor. The deration curve supplied by the rotor's manufacturer will indicate those speeds that could cause centrifugation media to precipitate and potentially damage the rotor and or centrifuge (Figure 4.4).

Is a Vertical Rotor the Right Angle for You?

Vertical rotors can purify DNA via cesium chloride centrifugation in three hours, as compared to the overnight runs using fixed-angle rotors. Vertical rotors re-orient your sample (Figure 4.5), so there is the small possibility that the RNA that pelleted against the outer wall of the tube will contaminate the DNA as the gradient re-orientes.

A near-vertical rotor from Beckman-Coulter eliminates this problem (Figure 4.6). The 9° angle of this rotor allows the RNA to pellet to the bottom of the tube without contaminating the DNA. The closeness to vertical keeps centrifugation times short. Triton X-100 was applied in this near-vertical system to improve the separation of RNA from plasmid DNA, although the impact of the detergent on the later applications of the DNA was not tested (Application Note A-1790A, Beckman-Coulter Corporation).

Vertical rotors also allow for the tube to be pulled out straight without the worry of disrupting the gradient. Fixed-angle rotors produce bands at an angle, requiring greater care when removing samples from the rotor.

What Can You Do to Improve the Separation of Supercoiled DNA from Relaxed Plasmid?

Centrifugation at lower *g* force will increase the resolution of supercoiled and relaxed DNA. Apply a step-run gradient, where high speed establishes the gradient, followed by lower speeds and *g* forces to better separate supercoiled and relaxed DNA. Rotor

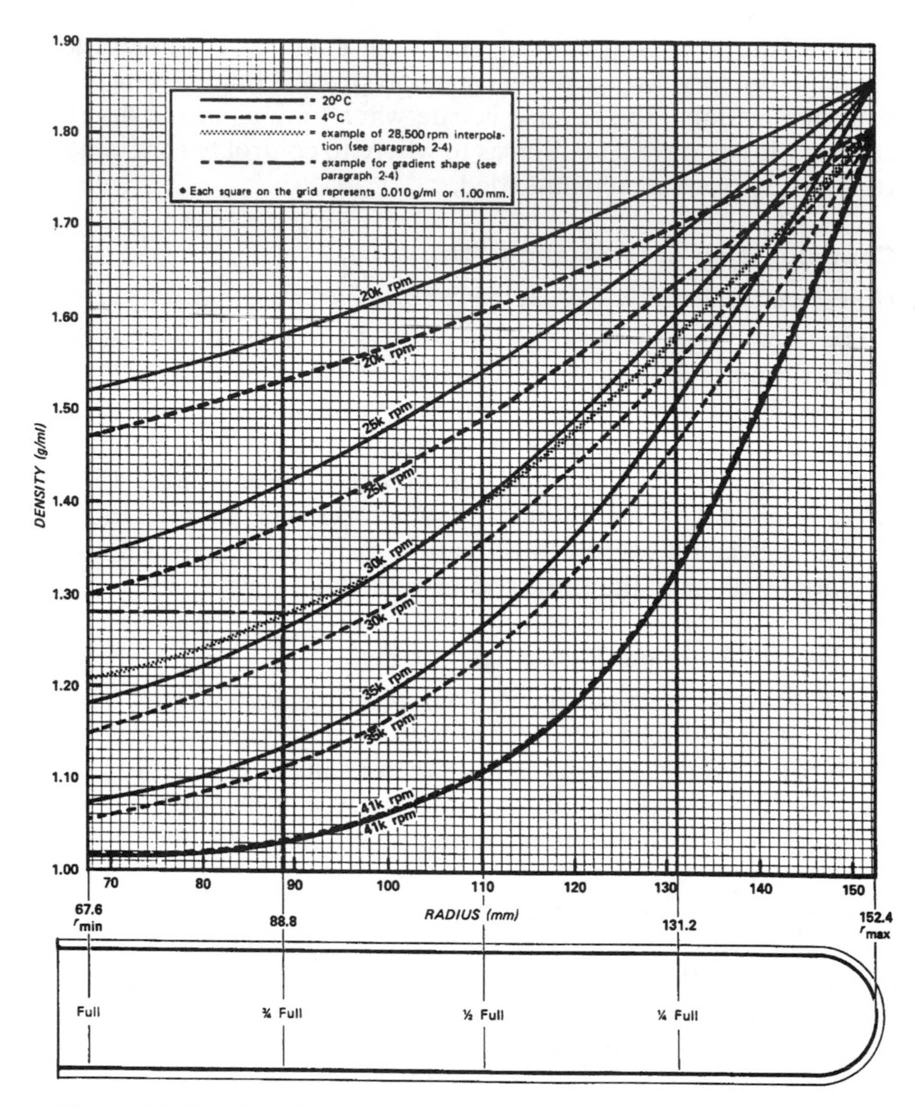

Figure 4.4 Deration Curve. Reproduced with permission of Kendro Laboratory Products.

manufacturers can provide the appropriate conditions for step-runs on your rotor-centrifuge combination.

Troubleshooting

How Can You Best Avoid Service Calls?

Too often operating manuals are buried in unmarked drawers, never to be seen again. This is a costly error because manuals contain information that can often solve problems without the expense and delay of a service call.

Older instruments may have brush motors. The more frequently

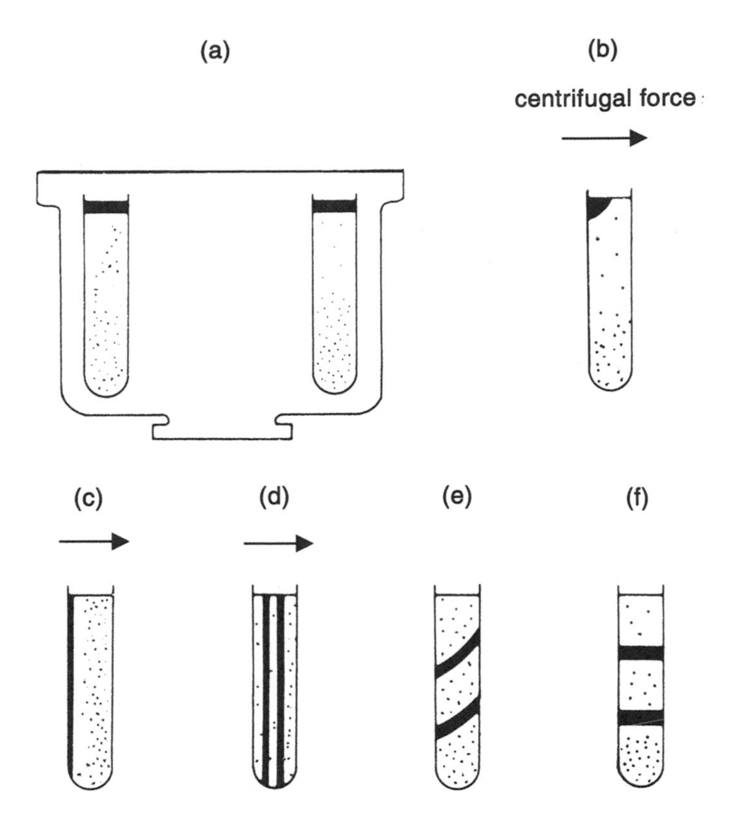

Figure 4.5 Operation of vertical rotors. (*a*) The gradient is prepared, the sample is layered on top, and the centrifuge tubes are placed in the pockets of the vertical rotor. (*b*) Both sample and gradient begin to reorient as rotor accelerates. (*c*) Reorientation of the sample and gradient is now complete. (*d*) Bands form as the particles sediment. (*e*) Bands and gradient reorient as the rotor decelerates. (*f*) Bands and gradient both fully reoriented; rotor at rest. From Centifugation: A Practical Approach (2nd Ed.). 1984. Rickwood, D., ed. Reprinted by permission of Oxford University Press.

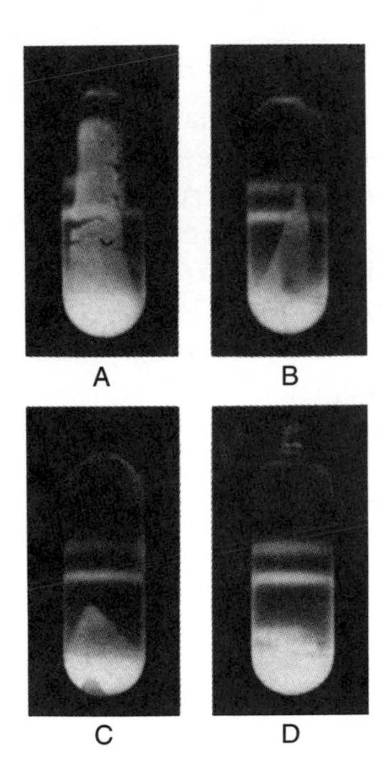

Figure 4.6 Effect of Triton X-100 on RNA pellet in the TLN-100 rotor. The final concentration of Triton X-100 in these tubes is (*a*) 0%, (*b*) 0.0001%, (*c*) 0.001%, and (*d*) 1%. As the Triton X-100 concentration increases, the pellet adhesion decreases. High Triton X-100 levels (0.1–1%) produce a visible reddish band at the top of the tube. Courtesy, Beckman Coulter, Inc., Fullerton, CA, application note A-1790A.

the instrument is used, the more frequently these brushes need to be changed. This is a procedure that you can do yourself, since brushes ordered form the manufacturer usually provide detailed instructions. Most instruments are equipped with an indicator light that signals a worn brush. As with any attempt to repair an instrument, disconnect any power source and consult the manufacturer for warnings on any hazards.

Cleanliness matters. Dirt and spilled materials can enter the motor compartment and cause failures. Clean any spills that occur inside the instrument as soon as possible.

When not in use, turn off the power to a refrigerated centrifuge, and open the chamber door to allow moisture to evaporate. If the instrument must be maintained with the power on, keep the chamber door shut as much as possible and check for frost before use.

Check the level of an instrument after it has been moved. An unleveled instrument can cause damage to the drive mechanism. A post-move preventative maintenance call can prevent problems and ensure that the machine is performing accurately.

What Is the Best Way to Clean a Spill within a Centrifuge?

Spills should be immediately cleaned using a manufacturer-approved detergent. Mild detergents are usually recommended and described in operating manuals. Avoid harsh solvents such as bleach and phenolics.

How Should You Deal with a Walking Centrifuge?

Older units are more prone to walking when imbalanced. If the instrument vibrates mildly, hit the stop button on the instrument. If there is major shaking, cut the power to the instrument by a means other than the instrument's power switch. You can't predict when or how the instrument is going to jump. An ultracentrifuge might require hours to come to a complete stop. Clear an area around the instrument and allow it to move if necessary. Vibration is going to be the greatest at a rotor-dependent critical speed, usually below 2000 rpm. Never attempt to open the chamber door while the rotor is spinning, and don't attempt to enter the rotor area even if the door is open. Don't attempt to use physical force to restrict the movement of the machine. Keep your hands off the instrument until it comes to a complete stop.

How Can You Improve Pellet Formation?

Fluffy pellets that form on the side of a wall are easily dislodged during attempts to remove the supernate. To form tighter pellets, switch to a rotor with a steeper angle, or spin harder and longer at the existing angle.

PIPETTORS (Michele A. Kennedy)

The accurate delivery of a solution is critical to almost all aspects of laboratory work. If the volume delivered is incorrect, the results can be compounded throughout the entire experiment. This section will discuss issues ranging from selecting the correct pipette from the start to ensuring that the pipette is working properly.

Which Pipette Is Most Appropriate for Your Application?

Different applications will require the use of different types of pipettes or different methods of pipetting. Prior to purchasing a pipette, one should decide which type of pipette will be required to address the needs of the lab. There are two main types of pipettes: air displacement and positive displacement. The air displacement pipette is the most commonly used pipette in the lab. In this type of pipette, a disposable pipette tip is used in conjunction with a pipette that has an internal piston. An air space, which is moved by the internal pipette piston, allows for the aspiration and dispensing of sample. This type of pipette is ideal for use with aqueous solutions.

The second type of pipette is the positive displacement pipette. In this pipette, the piston is contained within the disposable tip and comes in direct contact with the sample solution. Positive displacement pipettes are recommended for use with solutions that have a high vapor pressure or are very viscous. When pipetting solutions with a high vapor pressure, it is recommended to pre-wet the tip. This allows the small air space within a positive displacement system to become saturated with the vapors of the solution. Pre-wetting increases the accuracy of the pipetting because the sample will not evaporate into the saturated environment, which would normally cause a pipette to leak.

Once you have decided if an air or positive displacement pipette is the right choice for your lab, then the next thing to choose is the proper volume range. Determine what will be the most frequently used volume. This will then help you to decide on the style of pipette that will achieve the best accuracy. Fixed-volume pipettes provide the highest accuracy of manual pipettes, but they are limited to one volume. Adjustable-volume pipettes are slightly less accurate, but they allow for the pipetting of multiple volumes with one pipette. For example, an Eppendorf® Series 2100 10 to 100 μl adjustable pipette, set at 100 μl has an inaccuracy specification of ±0.8%, whereas a Series 2100 100 μl fixed-volume pipettes has an inaccuracy specification of ±0.6%. When choosing an adjustable-volume pipette, remember that all adjustable pipettes provide greater accuracy at the high end of their volume range. An Eppendorf Series 2100 10 to 100 μl adjustable pipette, set at 100 μl has an inaccuracy specification of ±0.8%, whereas a 100 to 1000 μl pipette set at 100 μl has an inaccuracy specification of ±3.0%.

What Are the Elements of Proper Pipetting Technique?

Once you have selected the correct pipette for your application, then you must ensure that the pipette is used correctly. Improper use of a pipette can lead to variations in the volume being dispensed. When working with a pipette, check that all movements of the piston are smooth and not abrupt. Aspirating a sample too quickly can cause the sample to vortex, possibly overaspirating the sample.

When aspirating a sample, it is important to make sure that the following guidelines are adhered to. First, the pipette tip should only be immersed a few mm into the sample to be aspirated (Figure 4.7).

This ensures that the hydrostatic pressure is similar during aspiration and dispensing. Next, the pipette should always be in a completely vertical position during aspiration. The result of holding a pipette at an angle of 30° could create a maximum of +0.5% inaccuracy (*Products and Applications for the Laboratory 2000*, Eppendorf® catalog, p. 161).

When dispensing the sample, the pipette tip should touch the side of the receiving vessel. This will ensure the even flow of the sample from the tip, without forming droplets. If a droplet remains inside the tip, the volume dispensed will not be correct.

Preventing and Solving Problems

A good maintenance and calibration schedule is the key to ensuring that your pipette is working properly. More often than not, the factory-set calibration on a pipette is changed before a proper inspection and cleaning has been performed. A good maintenance program can prevent unnecessary changes in the calibration, which saves money over the life of the pipette. A few maintenance suggestions are listed below:

- Always store pipettes in an upright position, preferably in a stand. This prevents the nose cones or pistons from being bent when placed in a drawer.

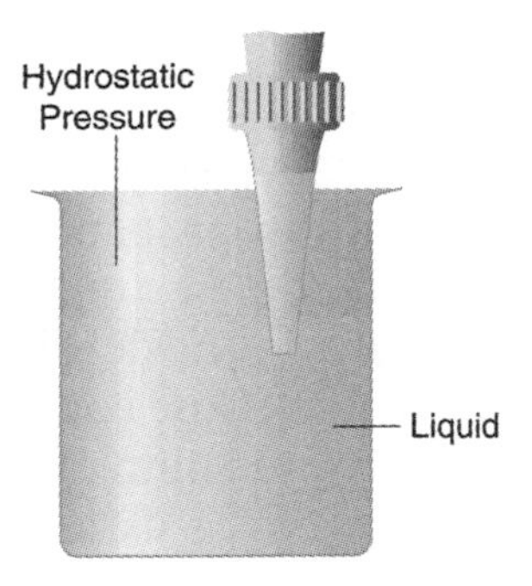

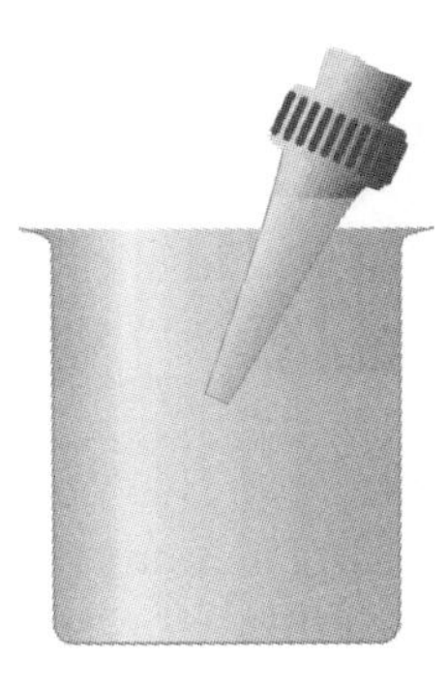

Figure 4.7 Proper placement of pipette tip. Reproduced with permission from Brinkmann™ Instruments, Inc.

• Always make sure that the pipette is clean. Dust or dirt on the nose cone can prevent the pipette tip from sealing properly, which in turn will affect the volume that is being aspirated and dispensed.

• Check that the nose cone of the pipette is not clogged or bent. If the nose cone is clogged or bent, it may change the air column that is aspirating the sample volume. This in turn will affect the volume that is aspirated.

• Make sure that the nose cone is securely fastened on the pipette. The process of placing tips on the pipette should not require the pipette to be twisted. This twisting method is often employed when tips that do not fit properly are used. This can loosen the nose cone and cause the pipette to aspirate incorrect volumes.

• It is strongly recommended that the leak tests described below are performed before you recalibrate a pipette. If a pipette is leaking, a new O-ring or seal might be all that is required to return the pipette to the original factory calibration. The last thing that you want to do is to change the factory calibration of a pipette.

How Should You Clean a Pipette?

Pipette cleaning should be conducted in a very methodical manner. Start from the outside and work your way in.

- Clean the external parts of the pipette with a soap solution. In most cases the pipette can also be cleaned with isopropanol, but check with the manufacturer. Next, remove the tip ejector to expose the nose cone. Ensure that the tip ejector contains no debris or residue. Thoroughly clean the external portion of the nose cone.
- Remove the nose cone, which is usually screwed into the pipette handle. In most cases this will expose the piston, seals, and O-rings. Clean the orifice at the tip of the nose cone to ensure that it is completely free of debris.
- Check the condition of the piston, O-ring, and seal. The piston should be free of any debris. Ceramic pistons (Eppendorf pipettes) should be cleaned and then lightly lubricated with silicone grease provided by the manufacturer. Stainless steel pistons (Rainin® pipettes) should be clean and corrosion-free, and should not be greased. The O-ring should fit on the seal and move freely on the piston.

Overall pipette design is similar from one brand to the next (Figure 4.8), but each pipette has a few slightly different features. Always refer to the manufacturer's instruction manual for recommended cleaning procedures.

How Frequently Should a Pipette Be Tested for Accuracy?

Protocols for testing a pipette may be dictated by several standards. The general market, which includes academic and research labs, follows ASTM, ISO, DIN, GMP, GLP, or FDA guidelines. Clinical laboratories may follow specific guidelines such as NCCLS, CAP, CLIA, and JCAHO.

Pipettes should be inspected and tested on a regular basis. The time interval between checks depends on the established guidelines in the facility and the frequency of use, but in general, a minimum of a quarterly evaluation is recommended.

The best way to establish the time interval between evaluations is to look at the work that you are conducting and determine how far back you would like to go to repeat your research if a problem is detected. For example, if a pharmaceutical company was at the halfway point of developing a new drug and had the calibration of their pipettes checked, only to find that they were all out of calibration—the course of action would be to repeat all of the work that was conducted using these out-of-calibration pipettes. By reducing the amount of time between calibrations,

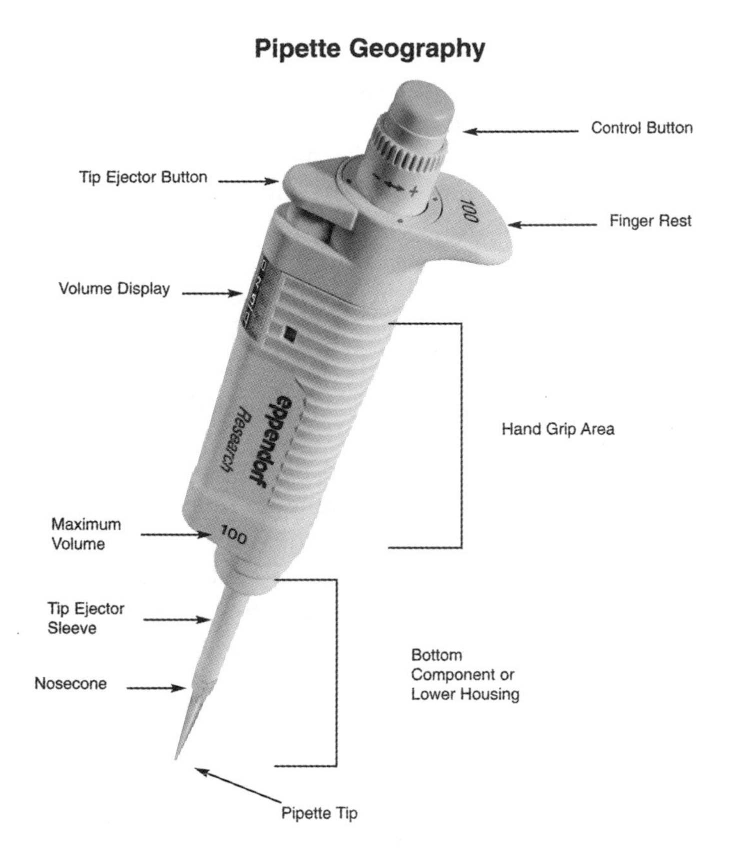

Figure 4.8 Pipette anatomy. Reproduced with permission from Brinkmann™ Instruments, Inc.

you can reduce the amount of time redoing experiments, if a problem is found.

How Can You Monitor the Performance of a Pipette?

All of the internal components of the pipette must be tested to determine that they are fully functional. The first thing that should be checked is the free movement of the piston. The piston should move up and down very smoothly. Next, verify that the internal parts are working properly by performing a leak test. Although this test does not measure accuracy and reproducibility, it is a quick and easy determination of the proper functioning of the internal parts. Please remember that this type of test will only ensure that the internal parts of the pipette are not contributing to a leak in the pipette or tip system. It does not test if the pipette is delivering the specified volume set on the volume display. Two methods to detect leaks are described below. These tests are appropriate for pipetted volumes greater than 10 μl; smaller volumes do not displace a sufficient amount of air to visually check the performance of the pipette.

The first, and easiest, approach is to set the pipette to the maximum rated volume, attach a pipette tip, aspirate liquid into the pipette tip, and hold the pipette in a vertical position for 15 seconds. If the liquid does not drip out, the fit of the seals and O-rings around the piston is good and there is no need to replace them. A leak will appear as a droplet on the pipette tip, which indicates that the pipette needs to be serviced.

A second method is to monitor the stability of a column of liquid in an attached 20 cm segment of PVC tubing. Hold the pipette vertically, aspirate the liquid (colored liquid can be used) and mark the meniscus level on the tubing. Wait one minute, then check if the meniscus level has changed. If a change in the level occurs, a leak exists and the pipette should be serviced. (*Eppendorf SOP Manual*, p. 26.) If the tubing is connected directly to the pipette, the internal parts are tested but not the interaction of the pipette and tip. Testing the direct connection to the pipette and then testing the pipette and tip connection ensures that internal parts are not leaking and that the tip is not causing any leaks. The size of the tubing depends upon the volume of the pipette to be checked.

Pipette Volume	Tubing Inner Diameter (mm)
10–100 μl	0.5–1
100–500 μl	1.5–2
>500 μl	5

How Can You Check If a Pipette Is Dispensing Accurate Volumes?

Gravimetric testing of pipettes refers to the technique of weighing a dispensed amount of liquid, changing the weight to a volume, and then determining if the volume is within the manufacturer's stated specifications. This is the most accepted form of testing the volume delivery of a pipette.

According to the Eppendorf standard operating procedure for pipette calibration, the following information details the equipment, the actual procedure, and the mathematical calculations needed to determine if the pipette is within the factory stated calibration specifications.

The following components are required for a measuring station for calibrating or adjusting pipettes:

1. Fine balance (tested by Board of Weights and Measures; e.g., Sartorius®, Mettler®, Ohaus®, or AnD). The resolution of the

balance depends on the volume of the pipette that is to be tested. The lower the volume, the better the resolution of the balance needs to be. The balance should be located in an area that is free of drafts and vibrations.

Nominal Volume of Pipette (μl)	Error Limits of Device to Be Tested (μl)	Required Accuracy to Be Tested (g)
1–50*	0.1–1.0	0.00001
100–1000	1.0–10	0.0001
>1000	>10	0.001

2. Evaporation protection. A moisture trap or other equipment that prevents evaporation, such as a narrow volumetric flask, are recommended for use. In addition to the narrow weighing vessel, it is advisable to use a moisture trap within the balance. This can be as simple as placing a dish filled with approximately 10 ml of distilled water within the balance. For pipettes with a maximum volume of 5000 μl and above, a moisture trap is not needed.

3. Room Temperature. Ambient temperature should be 20° to 25°C, ±0.5°C during measurement. Factors that affect the temperature of the pipettes and measuring station (e.g., direct sunlight) should be avoided. The ambient temperature and the temperature of the test liquid and pipettes must be the same as the temperature of the pipette tip. For example, if the sample is at 4°C and the pipette is at room temperature (22°C), this could result in a maximum error of –5.4% (Eppendorf catalog 2000, p. 161). It is advisable to equilibrate all components for approximately three to four hours prior to calibration.

4. Test Liquid. Degassed, bi-distilled, or deionized water which is at room temperature (20–25°C) should be used. The water in the liquid supply or in the weighing vessel must be changed every hour and must not be reused. The air humidity over the liquid surface of the weighing vessel should be maintained at a uniform value between 60% and 90% of the relative humidity.

**For volumes <1 μl, the balance is set with six decimal places, or where appropriate, a photometric test may be used.*

5. Instruction Manual. In view of the many different types of volume measuring devices, it is particularly important to refer to the manufacturer's instruction manual during testing.

6. Test Points. The number of test points is determined by the standard that is used. As a rule of thumb, a quick check involves 4 test points, a standard check involves approximately 8 test points, and a full calibration can involve 20 or more test points at each volume.

7. Test Volumes. Most standards test adjustable-volume pipettes at the following three increments:

a. The nominal volume (the largest volume of the pipette)
b. Approximately 50% of the nominal volume
c. The smallest adjustable volume, which should not be less than 10% of the nominal volume

When testing fixed-volume pipettes, only the nominal volume is tested. When testing multiple-channel pipettes, the same volumes are tested for each channel.

Perform The Gravimetric Test

1. Weigh the samples:

- Tare the balance.
- Pre-wet the tip.
- Aspirate and then dispense the set volume three times. Execute blow-out.

2. Aspirate the volume that is to be tested from the liquid supply as follows:

- Hold the pipette vertically in the liquid supply.
- Immerse the tip approximately 2 to 3 mm into the test liquid.
- Aspirate the test volume slowly and uniformly. Observe the waiting period of one to three seconds.
- Remove the pipette tip from the test liquid slowly and uniformly. Remove any remaining liquid by placing the pipette tip against the inside of the vessel.

3. Dispense the test volume into the weighing vessel as follows:

- Place the filled tip at an angle of 30° against the inside of the weighing vessel
- Dispense the test volume slowly and uniformly up to the first stop (measuring stroke) and wait for one to three seconds. (This applies to manual pipettes only.)

- Press the control button to the second stop (blowout) and dispense any liquid remaining in the tip
- Hold down the control button and pull the tip up along the inside of the weighing vessel. Release the control button.

4. Document the value that appears in the display of the balance immediately after the display has come to rest. Record the values from a measurement series as described above. Evaluate the inaccuracy and the imprecision as described below.

Determine Calibration Accuracy

In order to determine if the pipette is with in the factory calibration range, the mean volume, standard deviation, coefficient of variation, % inaccuracy and % imprecision must be determined. This involves completing the following calculations.

1. Mean Volume ($\bar{x}$). This is the sum of the number of weighs (at one volume setting) divided by the number of test points.

$$\bar{x} = \frac{X_1 + X_2 + X_3 + X_4}{\text{Number of weighings}}$$

where X_1, X_2, X_3, X_4, etc. are the actual measured weights

2. Adjustment for Z Factor. The Z factor accounts for the temperature and barometric pressure conditions during testing. (See Table 4.3.)

$$V = \bar{x} * Z$$

where Z = Z factor
$\bar{x}$ = mean of measured volume in μl
V = adjusted mean volume

3. Calculation of (In)Accuracy (A). Accuracy points to the amount of scatter that a pipette varies from its set point:

$$A = \frac{V - SV}{SV} * 100$$

where A = accuracy
V = adjusted mean volume
SV = set volume of pipette

4. Calculation of Standard Deviation (sd). The sd calculation points to the scatter of volume around the mean value:

Table 4.3 Factor Z (μl/mg) as a Function of Temperature and Air Pressure for Distilled Water (ISO DIS 8655/3)

Temperature (°C)	hPa(mbar)					
	800	853	907	960	1013	1067
15	1.0018	1.0018	1.0019	1.0019	1.0020	1.0020
15.5	1.0018	1.0018	1.0019	1.0020	1.0020	1.0020
16	1.0019	1.0020	1.0020	1.0021	1.0021	1.0022
16.5	1.0020	1.0020	1.0021	1.0022	1.0022	1.0023
17	1.0021	1.0021	1.0022	1.0022	1.0023	1.0023
17.5	1.0022	1.0022	1.0023	1.0023	1.0024	1.0024
18	1.0022	1.0023	1.0024	1.0024	1.0025	1.0025
18.5	1.0023	1.0024	1.0025	1.0025	1.0026	1.0026
19	1.0024	1.0025	1.0025	1.0026	1.0027	1.0027
19.5	1.0025	1.0026	1.0026	1.0027	1.0028	1.0028
20	1.0026	1.0027	1.0027	1.0028	1.0029	1.0029
20.5	1.0027	1.0028	1.0028	1.0029	1.0030	1.0030
21	1.0028	1.0029	1.0030	1.0030	1.0031	1.0031
21.5	1.0030	1.0030	1.0031	1.0031	1.0032	1.0032
22	1.0031	1.0031	1.0032	1.0032	1.0033	1.0033
22.5	1.0032	1.0032	1.0033	1.0033	1.0034	1.0035
23	1.0033	1.0033	1.0034	1.0035	1.0035	1.0036
23.5	1.0034	1.0035	1.0035	1.0036	1.0036	1.0037
24	1.0035	1.0036	1.0036	1.0037	1.0038	1.0038
24.5	1.0037	1.0037	1.0038	1.0038	1.0039	1.0039
25	1.0038	1.0038	1.0039	1.0039	1.0040	1.0041
25.5	1.0039	1.0040	1.0040	1.0041	1.0041	1.0042
26	1.0040	1.0041	1.0042	1.0042	1.0043	1.0043
26.5	1.0042	1.0042	1.0043	1.0043	1.0044	1.0045
27	1.0043	1.0044	1.0044	1.0045	1.0045	1.0046
27.5	1.0044	1.0045	1.0044	1.0045	1.0045	1.0046
28	1.0046	1.0046	1.0047	1.0048	1.0048	1.0049
28.5	1.0047	1.0048	1.0048	1.0049	1.0050	1.0050
29	1.0049	1.0049	1.0050	1.0050	1.0051	1.0052
29.5	1.0050	1.0051	1.0051	1.0052	1.0052	1.0053
30	1.0052	1.0052	1.0053	1.0053	1.0054	1.0055

$$sd = Z * \sqrt{\frac{(X_1 - SV)^2 + (X_2 - SV)^2 + (X_3 - SV)^2 + (X_4 - SV)^2}{\text{Number of weighings} - 1}}$$

where Z = Z factor

X_1, X_2, X_3, X_4, etc. are the actual measured weights

SV = set volume of pipette

5. Calculation of (Im) Precision with the Coefficient of Variation (CV). Calculate the standard deviation in percent:

$$CV = \frac{sd}{V} * 100$$

where sd = standard deviation

V = adjusted mean volume

After obtaining all of the preceding information, the results should be compared to the manufacturer's stated specifications. If the pipette is within the stated calibration specifications, it has passed the calibration test.

If the pipette does not meet the specifications, the calibration on the pipette must be changed. This can be accomplished in two different ways, depending on the brand and style of the pipette. In some pipettes, to change the calibration, you adjust the piston stroke length. This basically changes the amount of movement that the piston has during an aspiration/dispensing step, thus changing the volume that is aspirated to match the volume that should be aspirated. The other way to change the calibration of a pipette is to change the volume display to match the volume that was actually dispensed. Please refer to the manufacturer's instruction manual of your pipette to determine the correct way to adjust your pipette.

Once the pipette has been adjusted, the pipette should be retested to ensure that the pipette is now in proper working order.

Troubleshooting

Table 4.4 describes commonly found problems and possible solutions.

pH METERS (Jane Stevens)

What Are the Components of a pH Meter?

Sensing Electrode

This is described in greater detail later in the section "Which pH electrode is most appropriate for your analysis?"

Reference Electrode

The "reference" is the electrochemical industry term for the half-cell electrode whose constant potential is measured as E_0 in the Nernst equation (Figure 4.9). This half-cell is held under stable conditions generating a fixed voltage to which the pH-sensing electrode is compared. There are several types of reference electrode systems. Some such as the standard hydrogen electrode are important theoretically but not practical for actual use. The most commonly used reference electrode system is silver/silver chloride (Ag/AgCl). A silver wire is suspended in a solution of potassium chloride that has been saturated with silver to replenish the wire with silver ions. The calomel reference system uses mercury

Table 4.4 Pipette Troubleshooting Guide

Problem	Possible Cause	Solution
Pipette drips or leaks	Tip is loose or does not fit correctly	Use manufacturer recommended tips Use more force when putting the tip on the pipette
	Nose cone is scratched	Replace the nose cone
	Seal of the nose cone leaks	Replace the nose cone
	Piston is contaminated by reagent deposits	Clean and lubricate the piston (if recommended) Replace the seal
	Piston is damaged	Replace the piston and the piston seal
	Piston seal is damaged	Replace the piston seal and lubricate the piston (if recommended)
	Nose cone has been loosened	Retighten nose cone
Push button does not move smoothly	Piston is scraping due to contamination	Clean and lubricate the piston
	Seal is swollen due to reagent vapors	Open pipette and allow it to ventilate Lubricate the piston only if necessary
	Piston is visibly damaged or coated with insoluble solution	Replace piston seal and piston
Inaccurate volumes	Pipette is leaking	Verify that all of the above situations have been checked
	Pipette's calibration has been changed incorrectly	Recalibrate according to manufacturer's specifications
	Poor pipetting technique	Refer to section on pipetting technique

instead of silver; manufacturers also provide reference systems that lack metal ions altogether.

Junction

The junction is the means for the sample and electrode to contact electrically. The internal filling solution and the sample mix at the junction. The electrode should have a sufficient flow of filling solution that passes through the junction so that the sample and filling solution meet on the sample's side of the junction. This better protects the electrode from backflow of sample components. An electrical potential (the junction potential) due to the ion movement develops at the junction contributing a small elec-

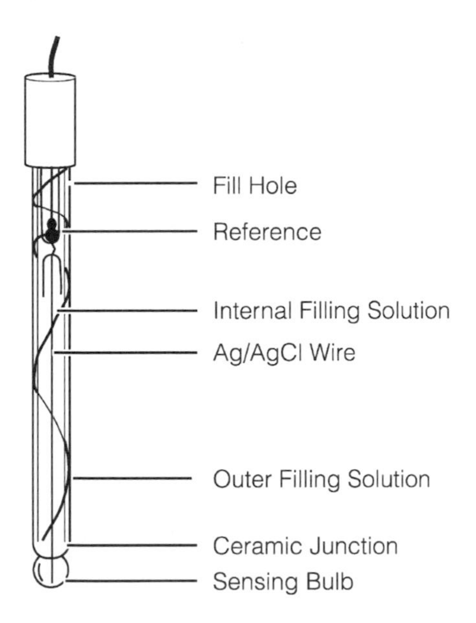

Figure 4.9 Double-junction combination electrode. Reproduced with permission from Thermo Orion Inc.

trical voltage to the overall measurement system. Generally this is a minor error. If the flow of filling solution is not adequate, backflow can cause this error to increase as ions moving at different rates cause an accumulation of charges. The filling solution should be equitransferent (the positive and negative ions can pass freely through the junction), thus minimizing charge accumulation and junction potential error. It is difficult to tell if the flow is adequate in some electrode junctions. A faster flowing junction such as the annular, flushable style (Figure 4.10) will reduce the chances of a poor junction between the sample and electrode. A poor junction will give erratic readings and thus erratic pH values as the additional charges are created in the dynamic solution. A change of 6mV is needed to change the pH by 0.1. It is very difficult to get reproducibility and accuracy without sufficient flow through a junction. Sluggish, drifting readings are indications that the flow may be impaired.

Fill Solution

Reference filling solution or internal filling solution is the electrolyte that is the contact point between the sample and the reference electrode. The filling solution completes the circuit to measure the voltage change due to the sample. It is comprised of salts that conduct electricity and allow the reference electrode to have a stable voltage for a period of time. The fill solution most often contains potassium chloride, but incompatibility with some

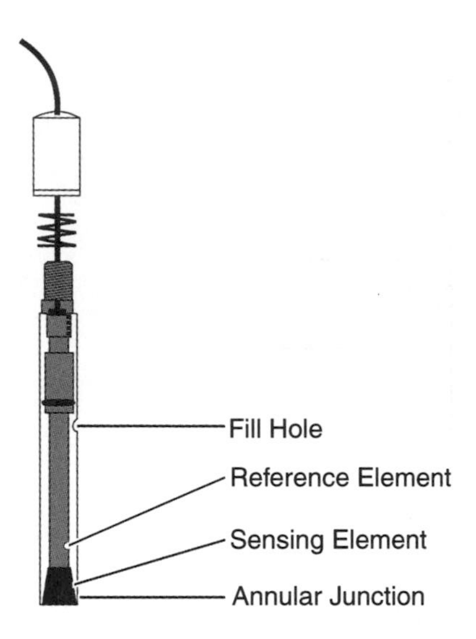

Figure 4.10 Annular pH electrode. Reproduced with permission from Thermo Orion Inc.

samples requires alternate solutions. An example where a different filling solution may be required is with ultra-pure, low ionic strength water. The concentrated KCl would cause the reading to drift as it mixed with the pure water. A lower ionic strength filling solution, such as 2.0 M KCl saturated with Ag^+, would produce faster, more accurate and reproducible readings.

Fill Hole

The filling hole cover on the electrode body must be removed for a positive flow through the junction.

How Does a pH Meter Function?

Theory

pH is an electrochemical measurement of the activity of the hydrogen ion, H^+, in a particular solution. The pH meter measures voltage, in millivolts (mV), from the "battery" created by the electrodes in an aqueous solution (Figure 4.11). The measured voltage is the difference between the electrical potentials of the reference and sensing electrodes. The sensing electrode is usually made of glass which is very sensitive to changes in hydrogen ion activity. Software in the pH meter makes the conversion to solution pH based on previous calibration data stored in its memory. The meter displays the calculated pH of the solution to the operator. Other pertinent information, such as temperature, time and date, and actual millivolts read from the sample are often displayed on more advanced meters.

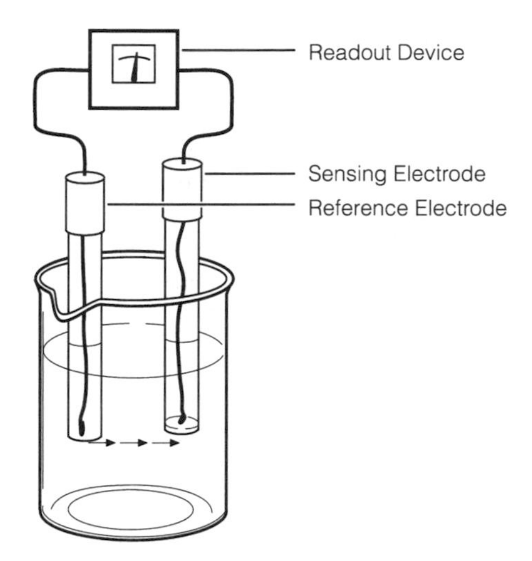

Figure 4.11 Mechanism of pH meter function. Reproduced with permission from Thermo Orion Inc.

Calibration

In order for the meter and electrode system to determine the pH of a sample, it must compare the sample to known values or standards. The standards are specially formulated buffers that have been carefully studied to determine the effect of temperature on their pH. These buffers are used to generate a calibration curve plotting the known pH value versus the measured millivolts to determine the pH of a sample. The millivolt measurement of the sample is plotted on the curve and the pH value is read from the curve.

The pH calibration curve is essentially linear. An average slope is determined at the end of the calibration and reported as percent, with 100% being the theoretical value at that temperature, or as a millivolts per decade of pH where 59.16 is theoretical at 25°C (Figure 4.12). This slope will change with temperature, but it will automatically be corrected by the meter if the correct temperature is entered.

Almost all laboratory pH electrodes are designed so that they read zero millivolts at pH 7. This zero point is called the isopotential point. Calibration curves at different temperatures, and therefore with different slopes, all pass through this point. Buffers are sold with the temperature-corrected pH values on their labels, or they can be found in the research literature.

How Does the Meter Measure the Sample pH?

The meter calculates the sample pH by measuring the mV of the sample, then using the Nernst equation to solve for pH. The Nernst equation is used to describe electrode behavior:

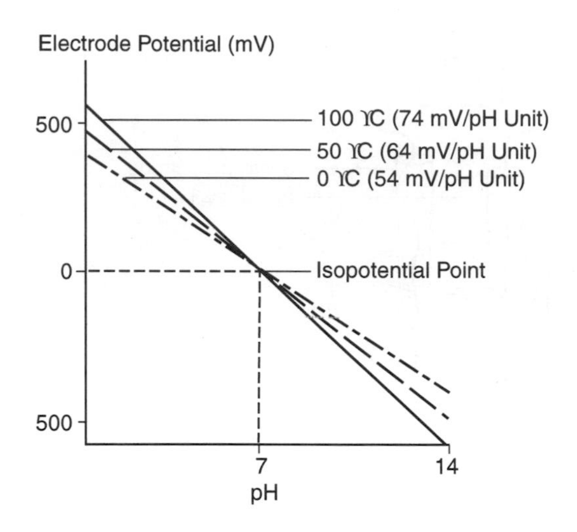

Figure 4.12 Effect of slope on pH determination. Reproduced with permission from Thermo Orion Inc.

$$E_{\text{measured}} = E_0 + S \log a_{\text{H+}}$$

where S is the slope, and E_{measured} is the electrical potential measured at the sensing electrode, and E_0 is the electrical potential measured at the reference electrode, and the log of the hydrogen activity is the pH. Both reference and sensing electrodes are present within a single probe in most pH meters. Due to the presence of the slope term, it is easy to understand the importance of calibration of the meter and electrode system. The meter uses the correct slope for the temperature of the sample, which gives more accurate pH readings. This is especially critical when the pH value varies dramatically from the isopotential point. The further away the pH is from the isopotential point, the greater is the effect a change in temperature produces on the result.

What Is the Purpose of Autobuffer Recognition?

Many meters have an autocalibration feature, autobuffer recognition, that simplifies calibration. This system enables the meter to identify buffers by the buffer's observed millivolt value. For example, a buffer with a millivolt reading of approximately 0 would be identified as pH 7.00, and a reading of around 167 mV would be identified as pH 4.01. If the meter is measuring the temperature, it can automatically adjust for the change in the value of the buffer caused by a change in the temperature.

Common calibration errors occur because of mistakes made in the operation of the autobuffer feature of their meter. Since most meters have pH versus temperature tables or algorithms in their software that will correct for the temperature, the user does not need to adjust the value displayed to account for the buffer pH change due to temperature as they would in a manual calibration. A second source of problems is not matching a buffer or group of buffers with the correct buffer set in the meter. Many meters allow the user to select the category of buffers they want to autorecognize. These could be the NIST (National Institute of Standards and Technology) buffers (4.01, 6.86, etc.), 4, 7, and 10 buffers, or other sets commonly used in different countries or required by government regulations, such as DIN buffers. Refer to the section "Should you use a non-NIST or non-NIST-traceable buffers to calibrate your pH meter?" for more details. Each set will have the pH versus temperature table or algorithm for that particular formulation, so errors will occur if a different buffer is used. For example, pH 7.00 and the NIST 6.86 buffer will both read close to 0 mV but will have different temperature effects due to their different chemical compositions. Errors in the calibration will give inaccurate sample results.

Which Buffers Are Appropriate for Your Calibration Step?

Proper Bracketing

It is essential that the calibration standards cover the entire range of your anticipated sample pH values and include a buffer near the isopotential point. If you will be measuring samples in the pH range of 5 to 6, then buffers 4 and 7 would be appropriate. If you have samples in the 9 to 10 range, buffers 4 and 7 would not generate accurate results as the meter would have to mathematically extrapolate the curve to determine the sample pH, but would not take into account other electrode variables. Calibration buffers with pH of 7 and 10 would be acceptable. Analyzing samples covering a wider range will be more accurately measured with a multiple-point calibration covering a range such as 4, 7, and 10, or even 1.68, 4, 7, 10, and 12.46. Advanced meters do not just draw a best-fit line in these calibrations and use that slope for all measurements. In order to generate the most accurate results, the sample is compared to the line segment that brackets the sample pH. Thus a sample with pH of 11 would be calculated using the slope of the line segment from pH 10 and pH 12.46. This multiple-point calibration strategy maximizes accuracy and saves time by reducing the number of calibrations performed when a wide range of samples is being analyzed.

NIST or Non-NIST-Traceable Buffers

The National Institute of Standards and Technology is a U.S. government agency that sells standards that have been extensively characterized for certain parameters. NIST sells the salts used to make a range of buffers and has generated tables of the temperature relationship to the pH in chemical references such as the *CRC Handbook of Chemistry* (Weast 1980). These buffer salts are primary reference standards and are very expensive due to the studies that these particular samples have undergone. Most buffers that are commercially available are compared to these buffer salts. Such commercial buffers are labeled NIST-traceable. The European community uses a set of buffers with different chemistries, and these are referred to as the DIN buffers. Most countries recognize other agencies' buffers, but it is best to verify this in cases where the measurements are for international regulatory requirements. As with any chemical measurement where standards are used to quantify a sample, the quality of the standard is critical to accurate sample results. Buffers that are traceable to a defined, accepted agency give more confidence in the sample results than results achieved by untested buffers or buffers of unknown quality. Sample measured against traceable buffers are more easily defended in laboratories where auditing occurs.

What Is Temperature Compensation and How Does One Choose the Best Method for an Analysis?

Temperature compensation is the term for the meter correction for the effect of temperature on the calibration curve. The sample temperature can be received by the meter in various ways. It can be measured by an automatic temperature compensation (ATC) probe, available separately or built directly into the electrode or, in certain new meter circuitries, by the pH electrode itself. Alternately, the operator can manually enter the temperature into the meter. The best method for the user is determined by sample characteristics and cost. An ATC probe or combination pH electrode with internal ATC probe is preferred if sample volume is sufficient. ATC probes are comprised of glass, epoxy, stainless steel, and other materials to provide compatibility with different sample types.

While it is possible for the meter to compensate for the effect of temperature on the calibration, the pH of a solution can itself change with temperature. The solution temperature can shift equilibria within the solution, which can create or scavenge the ions being measured. There is no way to predict or control this temperature effect, so the temperature at which the pH was determined should be recorded to ensure the data can be reproduced in the future. The internal reference of the electrode also requires time to reach a temperature equilibrium with the sensing bulb (the liquid inner filling solution and a solid metal wire change temperatures at different rates). This temperature effect can be minimized by using electrodes with nonmetallic reference systems.

How Does Resolution Affect pH Measurement?

Resolution is the number of decimal places past the decimal point shown by the meter. Some meters can be set to read, as, 9.0, 9.03, or 9.034. It is best to use only the number of places actually required for the measurement. For example, if a solution must be adjusted from pH 4.3 to 4.5, there is no point in trying to read pH 4.302. The more places that are used, the longer it takes to get a steady reading as the electrode gradually approaches the final value. The resolution of less expensive electrodes is not as high as refillable electrodes. Check the electrode specifications to ensure that the appropriate resolution is being used or the electrode may be slow or never fully reach stability. A stabilized pH reading may still "shift" on the display due to temperature fluctuation. This is a normal occurrence, but the effect will be more pronounced at higher resolution settings.

Why Does the Meter Indicate "Ready" Even as the pH Value Changes?

Because the electrode gradually approaches a final value, there is a trade-off between the time delay before a reading is recorded and getting sufficient precision. Meter design applies different criteria for deciding a sufficient wait period. Changes after the "ready" indicator comes on are usually small. A stabilized, temperature-compensated pH reading may still "shift" on the display due to the temperature fluctuations.

Which pH Electrode Is Most Appropriate for Your Analysis?

Sample Matrix

The sample matrix is the most significant factor to be considered. Proteins, Tris buffers, and other biological samples (agar plates, plasma, cells, fermentation vat samples) cause precipitates with silver ions. The best solution is an electrode without silver, but a free-flowing annular junction can reduce this effect because the greater outward flow of filling solution and ease of cleaning makes it less prone to clogging. Traditionally calomel electrodes have been used for protein and Tris buffered samples, but these electrodes use a mercury/mercuric chloride reference instead of silver/silver chloride. The health hazards associated with mercury and the expensive special disposal make them less desirable.

Some manufacturers now offer a nonmetallic, redox couple reference system which is not hazardous to biological samples or lab personnel. It contains no ions to cause precipitates with proteins. New electrodes are also available with isolated Ag wire references and AgCl inner filling solutions that do not come in contact with the sample. These electrodes use a polymer gel to keep the sample and silver separated. Electrode manufacturers can guide you to the best electrode for your sample.

Sample Volume and Format

Standard electrodes are typically 12 to 13 mm in diameter, but these might not contact the sample without first transferring a portion of the sample to a suitable container. A longer electrode will be required for samples that cannot be disturbed or transferred. Small sample volumes can be accommodated by semimicro electrodes with sensing bulb diameters of 4 to 6 mm, while flat surface electrodes can measure the pH of a sample surface, such as agar plates, cheese, or a drop of a limited quantity of sample. Microelectrodes are available, some with needles to pierce septa and measure pH inside a vial. This would benefit sample measurements where exposure to air is not desired. Microelectrodes can also measure pH in microtiter plates.

Temperature

Temperature of the sample is another consideration. Most glass electrodes are stable up to 100°C. Other electrodes are designed to be steam sterilizable or autoclavable. Epoxy bodied electrodes cannot be used at excessively high temperatures; some are stable up to 90°C but many are rated for 80°C.

Combination Electrodes

Combination electrodes are a single electrode that contains both the reference electrode and the sensing electrode within one body. The reference electrode is usually a silver wire with AgCl solution surrounding it, although it can also be a calomel, redox couple, or other reference system. The sensing electrode is usually pH glass but it can be a special transistor in the case of ion specific field effect transistor (ISFET or FET) electrodes. Combination electrodes requires smaller sample volumes than two-electrode systems, but lack their ability to isolate and vary the reference, which allows for more experimental control.

Refillable or Nonrefillable

Nonrefillable electrodes are less costly and require low maintenance. They may be submersible as there is no hole on the side that must have atmospheric pressure to cause the filling solution to flow. They usually have a fiber or wick junction for the gelled filling solution to leak outside and this flow cannot be stopped. Often these junctions exhibit a "sample memory" which is due to backflow of sample material into the electrode. Several double-junction, low-maintenance electrodes utilize a high-performance polymer for the internal filling solution. These electrodes have open junctions, where the polymer fills a hole in the glass and the silver/silver chloride reference is isolated from the sample by the polymer. These electrodes also can be used with commercial production samples since the silver never contacts the sample, just the gel which has no silver in it. While some polymer systems are prone to hydrolysis, these electrodes offer longer lifetimes, low maintenance and advanced features.

The refillable electrode offers the greatest flexibility and longest lifetime. Most are stored dry so they are better suited for infrequent users, such as classrooms. The filling solution may be altered as required by the samples and when sample contamination occurs. A refillable electrode costs more initially but can be more cost effective due to longer lifetime than a less expensive, nonrefillable electrode.

How Can You Maximize the Accuracy and Reproducibility of a pH Measurement?

New Systems

Prepare and Condition the Electrode

Electrodes need to be conditioned prior to use. The combination electrodes most commonly used in labs have several components that require stabilization for reliable operation. During long-term storage the electrode dries up to some extent, and refillable electrodes stored dry need to be filled with internal filling solution which itself must equilibrate chemically and thermally with the reference material. In addition the glass-sensing electrode needs to be hydrated to measure the pH. The junctions need to be flowing again so that the buffer and sample can create contact with the reference electrode. Refillable electrodes stored wet with their fill hole covers closed need the fill hole opened to create a positive flow at the junction. Junctions, particularly ceramic designs, become clogged as they dry out due to salt formation from KCl or other sample components, so they require soaking to flow properly. For all of these reasons, the electrode needs time to stabilize before use.

Conditioning is important for a good start to the analysis. The electrode should be soaked for approximately 15 minutes in a commercial storage solution to hydrate and equilibrate the electrode. If storage solution is not available, a mixture of 200 ml of pH 7 buffer and 1 gram of KCl may be used temporarily. Conditioning the electrode around pH 7 is the usual choice for the first calibration point, so it is the best choice for most electrodes. An exception is ion-selective field effect transistor (FET) electrodes which are better conditioned by pH 4 buffer. A calibration curve should then be performed on the reconditioned electrode. If the calibration fails and excessive drift or sluggish behavior is observed, ensure the junction is working and condition the electrode a little longer.

Proceed as per Existing Systems Below.

Existing Systems

Inspect the Reference Filling Solution

Optimally the solution should be filled to just below the fill hole at the start of the day or whenever the level lowers significantly. The electrode should be allowed to equilibrate with its new solution, which may vary slightly in temperature and composition from that already in the electrode.

The meter and electrode system should be calibrated at the beginning of each day of use and then approximately every two hours or when electrode performance is in question. Frequent calibration is recommended because sample components can migrate into the solution if the flow is not adequate or is interrupted for a time. If the pH response is slow, if the solution appears dirty or unusual, or if the samples are known to clog junctions, the filling solution should be changed. When routine maintenance is performed, the filling solution should be drained from the electrode and refilled with fresh solution.

Calibrate the System

The pH meter should be multiple-point calibrated at the beginning of each day of use and then about every two hours for accurate results. The electrode's sensing glass, filling solution, and junction change slightly during sample analysis and recalibration accounts for this. A one-point calibration is often used for the recalibration, since it will shift the millivolts intercept(E_0) of the line without altering the multiple-point calibration slope from the beginning of the day. The calibration should use buffers that bracket the pH of the samples to be measured, as discussed earlier.

The buffers and samples should be stirred using a magnetic stirring plate with an insulation barrier between the beaker and plate to prevent drift due to the heat generated by the stir plate. The first buffer should be pH 7 as the meter determines the E_0 point. The other pH buffers should be measured in order from lowest to highest, and the electrode rinsed with deionized water between buffers to reduce equilibration time. If the pH meter does not automatically monitor and incorporate temperature effects, these data should be manually entered into the instrument.

Verify the potency of your buffer standards. Buffer solutions expire over time because trace CO_2 from the air will leach into bottles, form carboxylic acid and cause shifting pH values and consuming buffer capacity. This is especially evident in pH 10 buffer where an open beaker will become more acidic over the course of a few hours. Because of varying environmental conditions, the lifetime of a buffer standard cannot be predicted. For this reason single-use buffer pouches have become very popular.

Measure the pH

The electrode must be immersed into the buffer or sample so that the sensor of the electrode and the junction are submersed. The filling solution of the electrode must be at least one inch above the sample surface. The buffers and samples should be stirred using a magnetic stirring plate with a thermal insulation barrier between the beaker and plate. The best reproducibility is achieved when the buffer and samples are measured at the same temperature. The electrode should be rinsed with deionized water, shaken or blotted dry, and rinsed with the next solution to be measured. The electrode should not be wiped dry because static discharge will build up on the electrode and give drifting readings until the discharge has dissipated. Ideally temperature should be recorded with every pH measurement, and a calibration should be run after the last sample to ensure the electrode still meets performance criteria. Quality control samples (i.e., a pH 5 buffer when using a meter calibrated at pH 4 and 7) can be interspersed among the samples.

How Do Lab Measurements Differ from Plant or Field Measurements?

Field measurements lack the controlled environment and sample handling of the laboratory. Field and plant measurements are often measured by placing the electrode directly into the total sample, rather than measuring an aliquot from a larger batch. This can mean less control of sample turbulence or homogeneity which can cause pH drift. The temperature should be recorded and calibration should occur at the site where the samples are measured.

Does Sample Volume Affect the Accuracy of the pH Measurement?

Generally, the sample volume itself doesn't directly affect pH accuracy, but reproducibility will suffer if the electrode sensor and junction are not submersed in the sample. As described below, smaller volumes are more prone to pH alteration from atmospheric CO_2. Temperature and cooling rates can vary with sample volume, so it is best to use the same amount of sample in each measurement. Any sample treatment (i.e., dilution or measuring a supernatant) should be performed for all samples; fewer variables generate more reliable data. Automated sampling systems will operate best with the same volume as the electrode will be lowered into the beaker a fixed amount.

How Do You Measure the pH of Viscous, Semisolid, Low Ionic Strength, or Other Atypical Samples?

Viscous and semisolid samples are best measured with a fast flowing flushable junction electrode or FET technology. Flushable junctions are designed for easy cleaning and will allow a better sample contact with the faster flow. FET sensing electrodes can be cleaned without the polarization issues of glass electrodes. Low ionic strength samples are best measured with refillable electrodes with low ionic strength filling solution.

Calibrating the system with buffers similar in ionic strength to the sample will increase reproducibility when measuring unusual samples. Small volumes of samples with low ionic strength can be affected by exposure to air. The greater ratio of surface area to volume for a smaller sample increases the potential for CO_2 to mix with the sample and cause a pH shift than with a larger sample. This is observed in ultrapure water and is part of the USP injection water testing protocol (USP 645, US Pharmacopeia).

How Can You Maximize the Lifetime of Your pH Meter?

Proper Usage

If the manufacturer's instructions are followed and product ratings adhered to, a quality meter should last many years. Protection of the meter from liquids, wiping up spills and respectful use gives long life. If the meter is to be used in harsh environments, use a meter rated rated for such work. For example, a waterproof system is better suited for work in the field or on ships.

Proper Cleaning

Precipitates at the Electrode Junction

The precipitate that forms at the electrode junction is crystallized potassium chloride from the inner filling solution. This "KCl creep" is created as the inner filling solution containing KCl comes through the junction when there is no sample or liquid on the external side of the junction. The water of the filling solution evaporates and crystals form. The creeping KCl should be rinsed away with deionized water and the filling solution height checked prior to putting the electrode back into service.

Clogged Electrode Junctions

There are several junction types, and they require different cleaning techniques. Consult the instruction manual for your particular electrode. In general, soaking or sonicating in a commercial cleaning solution for your sample type or a dilute hydrochloric acid solution can often remove sample buildup. Protein accumulation often requires a cleaning solution with pepsin for faster removal. After cleaning, the filling solution chamber of the electrode should be flushed with copious amounts of deionized water, then rinsed with filling solution several times. The filling solution rinses ensure that the electrode is put back into service with the proper concentration instead of a diluted filling solution. If your junction requires frequent cleaning, a different reference system or filling solution should be investigated. A junction cannot always be sufficiently cleaned; the electrode must be replaced.

Proper Storage

Precisely follow the manufacturer's recommendations for electrode storage. Some electrodes, such as gel-filled electrodes, should be stored in pH storage solution. They might be ruined if stored dry. The standard refillable electrodes are often stored with filling solution, and the fill hole cover closed. The sensing element is capped and kept moist with a few drops of pH storage solution. The filling solution can be emptied and then refilled when the electrode is returned to use. Some sleeve junction electrodes can be stored dry. Crystals may appear from evaporated residual filling solution, but they can be rinsed away with deionized water prior to returning the electrode to service.

Refillable electrodes offer a longer life as they are better designed for storage and do not have a fixed filling solution volume to dictate lifetime. Gel electrodes have a finite amount of continually leaking gel and when the gel is depleted, the electrode must be replaced. Refillable FET electrodes can be stored dry and refilled prior to use. The lifetime of any electrode is dependent upon level of care and maintenance, sample/application, type of filling solution and amount of filling solution if it is non-refillable.

TROUBLESHOOTING

Is the Instrument the Problem?

Meter

The meter alone, without the electrode, can be tested to verify performance. A quality pH meter can be tested easily by using a shorting cap over the electrode input to shunt or close off the BNC connector. This will allow the meter's internal diagnostics or self-test to check circuits and ensure that the electronics are functioning properly. There will be an error displayed on the meter if any tests have failed. Consult the instruction manual for details.

Slope

The best indicators of the electrode condition are the slope of the calibration curve and response time required to obtain a stable pH reading. A clean, well-performing electrode will produce a slope close to 100% or 59.16 mV/decade, which is the theoretical slope for pH determined by the Nernst equation. As any electrode ages, the percent slope decreases. This natural occurrence can be slowed by proper use and care of the electrode. The recommended operating range varies slightly by manufacturer but is usually 92% to 100% of the theoretical ideal above. The electrode should be replaced when the slope falls below the manufacturer's recommended operating range.

Response Time

The response time, or the time it takes until the reading stabilizes, will become longer as the sample components coat the sensing glass bulb. This can often be remedied with cleaning and/or replacing the filling solution. There is a point when the electrode may have damage that won't be recovered by cleaning. If the calibration data fall within the manufacturer's specifications, the sample may be causing the problem.

Is the Sample the Problem?

If the sample reading seems inappropriate, measure the pH of a buffer standard. A correct measurement of the standard points to a sample problem. If the electrode is sluggish or does not stabilize when measuring the standard, clean and recalibrate the electrode. Reanalyze the buffer standard with the cleaned electrode in the buffer to verify system operation. If this measurement is accurate, measure the sample again. If the sample still does not give a stable reading, further investigation into the measurement techniques and sample itself is recommended.

Sometimes the "expected" value is not obtained for a measurement but the correct value is. The problem is simply an incorrect perceived value. Competing ions, sodium ions at pH of 12 or above, or a sample that coats the electrode can affect pH measurements. pH sensing glass is optimized for hydrogen ions, but sodium ions are also detected to a lesser extent. This sodium error increases at high pH levels. A nomogram found in the electrode instruction manual can be used to correct the pH reading in samples with high sodium content. Other compounds or ions could be "complexed" out of solution or bound up or change its form so that it doesn't affect the sample any more.

Often the sample can be better analyzed using a different electrode design. Inexpensive gel electrodes with wick junctions—where the sample can migrate back into the gelled reference fill solution—are not as effective as refillable electrodes in complex matrices. Samples that may contaminate the filling solution are best analyzed with flushable electrodes. Samples need a sufficient amount of water to give a pH reading; a diluted sample may be measured more reproducibly. Samples and buffer should always be measured at the same temperature if possible. The sample may change its composition with temperature variation. pH is a relative measurement, so it might be necessary to optimize your sample preparation method.

Service Engineer, Technical Support, or Sales Rep: Who Can Best Help You and at the Least Expense?

The electrochemical measurement of hydrogen ion activity is simple, yet complex. Due to the many factors and interactions, many users increase errors in their measurements inadvertently. The best way to optimize your results is to educate yourself about your measurement system. Follow the instructions that the equipment manufacturer provides.

There are many versions of the standard glass pH electrode. Be sure that you are using the best electrode for your sample type. The sample only has contact with the electrode. So, if the electrode is not working properly, you cannot expect accurate results. The electrode preparation and conditioning steps are critical and vary by electrode type. Knowledge of your sample guides you to the proper measurement system and calibration procedures.

If you do need technical support for your analysis, it is best to call the company that manufactured your electrode. Due to the minimal cost of a pH meter as compared to other laboratory instruments and the replaceable electrode, service engineers are not a cost-effective option.

SPECTROPHOTOMETERS (Michael G. Davies and Andrew T. Dadd)

This overview addresses some of the basic aspects of UV-visible spectrophotometry and summarizes some of the standard operating procedures. It provides the reader with the fundamental background to select and operate a UV/Vis instrument addressing both specific and general requirements. This section also presents a number of methodologies that are currently available to successfully perform quantitative and qualitative analysis of macromolecules (e.g., proteins and nucleic acids) and small molecules including nucleotides, amino acids, or any UV/Vis-absorbing compounds

What Are the Criteria for Selecting a Spectrophotometer?

Most entry level instruments perform the most common applications involving the analysis of proteins and nucleic acids. The following information is provided to help you refine your choice of instrumentation.

What Sample Volumes Will You Most Frequently Analyze?

Advances in manufacturing of cuvettes has allowed greater flexibility both in terms of volumes and concentrations for the assessment of UV/Vis spectra or single/multiple wavelengths. Cell volumes as low as $10\,\mu l$ may be employed in some instruments, whereas special holders that position the cuvette in the light path might be required for others. Further details on cell types is provided later in this chapter. It is worth noting that continuous-flow as well as temperature-controlled cell holders are available for specific applications.

External Computer (PC)

PC control is especially beneficial for logging and archiving data via disk, LIMS (Laboratory Information Management System) and networks, and for producing customized reports. Spectrophotometers managed by an external PC will almost always provide more functions to analyze and manipulate data. Most free-standing instruments perform scanning, kinetics, quantitative analysis, and other functions, but the ability to store and manipulate data is usually limited. Some manufacturers sell software for use with an external PC that expands data manipulation and functionality. A combination of lower-cost instrument and supplemental software sometimes provides the most function for the least money. This may be offset by the extra bench space requirements and the cost of a PC.

Single Beam or Double Beam versus Diode Array

There are three modes of optical configuration available in UV-visible spectrophotometry.

Single-beam instruments with microprocessor control have good stability and simple optical and mechanical configurations. A light source is monochromated (a single wavelength is selected) usually by a diffraction grating (or a prism in older instruments) and then passed through the sample cuvette. Comparison between reference and sample is achieved by feeding the postdetector signal to a microprocessor that stores the reference data for subtraction from the sample signal prior to display of the result.

The light beam within a double-beam spectrophotometer is split or chopped and passed through both the sample and reference solutions to obtain a direct reading of the difference between them. This is useful in applications where the reference itself is changing and constant baseline subtraction can be employed for compensation, as can occur in enzyme analyses of biological systems. In a double-beam system a portion of the originating light energy is passed through the sample, and optically matched cuvettes need to be used for proper results.

A third optical configuration is the diode array. Here light is monochromated after passing through the sample. Transmitted light is then focused and measured by an assembly of individual detector elements arranged to collect a complete range of wavelengths. No sample compartment lid is necessary. Wavelength selection is dictated by the choice of detector elements (approximately 500 at 1 nm/diode), providing a more limited spectral range than single- or double-beam instruments.

Wavelength Range

Nucleic acids and proteins require the UV range 230 to 320 nm almost exclusively. Other compounds can be analyzed by monitoring specific wavelengths and scanning within the visible range. Until recently instruments were categorized into visible only (>320 nm) or UV and visible (190–1100 nm) primarily as a reflection of lamp technology. With improvements in lamp design and detector technology, another class of instrument can monitor absorbances between 200 and 800 nm with a single lamp. These compact instruments are designed mostly to measure the purity and concentration of nucleic acids and proteins, and some also possess basic scanning capabilities. For in-depth identification and verification studies or for a core facility, an instrument capable of scanning between 190 and 1100 nm is recommended.

Wavelength, Photometric Accuracy, and Stray Light

Wavelength accuracy describes the variation between the wavelength of the light you set for the instrument and the actual wavelength of the light produced. The variation in most instruments ranges from 0.7 to 2 nm. Should an instrument suffer from wavelength inaccuracy, the largest variation would be observed at wavelengths on either side of the absorbance maximum for a molecule, where there is a large rate of change of absorbance with respect to wavelength (Figure 4.13) and when working with dilute solutions. Note in Figure 4.13 the significant decrease in absorbance at wavelengths near 280 nm and above. Any wavelength variation by the instrument will produce very skewed data in these changeable regions of DNA's absorbtion spectra. This phenomenon also explains why A_{280} results in a very dilute sample have to be interpreted with caution.

Photometric accuracy describes the linearity of response over

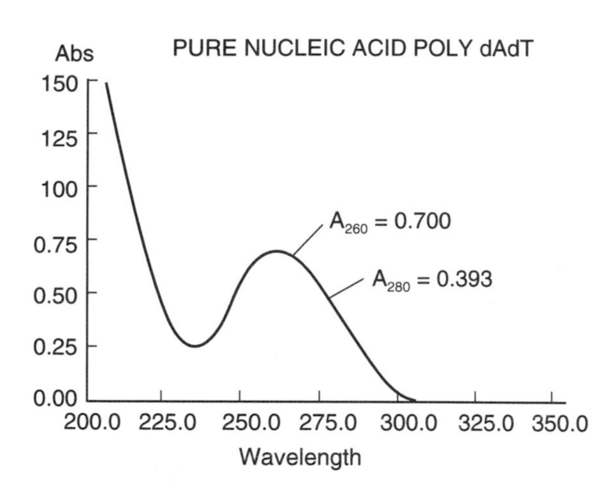

Figure 4.13 UV-visible absorption scan of DNA. Reproduced with permission from Biochrom Ltd.

the absorbance range. Normally this is expressed up to two absorbance units at specific wavelengths as measured against a range of calibrated standard filters from organizations such as the NIST. Typically it is within 0.5%. As most photodetectors are generally accurate to within 1%, the main factors compromising accuracy are errors in light transmission, most commonly stray light.

Stray light is radiation emerging from the monochromator other than the selected wavelength. This extra light causes the measured absorbance to read lower than the true absorbance, creating negative deviations from the Beer-Lambert law (Biochrom Ltd., 1997), ultimately ruining the reliability of subsequent concentration measurements. Stray light has a relatively large effect when sample absorbance is high, as in high concentrations of DNA measured at 260 nm. Dilution of concentrated samples or use of a smaller path length cell removes this effect.

Spectral Bandwidth Resolution

Bandwidth resolution describes the spectrophotometer's ability to distinguish narrow absorbance peaks. The natural bandwidth of a molecule is defined by the width of the absorbance curve at half the maximum absorbance height of a compound, and ranges from 5 to 50 nm for most biomolecules. The bandwidth of DNA is 51 nm, when measured from the spectrum in Figure 4.13. It can be shown that if the ratio of spectral to natural bandwidth is greater than 1:10, the absorbance measured by the spectrophotometer will deviate significantly from the true absorbance. A spectrophotometer with a fixed bandwidth of 5 nm or less is ideal for biopolymers, since there is no fine spectral detail, but for samples with sharp peaks such as some organic solvents, transition elements, and vapors like benzene and styrene, higher resolution is required.

Good Laboratory Practice

There has been an increase in laboratory requirements to conform with Good Laboratory Practice (GLP) techniques according to FDA regulations (1979). The FDA requires that results be traceable to an instrument and the instrument proved to be working correctly. Instrument performance criteria for spectrophotometers have been defined by the European Pharmacopoeia (1984) as being spectral bandwidth, stray light, absorbance accuracy, and wavelength accuracy. Standard tests are laid down and are checked against the appropriate filters and solutions to confirm instrument performance.

Beyond the Self-Tests Automatically Performed by Spectrophotometers, What Is the Best Indicator That an Instrument Is Operating Properly?

The Functional Approach

Measure a series of standard samples via your application(s) in your instrument and, if possible, a second spectrophotometer. Calibrated absorbance filters can be obtained from NIST and from commercial sources (Corion Corporation, Franklin, MA; the National Physical Laboratory, Teddington. London; and Starna, Hainault, U.K.). Quantitated nucleic acid solutions are commercially available (Gensura Corporation, San Diego, CA) but do not provide reproducible data over long-term use. Nucleic acid and protein standards prepared from solid material as required is recommended provided that the concentrations are carefully determined. Do not rely on the quantity of material indicated on the product label as an accurate representation of the amounts therein.

The Certified Approach

National and international standards organisations will likely require some or all of the following tests.

Bandwidth/Resolution

For external checks against a universally adopted method the Pharmacopoeia test is used. (*European Pharmacopeia*, 1984). The ratio of the absorbances at 269 and 266 nm in a 0.02% v/v solution of toluene *R* (*R* = reagent grade) in hexane *R* is determined as in the *European Pharmacopoeia* (2000).

Stray Light

Stray light is determined using a blocking filter that transmits light above a certain wavelength and blocks all light below that wavelength. Any measured transmittance is then due to stray light. The *European Pharmacopoeia* (2000) specifies that the absorbance of a 1.2% w/v potassium chloride *R* should be greater than 2.0 at 200 nm, when compared with water *R* as the reference liquid.

Wavelength Accuracy

This is determined using a standard that has sharp peaks at known positions. According to the *European Pharmacopoeia* (2000), the absorbance maxima of holmium perchlorate solution, the line of a hydrogen or deuterium discharge lamp, or the lines of a mercury vapor arc can be used to verify the wavelength scale.

Wavelength Reproducibility

Wavelength reproducibility is determined by repeatedly scanning a sharp peak at a known position, using the same standard as for wavelength accuracy.

Absorbance Accuracy (Photometric Linearity)

The absorbance of neutral density glass filters, traceable to NIST, NPL (National Physical Laboratory) *www.nist.gov*, *www.physics.nist.gov* or other internationally recognized standards, is measured for a range of absorbances at a stated wavelength. Neutral density filters provide nearly constant absorbances within certain wavelengths of the visible region, but measurements in the UV require metal on quartz filters or a liquid standard such as potassium dichromate *R* in dilute sulphuric acid *R* (*European Pharmacopoeia*, 2000). Metal on quartz filters can exhibit reflection problems, and dirt can contaminate the metal coating. The liquid must be prepared fresh for each use; sealed cells of potassium dichromate prepared under an argon atmosphere are commercially available.

Photometric Reproducibility

Photometric reproducibility is determined by repeatedly measuring a neutral density filter.

Noise, Stability, and Baseline Flatness

Noise is determined by repeatedly measuring the spectrum of air (no cuvette in the light path) at zero absorbance. This is achieved by setting reference on air. It is specified as the calculated RMS (root mean square) value at a single wavelength. The stability is the difference between the maximum and minimum absorbance readings at a specified wavelength (at constant temperature). The RMS (square root of $[a_1^2 + a_2^2 + a_3^2 + \ldots]$, where a represents the absorbance value at each wavelength) is calculated over the whole instrument wavelength range for the spectrum of air to provide the baseline flatness measurement.

Which Cuvette Best Fits Your Needs?

Small Volumes

Cuvettes with minimal sample volumes of 250 μl or greater usually do not require dedicated cuvette holders and are compatible with most instruments. Cuvettes with minimal sample volumes between 100 and 250 μl might require a manufacturer-specific, single-cell holder, and cuvettes requiring sample volumes below 100 μl almost always require specialized single-cell holders that are rarely interchangeable between manufacturers. These ultra-low-volume cuvettes have very small sample windows (2 × 2 mm) that require a specialized holder to align the window with the light beam. Some manufacturers recommend the use of masked cells to reduce overall light scatter.

If the light path length of your cuvette is less than 10 mm, check if the instrument automatically incorporates this when converting absorbance data into a concentration. The Beer-Lambert equation assumes a 10 mm path length. A double-stranded DNA sample that produces an absorbance at 260 nm of 0.5 in a cuvette of 10 mm path length produces a concentration of 25 μg/ml. The same sample measured in a cuvette with a 5 mm path length produces an absorbance of 0.25, and concentration of 12.5 μg/ml if the spectrophotometer does not take into account the cuvette's decreased path length. Capillaries of 0.5 mm path length can analyze very concentrated samples without dilution, but the quantitative reproducibility can suffer because of this extremely short light path.

Disposable Cuvettes

Plastic cuvettes are not recommended for quantitative UV measurements because of their reduced transmittance below 380 nm, which may seriously compromise accuracy and sensitivity of some quantitative methods. Polystyrene cuvettes may be replaced by a methacrylate-based version that supposedly allow higher transmittance values over the common plastic cuvettes. Cuvettes composed of novel polymers with superior absorbance properties are in development. However, caution should be exercised to ensure solvent compatibility using any material.

What Are the Options for Cleaning Cuvettes?

Dirty cuvettes can generate erroneous data, as they can trap air bubbles or sample carryover. Cuvettes made from optical glass or quartz should be cleaned with glassware detergent or dilute acid (e.g., HCl up to concentrations of 0.1 M) but not alkalis, which can etch the glass surface. When detergent is insufficient, first inspect your cuvette. If it is comprised of a solid block of glass or quartz and you see no seams within the cuvette, you can soak it in concentrated nitric or sulfochromic acids (but not HF) for limited periods of time. Then the cuvettes must be rinsed with copious amounts of water with the aid of special cell washers ensuring continuous water flow through the cell interior. Exposure to harsh acid must be of limited duration due to the possibility of long-term damage to the cuvette surface. Alternatively, polar solvents can also be employed to remove difficult residues. One cuvette manufacturer claims to provide a cleaning solution that is suitable for all situations (Hellmanex, Hellma, Southend, U.K.). Seams are indicative of glued joints and are more commonly present in low sample volume cuvettes. The interior sample chambers of seamed cuvettes can be treated with acid but not the seams. Cuvettes made from other materials or mixtures with glass should be treated with procedures compatible their chemical resistance.

How Can You Maximize the Reproducibility and Accuracy of Your Data?

Know Your Needs

Must your data be absolutely or relatively quantitative? If your situation requires absolute quantitation, your absorbance readings should ideally fall on the linear portions of a standard calibration curve. Dilute your sample if it's absorbance lies above the linear portion, or select a cuvette with shorter path length. If your absorbance values reside below the linear portion and you can't concentrate your samples, include additional calibration standards (to the original standard curve) that are similar to your concentration range. The objective is to generate curve-fitting compensation for values outside linear response.

Know Your Sample

What are the possible contaminants? Are you using phenol or chloroform to prepare DNA? Could the crushed glass from your purification kit be leaking out with your final product? If you can predict the contaminants, methods exist to remove them, as described in Chapter 7, "DNA Purification" and Chapter 8, "RNA

Purification." Many spectrophotometers also can compensate for contaminants by subtraction of reference or 3-point net measurements. If you can't predict the contaminant, scan your sample across the entire UV-visible spectrum, and compare these data to a scan of a purified sample control. The type of interference is indicated by the wavelengths of absorbance maxima that are characteristic of particular molecular groups and such information is available in Silverstein et al. (1967). Possible contaminants may be signified by comparison of outstanding absorbance peaks against an atlas of reference spectral data (e.g., commercially available from Sadtler, Philadelphia, PA). However, reference data sometimes do not give an accurate match, and it is more accurate and relevant to exploit the attributes of a fast scanning spectrophotometer and generate spectra of materials involved in the sample preparation procedure. This can give a direct comparison on the same instrument. Combined with the use of a PC for archiving, it is a convenient way to build up specific sample profiles for searching and overlays.

Cell suspension measurements at 600 nm (A_{600}) provide a convenient means of monitoring growth of bacterial cultures. Provided that absorbance is not above 1.5 units, A_{600} correlates quite well with cell numbers (Sambrook et al., 1989). The geometry of an instrument's optical system affects the magnitude of these absorbance measurements because of light scattering, so A_{600} values can vary between different instruments.

Opaque, solid, or slurried samples may block or scatter the light, preventing accurate detector response. A special optical configuration is required to deal with these samples to measure reflectance as an indicator of absorbance. This requires a specifically designed source and sample handling device, and costs can surpass the spectrophotometer itself.

Know Your Instrument's Limitations

Instruments costing the equivalent of tens of thousands of dollars might generate reproducible data between absorbance values of 0.001 and 0.01, but the scanning instruments found in most laboratories will not. Ultra-dilute samples are better analyzed using a long path length cell or a fixed wavelength monitor of high specification. A low sample volume cuvette might reduce or eliminate the need to dilute your sample.

How low an absorbance can your instrument reproducibly measure? Perform a standard curve to answer this question. Note that absorbance can be reproducible, but if the absorbance measurement does not fall on the linear part of the calibration curve, it might not correlate well with concentration.

What Can Contribute to Inaccurate A_{260} and A_{280} Data?

Instrument Issues

Aging, weakened UV lamps can generate inaccurate data, as can new deuterium lamps that were not properly warmed up (20–40 minutes for older instruments). Start-up is not an issue for most instruments produced within the last 10 years, which usually only require 10 minutes and may be accompanied by automatic internal calibration (required for GLP purposes). Lamp function is discussed in more detail below.

Sample Concentration

Measuring dilute samples that are near the sensitivity limits of the spectrophotometer is especially problematic for A_{280} readings. The sharp changes on either side of 280 nm (Figure 4.13) amplify any absorbance inaccuracy.

Contaminants

Contaminating salt, organic solvent, and protein can falsely increase the absorbance measured at 260 nm. Contaminants can be verified and sometimes quantitated by measuring absorbance at specific wavelengths. The additive effect on the spectrum is detected by alteration in the relevant absorbance ratio ($A\lambda_1/A\lambda_2$) as shown in Figure 4.14.

Absorbance at 230 nm

Tris, EDTA, and other buffer salts can be detected by their absorbance of light at 230 nm, a region where nucleotides and ribonucleotides generally have absorbance minima. At 230 nm this also is near the absorbance maximum of peptide bonds, indicating the presence of proteins. Therefore readings at 230 nm or preferably a scan incorporating wavelengths around 230 nm can readily show up impurities in nucleic acid preparations. High-absorbance values at 230 nm indicate nucleic acid preparations of suspect purity. In preparation of RNA using guanidine thiocyanate, the isolated RNA should exhibit an A_{260}/A_{230} ratio greater than 2.0. A ratio lower than this is generally indicative of contamination with guanidine thiocyanate carried over during the precipitation steps.

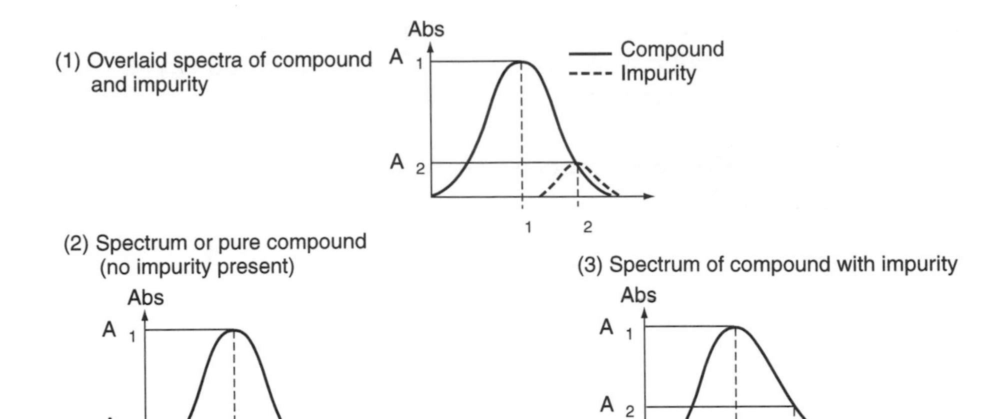

Figure 4.14 Detecting contaminants by absorbance ratio. Reprinted by permission of Biochrom Ltd.

Absorbance at 320 nm

Nucleic acids and proteins normally have virtually no absorbance at 320 nm, although absorbances between 300 and 350 nm may be indicative of aggregation, particularly in the case of proteins. Subtracting the absorbance at 320 nm from the absorbance detected at 260 nm can eliminate absorbance due to contaminants such as chloroform, ethanol, acetates, citrates, and particulates that cause turbidity. Background absorbance at 320 nm is more likely to skew the A_{260} readings of very dilute nucleic acid solutions or samples read in ultra-low-volume (<10 μl) cuvettes.

Does Absorbance Always Correlate with Concentration?

The Beer-Lambert law (Biochrom Ltd., 1997) gives a direct proportional relationship between the concentration of a substance, such as nucleic acids and proteins, and its absorbance. So a graph of absorbance plotted against concentration will be a straight line passing through the origin. Under straight line conditions, the concentration in an unknown sample can be calculated from its absorbance value and the absorbance of a known concentration of the nucleic acid or protein (or an appropriate conversion calibration factor).

When this Beer-Lambert relationship between absorbance and concentration is not linear, DNA and protein cannot be measured accurately using one factor (i.e., molar extinction coefficient) or concentration for calibration. For the greatest accuracy the absorbance readings have to be calibrated with known concentrations similar to those in the samples. The calibration standard range should cover the sample concentrations, which are measured to allow curve-fitting compensation for values outside linear response. Deviations from linearity result from three main experimental effects: changes in light absorption, instrumentation effects, and chemical changes.

Changes in light absorption can be produced by refractive index effects in the solution being measured. Although essentially constant at low concentrations, refractive index can vary with concentration of buffer salts, if above 0.001 M. This does not rule out quantitation as measurements can be calibrated with bracketing standard solutions or from a calibration curve.

Instrumentation effects arise if the light passing through the sample is not truly monochromatic, which was mentioned earlier in the section on spectral bandwidth. The Beer-Lambert law depends on monochromatic light, but in practice at a given spectrophotometer wavelength, a range of wavelengths, each with a different absorbance pass through the sample. Consequently the amount of light measured is affected and is not directly proportional to concentration, which results in a negative deviation from linearity at lower light levels due to higher concentrations. This effect only becomes apparent if absorbance peaks are narrow in relation to spectral bandwidth; it is not a problem with specifications set as discussed in that earlier section.

Chemical deviations arise when shifts occur in the wavelength maximum because of solution conditions. Some nucleotides are affected when there are pH changes of the buffer solvent, giving shifts of up to 5 nm. The magnitude of absorbance at 260 nm changes for DNA as it shifts from double-stranded to single-stranded, giving an increase in absorbance (hyperchromicity). In practice, frozen DNA solutions should be well thawed, annealed at high temperatures (80–90°C) and cooled slowly before measurements.

Why Does Popular Convention Recommend Working Between an Absorbance Range of 0.1 to 0.8 at 260 nm When Quantitating Nucleic Acids and When Quantitating Proteins at 280 nm?

Most properly functioning spectrophotometers generate a linear response (absorbance vs. concentration) between absorbance values of 0.1 and 0.8; hence this range is considered safe to quantitate a sample. If you choose to work outside this range, it is essential that you generate a calibration curve containing a sufficient number of standards to prove a statistically reliable correlation between absorbance and concentration. Such a calibration study must be performed with the cuvette to be used in your research. Cuvette design, quality, and path length can influence the data within such a calibration experiment. Calculations of protein and peptide concentration also require linearity of response and the same principles apply to their measurements.

Deuterium lamps can generate linear responses up to three units of absorbance; the linear response of xenon lamps decreases at approximately two units of absorbance.

Is the Ratio $A_{260}:A_{280}$ a Reliable Method to Evaluate Protein Contamination within Nucleic Acid Preparations?

The original purpose of the ratio $A_{260}:A_{280}$ was to detect nucleic acid contamination in protein preparations (Warburg and Christian, 1942), and not the inverse. This ratio can accurately describe nucleic acid purity, but it can also be fooled. The stronger extinction coefficients of DNA can mask the presence of protein (Glasel, 1995), and many chemicals utilized in DNA purification absorb at 260nm (Huberman, 1995). Manchester (1995) and Wilfinger, Mackey, and Chomczynski (1997) show the very significant effects of salt and pH on absorbance of DNA and RNA preparations at 260 and 280nm.

If you doubt the validity of your $A_{260}:A_{280}$ data, check for contaminants by monitoring absorbance between 200 and 240nm, a region where nucleic acids absorb weakly if at all, as described above. As discussed in Chapter 1, "Planning for Success in the Lab", a contaminant is problematic only if it interferes with your application. If contaminant removal is necessary but impractical, Schy and Plewa (1989) provide a method to assess the concentration and quality of impure DNA preparations by monitoring both diaminobenzoic acid fluorescence and UV absorbance.

What Can You Do to Minimize Service Calls?

Respect the manufacturers suggested operating temperatures and humidity levels, and avoid dust. Spills should be avoided and cleaned up immediately. This is because some materials not only attack instrument components but can also leave UV-absorbing residues and vapors.

How Can You Achieve the Maximum Lifetime from Your Lamps?

Deuterium

Older designs of deuterium lamps require that the lamp be powered up and kept on prior to sample measurement. The best indicator of vitality in these older designs is the hours of UV lamp use. As lamps approach the manufacturer recommended lifetimes, the light energy fades, producing erratic, irreproducible absorbance measurements. Deuterium lamps also lose effectiveness when stored unused and should not be kept longer than one year before use.

Should you automatically discard a deuterium lamp when it reaches the predicted lifetime? The answer is no. Deuterium lamps can generate accurate, reproducible data beyond their predicted lifetimes. Simply monitor the accuracy of an older lamp with control samples. Recently designed pulsed technology deuterium lamps turn on only when a sample is read (demand switching), resulting in lifetimes of five years or more.

Frequently switching the power on and off will prematurely weaken most deuterium lamps, but not the demand-switched lamps described above, which can last through thousands of switching cycles.

Tungsten

Tungsten lamps tend to give longer lifetimes—at least six months if left on continuously and several years when used during normal working hours. During long use, instruments tend to drift because of warming-up, while background noise decreases. It is better to leave instruments on during the working day and re-reference if lower noise measurements are required. Switching frequently may shorten total lamp lifetime unless the control circuits have been designed to minimize lamp wear on switching.

Xenon

Xenon lamps flash on only when a sample is read, resulting in lifetimes of 1000 to 2000 hours or more of actual use. Lifetime is not affected by frequent switching on and off.

The Deuterium Lamp on Your UV-Visible Instrument Burned Out. Can You Perform Measurements in the Visible Range?

With current internal calibration software, instruments can still self-calibrate and operate through the visible range without the deuterium lamp. Tungsten sources cover the range from 320 to 1100 nm, giving overlap at the lower end of the range into the UV. Likewise an instrument with a nonfunctional tungsten lamp will accurately generate UV absorbance data.

What Are the Strategies to Determine the Extinction Coefficient of a Compound?

The Beer-Lambert law defines absorbance A as equal to the product of molar absorptivity (extinction coefficient E) cell path length L and concentration C. The extinction coefficient defines the absorbance value for a one molar solution of a compound, and is characteristic of that compound.

$$A = ECL$$

An extinction coefficient can be empirically calculated from the absorbance measurement on a known concentration of a compound, as discussed in Chapter 10. Some extinction coefficients for nucleotides are shown in Table 10.2 of Chapter 10. Data for individual products can usually be found in manufacturers' information leaflets.

Issues of absorbance critical to the quantitation of nucleotides, oligonucleotides, and polynucleotides are discussed in greater detail in Chapter 10.

What Is the Extinction Coefficient of an Oligonucleotide?

A common approach applies a conversion factor of 33 or 37 μg per A_{260} for oligonucleotides and single-stranded DNA, respectively, and this appears sufficient for most applications. For a detailed discussion about the options to quantitate oligonucleotides and the limitations therein, refer to Chapter 10.

Is There a Single Conversion Factor to Convert Protein Absorbance Data into Concentration?

The heterogeneity of amino acid composition and the impact of specific amino acids on absorbance prevents the assignment of a single conversion factor for all proteins. The protein absorbance at 280 nm depends on contributions from tyrosine, phenylalanine, and tryptophan. If these amino acids are absent, this wavelength is not relevant and proteins then have to be detected by the peptide bond in the region of 210 nm. The Christian-Warburg citation provides a strategy to convert protein absorbance to concentration, but this requires modification based on composition (Manchester, 1996; Harlow and Lane, 1988).

Several methods are available in the literature, from which a relatively accurate extinction coefficient may be derived (e.g., Mach, Middaugh, and Lewis, 1992). Provided that the amino acid composition is known, an equation can be used to determine *E* that takes into account the number of tyrosines and tryptophans, as well as the number of disulfide bonds (if known); the latter less critical. It is sometimes imperative to conduct the measurements under denaturing conditions (e.g., 6M Guanidine-HCl) for accurate evaluation of the extinction of a protein, particularly when the majority of the aromatic residues are buried within the protein core. This may be revealed by comparing the normal or second derivative spectra in the presence and absence of the denaturing agent.

What Are the Strengths and Limitations of the Various Protein Quantitation Assays?

There are four main reagent-based assays for protein analysis:

1. Bradford (Coomassie Blue) has the broadest range of reactivity and is the most sensitive. The drawback is its variable responses with different proteins due to the varying efficiency of binding between the protein and dyestuff. The optimum wavelength for absorbance measurement is 595 nm. Sensitivity can be improved by about 15% for longer reaction times up to 30 minutes for microassays, and responses can be integrated over a longer period. Detergents give high background responses that require blank analyses for compensation.

2. BCA, measured at 562 nm, is about half the sensitivity of the Bradford method but has a more stable endpoint than the Lowry method. It also has a more uniform response to different proteins. There is little interference from detergents. It is not compatible with reducing agents.

3. Lowry, measured at 750 nm, is almost as sensitive as the Bradford assay, but it has more interference from amine buffer salts than other methods.

4. Biuret, measured at 546 nm, is in principle similar to the Lowry, but involving a single incubation of 20 minutes. Under alkaline conditions substances containing two or more peptide bonds form a purple complex with copper sulphate in the presence of sodium potassium tartrate and potassium iodide in the reagent. There are very few interfering agents apart from ammonium salts and fewer deviations than with the Lowry or ultraviolet absorption methods. However, it consumes much more material. In general, it is a good protein assay, though not as fast or sensitive as the Bradford assay.

Smith (1987) lists compounds that interfere with each assay and illustrates problems associated with the use of BSA as a standard (see also Harlow and Lane, 1988; Peterson, 1979).

BIBLIOGRAPHY

Amersham Pharmacia Biotech. 1995. *Percoll R Methodology and Applications*, 2nd ed., rev. 2. Uppsala, Sweden.

ASTM E1154-89 American Society for Testing Materials, 100 Barr Harbor Drive, West Conshohocken, PA 19428-2959.

Ausubel, F. M., Brent, R., Kingston, R. E., Moore, D. D., Seidman, J. G., Smith, J. A., and Struhl, K. (eds.), 1998. *Current Protocols in Molecular Biology*. Wiley, New York.

Beckman-Coulter Corporation. Application Note A—1790A. 1995.

Biochrom Ltd., 1997. *Basic UV/Visible Spectrophotometry.* Cambridge, England.

Biochrom Ltd., 1998. Spectrophotmetry Application Notes 52–55. Cambridge, England.

Biochrom Ltd., 1999. *GeneQuant Pro Operating Manual.* Cambridge, England.

DIN Standard 12650, Deutscites Institut für Normung, DIN/DQS Technorga GmbH, Kamekestr.8, D-50672 köh.

Eppendorf Catalog, 2000. Cologue, Germany, p. 161.

European Pharmacopoeia, 1984, V.6.19, 2nd ed. suppl., 2000.

GLP Standards FDA (HFE-88), Office of Consumer Affairs, 5600 Fisher's Lane, Rockville, MD 20857.

Good Laboratory Practice (GLP) Regulations, 21 CFR 58, 1979. FDA, USA.

Glasel, J. A. 1995. Validity of nucleic acid purities monitored by 260nm/280nm absorbance ratios. *Biotechniques* 18:62–63.

Harlow, E., and Lane, D. 1988. Protein quantitation—UV detection. *Antibodies: A Laboratory Manual.* Academic Press, New York, p. 673.

Huberman, J. A. 1995. Importance of measuring nucleic acid absorbance at 240nm as well as at 260 and 280nm. *Biotech.* 18:636.

ISO Guide 25, The International Organization for Standardization, 1, rue de Varembé, Case Postak 56, CH-1211 Genéve 20, Switzerland.

Mach, H., Middaugh, C. R., and Lewis, R. V. 1992. Statistical determination of the average values of the extinction coefficients of tryptophan and tyrosine in native proteins. *Anal. Biochem.* 200:74–80.

Manchester, K. L. 1995. Value of A_{260}/A_{280} ratios for measurement of purity of nucleic acids. *Biotech.* 19:209–210.

Manchester, K. L. 1996. Use of UV methods for measurement of protein and nucleic acid concentrations. *Biotech.* 20:968–970,

Peterson, G. L. 1979. Review of the Folin phenol protein quantitation method of Lowry, Rosebrough, Farr and Randall. *Anal. Biochem.* 100:201–220.

Rickwood, D. 1984. *Centrifugation: A Pracical Approach.* IRL Press, Washington, DC.

Sambrook, J., Fritsch, E. F., and Maniatis, T. 1989. *Molecular Cloning: A Laboratory Manual*, 2nd ed. Cold Spring Harbor Laboratory, Cold Spring Harbor, NY.

Schy, W. E., and Plewa, M. J. 1989. Use of the diaminobenzoic acid fluorescence assay in conjunction with UV absorbance as a means of quantifying and ascertaining the purity of a DNA preparation. *Anal. Biochem.* 180:314–318.

Silverstein, R. M., Bassler, C. G., and Morrill, T. C. 1967. *Spectrometric Identification of Organic Compounds.* Wiley, New York.

Smith, J. A. 1987. Quantitation of proteins. In Ausubel, F. M., Brent, R., Kingston,

R. E., Moore, D. D., Seidman, J. G., Smith, J. A., and Struhl, K., eds., *Current Protocols in Molecular Biology.* Wiley, New York, pp. 10.1.1–10.1.3.
US Pharmacopeia, USP645.
Voet, D., and Voet, J. 1995. *Biochemistry*, 2nd ed. Wiley, New York.
Weast, R. C. 1980. *CRC Handbook Of Chemistry and Physics*, 60th ed. CRC Press, Boca Raton, FL.
Warburg, O., and Christian, W. 1942. Isolation and crystallization of enolase. *Biochem. Z.* 310:384–421.
Wilfinger, W. W., Mackey, K., and Chomczynski, P. 1997. Effect of pH and ionic strength on the spectrophotometric assessment of nucleic acid purity. *Biotech.* 22:474–480.

Please note: Eppendorf® is a registered trademark of Eppendorf AG. Brinkmann™ is a trademark of Brinkmann Instruments, Inc.

5

Working Safely with Biological Samples

Constantine G. Haidaris and Eartell J. Brownlow

Biosafety 114
Is There Such a Thing as a Nonpathogenic Organism? 114
Do You Know the Biohazard Safety Level of Your Research Materials? 115
How Can You Learn More about the Genealogy of Your Host Cells? 117
Are You Properly Dressed and Equipped for Lab Work? 118
Are You Aware of the Potential Hazards during the Setup, Execution, and Cleanup of the Planned Experiment? 119
Are You Prepared to Deal with an Emergency? 122
What Are the Potential Sources of Contamination of Your Experiment and How Do You Guard against Them? 124
How Should You Maintain Microbial Strains in the Short and Long Terms? 125
How Do You Know If Your Culture Medium Is Usable? 126
Are Your Media and Culture Conditions Suitable for Your Experiment? How Significant Is the Genotype of Your Microbial Strains? 126

Molecular Biology Problem Solver, Edited by Alan S. Gerstein.
ISBN 0-471-37972-7

What Are the Necessary Precautions and Differences in Handling of Viruses, Bacteria, Fungi, and Protozoa? 126
What Precautions Should Be Taken with Experimental Animals? 128
What Precautions Should Be Considered before and during the Handling of Human Tissues and Body Fluids? 130
What Is the Best Way to Decontaminate Your Work Area after Taking down Your Experiment? 130
Is It Necessary to Decontaminate Yourself or Your Clothing? Is There Significant Risk of Contaminating Others? 132
Media Preparation and Sterilization 132
How Can You Work Most Efficiently with Your Media Preparation Group? 132
Which Autoclave Settings Are Appropriate for Your Situation? 133
What Is the Best Wrapping for Autoclaving? Aluminum Foil, Paper, or Cloth? 134
What Are the Time Requirements of Autoclaving? 135
What If the Appearance of the Indicator Tape Didn't Change during Autoclaving? 135
Why Is Plastic Labware Still Wet after Applying the Dry Cycle? Is Wet Labware Sterile? 135
Can Your Plastic Material Be Sterilized? 135
Requesting the Media Room to Sterilize Labware 136
Requesting the Media Room to Prepare Culture Media 136
Allow Sufficient Time 137
Autoclaving for the Do-It-Yourselfer 137
Bibliography 140

BIOSAFETY

Is There Such a Thing as a Nonpathogenic Organism?

The term "biohazard" is applied to any living agent that has the potential to cause infection and disease if introduced into a suitable host in an infectious dose. The "living agent" can include viruses, bacteria, fungi, protozoa, helminths (worms) and their eggs or larvae, and arthropods (insects, crustaceans) and their eggs or larvae.

We most commonly think of pathogenic microbes such as *Salmonella typhi* and *Leishmania donovani* as biohazards, but if introduced into a healthy body in large numbers (or an immunocompromised body in low numbers), organisms that are normally nonpathogenic can cause infection. The infectious dose will vary with the organism and the health of those infected. For example, *Shigella flexneri* requires the ingestion of only a few hundred organisms to cause intestinal disease. *Salmonella typhi* requires over a hundredfold more organisms to do so. Technically there is no such thing as a nonpathogenic microorganism.

Do You Know the Biohazard Safety Level of Your Research Materials?

Regardless of the type of work you will be doing with microorganisms, it is mandatory to know as much as possible about the safety precautions needed to handle the microbes you will be using, prior to entering the lab. Does your organism require special handling? Ask questions of the lab supervisor and your coworkers regarding the microbe itself, its safe handling and proper disposal. Know the location of first-aid kits, eyewash stations, and emergency lab showers.

In general, organisms used in the lab are classified in terms of the biosafety level (BSL) required to contain them, with BSL-1 being the lowest and BSL-4 being the highest levels, respectively. These classifications have been set by the U.S. government agencies such as the Centers for Disease Control and Prevention (CDC) in Atlanta, Georgia, and its associated institution, the National Center for Infectious Diseases (NCID), the World Health Organization (WHO), and the governments of the European Community (EC). The CDC and NCID are an excellent source of information on the biosafety level classification of individual organisms, and the methods suggested for their safe handling. The CDC Web site is at *www.cdc.gov*, which has links to the NCID site. Another excellent source of information is a book entitled *Laboratory Acquired Infections* (C. A. Collins and D. A. Kennedy, 1999).

BSL-1 and BSL-2

BSL-1 agents present no, or minimal, hazard under ordinary conditions of safe handling. The common host cells used in cloning experiments are classified as BSL-1, such as *E. coli* and *Saccharomyces cerevisiae*, and they can be handled at the benchtop. Simple disinfection of the workbench after handling and good hand-washing will be sufficient to eliminate organisms from any spillage. Low-level pathogens such as the fungus *Candida albicans* or the bacterium *Staphylococcus aureus* can also be safely handled at the benchtop, if the organism does not come into contact with the skin, or mucous membranes. Care should also be taken when handling sharps that may be contaminated with a microbe. A contaminated scalpel blade, needle, or broken glass can serve as a vehicle of entry to the body. As described more thoroughly below, wearing a lab coat, safety glasses, and disposable gloves is always a good idea when handling any quantity of microorganism. The handling of liquid culture could also lead to aerosolization of the microbial suspension, hence wearing a particle mask might be useful as a precaution. It is always best to take an approach that maximizes your own safety and the safety of those around you.

BSL-2 agents possess the potential for biohazard, and they may produce disease of varying degrees of severity as a result of accidental laboratory infections. Moderate level pathogens, such as *Neisseria gonorrhoeae*, are classified as BSL-2. A safe way to handle BSL-2 organisms is in a laminar airflow, biosafety hood. By generating a flow of air inside the cabinet, these hoods are designed to keep aerosols from leaving the hood and entering into the room's airspace. The hood's proper function should be certified yearly by professionals. The Environmental Health and Safety officers of your institution or outside contractors can perform this function.

Usually a germicidal UV light is used to disinfect the inside of the hood when not in use. As a precaution prior to use of the hood, the airflow should be on for 15 minutes while the germicidal light is on. A wipe-down of the hood's inside working surface with 70% ethanol is also a useful precaution after the UV light is turned off. The airflow should be kept on during the entire time the hood is in use, and the glass panel on the front of the hood should be raised only high enough to allow comfortable use of the worker's arms inside the hood. Most hoods are equipped with an alarm to warn the worker if the front panel is raised too high.

Basic microbiological techniques of sterility, a minimum of protective gear, disinfectant, and common sense are all that are required to safely handle BSL-1 microbes used as cloning or expression vectors in the laboratory. Under common sense and in accordance with safety regulations, there should be absolutely *no* eating or drinking by an individual during the handling of a microorganism in the lab. Those workers in a diagnostic microbiology lab or doing research on a BSL-2 pathogen will wish to use a biosafety hood when necessary, along with protective clothing.

BSL-3 and BSL-4

Any organism requiring BSL-3 or BSL-4 containment should only be handled by highly trained individuals, using extensive safety precautions. Training in a lab that has experience with the microbe is highly recommended.

BSL-3 pathogens pose special hazards to laboratory workers. They must be handled, at very least, in a biosafety hood. No open containers or those with the potential to break easily should leave the hood if they contain the BSL-3 agent. Eye protection, particle mask, lab coat, and gloves are mandatory. Centrifugation of such organisms requires sealed containers to prevent aerosol and spillage.

Virulent BSL-3 pathogens that cause disease in low numbers and are transmitted by aerosol, such as *Mycobacterium tuberculosis*, require environmentally sealed containment rooms and also require the worker to be completely protected by special clothing, colloquially referred to as a "moonsuit," because it resembles the type worn by astronauts. Disinfection and removal of the suit is required before the wearer can enter the open environment.

BSL-4 pathogens pose an extremely serious hazard to the laboratory worker. These are the "hottest" pathogens, such as Ebola virus. Only a few places in the United States and elsewhere in the world are equipped for such studies. Only highly trained professionals are qualified to handle these agents.

How Can You Learn More about the Genealogy of Your Host Cells?

If the cell is obtained from a commercial source, such as a biotechnology company or the American Type Culture Collection, then the background on the host cell is often provided with the cell stock or in the catalog of the company. For strains of *E. coli* commonly used for cloning, the catalogs of biotechnology companies often have appendixes that list phenotypes and original references for the given strains. If the cell comes from a personal contact (i.e., another scientist), then be sure to ask for references or technical material on the cultivation and use of the cell. If possible, try to reproduce the desired cellular activity in a small pilot experiment prior to using precious materials or resources in a large-scale study. A wealth of genetic information and links to *E. coli* resources is available at *http://cgsc.biology.yale.edu/*.

Are You Properly Dressed and Equipped for Lab Work?

Regardless of the level of one's experience in the laboratory, it is wise to be prepared for the worst in terms of accidents. Proper preparation starts with proper protection for yourself and your co-workers. A number of common accessories should be used as needed.

Lab Coat

Most laboratories require the wearing of a lab coat, but even when not mandatory, a lab coat is a good idea. Protection of clothing and, more important, the skin underneath, is worth the effort and (perceived) inconvenience. For those working with microbes, clothing can be permanently fouled by a spill of almost any microorganism. Even a small break in the skin can serve as a portal of entry for a seemingly innocuous microbe that can result in a serious infection if the microbes gain access to the circulation in sufficient numbers. For those working with flames, such as a Bunsen burner, some institutions require a lab coat of flame retardant material. This ensures an added level of protection should there be a spill of a flammable liquid followed by ignition at the lab bench.

Closed Footwear

The open-toed shoe, sandal, or "flip-flop," even while wearing socks, provides easy access to the foot for sharps, hazardous chemicals, and infectious agents. The protection afforded by a closed-toe shoe against these assaults could provide an important level of safety. When choosing between fashion, or even comfort, and protection, protect your feet!

Eye Protection

It is strongly suggested to wear some form of eye protection when handling large volumes of microorganisms to protect against splashing during handling. Even a nonpathogen can set up infection when introduced in large numbers on the conjunctiva. Eyeglass (but *not* contact lens) wearers are afforded reasonable protection in these circumstances, but more safety is provided with the larger protective surface afforded by safety goggles or glasses, which can fit over conventional eyeglasses if necessary. For individuals working with pathogenic organisms, it is essential to wear eye protection at all times.

Conventional safety glasses are suitable against BSL-1 and BSL-2 class pathogens. For BSL-3 pathogens, a full-length protective face mask will provide additional defense against accidental exposure of the face and eyes.

Latex Gloves

The wearing of protective gloves is a good idea if the skin on the hands is abraded or raw. It is very important to wear protective gloves if a pathogen is being handled. Some people's hands can be irritated by the powder on many types of latex gloves. Most can be obtained in powder-free form. Less fortunate individuals are allergic to the latex in the gloves. Cloth glove inserts are available that can prevent contact of the latex with skin.

Heavy-Duty Protective Gloves

For those individuals handling pathogens and sharps simultaneously, such as during inoculation of an experimental animal, it is a good idea to wear a heavier rubber glove over the latex glove. Removal of the last 0.5 to 1 inch of the finger tips of the heavy glove will afford the worker with the dexterity to handle instruments or other items with efficiency, and still provide protection to the bulk of the hand.

Safety Equipment and Supplies

The two most important pieces of safety equipment in any lab are the eyewash station and the emergency lab shower. These two equipment stations should be regularly tested (every 6–12 months) to be sure they are fully operative. Know their location. In the event of the splashing of a microbial suspension in the eye, if possible, go directly to the eyewash station and flush the eye thoroughly with water. Then seek medical attention immediately. Even small numbers of microbes can permanently damage the eye if it is left untreated. The lab shower is very useful in the case of a chemical spill, and a large spill of a serious pathogen on the body surface also could be removed in part by rinsing under the lab shower. Again, seek medical attention following such a circumstance. Finally, every lab should have a well-stocked first-aid kit for treatment of minor mishaps, and to provide intermediate care for more serious accidents. Take it upon yourself, or appoint a lab safety officer, to be sure the first-aid kit is stocked and the wash stations are operative.

Are You Aware of the Potential Hazards during the Setup, Execution, and Cleanup of the Planned Experiment?

In the microbiology lab there are physical, chemical, and microbial hazards. When handling a hazardous material or performing a hazardous procedure, the most important thing is to pay attention to what you are doing and, if transporting the material, where you are going. Wear the protective clothing as outlined in the previous section. Do not stop to answer the phone or chat. No clowning around; this is not a time for levity. Even an act as simple as sterilizing an inoculating loop can be hazardous if you get distracted.

The physical hazards include burns and cuts from sharps. Burns can result from a Bunsen burner or gas jet, an inoculating loop, a hot plate, or the autoclave. When handling items going in or coming out of the autoclave, heavy-duty cloth gloves designed for handling hot containers are essential. Materials heated on a hot plate or in a boiling water bath should also be handled with heavy duty protective gloves. Burns can also result from ignition of flammables like ethanol or acetone. *Always* keep containers with these liquids safely away from a heat source. Malfunctioning machinery can also be a source of a burn or a fire. A pump motor that has seized can cause a fire. If you smell smoke or other toxic gases emanating from a piece of equipment try to turn it off or unplug it immediately, and call the fire department if necessary. If this action seems unsafe, call the fire department immediately.

Needles, broken glass, scalpels, and razor blades are all potential hazards. Pay attention when handling them and dispose of them properly. Most all labs require disposal using a certified "sharps" container, and removal by housekeeping staff or health safety workers. *Never* throw a sharp into the everyday trash. This is a potential hazard and possible source of infection for the housekeepers.

Nearly all microbiology labs utilize corrosive acids, alkalis, and organic compounds that are toxic. The potential for toxicity can manifest itself through amounts as small as the fumes released by opening the container. Even a whiff of a concentrated acid or other corrosive liquid can cause tissue damage to the nasopharynx. A spill of even a few hundred milliliters of an organic chemical, like phenol, on the body can be life-threatening. Phenol vapors, in excessive amounts, can cause damage to the nasopharyngeal mucosa and to the mucous membranes of the eye. Brief exposure to phenol vapors can cause minor irritation of these mucosa as well. If one uses phenol frequently, it is sensible to perform the manipulations in a chemical fume hood if possible.

Sources of microbial contamination to you and others are aerosols formed by the handling of the inoculating loop, preparation of slides, plating of cultures, the pouring of microbial suspensions, and pipetting. These procedures can be serious sources of infection if a hazardous pathogen is being handled. Each of these actions will be discussed individually. For individuals whose body defenses are compromised by underlying disease or medical treatment, it is sensible for them to check with their physician as to the potential hazards to them of working in a microbiology lab where even organisms that are normally nonpathogenic are being handled.

The Inoculating Loop

Excessively long or improperly made loops can shed their inoculum, either by vibration or spontaneously. A film formed by a loopful of broth culture that is vibrated can break the surface tension that keeps the film in place, forming an aerosol. The longer the loop is, the more vibration that ensues from handling. An incompletely closed loop can also easily result in a break in surface tension of the film. The optimal size of loop is approximately 2 to 3 mm in diameter, and the loop should be completely closed. The length of the wire portion of the loop, or shank, should be approximately 5 to 6 cm. If a large flask is being inoculated, tilting the flask to bring the liquid closer to the neck may be a way to avoid the use of a very long wire. When loops become excessively bent or encrusted with carbonized material, they should be replaced. Pre-sterilized, single-use plastic loops are also available. They are not to be placed in contact with flame or solvents such as acetone, and should be discarded into a disinfectant solution.

The discharge of proteinaceous or liquid material that often follows flaming a loop has been suspected as a source of contamination, but there is little evidence to support this contention. Nonetheless, it is best to decontaminate the wire loop by placing it into the apex of the internal blue flame of the burner so that any discharged material has to pass through the bulk of the flame as it leaves the loop.

Preparation of Slides

The production of aerosols by spreading of a bacterial suspension on a slide is minimized by gentle movements, especially when removing the loop from the spread suspension. For pathogens this activity is best performed in a biosafety hood, and the slides should not be removed from the hood until completely dried.

Streaking of Plates

In general, aerosol production is minimized by using smooth plates. Rough surfaces or bubbles in the agar can lead to excessive vibration of the loop. Spreading of samples with a sterile glass rod may minimize the production of aerosols on agar plates with rough surfaces.

Pouring of Microbial Suspensions

Following centrifugation, pouring off the supernatant from a microbial suspension can result in aerosols. One safe way to minimize aerosol production is to use a funnel with the narrow end submerged into disinfectant in a large container. Use enough disinfectant to mediate the decontamination of the supernatant, and a container large enough to avoid overflow. A volume of disinfectant equal to, or greater than, the amount of supernatant to be decontaminated should suffice. Rinse the funnel with more disinfectant to handle any residual supernatant clinging to the funnel wall.

Pipetting

Even for solutions deemed to be safe, it is good practice to *never* mouth pipette. It is mandatory not to mouth-pipette a microbial solution, even with a cotton-plugged pipette. Aspiration of organisms into the mouth can occur despite the cotton plug. Blowing out the last few droplets from a pipette can also form aerosols. Use either a manual or automatic pipetting aid to pipette. The discharge of the last few droplets using either manual or automatic pipette aids can result in aerosols, so avoid this if possible. If it is necessary to discharge the entire contents of the pipette, try to avoid spraying. Again, for serious pathogens, pipetting should be performed in a biosafety hood. Contaminated pipettes should be discarded into a container containing a sufficient volume of disinfectant to permit the *complete* immersion of the pipette.

Are You Prepared to Deal with an Emergency?

Adequate preparation for an emergency requires both an appreciation for the potential hazards involved and knowledge of the resources available to handle the emergency. Such preparation should precede actual lab work, although this is rarely done. We have discussed the appropriate types of laboratory clothing that should be worn in the microbiology laboratory, important safety equipment and supplies, and potential sources of harm.

There are several potential sources of an emergency situation in the lab.

Flammable Liquid and Microbial Culture Spills

In case of a flammable liquid (i.e., ethanol) spill, turn off any sources of ignition immediately, such as a Bunsen burner or hot plate. Then try to contain the spill. If the ethanol spilled onto your clothing, be careful not to ignite the clothing while trying to turn off potential sources of ignition. Remove the wet clothing as soon as possible.

For the microbial spill, contain the spill with disposable cloth or paper and decontaminate the area with a disinfectant. If any microbial suspension has contaminated your clothing, remove and disinfect the clothing. Any spill onto the skin should be washed off as soon as possible. If necessary, use the eyewash station or emergency shower.

Accidental Inoculation of Self with Microbe; Nonpathogen versus Pathogen

Accidental inoculation with any microbe should be treated as a serious situation, regardless of whether the organism is considered pathogenic or not. Even "cloning strains" of *E. coli* can cause septicemia if inoculated in substantial numbers. Seek immediate medical attention. Accidental inoculation of oneself with a syringe containing blood or other human tissue fluid should also be considered a medical emergency. Animal blood or tissues can also contain serious pathogens. Do not let accidental inoculation pass unattended; go immediately to seek attention from a medical professional. Be prepared to provide as much information as possible to the health care provider regarding the microbial agent or source of biological fluid.

Fire and Burn to Self or Fire in Surroundings

A direct burn from a Bunsen burner flame or hot plate can be serious. Even a minor burn can become infected. Seek medical attention if necessary. If one's clothing catches fire, try to remove it immediately or roll around on the floor to extinguish the flame. This is no time for modesty! Burning clothing can result in a life-threatening or fatal burn. Fire in one's surroundings can also develop into a life-threatening situation. If the fire is of a small, manageable scale, and you know the proper technique for extinguishing the different types of fires, you might attempt to stop the fire yourself. However, many fires are not easy to extinguish, and can be spread by incorrect use of extinguishing equipment. If you have any doubts, immediately leave the area and pull the fire alarm or call the fire department.

Electrical Hazard from Malfunctioning Equipment

Be aware of frayed electrical cords, which should be repaired or replaced immediately. Any piece of electrical equipment that produces sparks or smoke has the potential to cause harm by electric shock or fire. If in doubt, leave the room and call the fire department.

What Are the Potential Sources of Contamination of Your Experiment and How Do You Guard against Them?

Two prominent sources of "contamination" are (1) the introduction of microbes from the environment and (2) a mix-up of two or more closely related strains of organisms that you are working with simultaneously.

Environmental contaminants are most commonly fungi from the air that grow rapidly as fuzzy colonies on a bacterial plate. They are not harmful to the healthy researcher but can wreak havoc on your bacterial or mammalian cell cultures. Once a fungal contamination of this type occurs, it is nearly impossible to get rid of it. The best course of action is to get a fresh aliquot of the lab strain you're working with and start over again. Once the contaminant is visible by the naked eye, it has already produced numerous spores that will be spread as soon as the Petri dish lid is opened, making it difficult to go into that plate in order to pass the bacterial strain onto a fresh plate. You will only carry the spores from the contaminant with you. This speaks to the necessity of backup cultures of important organism strains, whether they are viruses, bacteria, fungi, protozoa, or higher eukaryotic cells. If the only copy of an important strain is contaminated by airborne fungi, heroic measures including treatment of the culture with antifungals such as amphotericin B or nystatin can be attempted, but this approach is usually not successful since the entire fungal population is not killed. Seal the contaminated plate with either tape or paraffin film, and discard it in the biological waste for autoclaving. For contaminated cell culture flasks, keep them sealed and dispose of them by autoclaving.

To prevent contamination from environmental sources, adhere to good sterile technique. Flame the inoculating loop thoroughly, leave the lid of the Petri dish or medium bottle open for as short a period of time as possible, and avoid working in a drafty area.

If environmental contamination persists, work in a laminar flow biosafety hood. If a contamination problem persists in the lab, consider an examination of the lab's air delivery system. Placing filters on the air ducts may reduce levels of contamination. Also consider lab clothing or dirty hands as a possible source of contamination. If circumstances dictate that microbial cultures and mammalian cell cultures be handled in the same hood, the working surfaces of the hood should be decontaminated with disinfectant after working with bacteria, and the germicidal lamp should be on at least 30 to 60 minutes before using the hood for cell culture.

The second possible source of contamination is by introducing a closely related strain to your culture while working with both strains simultaneously or consecutively. Such a mix-up is quite possible unless you are very attentive. It is difficult, if not impossible, to distinguish different *E. coli* strains by the naked eye at the level of colony morphology or microscopically by Gram staining. It may only be possible to do so on the basis of genotypic or phenotypic markers. Try to avoid handling more than one strain at a time. Most important, label all plates, tubes, and flasks thoroughly. If multiple strains are to be handled simultaneously, using different colors for the labeling of different strains is helpful. If a mix-up is suspected, put the strain of interest through as many types of tests as necessary to confirm its identity. This will save time and money in the long run compared to continuing to work up a contaminant, in which case you will need to start over anyway.

How Should You Maintain Microbial Strains in the Short and Long Terms?

For the short term, most bacterial strains can be maintained on plates without subculturing at 4°C for two weeks. Store plates inverted (with the half containing the agar facing you), because condensation on the plate lid of water evaporating from the agar can drip down onto the plate, and cross-contaminate isolated colonies as the liquid spreads. It is possible to remove excess liquid from the lid by a flick of the wrist into a lab sink and then using a disposable tissue to blot the excess liquid. This may introduce a contaminant, but it's a trade-off between that and soaking the plate with the condensation.

In the clinical microbiology lab, blood agar plates that contain cultures under evaluation are stored at the benchtop for up to 7 to 10 days after the initial overnight incubation at 37°C.

The concern here is desiccation of the agar over time and subsequent death of the microbe. Wrapping the plates in plastic wrap or sealing the edge with Parafilm® will help prevent desiccation.

For the long term, most microbial strains can be stored in glycerol (10–15% final concentration) at either –20°C (1–2 years) or –80°C (>2 years). There is no need to thaw the entire culture to recover the strain. A scraping of the frozen stock with a sterile toothpick or inoculating loop and inoculation of a plate is sufficient to recover the culture. Repeated freeze–thaw cycles will decrease the longevity of the frozen culture.

How Do You Know If Your Culture Medium Is Usable?

If liquid microbiological media remains sterile, it is usable for long periods of time (years) for most strains. Some fastidious microbes, such as the streptococci, are more sensitive and require freshly prepared media due to the lability of critical nutrients. In a clinical microbiology lab, where quality control is important, media should not be used past the expiration date for diagnostic purposes, but it may be used for less critical tasks.

Any medium that shows signs of contamination should not be used and should be discarded appropriately. Do not try to resterilize the medium and use it, since the contaminating microbe has not only depleted nutrients from the medium but has released or shed products that could interfere with the growth of your strain. It is easier and cheaper to simply make fresh medium.

Are Your Media and Culture Conditions Suitable for Your Experiment? How Significant Is the Genotype of Your Microbial Strains?

These two questions are related since the genotype of the organism will influence medium composition and growth conditions. Knowing the genotype of the organism in relationship to auxotrophic markers is critical, since essential nutrients, such as an amino acid or nucleoside, may need to be added to the basal medium to permit growth of the strain. Knowing the genotype of the organism in relationship to antibiotic resistance markers is critical to allow selection of the correct strain in the presence of organisms sensitive to the specific antibiotic.

What Are the Necessary Precautions and Differences in Handling of Viruses, Bacteria, Fungi, and Protozoa?

Since each of these four main groups of microbes have members that are either overt or opportunistic pathogens, a preliminary understanding of the pathogenic potential of the specific organism you will be working with is warranted. Part of this understanding is that of your own susceptibility to infection by the organism with which you are working. If you are in good health, then those organisms classified as opportunists should pose no threat unless accidental inoculation occurs. Individuals who are immunocompromised are at risk from infection from organisms that cause no harm to the healthy person. For overt pathogens, the appropriate precautions of laboratory apparel and, if necessary, a biosafety cabinet, applies to all workers.

Viruses of bacteria, bacteriophages, are not believed to pose a threat to humans and can be handled at the benchtop. Accidental ingestion of phage could, potentially, perturb the normal bacterial flora of the gut and possibly lead to diarrheal disease. As always, avoid mouth-pipetting to eliminate this possibility. Animal viruses, and those that infect plants as well, are propagated inside their respective eukaryotic cells. Again, plant viruses pose no health threat to humans. Many animal viruses that are conventionally handled using BSL-1 and BSL-2 containment can pose a threat to one's health, however, and should be handled carefully. This includes using the appropriate lab safety gear, taking care while handling sharps or glassware that have come in contact with the virus, and avoiding pipetting by mouth. Cell culture is highly susceptible to contamination from organisms in air and water droplets, hence is routinely performed in a biosafety cabinet with laminar airflow. This protects the cell culture from environmental contamination and has the benefit of protecting the worker from exposure to the virus. There is always the potential for accidental inoculation via a sharp object, or by a spill that occurs while handling viral stocks or infected cells.

Many of the precautions for handling bacteria have been described in the earlier discussion of biosafety. Inoculation via sharps, aerosols, and spills are the most common means of infection during handling.

Spore-forming filamentous fungi, like *Aspergillus*, pose the risk of infecting the lab worker by the release of the spores that are an integral part of their life cycle. They are easily made airborne and inhaled. They should be handled in a biosafety cabinet with laminar airflow. Inoculating loops and needles for passing fungal strains are sterilized and decontaminated with special electric heating coils shielded with a metal or ceramic covering to protect the worker from accidental burn. This mode of sterilization replaces the flame burner, which can be difficult to manage in a biosafety cabinet with laminar airflow. Non-spore formers such as *Saccharomyces cerevisiae* and *Candida albicans* can be handled at the benchtop using sterile technique. More serious pathogens in the non-spore former category that cause respiratory disease, such as *Coccidioides immitis* and *Histoplasma capsulatum*, should always be handled in a biosafety cabinet with laminar airflow.

Nonpathogenic protozoa that can be grown in axenic (pure culture; no feeder or accessory cells) culture can be handled at the benchtop using sterile technique. All pathogenic protozoa, grown in either axenic culture or in cell culture, should be handled in a biosafety cabinet with laminar airflow. They can become airborne in aerosols and make contact with the soft tissues of the eye, nose, oral cavity, and throat. Pathogenic hemoflagellates, such as those of the genera *Leishmania* and *Trypanosoma*, can penetrate the mucosal epithelium and establish infection.

While many pathogens pose the threat of infection, proper lab gear, safety equipment, and training permit their safe handling. Know the potential hazards before beginning work. Be sure the necessary supplies and equipment are available and accessible to deal with an emergency, should one arise. Above all, don't panic, and seek help immediately should an accident occur.

What Precautions Should Be Taken with Experimental Animals?

The Infection of Experimental Animals by a Natural Route or by Inoculation

Several techniques are used to infect experimental animals, depending on the nature of the infection to be induced. Each procedure has its attendant hazards for the investigator. Aerosols, contact of skin or mucosal surfaces by contaminated clothing and gloves, and inoculation by sharps are common hazards.

Respiratory infections are commonly initiated by (1) natural acquisition of the infection by co-housing the recipient animal with an infected donor animal, (2) direct introduction of the infectious agent via intubation of the lung, and (3) exposure of the animal to the infectious agent via aerosol in a sealed chamber. Aerosolization via mode (3) is performed with serious pathogens such as *Mycobacterium tuberculosis* and requires special containment rooms as well as airtight suits on the part of the investigators.

Mucosal pathogens that infect the nasopharynx, oral cavity, eye, gastrointestinal tract, or genitourinary tract are applied by direct inoculation, using a swab containing the infectious agent in buffer or medium. Gastrointestinal pathogens can also be applied by feeding the infectious agent to the animal or by gavage (direct intubation and inoculation). Accidental inoculation of the mucous membranes by the contaminated glove or sleeve of the worker is a common mode of transmission.

In particular, the inoculation of an animal by injection with a needle is particularly hazardous if the animal is not anesthetized or restrained during the procedure. Wriggling, other rapid movement of the animal, or an unrelated distraction during inoculation can readily result in accidental inoculation of the handler. If the animal cannot be anesthetized or restrained, assistance by another person in either the holding or inoculation of the animal can help minimize accidental inoculation of the investigator. Avoid unnecessary distractions and pay attention to what you are doing.

Safe Handling of Infected Experimental Animals

A reasonable precaution for the safe handling of infected experimental animals is to learn the techniques for safe handling of the same animal species when it is uninfected. This always involves minimizing the aggravation of the animal. First and foremost, be calm. Like pets, lab animals can sense nervousness and fear. Be firm but gentle. Hold the animal in a way that is not painful or hazardous to the animal. For example, improper handling of rabbits can lead them to contort themselves in a way that they damage their own spine, which results in having to sacrifice the rabbit. Trained laboratory animal medicine personnel are the best source of information and technique for the novice. In the infected animal, the hazards are increased by the fact that the animal may feel poorly or be irritable as a result of the infection. Those animals that have been injected will remember that experience and may be reluctant to be handled again. In some cases a bite or a scratch by the animal could serve as a means to transmit the primary infection, or another microorganism, to the handler.

Disposal of Carcasses, Tissues, and Body Fluids of Infected Experimental Animals

First and foremost, *never* place these materials in the regular trash. Wrap carcasses and tissues in a plastic bag and seal the bag. Do not autoclave carcasses or large amounts of tissue. For temporary storage, freezing the tissue or carcasses in the plastic bag at –20°C is acceptable until proper disposal can be performed. Most institutions that undertake animal work have specific facilities in their vivarium for disposal of tissues by incineration. Check with the vivarium personnel for the proper protocols. Fluids such as blood from an infected animal, and very small amounts (a few grams) of tissue can usually be autoclaved with the other infectious waste, and disposed of properly.

What Precautions Should Be Considered before and during the Handling of Human Tissues and Body Fluids?

Any fresh or fresh-frozen human tissues or body fluids should be considered to be potentially infectious, and handled with the utmost care. When handling human tissues, there are always risks of exposure to hepatitis, HIV, and *Mycobacterium tuberculosis*, to mention only a few pathogens. Proper lab attire and use of a biosafety cabinet with laminar airflow, is essential. Further containment in a BSL-3 facility may be warranted under the appropriate circumstances. Get as much information as possible as to the origin of the material and potential risk factors before handling the samples.

Any materials or equipment that comes into contact with the tissues should be decontaminated by treatment with disinfectant. Any extraneous tissue fragments or fluids left over from the work should be sterilized by autoclaving. Disposal of large pieces of tissues should be handled under the supervision of the institution's Department of Environmental Health and Safety, which will likely incinerate the material much like is done with animal carcasses. They will also be able to answer general questions about risk factors, and other disposal procedures. In the absence of institutional procedures, use common sense and approach the situation as if the biohazard risk is high.

If the tissue is properly and thoroughly fixed, then the risk of infection is eliminated and the material can be handled at the benchtop. Avoid accidental ingestion of the material via eating and drinking in the lab. The preparation of fixed tissues and safety considerations for the handling of contaminated tissue specimens is comprehensively described in the text by Prophet et al. (1992).

What Is the Best Way to Decontaminate Your Work Area after Taking down Your Experiment?

No matter how harmless the microbial agent that you have worked with may be, it is always a good idea to decontaminate your work area when finished. While many microbes do not survive on dry surfaces for prolonged periods, others such as

Staphylococcus aureus, do. If you have used absorbent bench paper to cover the lab bench, dispose of it in the biohazardous waste.

If you have worked directly on the benchtop, there are numerous disinfectants suitable for use on these surfaces. A wide variety of additional disinfectants are commercially available. They are inexpensive; use them liberally. Not only does this reduce the risk of transmitting infection to yourself and others, but it also reduces the risk of contaminating your next experiment with organisms left at the benchtop.

There are several groups of disinfectants, each with their specific uses and advantages (Jensen, Wright, and Robinson, 1997).

- *Alchohols.* Wiping the bench with a solution of 70% ethanol is commonly done. Ethanol and isopropanol both kill microbes by disrupting cytoplasmic membranes and denaturing proteins. They kill most vegetative bacteria (but not endospores), viruses, and fungi. Ethanol and isopropanol concentrations of 70% to 90% are more microbicidal than concentrations above 90%, since they require some hydration for antimicrobial activity. Ethanol is also very effective as a skin antiseptic.
- *Heavy metals.* Various mercury-containing compounds were used as disinfectants in the past, but are rarely used now because of their toxicity. Dilute (1%) silver and copper ion-containing solutions are also used as disinfectants but are commonly used in water purification. They act by combining with proteins, such as enzymes and inactivating them.
- *Phenol and its derivatives.* These act by damaging cell walls and membranes and precipitating proteins. These are toxic in concentrated form, but are often mixed at low concentration (3%) with detergents or soaps to provide added disinfectant capacity on skin. Their advantage is that they are not inactivated by organic matter. Dilute solutions of phenols are found in many of the commercial disinfectants used on lab benchtops. Lysol™ spray, a very effective surface disinfectant, is a mixture of 79% ethanol and 2% *o*-phenylphenol.
- *The halogens.* The halogens iodine and chlorine are useful chemical disinfectants that oxidize cell proteins. Hypochlorites (0.5%) are the active agents of household bleach. The halogens are inactivated by organic matter and can lose their effectiveness when excessive organic matter is present in solutions and on surfaces. Iodine (2%) is an effective skin disinfectant, and is a component of some disinfectant soaps.

- *Quaternary ammonium compounds.* These compounds have detergent activity and solubilize cell membranes. They are commonly used at concentrations of 1% or less to disinfect floors, benchtops, and other inanimate objects. They are inexpensive, odorless, and nontoxic, but they are not as microbicidal as the aforementioned compounds.

Is It Necessary to Decontaminate Yourself or Your Clothing? Is There Significant Risk of Contaminating Others?

The answer to both of the above questions is a resounding *yes*! This is particularly important in a hospital setting where you will likely encounter patients that may be highly susceptible to infection. Recent studies (Neely and Maley, 2000) have shown that gram-positive bacteria, particularly multidrug-resistant *Staphylococcus aureus* and vancomycin-resistant enterococcus, can survive for prolonged time periods on common hospital and laboratory fabrics. Of 22 different organisms tested, all survived at least a day and some for more than 90 days! This study alone provides a good reason not to wear your lab coat to the cafeteria. Leave your lab wear, including lab coats, in the lab. Get your coat laundered with regularity. Any article of clothing, whether lab attire or personal clothing, should be washed very soon after an accidental spill of a microbial suspension or contaminated material onto it. Even if you have been wearing latex gloves, wash your hands when finished working. These simple control procedures will go a long way to stop the spread of infection to yourself and to those around you.

MEDIA PREPARATION AND STERILIZATION

How Can You Work Most Efficiently with Your Media Preparation Group?

Learn the Capabilities and the Guidelines

Media preparation facilities serve a crucial role for large numbers of researchers, and they are usually extremely busy. The more familiar you become with the operational guidelines and functional capabilities of the facility, the more likely you will get the materials you need when you need them.

Get to Know the Staff

Learn the media group's supervisory structure, and let the staff get to know you. As with any situation where a facility has to serve the needs of many different people, getting along with those that do the work will benefit you in the long run. A courteous, considerate, and respectful approach is always rewarded by an extra effort on the part of the staff to get your work done in a timely manner. Media preparation is physically demanding and requires more technical skill than you might initially imagine. "Thank you, I appreciate your help" goes a very long way. When things go wrong, seek out the responsible individuals and make your concerns known in a civil manner.

Show Consideration for Their Safety

Notify the personnel of any potential hazard in the job you want them to do. If there is broken glass or other sharps in material to be decontaminated, advise them of the hazard. If the material they will be handling contains an agent that is considered a biohazard, inform them and discuss ways that the material can be handled safely. It is your responsibility to learn which chemical or biological materials require special disposal. Organic reagents such as phenol, or animal parts and bedding, are examples of materials that should not be sent to the typical media preparation facility.

Radioactive material should never be included with material destined for decontamination by media room personnel. The disposal of this material should be handled by authorized personnel from your institute's Radiation Safety or Health Physics office.

Which Autoclave Settings Are Appropriate for Your Situation?

Liquids

For liquids, use the liquid cycle with slow exhaust (exhaust rate describes the speed with which steam exits the sterilizing chamber). Liquids need a slow exhaust to ensure that the steam in the chamber does not depressurize so fast as to cause the liquids to boil over or evaporate excessively, as occurs with a fast exhaust setting. However, if the liquids and other materials are to be discarded, a fast exhaust rate can be applied.

Nonliquids

For nonliquid materials, use the wrapped, or gravity, cycle with fast exhaust. For sterilization of dry items (pipettes, instruments, test tubes, etc.) you would use the wrapped or gravity cycle. The fast exhaust serves to remove most of the excess moisture that accumulates during steam autoclaving. Sometimes a period of drying in a warm room (usually overnight) is required to evaporate any excess moisture that remains after the fast exhaust cycle.

Some of the newer autoclaves don't have a cycle that provides only fast exhaust; here you would have to use either the liquid cycle or the gravity cycle, which would have the dry cycle built in.

What Is the Best Wrapping for Autoclaving? Aluminum Foil, Paper, or Cloth?

For dry materials such as surgical instruments, the best wrapping is cloth; paper is the second choice. However, as long as the aluminum foil is free of holes, it is the easiest and fastest method of wrapping. If wrapped properly, items in cloth and paper will stay sterile as long or longer than items in foil. They can be re-sterilized without repackaging. Although foil can be used several times to cut down on waste, repeated autoclaving can break it down. Reused foil should be checked for pinpoint holes or cracks by holding it to the light. Pipettes and surgical instruments can also be placed in metal canisters, metal pans with lids, or glass tubes with a metal cap for sterilization. Paper sterilizing bags are available, including those with see-through plastic on one side, so you can see the contents of the bag.

For liquids, use a bottle with a screwcap or a flask with a covering. In bottles, leave approximately 25% to 30% of the bottle volume empty to allow for liquid expansion during autoclaving. *Never* autoclave a tightly sealed bottle; it could crack or break. Leave the cap slightly loosened. If you are worried about excessive evaporation of a precious solution, or of a solution with a small volume, use a permanent black marker to mark the initial fluid level on the outside of the bottle. It is always wise to autoclave pieces of glassware on a tray or a container with a low side; this simplifies clean-up if there is accidental breakage.

For autoclaving liquids in a flask, cover the opening with a loose metal cap or a cotton/gauze plug. Most commonly used, however, is a double layer of heavy duty aluminum foil, squeezed firmly, but not too tightly around the opening of the container. This double-layer foil wrap still allows adequate release of pressure that may build up in the flask during autoclaving. Single-layer foil wrap runs the risk of tearing and subsequent contamination of the contents.

It is best to sterilize liquids and dry items during separate cycles.

What Are the Time Requirements of Autoclaving?

The minimum requirements for sterilization are 15lb of pressure per square inch at 115°C for 15 minutes of sterilization time (not including the exhaust cycle). Slightly longer sterilization times are usually not harmful and sometimes necessary. Large volumes of liquid require more sterilization time to ensure that the conditions reach the appropriate levels in the center of the liquid. Consult the media preparation staff for guidance when sterilizing unusual items.

What If the Appearance of the Indicator Tape Didn't Change during Autoclaving?

Indicator tape is often used to confirm material has been sterilized. A pale white striping or lettering on the tape turns black, indicating that that the autoclave reached the desired temperature. If the autoclaving proceeded normally but the tape didn't change color, the tape might be old, or the first two or three feet may have dried out due to improper storage. Autoclave tape should be stored in the cold and allowed to come to room temperature before use. Also remember that masking tape looks like indicator tape. If all of the above suggestions have been tested, the autoclave is not sterilizing properly.

Why Is Plastic Labware Still Wet after Applying the Dry Cycle? Is Wet Labware Sterile?

The plastic and some glassware will show condensation after a dry cycle due to the length of the cycle, or due to the cool temperature in the room that the load is brought out into. Despite the condensation, the material is still sterile. If absolute dryness is required, incubate the material, still wrapped and sealed, in a warm room for several hours to overnight. Alternatively, drying ovens set at a compatible temperature (<123°C) can dry plastic faster.

Can Your Plastic Material Be Sterilized?

Depending on the structure of the plastic, it can be sterilized by steam, gas, dry, or chemical means. The Nalgene Company catalog contains comprehensive information regarding sterilization of different plastics.

Requesting the Media Room to Sterilize Labware

Identify Your Goods

Use black permanent ink to mark your objects. The colored ink from many marker pens will wash away during autoclaving. Don't use pieces of adhesive tape to identify your labware; these will dislodge during sterilization and clog the autoclave drains.

Allow Sufficient Time

A liquid cycle (20 minutes of sterilization time) usually requires approximately 35 minutes. A wrapped (dry with fast exhaust) cycle (20 minutes of sterilization time) lasts approximately 45 minutes. On dry cycle, there is a 15 minute period beyond sterilization when the steam is exhausted, and the heat within the autoclave dries the contents. However, the actual time required to prepare your materials depends on the size of the facility, the number of autoclaves, the time of your request, the facility's workload, and the number of people on duty.

Sometimes your materials can be sterilized quickly in an emergency, but at other times you must wait your turn. Lack of planning on your part does not constitute an emergency to media room personnel with responsibilities to several laboratories. In addition autoclave cycles are pre-set, and can't be rushed.

Requesting the Media Room to Prepare Culture Media

Document Your Needs

A written request will prevent misinformation as to what was ordered and when, by whom, and when it is needed. It is crucial to place your request according to the facility's guidelines. With many media requests coming in, mistakes can happen when procedure isn't followed.

Media room personnel can provide you with information for the most commonly used media, but ultimately you should provide the details on your media needs. Does the media require a low sodium content? What pH is required? Any unusual nutrients needed?

Indicate the specific amount of media requested, and the concentration of any required antibiotic. The definition of "standard amount of antibiotic" will vary between research groups and occasionally between applications. Specify the design and size of the preferred bacteriological plate.

Allow Sufficient Time

It is best to allow at least two days for the request to be filled. File the request on Monday, the plates will be made on Tuesday, incubated overnight to test for contamination, and delivered to you on Wednesday.

Autoclaving for the Do-It-Yourselfer

Extensive rinsing and sterilization will make your washed glassware suitable for protein or DNA work. If special treatment is required, such as making the glassware endotoxin- or RNase-free, then this will require special protocols above and beyond conventional washing.

Volumes of Media and Vessels

As a general rule, add considerably less material in the container than it is designed to hold. For liquids without agar, it is safe to autoclave 900 ml in a liter bottle without losing much of the volume. The smaller the container, the safer it is to autoclave a volume of liquid approaching the maximum volume of the vessel.

The shape of the container also affects the recommended amount of media for autoclaving. For example, when autoclaving media with agar in an Erlenmeyer flask never fill the container over half of the flask volume (i.e., 500 ml in a liter flask).

Preventing Boil-Over

The most common cause of boil-over is a cap secured too tightly on a container when placed into the autoclave. The cap should be loosened at least one-half turn before sterilizing. Also don't crimp the foil too tightly around the opening. Remember to employ slow exhaust rates when autoclaving liquids.

Preventing Lumps of Powdered Stocks in Liquid or Agar Media

Never add powder or agar to a dry container. Always add some water to the container ($\frac{1}{4}$ to $\frac{1}{3}$ of the final volume of the media) before adding powdered media components. It is more accurate to do this procedure in a graduated cylinder than a flask or beaker. If your medium requires agar, do not add it yet. After your dry ingredients (except agar) are in, add a magnetic stir bar and place the container on a stir plate. This will bring most of the powder off the bottom. If the stir bar is large and the container is glass, add the stir bar carefully and gently to the container so as not to crack the bottom. It is usually not necessary to apply heat to microbiological medium to get the powdered components into solution. If needed, place the container in a warm water bath to help dissolve the powders. Add more water until approximately 80% of the final volume of the medium is in the container. This allows a bit of volume to adjust the pH of the medium, but don't adjust the pH until all the powder components are completely dissolved. After pH adjustment, bring the solution to its final volume with water, and mix and prepare for dispensing.

If the medium is to remain as liquid, distribute it in the desired containers, leaving adequate airspace in each container as described above. If you need to add agar, follow the same principle as with preparation of the liquid medium. Never add dry agar to a dry container before adding the medium. Determine the final volume your container will hold and add about half the liquid medium to it. Add the desired amount of dry agar to the medium in the container, and use the remainder of the appropriate volume of medium to wash down the agar that has stuck to the sides of the container. If the agar powder remains on the sides of the flask, it will turn to a sticky residue during autoclaving that will never come off. The agar will not dissolve until autoclaving is completed.

Before pouring plates with the agar, gently swirl the contents to adequately mix the agar with the medium. *Warning*: Swirling a hot container immediately out of the autoclave will cause the solution to boil over, causing losses and possibly burning the handler. The agar will not sufficiently cool to solidify for quite some time, so allow the flask to cool slightly before swirling. If the plates are not to be poured immediately, the agar will stay molten in a 65°C water bath. If antibiotic is to be added, the agar must be cooled to approximately 48°C before adding the antibiotic, or else the antibiotic will be degraded.

Lumps of Agar

If you inadvertently cook the agar into a large lump in the flask, it is unusable, and must be properly disposed. After rinsing the flask with hot water to remove most of the agar, add enough undiluted bleach to cover the remaining lump and let it sit at room temperature until the lump dissolves. Large clumps of agar can clog pipes, so keep them out of the sink. Dispose of these clumps before you send the flask to be washed.

Use a Secondary Container

Glassware does "age" and can break over time during the washing and sterilization process. Bottles or flasks with liquid to be autoclaved should be protected from the hot metal of the autoclave surface by placing them in a leak-proof container with enough water to cover the bottom of the bottle, jug, or glassware. Then, if the piece breaks, the floor of the autoclave is protected, and cleanup is simplified. Spilled agar or broth will bake on the surface of the autoclave floor, causing materials autoclaved subsequently to become covered with the residue of the spilled material.

Be aware of the way you leave the autoclave. If there is an agar spill in the autoclave, never turn the steam off to clean the floor of the autoclave, as this will cause the agar to harden in the autoclave drain. Clean the spill by pouring small amounts of hot water in the chambers to gradually and gently wash the agar down the drain. Stand away from the door as the steam develops from the hot water touching the chamber surface.

Always protect yourself adequately when handling the autoclave or recently autoclaved materials. Use protective gloves, wear a lab coat, and guard your eyes with safety glasses. Always open the autoclave door slowly, keeping your face away from the door opening. Steam escaping from a newly opened autoclave door can scald the face and damage the eyes.

Carbohydrates and Other Atypical Solutions

Some amino acids and carbohydrates can be safely autoclaved at 121°C; others cannot. The minimum temperature for autoclaving carbohydrates is 110°C. Before sterilization, read any information you can obtain regarding the product. For small volumes of solutions that will be used in analytical procedures, filtration is also an effective sterilization method. Elimination of particulate matter by centrifugation and a prefiltration step can help prevent clogging of the 0.45 or 0.22 micron filters used to remove bacteria and fungi. These filters will not remove mycoplasma or viruses, however.

Decontamination of Waste

First and foremost, don't dispose of biological materials in the normal trash; these materials must be decontaminated first. Some liquid cultures of microorganisms can be killed by addition of an iodine solution, such as Betadine™. Add enough to make your suspension turn deep brown in color. Organisms on agar plates or cells in tissue culture flasks should be autoclaved to decontaminate them. Second, determine the biohazard level of the material to be decontaminated. Consult your institutional guidelines on what can and cannot be decontaminated at a given facility. These rules are often determined by the Occupational Health and Safety Office (OSHA), and they must be followed to the letter.

Autoclave bags are commonly used to dispose of biological materials. Orange bags contain biological materials that are thought to pose no threat to humans, but need to be decontaminated nonetheless. In most cities, waste of this type is safe for disposal in landfills after decontamination. Red autoclave bags are for biohazardous waste that could be infectious for humans or animals. *Never* put glass or other sharps in these bags. Broken glass can be decontaminated in a sealed cardboard box clearly marked with a description of the contents. Glass and metal sharps can also be disposed of and decontaminated by placing them in special plastic containers (red plastic) that are designed for this purpose. These special containers can be obtained from most large lab supply houses. Consult OSHA or the media room personnel for information on these containers.

To summarize, hazardous chemicals, radioactive waste, animal parts, and volatile substances should never be autoclaved. There are special facilities and conditions for their disposal. If you need to dispose of some of these materials, consult the media room personnel or other regulatory offices at your institution.

BIBLIOGRAPHY

Collins, C. A., and Kennedy, D. A. 1999. *Laboratory Acquired Infections*, 4th ed. Butterworth-Heinemann, Oxford, U.K.

Prophet, E. B., Mills, B., Arrington, J. B., and Sobin, L. H., eds. 1992. *Laboratory Methods in Histotechnology*. American Registry of Pathology, Armed Forces Institute of Pathology, Washington, DC.

Jensen, M. M., Wright, D. N., and Robinson, R. A., eds. 1997. Microbiology for the Health Sciences, 4th ed. Prentice Hall, Englewood Cliffs, NJ.

Neely, A. N., and Maley, M. P. 2000. Survival of enterococci and staphylococci on hospital fabrics and plastic. *J. Clin. Microbiol.* 38:724–726.

6

Working Safely with Radioactive Materials

William R. J. Volny Jr.

Licensing and Certification 143
Do You Need a License to Handle Radioactive Materials? 143
Who Do You Contact to Begin the Process of Becoming Licensed or Certified to Use Radioactivity? 144
Selecting and Ordering a Radioisotope 144
Which Radiochemical Is Most Appropriate for Your Research? 144
What Quantity of Radioactivity Should You Purchase? 147
When Should You Order the Material? 148
How Do You Calculate the Amount of Remaining Radiolabel? 148
How Long after the Reference Date Can You Use Your Material? 149
Can You Compensate by Adding More Radiochemical If the Reference Date Has Long Passed? 150
Handling Radioactive Shipments 150
What Should You Do with the Radioactive Shipment When It Arrives? 150
A Wipe Test Detected Radioactivity on the Outside of the Vial. Does This Indicate a Problem? 151

Molecular Biology Problem Solver, Edited by Alan S. Gerstein.
ISBN 0-471-37972-7

Your Count-Rate Meter Detects Radiation on the Outside of a Box Containing 1 mCi of a ^{32}P Labeled dATP. Is It Contaminated? 152
You Received 250 μCi of ^{32}P and the Box Wasn't Labeled Radioactive. Isn't This a Dangerous Mistake? 153
Designing Your Experiments 153
How Do You Determine the Molarity and Mass in the Vial of Material? 153
How Do You Quantitate the Amount of Radioactivity for Your Reaction? 155
Storing Radioactive Materials 156
What Causes the Degradation of a Radiochemical? 156
What Can You Do to Maximize the Lifetime and Potency of a Radiochemical? 156
What Is the Stability of a Radiolabeled Protein or Nucleic Acid? 157
Radioactive Waste: What Are Your Options and Obligations? 158
Handling Radioactivity: Achieving Minimum Dose 159
How Is Radioactive Exposure Quantified and What Are the Allowable Doses? 159
Monitoring Technology: What's the Difference between a Count-Rate Counter and a Dose-Rate Meter? 159
What Are the Elements of a Good Overall Monitoring Strategy? 161
What Can You Do to Achieve Minimum Radioactive Dose? 162
How Can You Organize Your Work Area to Minimize Your Exposure to Radioactivity? 164
How Can You Concentrate a Radioactive Solution? 164
Bibliography 166
Appendix A: Physical Properties of Common Radionuclides 166

The information within this chapter is designed as a supplement, not a replacement, to the training provided by your institutional rules and/or radiation safety officer. At the very least, there are some 10 fundamental rules to consider when working with radioactivity (Amerhsam International, 1974):

1. Understand the nature of the hazard, and get practical training.
2. Plan ahead to minimize time spent handling radioactivity.

3. Distance yourself appropriately from sources of radiation.
4. Use appropriate shielding for the radiation.
5. Contain radioactive materials in defined work areas.
6. Wear appropriate protective clothing and dosimeters.
7. Monitor the work area frequently for contamination control.
8. Follow the local rules and safe ways of working.
9. Minimize accumulation of waste and dispose of it by appropriate routes.
10. After completion of work, monitor yourself; then wash and monitor again.

LICENSING AND CERTIFICATION

Do You Need a License to Handle Radioactive Materials?

Whichever type of license is granted by the Nuclear Regulatory Commission (NRC), it tends to be a single license issued to the institution itself, to regulate its entire radioisotope usage. Separate licenses are not normally granted to the various departments or to individuals at that institution. However, everyone who works with radioactive materials at a licensed institution must be trained and approved to use the radioactive materials. Keep in mind that some states may have control over radioactive materials not controlled by the NRC. In addition, in "agreement states," the NRC requirements are regulated and controlled by a state agency.

Universities, governmental institutions, or industry are usually licensed to use radionuclides under a Type A License of Broad Scope (U.S. Nuclear Regulatory Commission Regulatory Guide, 1980). This is the most comprehensive license available to an institution. It requires that the institution have a radiation safety committee, an appointed radiation safety officer (RSO), and detailed radiation protection and training procedures. Researchers who want to use radionuclides in their work must present the proposal to the radiation safety committee and have it approved before being able to carry out the experiments.

There are other types of licenses issued by NRC or by agreement states. For example, these may be specific by-product material licenses of limited scope, specific licenses of broad scope, licenses for source or special nuclear materials, or licenses for kilocurie irradiation sources (U.S. Nuclear Regulatory Commission Regulatory Guide, 1979, 1976). By-product materials are the radionuclides that form during reactor processes. The most commonly used radionuclides, ^{32}P, ^{33}P, ^{35}S, ^{3}H, ^{14}C, and ^{125}I are all by-product materials. The licensing of by-product material is covered in detail under Title 10, Code of Federal Regulations (CFR), Part 30, *Rules of General Applicability to Licensing of Byproduct Material* (10CFR Part 30), and 10CFR Part 33, *Specific Domestic Licenses of Broad Scope for Byproduct Material* (10CFR Part 33).

For more information, a recent publication by the NRC is now available entitled: *Consolidated Guidance about Materials Licenses. Program-Specific Guidance about Academic Research and Development, and other Licenses of Limited Scope*. Final Report U.S. Regulatory Commission, Office of Nuclear Material Safety and Safeguards. NOREG-1556, Vol. 7. M. L. Fuller, R. P. Hayes, A. S. Lodhi, G. W. Purdy, December 1999. You can also find information on the NRC Web site *www.NRC.gov*. The Atomic Energy Control Board, or AECB, governs radioactive use in Canada. Their Web site is *www.aecb-ccea.gc.ca*.

Who Do You Contact to Begin the Process of Becoming Licensed or Certified to Use Radioactivity?

If you want to use radioactivity in your research, you may need to become an authorized user at your institution. First, decide what type of isotope or isotopes will be used in your research, the application, how much material you will need, disposal methods, and for how long you will use it. Then, present this information to your radiation safety officer or radiation safety committee so that they can determine whether such radionuclide use is possible under your institution's license. If the request is approved, carry out the requirements stated on your institution's license to become an authorized user operating in an approved laboratory.

SELECTING AND ORDERING A RADIOISOTOPE

Which Radiochemical Is Most Appropriate for Your Research?

The Institution's Perspective

Your institution's license defines specific limits to the type and amount of radionuclide allowable on site (this includes on-site waste). Before determining how much material you think you'll need, find out how much you'll be allowed to have in your lab at any one time. You can then get an idea about how or if you'll need to space out the work requiring radioactivity.

Your Perspective

These are some of the most important parameters to consider when deciding which isotope to use.

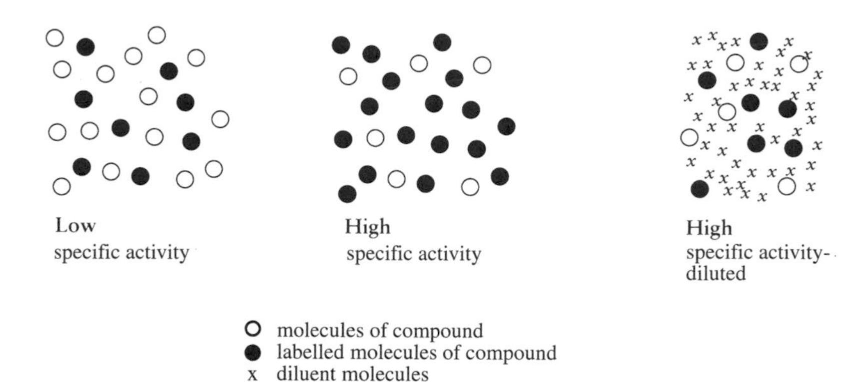

Figure 6.1 Diagrammatical representation of radiochemicals at low and high specific activity, and at high specific activity in a diluent. From *Guide to the Self-decomposition of Radiochemicals*, Amersham International, plc, 1992, Buckinghamshire, U.K. Reprinted by permission of Amersham Pharmacia Biotech.

Radionuclide, Energy, and Type of Emission (Alpha, Beta, Gamma, X ray, etc.)

In most cases you won't have the choice. You will choose the radionuclide because of its elemental properties, and its reactivity in reference to the experiment, not its type of emission. Each radionuclide has its unique emission spectrum. The spectra are important in determining how you detect the radioactivity in your samples. This is discussed more fully later in the chapter.

Specific Activity and Radioactive Concentration

The highest specific activity and the highest radioactive concentration tend to be the best since it means that there will be the greatest number of radioactive molecules in a given mass and volume (Figure 6.1). But there are two caveats to this ideal. The first is that as you increase the specific activity, you decrease the molar concentration of your desired molecule. This molecule will become the limiting reagent and possibly slow down or halt the reaction. The second danger is that at high specific activities and/or radioactive concentrations, the rate of radiolytic decomposition will increase. These parameters are discussed in more detail in Chapter 14, "Nucleic Acid Hybridization."

To take an example, a standard random priming labeling reaction requires 50 μCi (1.85 MBq)* of ^{32}P dNTP (Feinberg

**In the United States the unit of activity of "Curie" is still used. The unit of common usage is the Becquerel (Bq). Whereas 1 Curie = 3.7×10^{10} disintegrations per second (dps), the Bq = 1 dps. For example, to convert picocuries (10^{-12} Curies) to*

and Vogelstein, 1983). At a specific activity of 3000 Ci/mmol, that 50 μCi translates to 16.6 femtomoles of ^{32}P dNTP being added to the reaction mix, while 50 μCi of a ^{32}P labeled dNTP at a specific activity of 6000 Ci/mmol will add only 8.3 femtomoles to the reaction. Unless sufficient unlabeled dNTP is added, the lower mass of the hotter dNTP solution added might end up slowing the random prime reaction down, giving the resulting probe a lower specific activity than the probe that used the 3000 Ci/mmol material.

Label Location on the Compound

Consider the reason for using a radioactive molecule. Is the reaction involved in the transferring of the radioactive moiety to a biomolecule, such as a nucleic acid, peptide, or protein? Is the *in vivo* catabolism of the molecule being studied, perhaps in an ADME (absorption, distribution, metabolism, and excretion) study? Or perhaps the labeled molecule is simply being used as a tracer. For any situation, it's worthwhile to consider the following impacts of the label location: First, will the label's location allow the label or the labeled ligand to be incorporated? Next, once incorporated, will it produce the desired result or an unwanted effect? For example, will the label's presence in a nucleic acid probe interfere with the probe's ability to hybridize to its target DNA? The latter issue is also discussed in greater detail in Chapter 14, "Nucleic Acid Hybridization."

There are some reactions where the location of label is not critical. A thymidine uptake assay is one such case. The labeling will work just as effectively whether the tritium is on the methyl group or on the ring.

The Form and Quantity of the Radioligand

The radionuclide is usually available in different solvents. The two main concerns are the effect (if any) of the solvent on the reaction or assay, and whether the radioactive material will be used quickly or over a long period. For example, a radiolabeled compound supplied in benzene or toluene cannot be added directly to cells or to an enzyme reaction without destroying the biological systems; it must be dried down and brought up in a compatible solvent. Likewise a compound shipped in simple aqueous solvent might be added directly to the reaction, but might not be

Becquerels, divide by 27 (27.027): $50\,\mu Ci = 50 \times 10^{-6}\,Ci = (50 \times 10^{-6}\,Ci \times 3.7 \times 10^{10}\,dps/Ci) = 1.85 \times 10^{6}\,dps = 1.85 \times 10^{6}\,Bq = 1.85\,MBq$.)

the best solvent for long-term storage. From a manufacturing perspective, the radiochemical is supplied in a solvent that is a compromise between the stability and solubility of the compound and the investigator's convenience.

Some common solvents to consider, and the reasons they are used:

- *Ethanol, 2%.* Added to aqueous solvents where it acts as a free radical scavenger and will extend the shelf life of the radiolabeled compound.
- *Toluene or benzene.* Most often used to increase stability of the radiolabeled compound, and increase solubility of nonpolar compounds, such as lipids.
- *2-mercaptoethanol, 5 mM.* Helps to minimize the release of radioactive sulfur from amino acids and nucleotides in the form of sulfoxides and other volatile molecules.
- *Colored dyes.* Added for the investigator's convenience to visualize the presence of the radioactivity.

When not in use, the "stock" solution of the radioactive compound is capped and usually refrigerated to minimize volatilization/evaporation.

What Quantity of Radioactivity Should You Purchase?

There are three things to consider when deciding how much material to purchase:

1. How much activity (radioactivity) will be used and over what period?
2. What are the institutional limits affecting the amounts of radioisotope chosen that your lab may be authorized to use?
3. What are the decomposition rate of your radiolabeled compound and its half-life.

In general you will want to purchase as large a quantity as possible to save on initial cost, while at the same time not compromising the quality of the results of the research by using decomposed material. For example, certain forms of tritiated thymidine can have radiolytic decomposition rates (thymidine degradation) of 4% per week. This decomposition rate is not to be confused with tritium's decay rate, or half-life, which is over 12 years. Stocking up on such rapidly decomposing material, or by using it for more than just a few months could compromise experiments carried out later in the product's life.

When Should You Order the Material?

Analysis Date

Ideally you will want to schedule your experiments and your radiochemical shipments such that the material arrives at its maximum level of activity and lowest level of decomposition. This will tend to be when the product is newer, or nearer its analysis date (the date on which the compound passes quality control tests and is diluted appropriately so that the radioactive concentration and specific activity will be as those stated on the reference date). Some isotopes and radiochemicals decompose slowly, so it is not always necessary to take this suggestion to the extreme. As you use a radiolabeled product, you'll come to know how long you can use it in your work. An ^{125}I labeled ligand will not last as long as a ^{14}C labeled sugar. An inorganic radiolabeled compound, such as $Na^{125}I$ or sodium ^{51}chromate, will decompose at the isotope's rate of decay, whereas a labeled organic compound, such as the tritiated thymidine alluded to earlier, will decompose at a much faster rate than the half-life of the isotope would indicate. Manufacturers take this into account by having a terminal sale date. The material will only be sold for so long before it is removed from its stores. Up until this date you will be able to purchase the material and still expect to use it over a reasonable period of time.

Reference Date

The reference date is the day on which you will have the stated amount of material. If you purchased a 1 mCi vial of ^{32}P dCTP, you will have greater than 1 mCi (37 MBq) prior to the reference date, 1 mCi on that date, and successively less beyond the reference date. (Note that since you will most likely receive your radioactive material prior to reference, it is possible to exceed possession limits; consider this when determining limits on your radiation license.) In the case of longer-lived radioisotopes, such as ^{3}H and ^{14}C, the analysis date will also serve as the reference date.

How Do You Calculate the Amount of Remaining Radiolabel?

The most straightforward way of calculating radioactive decay is to use the following exponential decay equation. For convenience's sake, most manufacturers of radiochemicals provide decay charts in their catalogs for commonly used isotopes. This equation comes in handy for the less common isotopes.

$$A = A_0 e^{-0.693t/T}$$

where

A_0 is the radioactivity at reference date,
t is the time between reference date and the time you are calculating for,
T is the half-life of the isotope (note that both t and T must have the same units of time).

It is easy to use the aforementioned decay charts as shown in the following two examples.

Say you had 250 μCi of ^{35}S methionine at a certain reference date, and the radioactive concentration was 15 mCi/ml. Now it is 25 days after that reference date. You calculate your new radioactive concentration and total activity in the vial by looking on the chart to locate the fraction under the column and row that corresponds to 25 days postreference. This number should be 0.820. Multiply your starting radioactive concentration by this fraction to obtain the new radioactive concentration:

$$15\,\text{mCi/ml} \times 0.820 = 12.3\,\text{mCi/ml}$$

The total amount of activity can be likewise calculated for ^{35}S with a half-life of 87.4 days; namely $A = A_0e^{-0.693t/T} = 15\exp(-0.693 \times 25/87.4) = 12.3\,\text{mCi/ml}$.

For the second example you can find out how much activity you had before the reference date. Some decay charts only have postreference fractions, but if you have a 1 mCi vial of ^{33}P dUTP at 10 mCi/ml, and it is 5 days prior to the reference date, how do you figure out how much you have? Go to the column and row on the ^{33}P decay chart corresponding to 5 days postreference. There you will see the fraction 0.872. You will divide your reference activity and radioactive concentration by this number to obtain the proper amount of activity present, or 1/0.872 = 1.15 mCi. Note that the values should be greater than the stated amounts of activity and the referenced radioactive concentration. For the calculation method you are now looking for A_0. Therefore $A_0 = Ae^{0.693t/T} = 10\exp(0.693 \times 5/25) = 11.5\,\text{mCi/ml}$, using a half-life of 25 days for ^{33}P.

How Long after the Reference Date Can You Use Your Material?

Radioactively labeled compounds do not suddenly go bad after the reference date. It isn't an expiration date. It is used as a benchmark by which you can anchor your decay calculations as described above.

Table 6.1 Shelf Lives for Commonly Used Isotopes

Isotope	Shelf life
^{32}Phosphorous	1–3 weeks
^{33}Phosphorous	4–12 weeks
^{35}Sulfur	2–6 weeks
^{125}Iodine	3–12 weeks
^{3}Hydrogen	1–12 months
^{14}Carbon	1–2 years

Only you can determine how long you can use your radioisotope after the reference date. The answer depends on the isotope, the compound it's bound to, the experiment, storage, the formulation of the product, and the like. Table 6.1 lists the general ranges for the most commonly used radioisotopes, which is a guideline only. As you carry out your work, you will discover when your material starts to give poorer results.

Can You Compensate by Adding More Radiochemical If the Reference Date Has Long Passed?

Sometimes it is not that simple. As an example of the complexities involved with radiolytic decomposition, suppose you had a vial of ^{32}P gamma labeled ATP that you routinely use to label the 5′ end of DNA via T4 Polynucleotide kinase. If one half-life has passed since the reference date (14.28 days), you will have 50% of the stated radioactivity remaining. You might still achieve satisfactory 5′ end labeling with T4 Polynucleotide kinase if you double the amount of the ^{32}P added to the reaction. Often, however, you may find that though you have compensated for the radioactive decay by adding more material, you have also introduced more of the decomposition products, which will be fragments of the original labeled compound and free radicals. You also will have added more of the solute that might be present in the stock vial. These contaminants and decomposition products can significantly interfere with the reaction mechanism and compromise your results.

HANDLING RADIOACTIVE SHIPMENTS

What Should You Do with the Radioactive Shipment When It Arrives?

The radiation safety officer is responsible for ensuring that radioactive materials are received in satisfactory condition, but procedures may vary within the institution. Sometimes the RSO will check the shipping box for contamination and then discard the outer box, forwarding only the radioactive container to the researcher. At other facilities the receiving group will do a wipe test on the outer shipping container only, and if found to be uncontaminated, forward the entire package to the researcher. Upon receipt, you will want to carry out a final wipe test on the vial of radioactive material before opening, to make sure there is no gross contamination.

The Wipe Test

The manner in which a wipe test is to be carried out will be described in your institution's radioactive use license, the details of which can be explained to you by your RSO. A wipe test involves dragging or rubbing a piece of absorbent paper, or cotton swab across a portion of a vial, package, or surface (the standard area being $100\,cm^2$). You are testing for the presence of removable radioactive contamination. The paper or swab may be dry or wetted with methanol or water. Your RSO will let you know which way is preferred by the institution. After wiping the surface, the paper or swab may be placed into a liquid scintillation counter (LSC) to detect if any contamination was removed. It is usually best to count the wipes in an LSC rather than use a count-rate meter because some isotopes are not detected with a count-rate meter (e.g., tritium). Knowing the radioisotope and its decay products will help to determine the best detection method. The count-rate meter will be described more fully below.

A Wipe Test Detected Radioactivity on the Outside of the Vial. Does This Indicate a Problem?

If you detect contamination on the outside of your vial, contact your RSO. She will tell you, based on experience and institutional norms, whether the amount of contamination you have found is of concern, and whether the counts detected on the LSC may be artifactual (caused by chemiluminescence), or if they are being caused by radioactive contamination.

While the ideal is to have no detectable counts on the outside of the primary container, the act of packaging, shipping, and handling can work together to make this difficult to achieve. Then there are some radioisotopes, most notably, ^{35}S and ^{3}H, that are volatile, and can leach through the crimped overseals. This is one of the reasons why radioactive materials are shipped in secondary containers. The vial of ^{35}S labeled methionine you might receive is first dispensed into a primary container, which is sealed, then placed into a secondary container. The secondary container usually has absorbent material placed in it that will absorb any liquid should there be a spill. It is always prudent to wear some thin plastic gloves when dealing with radioactive materials, especially when they first arrive and you have no indication on whether or not they are contaminated.

The NRC has set action limits to contaminated surfaces of outer packages and containers, and the RSO is required to contact the NRC when these levels are surpassed. The amount of contamination considered significant will differ, depending on whether the activity is on the primary container, secondary container, or the outer package. Based on these action levels, an institution sometimes sets its own, lower, contamination limits. Contact your RSO for more information on contamination limits.

Your Count-Rate Meter Detects Radiation on the Outside of a Box Containing 1 mCi of a ^{32}P Labeled dATP. Is It Contaminated?

When you put a count-rate meter up to the outer package containing the ^{32}P substance, you will hear clicking sounds, indicating the presence of radioactivity. To determine whether the radiation is coming from contamination on the outside of the package, or emanating from the vial of material, it is necessary to carry out a wipe test on the package. In the overwhelming majority of cases, what the instrument is detecting is called *Bremsstrahlung*.

Bremsstrahlung

Bremsstrahlung, or "braking radiation," is created when a beta particle interacts with the shielding material to produce X rays. The Plexiglas™ vial that contains the radioactive material is sufficient to block essentially all beta emissions but not the X rays. Is *Bremsstrahlung* dangerous? The dose rate detected on the surface of a vial with 1 mCi of ^{32}P tends to be between 1 and 5 mrem/h, while there will be no detectable dose rate three feet away. This level of dose rate is considered low for those working in occupations that use radioactivity. It is important to remember that dose rate decreases as you move away from the source. Doubling the distance from the source will quarter the radiation dose. This is known as the inverse square law, and it is applicable whenever the source can be considered a point source. You may wish to discuss dose rates in more detail with your RSO.

You Received 250 μCi of ^{32}P and the Box Wasn't Labeled Radioactive. Isn't This a Dangerous Mistake?

Both the Department of Transportation (DOT) and The International Air Transport Association (IATA) have regulations concerning the labeling of packages containing limited quantities of radioactive materials (International Air Transport Association [IATA] Dangerous Goods Regulations, 6.2, and Code of Federal Regulations [CFR] 173.421, 173.422, 173.424, and 173.427). A package is defined as containing a limited quantity of an isotope if it conforms both to a certain physical amount of radioactive substance, and if the dose rate on the outside surface of the package is less than 0.5 mrem/h. For example, the isotope ^{32}P has a limited quantity of 3.0 mCi if it is in liquid form. This means that if the package contains less than 3.0 mCi, and if the dose rate is less than 0.5 mrem/h on the package's surface, then it is considered to be a package of limited quantity. So the package does not require an external label which bears the marking "radioactive." The regulations do require, however, that there be such labeling somewhere inside of the package, and that the packaging itself prevent leakage of the radioactive material under "conditions likely to be encountered during routine transport (incident-free conditions). . . ." (International Air Transport Association [IATA] Dangerous Goods Regulations, 6.2).

DESIGNING YOUR EXPERIMENTS

How Do You Determine the Molarity and Mass in the Vial of Material?

Let's start with some definitions.

Specific Activity

The definition of specific activity is the amount of radioactivity per unit mass, and is usually reported as curies per millimole (or Becquerels per millimole), abbreviated Ci/mmol (Bq/mmol). Specific activity is a quantitative description for how many molecules in a sample are radioactively labeled.

In order to determine the ratio of labeled molecules in the total molecule population, the specific activity of the material is divided by the theoretical maximum specific activity. The theoretical maximum specific activity is defined as the greatest amount of radioactivity that can be achieved if there were 100% isotopic abundance at a single location. This number is specific to the type of radionuclide.

As a simple example, the theoretical maximum specific activity for ^{32}P is 9131 Ci/mmol. If the percentage of radioactive molecules in a 3000 Ci/mmol product is desired, it simply is a matter of dividing 3000 by 9131 to find that approximately one-third of the molecules in that sample are radioactive. This will give the investigator an idea of how many radioactive molecules may get incorporated into the final product.

Molarity

The molarity of a labeled compound in solution can be calculated by dividing the radioactive concentration of your radiochemical by its specific activity:

$$\frac{\text{Radioactive concentration}}{\text{specific activity}}$$

For example, a vial of ^{32}P-labeled gamma ATP at a radioactive concentration of 10 mCi/ml and specific activity of 3000 Ci/mmol will have a molarity of

$$\frac{10\,\text{mCi/ml}}{3000\,\text{Ci/mmol}} = 3.33\,\mu\text{M}$$

Moles

Once you have the molarity of your stock solution, simply multiply that by the volume of stock you'll be adding to your reaction in order to obtain the number of moles you have

$$\text{Molarity} \times \text{volume} = \text{moles}$$

Continuing with this example, if you are adding 5 μl of the ^{32}P-ATP to your reaction, you will be adding

$$3.33\,\text{pmol}/\mu\text{l} \times 5\,\mu\text{l} = 16.7\,\text{pmols}$$

to the reaction vessel. After calculating the molarity of your radioactive stock solution, you might be shocked to learn how low it is. You might even think that it's too low for the reaction to run, or perhaps your protocol states that you should start out with a higher molarity. The solution is as follows:

Make up a stock solution of the cold compound in question, at the appropriately higher molarity. To this stock or to the reaction mix itself, add the amount of radioactivity required. The number of picomoles of radiolabeled compound that you'll be adding to your reaction mix will, in most cases, be so low that it will make no practical difference to the overall molarity of the compound. Note, however, that by adding cold compound, you will be dramatically lowering the specific activity of the radioactive label, which is in the reaction vessel. You may find that you get lower incorporation rates.

How Do You Quantitate the Amount of Radioactivity for Your Reaction?

DPM, CPM, and μCi

One curie of activity is 3.7×10^{10} disintegrations per second (dps) or 2.22×10^{12} disintegrations per minute (dpm). Thus the definition of a μCi is 2.22×10^{6} dpm, regardless of the isotope involved. Disintegrations per minute (dpm) are a function of nature; counts per minute (cpm) are a function of a detection device. Cpm/dpm is a measure of the instrument's efficiency to detect an isotope's decay event. A liquid scintillation counter (LSC), because of its photomultiplier tubes, cannot be 100% efficient. The instrument will give values in cpm, which will always be lower than the true number of disintegrations occurring in the sample. This counting efficiency can vary with the isotope and even the type of solvent and scintillation fluid (if a liquid scintillation method is used) that your samples are in. The counting efficiency should be determined if quantitative results are needed in your work. The procedure is to measure the cpm detected from a sealed calibration source that contains a known number of dpm. The percent efficiency of your counter is calculated by dividing the counted cpm by the dpm as indicated on the vial, then multiplying this quotient by 100. Typical examples of counting efficiencies for some commonly used isotopes are 75 to 85% for ^{14}C, ^{35}S, and ^{33}P; 35 to 55% for ^{3}H; 70 to 80% for ^{125}I (this radioisotope, while being a gamma ray emitter, is actually more efficiently counted on an LSC); and almost 100% for ^{32}P. These efficiencies are approximations only. Efficiencies can vary widely, however, depending on instrument, isotope, and sample type.

The Becquerel (Bq)

A Becquerel, or Bq, is a *Systeme Internationale* unit of measure for radioactivity. One Bq is one disintegration per second (dps). One dpm will be 60 Bq. One μCi is defined as 37 kBq. The Bq is a defined value of radioactivity that is small, whereas the Ci is very large. In the United States, the Ci is still a common unit. Most other countries have converted to using the Bq.

STORING RADIOACTIVE MATERIALS

As you gain experience with your radioactive materials, you will gain insight into two of their important but not intuitive physical properties. First, their lifetime is shorter than their unlabeled counterparts (because attached to them is a huge ball of energy waiting to blow). Second, the compound's shelf life is often dramatically less than the half-life of the isotope used to label it.

What Causes the Degradation of a Radiochemical?

The mechanisms of radiolytic decomposition are fairly complex but can be divided into primary and secondary decomposition (Amersham International, 1992). Internal primary degradation is caused by the release of energy from the radioactive atom's unstable nucleus. This energy release in turn is thought to break up the bonds of the parent molecule, destroying it (for very large molecules, e.g., proteins, it is unlikely to destroy the entire molecule). The rate of primary degradation is identical to the radioactive decay rate.

Another mode of primary decomposition is external, arising when ionizing radiation emissions hit nearby molecules. The energy transferred to the molecule is often enough to break chemical bonds within the molecule producing random fragments.

Secondary decomposition is caused by free radicals generated by the interaction of beta particles with the solvent. It is the most insidious form of decomposition. Free radicals can potentially interact with any compound within the solvent, generating innumerable contaminants and breakdown products. Some reactions generate more free radicals, leading to exponential rates of breakdown and contaminant production.

What Can You Do to Maximize the Lifetime and Potency of a Radiochemical?

1. Do not alter the recommended storage conditions.

Colder is not always better. If the solvent containing the radioactive material is stored at a temperature that allows the solvent to freeze slowly, an event called molecular clustering will occur (Figure 6.2). The freezing solvent pushes nonsolvent molecules into pockets or clusters. This results in extremely high radioactive concentrations, which in turn will cause extremely high rates of radiolytic decomposition. Examples of solvents freezing slowly will be: water at –20°C, or ethanol at –70°C. If the solution is quick-frozen (in liquid nitrogen), you will avoid the effects of molecular clustering.

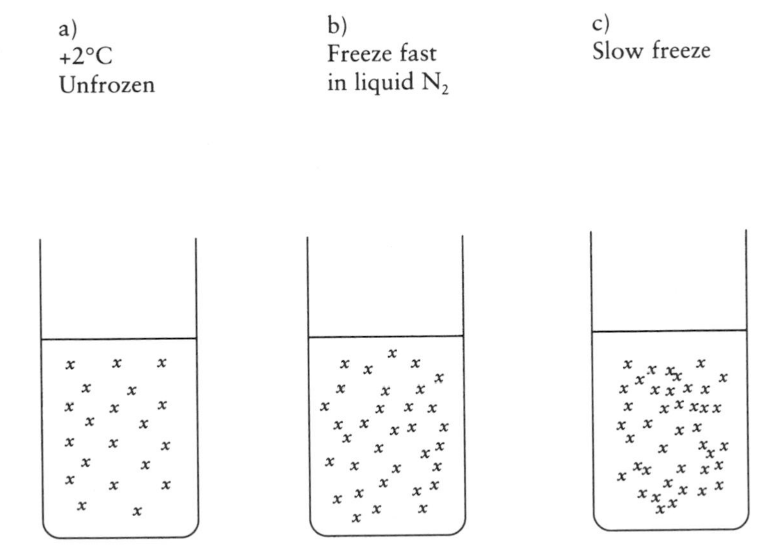

Figure 6.2 Molecular clustering effects. From *Guide to the Self-decomposition of Radiochemicals*, Amersham International, plc, 1992, Buckinghamshire, U.K. Reprinted by permission of Amersham Pharmacia Biotech.

2. Keep the radioactive concentration as low as possible to minimize primary external and secondary decomposition.
3. Minimize the number of freeze-thaws, which may increase the decomposition rate
4. Don't alter the recommended solvent.

Some solvents will cause greater rates of radiolytic decomposition. It cannot be predicted which ones will be better or worse until they have been tested. Manufacturers will have chosen the most appropriate solvent for the radioactive compound.

5. Schedule experiments to consume your store of radioisotope as quickly as possible.

What Is the Stability of a Radiolabeled Protein or Nucleic Acid?

After labeling or incorporation of radioactivity into your molecule of interest, radiolytic decomposition occurs. As the isotope decays into the surrounding solution, there will be primary decomposition, giving rise to nicked, or broken strands in your labeled nucleic acid, as well as the less predictable secondary decomposition, which might break the chemical bonds comprising that molecule. It is best to use your labeled molecule as soon as possible, or to store in as dilute a concentration as is reasonable for your work.

In this regard, ^{32}P is the most offending of the three most commonly used radioisotopes (^{32}P, ^{33}P, and ^{35}S). Compounds labeled with ^{32}P can have extremely high specific activities, and the energy of the beta is also extremely high. On the other hand, ^{33}P and ^{35}S have similar energies to each other. They have much lower emission energies and thus are less destructive to surrounding molecules. This issue is discussed in greater depth in Chapter 14, "Nucleic Acid Hybridization."

Radioactive Waste: What Are Your Options and Obligations?

It is essential to keep accurate records of the amount and type of radioactive waste that you generate. The RSO keeps track of all incoming radioactivity and all outgoing radioactive waste, so it is important to keep track of the material you use, store, and dispose of for the RSO's records. When the NRC or governing body inspects your institution, it will check its receipt and disposal records. If they are not in order, there is the possibility of suspension of your institution's license.

Obligations

Consult your RSO. Your institution has in place a detailed waste management program, and your radiation license requires that you follow your institution's waste handling procedure without variance. Minimally you must separate the waste of different nuclides, and you will probably be required to separate liquid and solid wastes and to minimize the creation of mixed waste, which is discussed further below.

Options

Generate no more radioactive waste than is absolutely necessary. Most countries have few sites that accept radioactive waste, so the costs per pound are outrageously expensive, forcing more institutions to store waste locally. Although there have been major advances in radioactive waste processing, these new technologies may be years away from being commonly available to any but the largest producers of radioactive waste.

Radioactive waste can be treated as nonradioactive after 10 half-lives. This is convenient for isotopes with short half-lives, such as ^{32}P and ^{33}P (10 half-lives is 250 days for ^{33}P, 143 days for ^{32}P), but the very long half-lives of ^{14}C (5730 years) and ^{3}H (12.41 years) more urgently illustrate the need for waste reduction. Your institution will have a policy on which radioisotopes will be disposed of by "decay in storage" or dumping into the sanitary sewer.

Limit the production of mixed waste, which is defined as a combination of two or more hazardous compounds, such as scintillation fluid and radioactivity. This waste is especially expensive to process and it will be worth your time to investigate if this type of waste can be avoided or minimized.

HANDLING RADIOACTIVITY: ACHIEVING MINIMUM DOSE

How Is Radioactive Exposure Quantified and What Are the Allowable Doses?

Radiation exposure is defined in REM, or "radiation equivalent man," and mrem, or millirem. In the United States, the maximum annual allowable dose is 5000 mrem to the internal organs, and 50,000 mrem to the extremities for those individuals working with radioactivity. (Note: Most other countries use a similar level, but the units are the international units of Sieverts, 5000 mrem = 50 mSv.) For comparison, the average person who doesn't work with radioactivity receives between 300 and 500 mrem per year. They receive this exposure from sources such as ^{40}K (potassium-40) and other naturally occurring radioactive isotopes found in foods, soil and rock, radon gas, cosmic rays, medical and dental X rays, and so forth.

Monitoring Technology: What's the Difference between a Count-Rate Counter and a Dose-Rate Meter?

Count-Rate Meter

A count-rate meter, generally configured with a Geiger-Müller detector, is used to detect small amounts of surface contamination. It is a common laboratory instrument. The unit is small and hand-held with an attached probe. When the probe face is directed toward an appropriate radiation field (most beta or gamma emitters), the count-rate meter produces the familiar clicking sound made so famous in science fiction movies of the 1950s. On the body of the meter is either an analog needle-type gauge or a digital readout, which will indicate the counts per second (cps) or counts per minute (cpm) of the field based on where the probe is located. In general, the efficiency of the count-rate meter versus a liquid scintillation counter, for example, is quite low. It has been designed to provide a quick, qualitative means of determining the presence of minute quantities of radioactivity (most instruments will detect between 50 and 5000 cps).

In the presence of a significant radiation field, the count-rate meter will be overloaded and cease to "click" or give a reading on the needle gauge. You can misinterpret the lack of sound as meaning that no dose field is present, when in fact what you need is another type of instrument to detect dose rates. The count-rate meter is best used to detect nanocurie or less quantities of contamination on gloves, benchtop, and other equipment.

Dose-Rate Meters

The dose-rate instrument should be used to detect larger quantities of ionizing radiation. It measures radiation fields in units of mrem/h. The dose rate meter also has a probe, generally an ionization chamber, and registers values on an analog needle gauge, or digital reading. A dose-rate meter does not aurally indicate the presence of radioactivity with clicks, however. It converts detected nuclear events into units that can be related to how much radioactive dose is present. This conversion is dependent on type and energy of emission, as well as on the distance from the radiation source. A count-rate meter detects an event, while a dose-rate meter converts that event into a meaningful energy reading. It is not a simple matter to convert cpm into a dose rate of mrem/h in our heads or by use of a chart because of the number of variables involved. Where a count-rate meter will go off scale, or become overloaded in a modest radiation field of 10 mrem/h, a dose-rate meter can measure much greater readings, depending on the particular instrument. Dose-rate meters are generally more expensive and not normally present in a lab, but your RSO will have them on hand when one is needed.

An illustration of the difference between dose rate and counts per second (or per minute) is seen in the following example. If you place a 1 mCi vial of ^{32}P-labeled dCTP right next to the probe face of a count-rate meter, there would suddenly be an "alarming" clicking sound. If you were to then open the vial, face the probe vertically down toward the open solution of ^{32}P, the meter would almost immediately become overloaded and stop giving off any sound, and fail to register a value on the needle gauge (The author does not recommend that you actually do this as it will needlessly increase both your exposure and the editor's legal liability.)

If you were to place the same sealed vial directly next to a dose-rate meter, you will detect an exposure rate of 2 to 5 mrem/h, which is typically considered to be a low or modest exposure dose field. If the dose-rate meter's probe is one inch directly above the open vial, you will read in excess of 1.0 rem/h, or 1000 mrem/h, a dramatic increase in dose. This is a significantly higher dose rate, yet the count-rate meter would not provide you with any warning.

To relate this scenario into the amount of exposure a dose film badge or thermoluminescent dosimeter (TLD) used for personnel monitoring might detect, suppose that your hand with a finger badge were placed directly over the open vial for one minute. Your finger might receive close to 17 mrem, which can quickly add up if it is part of your routine. On the other hand, if you held the closed vial in your hand for one hour, the finger dosimeter would register only 5 mrem, or 0.01% of the annual allowable dose.

What Are the Elements of a Good Overall Monitoring Strategy?

Identify the Hot Spots

Consider inviting your RSO to inspect the organization of your radioactive work area and to monitor your laboratory with a dose-rate meter to identify locations of significant exposure. This step is especially relevant when working with strong emitters such as ^{32}P and ^{125}I.

Short Term, or Contamination Monitoring

At the start of each workday, use a count-rate meter to check any work surface you plan to encounter, such as the benchtop and the lip of the hood. Next apply a count-rate meter to monitor the entire front part of your body and legs; pay special attention to your gloves and lab coat, especially the sleeves.

In all cases of contamination of yourself or if a serious spill occurs, your institution will have a very clear procedure on what steps to take to resolve it. You must know this procedure before working with radioactivity in your lab.

Long Term, or Dose Monitoring

Whole body dosimeters, often referred to as "badges" should be worn on the chest or abdomen to estimate exposure to critical organs. Ring badges worn on fingers are recommended to monitor extremity exposure. In some cases, and with particular radioisotopes in use, the radiation safety officer may require more specific monitoring techniques in order to test for the presence of radioactive contamination. A common example is the requirement of urine samples from those investigators working with tritium, and thyroid monitoring for those working with radioactive iodine.

What Can You Do to Achieve Minimum Radioactive Dose?

Attitude

Consider the benefits of an attitude whereby everyone working with radioactivity continuously ponders if they are working in the safest, most efficient manner. Is a particular radioactive experiment necessary? Can the amount of radioactivity in an experiment be reduced? Is there a faster, safer way to carry out the work? Questions like these will reduce cost and radioactive exposure. An institution's RSO is also required to implement a continuing education program regarding the principles of keeping personnel exposure dose low.

If you find yourself becoming stressed while handling radioactivity, or if that "incessant clicking sound" of the count-rate meter is causing a heightened sense of alarm, you can always step away from the bench to put things into perspective. Estimate how much dose you are receiving from your activities and relate those back to your annual allowable dose.

Time

Work quickly and neatly. In the example above, a finger lingering for 1 minute over an open 1 mCi vial of ^{32}P will receive 17 mrem, whereas a 10 second exposure receives a sixfold lower dose.

Practicing the manipulations of your experiments with non-radioactive materials will identify problem areas and ultimately enable you to work faster and safer. Working with radioactivity while feeling panicked or rushed will slow you down or cause an accident. If you can't smoothly do the motion in 10 seconds, take 20. You'll improve through time and all along will be well aware of your estimated dose. You'll automatically be striving to lower your dose.

Distance

Dosage decreases with distance. Why? A radiation source is like a light bulb. As the rays radiate outward in a sphere, they cover a wider area but become less potent at any single point. Use the inverse square law to your advantage. Can you pipette with a longer pipettor? Can you place the reaction vial even a few inches farther from you and others in the lab? Can you place a film cassette containing a radioactive membrane farther away from your work area? Small steps such as these can go a long way in reducing dose to you and your colleagues.

Shielding

A good shielding strategy will effectively reduce dose rate without preventing you from working smoothly and safely. It will not force you to get closer or stay longer in high radiation areas. If the use of lead-lined gloves makes you feel like you're working in a vat of honey and increase the likelihood of a spill, you might want to consider alternative shielding.

Shielding for Beta Emitters

Acrylic plastic (Plexiglas™) is used for the pure beta emitters, like ^{32}P, ^{33}P, and ^{35}S. A half inch thick piece of acrylic will stop essentially 100% of all betas, even for strong emitters, such as ^{32}P.

Shielding for Gamma Emitters

Lead will attenuate rather than completely obstruct gamma or X radiation. You may see in some literature that for a particular gamma-emitting isotope, a certain thickness of lead is required to "reduce the dose rate by a factor of 10." This means that if a source is giving off a field of 100 mrem/h without shielding, the dose rate with that particular thickness of lead will be brought down to 10 mrem/h. For example, ^{125}I needs to have 0.25 mm of lead shielding in order to reduce the dose rate by a factor of 10. Each successive layer of 0.25 mm will continue to decrease the dose rate by a factor of 10.

Lead is best used as shielding for an isotope giving off both gamma radiation and beta particles, rather than a combination of acrylic and lead.

Volatile Nuclides

The three isotopes you are likely to encounter with volatile properties are ^{3}H, ^{35}S, and ^{125}I. Their chemical properties and the incredibly complex reactions involved with radiolytic decay cause these two isotopes to form gaseous by-products. If you work with any of these isotopes, your RSO and institution may have approved fume hoods for their use.

Isotopes That Do Not Require Shielding

Tritium, being a very weak beta emitter, travels only a few microns in air. Acrylic shielding would be of no use. What you do not want to do is to ingest tritium. Tritium in an aqueous form is 25,000 times more radiotoxic than tritium in a gaseous form.

How Can You Organize Your Work Area to Minimize Your Exposure to Radioactivity?

If feasible, select bench space at the corner of the room, rather than in a central location, to reduce unnecessary traffic. Clearly delineate this work area as radioactive. Although it is not always possible due to space restrictions, it is recommended that if your lab is working with different radioisotopes that there be separate work areas for each radioisotope. Check with your RSO about any additional requirements listed on the institution's license.

Of main importance will be arranging your work space. Begin with absorbent material, perhaps a double thick section, taped onto the bench. A waste container that shields against the radioactivity should be placed in a location that makes it easy, quick, and safe to dispose of pipette tips, hot gloves, and the like. A box made of acrylic with a lid is sufficient for ^{32}P, ^{33}P, and ^{35}S, while for ^{125}I, lead-impregnated acrylic will help attenuate the gamma rays. Each radioisotope may need its own separate container for waste, depending on your institution's disposal protocols.

If you are using ^{32}P, acrylic shielding between you and the source is strongly recommended. There are many commercially available shields that will meet your needs. Once you establish your radioactive area, do a couple of practice runs to make sure that your work area is properly organized. Bring the RSO in so that she/he can approve your radioactive area and perhaps make further suggestions.

You'll want to examine closely any areas or actions that have the potential for high doses. An open vial of ^{32}P, an Eppendorf tube with 50 μl of ^{32}P, and a tray containing your blot with hybridization solution mixed with radioactive probe will all be obvious areas where you'll need to pay close attention. The open vial may simply need an acrylic pipette guard on your pipettor in order to bring the dose down 10,000 fold. The Eppendorf incubation tube can be kept in an acrylic box, or behind acrylic shielding while the labeling reaction is going on. You may devise a way of not picking the reaction tube up with your fingers while you remove the reaction mix with a pipettor. The blotting container may present a potential spill. Finding a safe, out-of-the-way place, preferably in a fume hood and behind some acrylic shielding will go a long way toward reducing dose.

How Can You Concentrate a Radioactive Solution?

Three convenient approaches are lyophilization, a spinning vacuum chamber, and drying with a gentle stream of nitrogen gas.

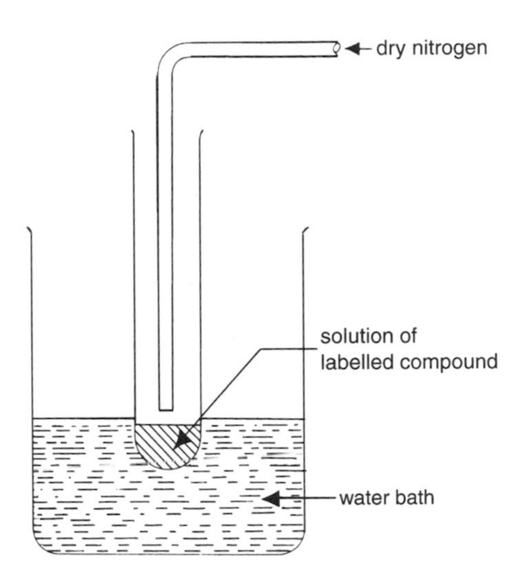

Figure 6.3 Removal of solvent from a non-volatile radiochemical using dry nitrogen. From *Guide to Working Safely with Radiolabelled Compounds*, Amerhsam International, plc, 1974, Buckinghamshire, U.K. Reprinted by permission of Amersham Pharmacia Biotech.

There is significant risk of contamination when using a lyophilizer or spinning vacuum chamber, so most facilities dedicate specific equipment for radioactive work. Blowing a very gentle stream of nitrogen gas over the solution works efficiently, but practice is required to avoid blowing the radioactive solution out of its container.

The nitrogen stream method is straightforward (Figure 6.3). Attach a small glass pipette/dropper tip to tubing that is attached to the gas regulator of a tank of dry nitrogen gas, being careful not to break the top of the pipette into your hand. Turn on the gas flow, keeping the gas flow as gentle as possible. This procedure requires very little nitrogen flow. Before impinging upon the surface of the radioactive material, test the gas flow on a vial containing a like amount of water. Adjust the flow so that there is no splashing of the liquid but only a noticeable indentation of the liquid's surface. Once you are satisfied that it is safe, gently direct the stream of gas onto the surface of the radioactive liquid, ensuring no splashing. Do all of this in a hood and in a location that is safe and will be able to contain accidental spills.

Continue blowing off the solution until dryness. Overdrying can sometimes be of concern, so it is best not to leave the area and come back to it after an extended period. It is also best to bring the solution to complete dryness so that when you bring it up into a known amount of solution, you will have an accurate idea of the concentration.

BIBLIOGRAPHY

Feinberg, A. P., and Vogelstein, B. 1983. A technique for radiolabeling DNA restriction endonuclease fragments to high specific activity. *Anal. Biochem.* 132:6–13.

U.S. Nuclear Regulatory Commission Regulatory Guide, Office of Standards Development. Regulatory Guide 10.5, *Applications for Type A Licenses of Broad Scope.* Revision 1, December 1980.

U.S. Nuclear Regulatory Commission Regulatory Guide, Office of Standards Development. Regulatory Guide 10.7, *Guide for the Preparation of Applications for Licenses for Laboratory and Industrial Use of Small Quantities of Byproduct Material*, Revision 1, August, 1979.

U.S. Nuclear Regulatory Commission Regulatory Guide, Office of Standards Development. Regulatory Guide 10.2, *Guidance to Academic Institutions Applying for Specific Byproduct Material Licenses of Limited Scope*. Revision 1, December 1976.

Guide to the Self-decomposition of Radiochemicals. Amersham International, plc, Buckinghamshire, U.K., 1992.

Guide to Working Safely with Radiolabelled Compounds. Amersham International, plc, Buckinghamshire, U.K., 1974.

International Air Transport Association (IATA) Dangerous Goods Regulations, 6.2, *Packing Instructions*.

Code of Federal Regulations (CFR) 173.421, 173.422, 173.424, and 173.427. *Additional Requirements for Excepted Packages Containing Class 7 (Radioactive) Materials.*

Appendix A. Physical Properties of Common Radionuclides

Radionuclide	Halflife	Beta Energy, max(MeV)	Specific Activity, max
Tritium (hydrogen-3)	12.4 years	0.0186	28.8 Ci/matom[a] 1.06 TBq/matom[a]
Carbon-14	5730 years	0.156	62.4 mCi/matom 2.31 GBq/matom
Sulfur-35	87.4 days	0.167	1.49 kCi/matom 55.3 TBq/matom
Phosphorus-32	14.3 days	1.709	9.13 kCi/matom 338 TBq/matom
Phosphorous-33	25.3 days	0.249	5140 Ci/matom
Iodine-125	59.6 days	Electron capture[b]	2.18 Ci/matom 80.5 TBq/matom[a]

Source: Data reproduced from *Guide to Working Safely with Radiolabelled Compounds* (Amerhsam International, 1974).

[a] A milliatom is the atomic weight of the element in milligrams.

[b] Electron capture is a radioactive transformation in which the nucleus absorbs an electron from an inner orbital. The remaining orbital electrons re-arrange to fill the empty electron shell and in so doing energy is released as electromagnetic radiation at X-ray wavelengths and/or electrons.

7

DNA Purification

Sibylle Herzer

What Criteria Could You Consider When Selecting a Purification Strategy? 168
How Much Purity Does Your Application Require? 168
How Much Nucleic Acid Can Be Produced from a Given Amount of Starting Material? 168
Do You Require High Molecular Weight Material? 168
How Important Is Speed to Your Situation? 168
How Important Is Cost? 169
How Important Is Reproducibility (Robustness) of the Procedure? 169
What Interferes with Nucleic Acid Purification? 169
What Practices Will Maximize the Quality of DNA Purification? 171
How Can You Maximize the Storage Life of Purified DNA? 172
Isolating DNA from Cells and Tissue 172
What Are the Fundamental Steps of DNA Purification? 172
What Are the Strengths and Limitations of Contemporary Purification Methods? 174
What Are the Steps of Plasmid Purification? 180
What Are the Options for Purification after In Vitro Reactions? 184
Spun Column Chromatography through Gel Filtration Resins 184

Molecular Biology Problem Solver, Edited by Alan S. Gerstein.
ISBN 0-471-37972-7

Filter Cartridges 185
Silica Resin-Based Strategies 186
Isolation from Electrophoresis Gels 187
What Are Your Options for Monitoring the Quality of Your DNA Preparation? 190
Bibliography 191

WHAT CRITERIA COULD YOU CONSIDER WHEN SELECTING A PURIFICATION STRATEGY?

How Much Purity Does Your Application Require?

What contaminants will affect your immediate and downstream application(s)? As discussed below and in Chapter 1, "Planning for Success in the Laboratory," time and money can be saved by determining which contaminants need not be removed. For example, some PCR applications might not require extensively purified DNA. Cells can be lysed, diluted, and amplified without any further steps. Another reason to accurately determine purity requirements is that yields tend to decrease as purity requirements increase.

How Much Nucleic Acid Can Be Produced from a Given Amount of Starting Material?

While it is feasible to mathematically calculate the total amount of nucleic acid in a given sample, and values are provided in the research literature (Sambrook et al., 1989; Studier and Moffat, 1986; Bolivar et al., 1977; Kahn et al., 1979; Stoker et al., 1982), the yields from commercial purification products and noncommercial purification strategies are usually significantly less than these maxima, sometimes less than 50%. Since recoveries will vary with sample origin, consider making your plans based on yields published for samples similar if not identical to your own.

Do You Require High Molecular Weight Material?

The average size of genomic DNA prepared will vary between commercial products and between published procedures.

How Important Is Speed to Your Situation?

Some purification protocols are very fast and allow isolation of nucleic acids within 30 minutes, but speed usually comes at the price of reduced yield and/or purity, especially when working with complex samples.

How Important Is Cost?

Reagents obviously figure into the cost of a procedure, but the labor required to produce and apply the reagents of purification should also be considered.

How Important Is Reproducibility (Robustness) of the Procedure?

Some methods will not give consistent quality and quantity. When planning long-term or high-throughput extractions, validate your methods for consistency and robustness.

What Interferes with Nucleic Acid Purification?

Nuclease

One of the major concerns of nucleic acid purification is the ubiquity of nucleases. The minute a cell dies, the isolation of DNA turns into a race against internal degradation. Samples must be lysed fast and completely and lysis buffers must inactivate nucleases to prevent nuclease degradation.

Most lysis buffers contain protein-denaturing and enzyme-inhibiting components. DNases are much easier to inactivate than RNases, but care should be taken not to reintroduce them during or after purification. All materials should be autoclaved or baked four hours at 300°F to inactivate DNases and RNases, or you should use disposable materials. Use only enzymes and materials guaranteed to be free of contaminating nucleases. Where appropriate, work on ice or in the cold to slow down potential nuclease activity.

Smears and lack of signal, or smeared signal alone, and failure to amplify by PCR are indicative of nuclease contamination. The presence of nuclease can be verified by incubating a small aliquot of your sample at 37°C for a few hours or overnight, followed by evaluation by electrophoresis or hybridization. If nuclease contamination is minor, consider repurifying the sample with a procedure that removes protein.

Shearing

Large DNA molecules (genomic DNA, bacterial artificial chromomoses, yeast artificial chromosomes) can be easily sheared during purification. Avoid vortexing, repeated pipetting (especially through low-volume pipette tips), and any other form of mechanical stress when the isolate is destined for applications that require high molecuar weight DNA.

Chemical Contaminants

Materials that interfere with nucleic acid isolation or downstream applications involving the purified DNA can originate from the sample. Plants, molds, and fungi can present a challenge because of their rigid cell wall and the presence of polyphenolic components, which can react irreversibly with nucleic acids to create an unusable final product.

The reagents of a DNA purification method can also contribute contaminants to the isolated DNA. Reagents that lyse and solubilize samples, such as guanidinium isothiocyanate, can inhibit some enzymes when present in trace amounts. Ethanol precipitation of the DNA and subsequent ethanol washes eliminate such a contaminant. Phenol can also be problematic. If you experience problems with DNA purified by a phenol-based strategy, apply chloroform to extract away the phenol. Phenol oxidation products may also damage nucleic acids; hence re-distilled phenol is recommended for purification procedures.

A mixture of chloroform and phenol is often employed to maximize the yield of isolated DNA; the chloroform reduces the amount of the DNA-containing aqueous layer at the phenol interphase. Similar to phenol, residual chloroform can be problematic, and should be removed by thorough drying. Drying is also employed to remove residual ethanol. Overdried DNA can be difficult to dissolve, so drying should be stopped shortly after the liquid can no longer be observed. Detailed procedures for the above extraction, precipitation and washing steps can be found in Sambrook, Fritsch, and Maniatis (1989) and Ausubel et al. (1998).

Ammonium ions inhibit T4 polynucleotide kinase, and chloride can poison translation reactions (Ausubel et al., 1998). The common electrophoresis buffer, TBE (Tris, borate, EDTA) can inhibit enzymes (Ausubel et al., 1998) and interfere with transformation due to the increased salt concentration (Woods, 1994). Phosphate buffers may also inhibit some enzymes, namely T4 Polynucleotide kinase (Sambrook et al., 1989), alkaline phosphatase (Fernley, 1971), Taq DNA polymerase (Johnson et al., 1995), and Poly A polymerase from *E. coli* (Sippel, 1973). Agarose can also be a problem but some enzyme activity can be recovered by adding BSA to 500 μg/ml final concentration (Ausubel et al., 1998). EDTA can protect against nuclease and heavy metal damage, but could interfere with a downstream application.

The anticoagulant heparin can contaminate nucleic acids isolated from blood, and should be avoided if possible (Grimberg et al., 1989). Taq DNA polymerase is inhibited by heparin, which can be resolved by the addition of heparinase (Farnert et al., 1999). Heparin also interacts with chromatin leading to release of denatured/nicked DNA molecules (Strzelecka, Spitkovsky, and Paponov, 1983). Narayanan (1996) reviews the effects of anticoagulants.

What Practices Will Maximize the Quality of DNA Purification?

The success of DNA purification is dependent on the initial quality of the sample and its preparation. It would be nice to have a simple, straightforward formula that applies to all samples, but some specimens have inherent limitations. The list below will help guide your selection and provide remedies to nonideal situations:

1. Ideally start with fresh sample. Old and necrotic samples complicate purification. In the case of plasmid preparations, cell death sets in after active growth has ceased, which can produce an increase in unwanted by-products such as endotoxins that interfere with purification or downstream application.

   The best growth phase of bacterial cultures for plasmid preparations may be strain dependent. During the log phase of bacterial culture, actively replicating plasmids are present that are "nicked" during replication rather than being supercoiled. Still some researchers prefer mid to late log phase due to the high ratio of DNA to protein and low numbers of dead cells. Others only work with plasmids that have grown just out of log phase to avoid co-purification of nicked plasmid.

   If old samples can't be avoided, scaling up the purification can compensate for losses due to degradation. PCR or dot blotting is strongly recommended to document the integrity of the DNA.

2. Process your sample as quickly as possible. There are few exceptions to this rule, one being virus purification. When samples can't be immediately purified, snap freeze the intact sample in liquid nitrogen or hexane on dry ice (Franken and Luyten, 1976; Narang and Seawright, 1990) or store the lysed extract at –80°C. Commercial products, such as those from Ambion, Inc., can also protect samples from degradation prior to nucleic acid purification. Samples can also be freeze-dried, as discussed below in the question, *How Can You Maximize the Storage Life of Purified DNA?*.

3. Thorough, rapid homogenization is crucial. Review the literature to determine if your sample requires any special physical or mechanical means to generate the lysate.

4. Load the appropriate amount of sample. Nothing will impair the quality and yield of a purification strategy more than overloading the system. Too much sample can cause an increase in the viscosity of the DNA preparation and lead to shearing of genomic DNA. If you do not know the exact amount of starting material, use 60 to 70% of your estimate.

How Can You Maximize the Storage Life of Purified DNA?

The integrity of purified DNA in solution could be compromised by nuclease, pH below 6.0 and above 9.0, heavy metals, UV light, and oxidation by free radicals. EDTA is often added to chelate divalent cations required for nuclease activity and to prevent heavy metal oxidative damage. Tris-based buffers will provide a safe pH of 7 to 8 and will not generate free radicals, as can occur with PBS (Miller, Thomas, and Frazier, 1991; Muller and Janz, 1993). Free-radical oxidation seems to be a key player in breakdown and ethanol is the best means to control this process (Evans et al., 2000).

Low temperatures are also important for long-term stability. Storage at 4°C is only recommended for short periods (days) (Krajden et al., 1999). Even though some studies have shown that storage under ethanol is safe even at elevated temperatures (Sharova, 1977), better stability is obtained at –80°C. Storage at –20°C can lead to degradation, but this breakdown is prevented by the addition of carrier DNA. RNA stored in serum has also been shown to degrade at –20°C (Halfon et al., 1996).

Another approach for intermediate storage is freeze drying DNA-containing samples intact (Takahashi et al., 1995). The DNA within freeze-dried tissue was stable for 6 months, but RNA began degrading after 10 weeks of storage. The control of moisture and temperature had a significant effect on shelf life of samples. The long term stability of DNA-containing samples is still being investigated (Visvikis, Schlenck, and Maurice, 1998), but some companies offer specialized solutions (e.g., RNA Later™ from Ambion, Inc.) allowing storage at room temperature.

ISOLATING DNA FROM CELLS AND TISSUES

What Are the Fundamental Steps of DNA Purification?

The fundamental processes of DNA purification from cells and tissues are sample lysis and the segregation of the nucleic acid away from contaminants. While DNA is more or less universal to all species, the contaminants and their relative amounts will differ considerably. The composition of fat cells differs significantly from muscle cells. Plants have to sustain high pressure, contain chloroplasts packed with chromophores, and often have a very rigid outer cell wall. Bacteria contain lipopolysaccharides that can interfere with purification and cause toxicity problems when present in downstream applications. Fibrous tissues such as heart and skeletal muscle are tough to homogenize. These variations have to be taken into consideration when developing or selecting a lysis method.

Lysis

Detergents are used to solubilize the cell membranes. Popular choices are SDS, Triton X-100, and CTAB(hexadecyltrimethyl ammonium bromide). CTAB can precipitate genomic DNA, and it is also popular because of its ability to remove polysaccharides from bacterial and plant preparations (Ausubel et al., 1998).

Enzymes attacking cell surface components and/or components of the cytosol are often added to detergent-based lysis buffers. Lysozyme digests cell wall components of gram-positive bacteria. Zymolase, and murienase aid in protoplast production from yeast cells. Proteinase K cleaves glycoproteins and inactivates (to some extent) RNase/DNase in 0.5 to 1% SDS solutions. Heat is also applied to enhance lysis. Denaturants such as urea, guanidinium salts, and other chaotropes are applied to lyse cells and inactivate enzymes, but extended use beyond what is recommended in a procedure can lead to a reduction in quality and yield.

Sonication, grinding in liquid nitrogen, shredding devices such as rigid spheres or beads, and mechanical stress such as filtration have been used to lyse difficult samples prior to or in conjunction with lysis solutions. Disruption methods are discussed at *http://www.thescientist.com/yr1998/nov/profile2_981109.html.*

Segregation of DNA from Contaminants

The separation of nucleic acid from contaminants are discussed below within the question, *What Are The Strengths and Limitations of Contemporary Purification Methods?*

DNA Precipitation

To concentrate nucleic acids for resuspension in a more suitable buffer, solvents such as ethanol (75–80%) or isopropanol (final concentration of 40–50%) are commonly used in the presence of salt to precipitate nucleic acids (Sambrook, Fritsch, and Maniatis,

1989; Ausubel et al., 1998). If volume is not an issue, ethanol is preferred because less salt will coprecipitate and the pellet is more easily dried. Polyethylene glycol (PEG) selectively precipitates high molecular weight DNA, but it is also more difficult to dry and can interfere with downstream applications (Hillen, Klein, and Wells, 1981). Trichloroacetic acid (TCA) precipitates even low MW polymers down to (5kDa) (*http://biotech-server.biotech.ubc.ca/biotech/bisc437/lecture/e-na-isoln/na-isoln3.html*), but nucleic acids cannot be recovered in a functional form after precipitation.

Salt is essential for DNA precipitation because its cations counteract the repulsion caused by the negative charges of the phosphate backbone. Ammonium acetate is useful because it is volatile and easily removed, and at high concentration it selectively precipitates high molecular weight molecules. Lithium chloride is often used for RNA because Li^+ does not precipitate double-stranded DNA, proteins, or carbohydrates, although the single-stranded nucleic acids must be above 300 nucleotides. To efficiently precipitate nucleic acids, incubation at low temperatures (preferably ≤–20°C) for at least 10 minutes is required, followed by centrifugation at 12,000 × *g* for at least five minutes. Temperature and time are crucial for nucleic acids at low concentrations, but above 0.25mg/ml, precipitation may be carried out at room temperature. Additional washing steps with 70% ethanol will remove residual salt from pelleted DNA. Pellets are dried in a speed vac or on the bench and are resuspended in water or TE (10mM Tris, 1mM EDTA). Do not attempt to precipitate nucleic acids below a concentration of 20ng/ml unless carrier such as RNA, DNA, or a high molecular weight co-precipitant like glycogen is added. In the range from 20ng/ml to 10μg/ml, either add carrier or extend precipitation time, and add more ethanol. Polyethylene glycol (PEG) precipitation is even more concentration dependent and will only work at DNA concentrations above 10μg/ml (Lis and Schleif, 1975). Pellets will dissolve better in low-salt buffers (water or TE) and at concentrations below 1mg/ml. Gentle heating can also help to redissolve nucleic acids

What Are the Strengths and Limitations of Contemporary Purification Methods?

Salting out and DNA Precipitation

Mechanism

Some of the first DNA isolation methods were based on the use of chaotropes and cosmotropes to separate cellular components based on solubility differences (Harrison, 1971; Lang, 1969). A chaotrope increases the solubility of molecules ("salting-in") by changing the structure of water, and as the name suggests, the driving force is an increase in entropy. A cosmotrope is a structure-maker; it will decrease the solubility of a molecule ("salting-out"). Guanidium salts are common chaotropes applied in DNA purification. Guanidinium isothiocyanate is the most potent because both cation and anion components are chaotropic. Typical lyotropes used for salting out proteins are ammonium and potassium sulfate or acetate. An all solution based nucleic acid purification can be performed by differentially precipitating contaminants and nucleic acids.

Cells are lysed with a gentle enzyme- or detergent-based buffer (often SDS/proteinase K). A cosmotrope such as potassium acetate is added to salt out protein, SDS, and lipids but not the bulk of nucleic acids. The white precipitate is then removed by centrifugation. The remaining nucleic acid solution is too dilute and in a buffer incompatible with most downstream applications, so the DNA is next precipitated as described above.

Features

Protocols and commercial products differ mainly in lysis buffer composition. Yields are generally good, provided that sample lysis was complete and DNA precipitation was thorough. These procedures apply little mechanical stress, so shearing is generally not a problem.

Limitations

If phenolic contaminants (i.e., from plants) are a problem, adding 1% polyvinylpyrrolidine to your extraction buffer can absorb them (John, 1992; Pich and Schubert, 1993; Kim et al., 1997). Alternatively, add a CTAB precipitation step to remove polysaccharides (Ausubel et al., 1998).

Extraction with Organic Solvents, Chaotropes, and DNA Precipitation

Mechanism

Chaotropic guanidinium salts lyse cells and denature proteins, and reducing agents (β-mercaptoethanol, dithiothreitol) prevent oxidative damage of nucleic acids. Phenol, which solubilizes and extracts proteins and lipids to the organic phase, sequestering them away from nucleic acids, can be added directly to the lysis buffer, or a phenol step could be included after lysis with either

GTC- or SDS-based buffers as above. GTC/phenol buffers often require vortexing or vigorous mixing.

The affinity of nucleic acids for this two-phase extraction system is pH dependent. Acidic phenol is applied in RNA extractions because DNA is more soluble in acidic phenol; smaller DNA molecules (<50 kb) will be found in the organic phase and larger DNA molecules (>50 kb) in the interphase. When purifying RNA via this procedure, it is essential to shear the DNA to ensure a light interphase.

Phenol titrated to a pH of 8 is used to separate DNA from proteins and lipids, since DNA is insoluble in basic phenol. Whether protocols call for a GTC/phenol, a GTC, or an SDS based step followed by phenol, it is best to follow a phenol extraction with chloroform in order to extract residual phenol from the aqueous phase. Phenol is highly soluble in chloroform, and chloroform is not water soluble. Remaining lipids may also be removed by this step. Phenol extractions are followed by nucleic acid precipitation steps as described above.

Features

Though caustic and toxic, this strategy still has wide use because yield, purity, and speed are good, and convenient for working with small numbers of samples.

Limitations

If lysis is incomplete, the interphase between organic and aqueous layers becomes very heavy and difficult to manipulate, and may trap DNA. Phenol is not completely insoluble in water, so if chloroform steps are skipped, residual phenol can remain and interfere with downstream applications. High salt concentrations can also lead to phase inversion, where the aqueous phase is no longer on top (problematic if colorless phenol is used). Diluting the aqueous phase and increasing the amount of phenol will correct this inversion. When working with GTC/phenol-based extraction buffers, cross-contamination of RNA with DNA, and vice versa, is frequent.

Glass Milk/Silica Resin-Based Strategies

Mechanism

Nucleic acids bind to glass milk and silica resin under denaturing conditions in the presence of salts (Vogelstein and Gillespie, 1979). Recent findings indicate that binding of some nucleic acids might even be feasible under nondenaturing conditions (Neudecker and Grimm, 2000). The strong, hydrophobic interaction created in the presence of chaotropic substances can be easily disrupted by removal of salt. The adsorption is followed by wash steps, usually with salt/ethanol which will not interfere with the strong binding of nucleic acids but will wash away remaining impurities and excess chaotrope. Depending on the protocol, this can be followed by a low salt/ethanol wash step that can lead to a reduction in yield. Finally nucleic acids are eluted from the glass in a salt or TE buffer. Nucleic acids are then ready for use.

Most methods create a denaturing adsorption environment by using guanidium salts for one-step lysis and binding. The strength of the binding depends on the cation used to shield the negative charges of the phosphate backbone and the pH (Romanowski et al., 1991). Slightly acidic pH and divalent cations, preferably magnesium, seem to work best.

Differences between glass milk, silica resin, and powdered glass consist mainly in capacity and adsorption strength, a function of impurities present in the binding resins. Diatomaceous earths seem to have an especially high binding capacity (*http://www.nwfsc.noaa.gov/protocols/dna-prep.html*). Pure silica oxide has the lowest affinity to nucleic acids (Boom et al., 1990), but this can improve recovery even though initial binding capacity is lower.

Glass milk is silica presuspended in chaotropic buffer, whereas the silica resin is a solid, predispensed matrix usually found in spin or vacuum flow-through format. Glass milk gives more flexibility for scale of prep, predispensed resin is more convenient for high-throughput applications. Glass milk or silica-based kits are available from numerous vendors, and even though the basic principle is the same, there can be significant differences in efficiency, purity, and yield.

Features

DNA purification based on hydrophobic adsorption to glass or silica is fast, simple, straightforward, and scalable. No additional time-consuming and yield-reducing precipitation steps are required. Depending on binding and wash buffer composition, very good yield and purity values are obtained. This purification approach can also allow restriction digestion/ligation reactions directly on the glass surface, improving transformation efficiency of complex ligation mixtures (Maitra and Thakur, 1994).

Limitations

One of the dangers of silica-based strategies is underloading the sample. Even though yields are good, there is always sample loss due to some remaining material on the resin or filter. The smaller the DNA fragment, the tighter is the interaction. Oligonucleotide primers are actually removed because binding to the glass becomes virtually irreversible. Underloading can become a critical issue when working with small samples and large volumes of glass milk or silica filter.

Some of the older methods utilized unstable buffer components, such as NaI, that tended to oxidize over time, leading to very poor recoveries. Some procedures required the addition of reagents to produce functional wash or elution buffers. If the concentrations were incorrect, or if volatile reagents (i.e., ethanol) were added and the buffers stored long term, these buffers lost their effectiveness. Incomplete sample lysis can be problematic because intact cells may also bind to silica and lyse under low- or no-salt elution conditions, leading to degradation of nucleic acids. Incomplete ethanol removal after wash steps will cause the problems described earlier for ethanol precipitation (discussed below under the question *What Are The Fundamental Steps Of DNA Purification?*). Ethanol must be completely removed from the samples after wash steps to avoid problems such as diffusion out of agarose gel wells ("unloadable" DNA/RNA) or undigestable DNA. Overdrying will lead to irreversible binding of nucleic acids to the resin severely impairing yields.

Anion Exchange (AIX) Based Strategies

Mechanism

Nucleic acids are very large anions with a charge of –1/base and –2/bp; hence they will bind to positively charged purification resins (commonly referred to as anion exchangers). After washes in low-salt buffers, the DNA is eluted in a high-salt buffer. AIX strategies are applied to purify genomic and plasmid DNA.

Logic might suggest that the greater the strength of the anion exchanger, the more DNA it would bind (and more tightly), which would make for superior DNA purification. In practice, however, if an anion exchanger is too strong, most DNA is never recovered. This is especially problematic when working with small samples and with spun column formats. Forcing liquid through porous chromatography resins via centrifugation does not allow for even flow rates, hence resolution is poor. For this reason some spun column plasmid purification procedures advise the recovery of only a portion of the potential total material to avoid contamination by genomic DNA. Procedures where the buffer flows through columns packed with AIX resins under the force of gravity (as in standard column chromatography) can overcome this problem, but are slower. Gravity flow-based columns can clog if lysis is incomplete or if removal of protein or lipid is incomplete. Resolution is very much flow rate dependent, and tight control of linear flow rates on HPLC or FPLC™ systems are superior to gravity flow and/or spun column formats when it comes to resolution and scale-up.

Features

These methods can produce very pure DNA, but the yields in small-scale applications tend to be low, especially in spun column formats.

Limitations

Not the most robust method, and recoveries tend to be lower, and the final elution step of AIX protocols involves high-salt buffers. The 0.7 to 2M sodium chloride eluate needs to be desalted, usually by a precipitation step, which decreases recovery and increases the overall procedure time. The binding capacities tend to be low (0.25–2 mg/ml of resin), increase with pH, and decrease with increasing size of the DNA. The amount of RNA present in the sample will also affect binding capacity because RNA will compete with DNA for binding.

Hydroxyapatite (HA) Based Strategies

Mechanism

Nucleic acids bind to crystalline calcium phosphate through the interaction of calcium ions on the hydroxyapatite and the phosphate groups of the nucleic acids. An increase in competing free phosphate ions from 0.12 to 0.4 M will elute nucleic acids, with single-stranded nucleic acids eluting before double-stranded DNA. The entire experiment needs to be run at 60°C for thermal elution (Martinson and Wagenaar, 1974) or in the presence of formamide at room temperature (Goodman et al., 1973).

Sodium phosphate buffers are most commonly used; the phosphate salt affects the selectivity of the resin (Martinson and Wagenaar, 1974). Nucleic acids may also be eluted by increasing the temperature until nucleic acid strands melt and elute from the column.

Features

Excellent separation of single-stranded from double-stranded DNA molecules.

Limitations

The quality and performance of hydroxyapatite can vary from batch to batch and between manufacturers. Thermal elution procedures require reliable temperature control, but fluctuations occur because of lack of heat-regulated chromatography equipment. These elevated temperatures can also produce bubbles in the buffer that can interfere with the separation. Hydroxyapatite has poor mechanical stability. Hydroxyapatite procedures often employ high-salt buffers and lead to sample dilution, requiring an additional precipitation step.

For these reasons hydroxyapatite is not extensively referenced. It is mostly limited to subtractive cDNA cloning (Ausubel et al., 1998), removal of single-stranded molecules, and DNA reassociation analysis (Britten, Graham, and Newfeld, 1974).

What Are the Steps of Plasmid Purification?

Alkaline Lysis and Boiling Strategies

Mechanism (Small Scale)

Plasmid purification holds a special challenge because the target DNA must be purified from DNA contaminants. Isolation strategies take advantage of the physical differences between linear, closed, and supercoiled DNA. Alkaline lysis (Birnboim and Doly, 1979), boiling, and all other denaturing methods exploit the fact that closed DNA will renature quickly upon cooling or neutralizing, while the long genomic DNA molecules will not renature and remain "tangled" with proteins, SDS, and lipids, which are salted out. Whether boiling or alkaline pH is the denaturing step, the renaturing step is usually performed in the cold to enhance precipitation or salting-out of protein and contaminant nucleic acids.

Buffer 1 of an alkaline lysis procedure contains glucose to buffer the effects of sodium hydroxide added in step 2, and lysozyme, to aid cellular breakdown which prevents plasmid from becoming trapped in cellular debris. Buffer 2 contains SDS and NaOH. SDS denatures proteins and NaOH denatures DNA, both plasmid and genomic, and proteins, and partially breaks down RNA. Buffer 3 contains an acidic potassium acetate solution that will salt out proteins by complexing SDS with potassium and precipitating out a mix of SDS, K^+, proteins, and denatured genomic

DNA. Supercoiled plasmids and RNA molecules will remain in solution.

Another method lyses cells by a combination of enzymatic breakdown, detergent solubilization, and heat (Holmes and Quigley, 1981). The lysis buffer usually contains lysozyme, STE, and Triton X-100 or CTAB. Bacterial chromosomal DNA remains attached to the membrane and precipitates out. Again, the aqueous supernatant generated by this method contains plasmid and RNA.

Polyethylene glycol (PEG) has been used to separate DNA molecules by size, based on it's size-specific binding to DNA fragments (Humphreys, Willshaw, and Anderson, 1975; Hillen, Klein, and Wells, 1981). A 6.5% PEG solution can be used to precipitate genomic DNA selectively from cleared bacterial lysates. Trace amounts of PEG may be removed by a chloroform extraction.

Isolation of plasmid DNA by cesium chloride centrifugation in the presence of ethidium bromide (EtBr) is especially useful for large-scale DNA preparations. The interaction of EtBr with DNA decreases the density of the nucleic acid; because of its supercoiled conformation and smaller size, plasmid incorporates less EtBr than genomic DNA, enhancing separation on a density gradient.

Chromatographic methods such as anion exchange and gel filtration may also be used to purify plasmids after lysis. For chromatography, RNA removal prior to separation is essential because the RNA will interfere with and contaminate the separation process. RNase A treatments (Feliciello and Chinali, 1993), RNA-specific precipitation (Mukhopadhyay and Mandal, 1983; Kondo et al., 1991), tangential flow filtration (Kahn et al., 2000), and nitrocellulose filter binding (Levy et al., 2000a, 2000b) have been employed to desalt, concentrate, and generally prepare samples for column purification.

Limitations

The efficiency of plasmid purification will vary with the host cell strain due to differences in polysaccharide content and endonuclease—End A+ strains such as HB101 (Ausubel et al., 1998). Recombination impaired hosts are often selected when producing plasmids prone to deletion and rearrangement of cloned inserts (Summers and Sherratt, 1984; Biek and Cohen, 1986). The University of Birmingham's Web site gives useful links to research strain genotypes and characteristics at *http://web.bham.ac.uk/bcm4ght6/res.html*, as does the *E. coli* Genetic Stock Center at Yale Univeristy (*http://cgsc.biology.yale.edu*).

Whenever nucleic acids are denatured, there is a risk of irreversible denaturation. Never increase the denaturation time beyond what is recommended, and ensure that pH values are accurate for neutralization. Prolonged high pH or heat exposure may lead to more contamination with genomic DNA (Liou et al., 1999) and nicked, open, and irreversibly denatured plasmid. The pH of solution 3 of an alkaline lysis procedure needs to be pH 5.5 to precipitate out SDS/protein/genomic DNA. Effects of changing criticial parameters have been studied in detail (Kieser, 1984). These protocols have been modified to purify cosmids, but larger DNA molecules will not renature as well as small plasmids. Most methods work well for plasmids up to 10 kb; above 10 kb, denaturation has to be milder (Hogrefe and Friedrich, 1984; Azad, Coote, and Parton, 1992; Sinnett, Richer, and Baccichet, 1998).

The yield of low copy number plasmids can be improved dramatically by adding chloramphenicol (Norgard, Emigholz, and Monahan, 1979) or spectinomycin (300 μg/ml; Amersham Pharmacia Biotech, unpublished observations), which prevent replication of chromosomal but not plasmid DNA. However, extended exposure to such agents have also been shown to damage DNA in vitro (Skolimowski, Knight, and Edwards, 1983).

Resources

Plasmid purification methodology could fill an entire book of its own. Traditional chromatography has been applied to isolate large- and small-scale preparations of plasmid from a variety of hosts. Techniques include gel filtration, anion exchange, hydrophobic interaction chromatography, single-strand affinity matrix (Pham, Chillapagari, and Suarez, 1996; Yashima et al., 1993a, b), triple helix resin, silica resin, and hydroxyapatite in a column as well as microtiter plate format. Plasmid purification procedures are reviewed in O'Kennedy et al. (2000), Neudecker and Grimm (2000), Monteiro et al. (1999), Ferreira et al., 2000. Ferreira et al. (1999), Ferreira et al. (1998), Huber (1998), Lyddiatt and O'Sullivan (1998), and Levy et al. (2000a).

CsCl Purification

Mechanism

The separation of DNA from contaminants based on density differences (isopycnic centrifugation) in CsCl gradients remains an effective if slow method. High *g* forces cause the migration of dense Cs^+ ions to the bottom of the tube until centripetal force and force of diffusion have reached an equilibrium.

Within a CsCl gradient, polysaccharides will assume a random coil secondary structure, DNA a double-stranded intermediate density conformation, and RNA, because of its extensive secondary structure, will have the highest density. Dyes that bind to nucleic acids and alter their density have been applied to enhance their separation from contaminants. The binding of EtBr decreases the apparent density of DNA. Supercoiled DNA binds less EtBr than linear DNA, enhancing their separation based on density differences. CsCl centrifugation is most commonly applied to purify plasmids and cosmids in combination with EtBr. Ausubel et al. (1998) also provides protocols for the isolation of genomic DNA from plants and bacteria.

Features

Cesium gradient formation requires long periods (at least overnight) of ultracentrifugation and are caustic, yet remain popular because they produce high yield and purity and are more easily scaled up.

Limitations

GC content of DNA correlates directly to its density. Equilibrium density of DNA can be calculated as $1.66 + 0.098 \times \%GC$ (Sambrook, Fritsch, and Maniatis, 1989). The density of very GC-rich DNA can be sufficiently high as to cause it to migrate immediately adjacent to RNA in a CsCl gradient. If too much sample is loaded onto a gradient, or if mistakes were made during preparation of the gradient, separation will be incomplete or ineffective.

Affinity Techniques

Triple helix resins have been used to purify plasmids and cosmids (Wils et al., 1997). This approach takes advantage of the adoption of a triple rather than a double helix conformation under the proper pH, salt, and temperature conditions. Triple helix affinity resins are generated by insertion of a suitable homopurine sequence into the plasmid DNA and crosslinking the complement to a chromatographic resin of choice. The triple helix interaction is only stable at mild acidic pH; it dissociates under alkaline conditions. The interaction at mildly acid pH is very strong (Radhakrishnan and Patel, 1993). This strong affinity allows for extensive washing that can improve the removal of genomic DNA, RNA, and endotoxin during large-scale DNA preparations.

A radically different approach applies covalent affinity chromatography to trap contaminants. Some of the examples include a chemically modified silica resin that irreversibly binds protein via an imide bond (Ernst-Cabrera and Wilchek, 1986), and a modified silica resin that covalently binds to polysaccharides via a cyclic boric acid ester, trapping proteins in the process. This latter reaction was initially applied to purify tRNA (McCutchan, Gilham, and Soll, 1975); it is described in greater detail by O'Neill et al. (1996). Some commercial products use salts to generate an irreversible protein precipitate that forms a physical barrier between the aqueous nucleic acid and the solid protein phase. Affinity-based technologies are also described at *http://www.polyprobe.com/about.htm* and at *http://www.edgebio.com*.

Features

Affinity techniques can produce excellent yields. Impressive purity is achieved if the system is not overloaded; if need be, the affinity steps can be repeated to further enhance purity. These methods are especially recommended when sample is precious and limited or purity requirements are very high.

Limitations

Cost, which may be minimized by reuse of resin. However cleaning of resin and its validation may be problematic.

WHAT ARE THE OPTIONS FOR PURIFICATION AFTER IN VITRO REACTIONS?

Spun Column Chromatography through Gel Filtration Resins

Mechanism

As in standard, column-based gel filtration (size exclusion) chromatography, a liquid phase containing sample and contaminant passes through a resin. The smaller molecules (contaminant) enter into the resin's pores, while the larger molecules (desired product) will pass through without being retained. Properly applied, this procedure can accomplish quick buffer exchange, desalting, removal of unincorporated nucleotides, and the elimination of primers from PCR reactions (gel filtration spin columns will *not* remove enzyme from a reaction; this requires organic extraction) to name a few applications.

Features and Limitations

These procedures are fast, efficient, and reproducible when the correct resins and centrifugation conditions are applied to the appropriate samples. Viscous solutions are not compatible with this technique.

One should not approach spun column, size exclusion chromatography with a care-free attitude. The exclusion limits based on standard chromatography should not be automatically applied to spun columns. Spinning makes such standard chromatography data obsolete. Before you apply a resin or a commercial spun column in an application, verify that the product has been successfully used in your particular application. Just because a resin has a pore size that can exclude a 30 nucleotide long oligo isn't a guarantee that a column with this resin will remove all or even most of the primer from a PCR reaction.

Manufacturers will optimize the columns and/or the procedures to accomplish a stated task. The presence of salt (100–150 mM NaCl) improves the yield of radiolabeled probes from one type of spun column, but the presence of Tris can interfere with the preparation of templates for automated sequencing (Amersham Pharmacia Biotech, unpublished observations, and *Nucleic Acid Purification Guide*, 1996). Too much *g* force, and the contaminants can elute with the desired product; too little *g* force, and the desired product is not eluted. If the volume you're eluting off the spun column is much greater or less than the volume you've loaded, the applied *g* force is no longer correct.

If you plan to create a spun column from scratch, consider the following:

- Sample volumes should be kept low with respect to the volume of resin, usually below a tenth to a twentieth of the column volume to allow for good resolution.
- Gel filtration resin will not resolve components efficiently (purity >90%) unless the largest contaminant is at least 20 times smaller than the smallest molecule to be purified.
- Desalting, where the size difference between ions and biomolecule is >>1:20, works well even at high flow rates.

Filter Cartridges

Mechanism

Filtration under the influence of vacuum suction or centrifugation operates under principles similar to gel filtration. Semipermeable membranes allow passage of small molecules such as salts, sugars, and so forth, but larger molecules such as DNA are retained. Since the retentate rather than an eluate is collected, samples will be concentrated. Ultrafiltration and microfiltration are reviewed by Munir (1998) and Schratter et al., (1993).

Features and Limitations

Filtration procedures are fast and reproducible provided that the proper *g* force or vacuum are applied. Membranes can clog from debris when large molecules accumulate at the membrane surface (but don't pass through), forming a molecule-solute gel layer that prevents efficient removal of remaining contaminants. As with gel filtration spun columns, filtration will not remove enzymes from reaction mixes unless the enzyme is small enough to pass through the membrane, which rarely is the case.

Silica Resin-Based Strategies

Mechanism

The approach is essentially identical to that described for silica resins used to purify DNA from cells and tissue, as described above.

Features and Limitations

Advantages and pitfalls are basically the same. Recoveries from solutions are between 50 to 95% and from agarose gels, 40 to 80%. Fragments smaller than 100 bp or larger than 10 kb (gel), or 50 kb (solution), are problematic. Small fragments may not elute unless a special formulation of glass milk is used (e.g., Glass Fog™ by 5′-3′ Eppendorf), and large fragments often shear and give poor yield. Depending on the capture buffer formulation, RNA and single-stranded molecules may or may not bind.

When using silica resins to bind nonradioactively labeled probes, investigate the stability of the label in the presence of chaotrope used for the capture and washing steps. Chaotropes create an environment harsh enough to attack contaminants such as proteins and polysaccharides, so it would be prudent to assume that any protein submitted to such an environment will lose its function. Nucleic acids covalently tagged with horseradish peroxidase or alkaline phosphatase are less likely to remain active after exposure to harsh denaturants. The stability of the linker connecting the reporter molecule to the DNA should also be considered prior to use.

Also consider the effect of reporter molecules/labels on the ability of DNA to bind to the resin. Nucleic acids that elute well in the unlabeled state may become so tightly bound to the resin by virtue of their label that they become virtually "sorbed out" and hence are unrecoverable. This is a notable concern when the reporter molecule is hydrophobic.

Isolation from Electrophoresis Gels

This subject is also addressed in Table 8.4 of Chapter 8, "Electrophoresis." Purification through an electrophoresis gel (refered to hereon as *gel purification*) is the only choice if the objective is to simultaneously determine the fragment size and remove contaminants. It could be argued that gel purification is really a two-step process. The first step is filtration through the gel and separation according to size. The second step is required to remove impurities introduced by the electrophoresis step (i.e., agarose, acrylamide, and salts). There are several strategies to isolate DNA away from these impurities, as summarized in Table 7.1 and discussed in detail below.

All these procedures are sensitive to the size and mass of the amount of gel segment being treated. The DNA should appear on the gel as tight bands, so in the case of agarose gels, combs must be inserted straight into the gel. When isolating fragments for cloning or sequencing, minimize exposure to UV light; visualize the bands at 340 nm. Any materials coming in contact with the gel slice should be nuclease free. Crush or dice up the gel to speed up your extraction method.

Polyacrylamide Gels

Crush and Soak

With time, nucleic acids diffuse out of PAGE gels, but recovery is poor. The larger the fragment size, the longer is the elution time required for 50% recovery. Elevated temperatures (37°C) accelerate the process. A variation of the crush and soak procedure is available at *http://www.ambion.com/techlib/tb/tb_171.html.* A procedure for RNA elution is provided at *http://grimwade.biochem.unimelb.edu.au/bfjones/gen7/m7a4.htm.*

Electroelution

Depending on the instrumentation, electroelution can elute DNA into a buffer-filled well, into a dialysis bag, or onto a DEAE cellulose paper strip inserted into the gel above and below the band of interest. Inconsistent performance and occasionally difficult manipulations make this approach less popular.

Specialized Acrylamide Crosslinkers

These are discussed in Chapter 8, "Electrophoresis."

Table 7.1 Comparison of Nucleic Acid Punfication Methods from Gel and/or Solution

Method	Used for	Yield	Speed	Benefits	Limitations
l.m.p. agarose with or without agarase, or phenol[a] (Ausubel et al., 1998, Hengen, 1994)	DNA fragments and/or plasmids	Up to 70%; typically 50%	From 0.5 to 2h depending on downstream purification method chosen	Agarase especially useful for large fragments or cosmids, since treatment is very gentle; some applications may allow treatment directly in melted gel slice (e.g., ligation, labeling with Klenow)	Requires an additional purification step; carrier often required for precipitation because solutions are dilute; extraction with phenol is caustic, especially hot phenol
"Freeze and squeeze" (Benson and Spencer, 1984, Ausubel et al., 1998)	DNA, RNA fragments	40–60% (for fragments up to 5kb, above that, lower)	Slow; freeze for at least 15′ or up to 2h, then follow with precipitation	Very gentle; good for larger molecules; very inexpensive	Low yield
"Crush and soak" (for acrylamide gels) Sambrook, Fritsch, and Maniatis, 1989	Mostly RNA, but works for any nucleic acid	40–70% depending on elution time, concentration, etc.	2–4h depending on fragment size	Allows high sample loads; best for reactions generating larger quantities of probe (in vitro transcription) to compensate for low recoveries	Significant chance of contamination when working with radioactivity; poor recovery
Gel filtration, desalting Ausubel et al., 1998; Sambrook, Fritsch, and Maniatis, 1989	DNA and RNA; fragment size must be well above exclusion limit of resin	>90% for fragments above exclusion limit	3–15min, depending on column format (gravity flow vs. spun column); primer removal protocols might require 30min	Fast, with high purity and yield	Often leads to dilution of sample; only removes small contaminants, difficult to monitor separation of noncontaminants without radioactivity
Glass milk/Na I (Ausubel et al., 1998; Hengen, 1994)	Usually DNA fragments and plasmids from agarose gel or solution	50–75% from solution, 40–70% from gel	0.25–1.5h	Fast, versatile; removes most major contaminants (proteins, primers, salts); efficient one-step purification	Yield; Na I stability; shearing

Silica/guanidinium salts (Ausubel et al., 1998; Gribanov et al., 1996; Boom et al., 1990; Vogelstein and Gillespie, 1979)	Usually DNA fragments and plasmids from agarose gel or solution	80–90% from solution, up to 80% from gel	5 min from solution; up to 1 h from gel	Fast, versatile; removes most major contaminants (proteins, primers, salts); efficient one-step purification	Shearing; yields very much resin/protocol-dependent
Filter cartridges in combination with freezing or without freezing (Leonard et al., 1998; Blattner et al. 1994; Li and Ownby, 1993; Schwarz and Whitton, 1992)	Concentration and desalting of DNA/RNA samples; desalting of freeze-squeeze eluted agarose gel slices	Up to 95% depending on fractionation range of membrane and nonspecific interaction with membrane	Often 2–5 min; depends on required concentration and salt tolerance of downstream applications	Fast; simultaneous desalting and concentration possible	Molecular size cutoff is not always well defined; not recommended for primer removal unless size cutoff well above primer size; will not remove large contaminants like proteins
Ethanol or isopropanol precipitation (Sambrook, Fritsch, and Maniatis, 1989, Ausubel et al., 1998)	Any nucleic acid as long as concentration is >10 μg/ml and at least 0.1 M monovalent cations are present	Up to 95% depending on protocol	20 min-overnight depending on sample concentration	Easy to monitor (visible pellet); noncaustic, robust, high yields, versatile in combination with different precipitation salts	More time-consuming; difficult for multiple samples, pellet may be lost; may not remove protein contaminants
Electroelution (Ausubel et al., 1998; Bostian, Lee, and Halvorson, 1979, Dretzen et al., 1981, Girvitz et al., 1980; Henrich, Lubitz, and Fuchs, 1982; Smith, 1980; Strongin et al., 1977; Tabak and Flavell, 1978)	Mostly DNA fragments from gel; elution onto DEAE membrane does not work well for fragments >2 kb	Up to 90% for fragments <1 kb, very small fragments between 50–60%, large fragments as low as 20%	2–4 h; or 1–3 h for DEAE elution	Few reagents required; not caustic/toxic; yields for fragments up to 1 kb are quite high	More difficult to monitor; only for fragments from 0.05–20 kb; need to be combined with a second method

Source: Data in table aside from references also based on average values found in catalogs and online of the following manufacturers: Ambion, Amersham Pharmacia Biotech, Amresco, Bio101, BioRad, Biotronics, BioTecx, Bioventures, Boehringer Mannheim/Roche, Clontech, CPG, Dynal, Edge Biosystems, Epicentre, FMC, Genhunter, Genosys, Gentra Systems, GIBCO Life Technologies, Invitrogen,Ligochem, 5′ 3′ (Eppendorf), Macromolecular Resource Center, Maxim Biotech, MBI Fermentas PerSeptive (now Life Technologies), Nucleon, Promega, Qiagen, Schleicher & Schuell, Sigma, Stratagene, USB, Worthington. For additional data, see DeFrancesco (1999), who provides a fragment purification products table and a comparison of size, agarose limitiations, buffer compatibility, time requirements, yield, capcity, and volume for isolation of DNA from agarose gels.

[a] If hot phenol is used, avoid phenol chloroform which can severly impair yields (Ausubel et al., 1998).

Agarose

Detailed procedures regarding the methodology discussed below are available at *http://www.bioproducts.com/technical/dnarecovery.shtml#elution*.

Freeze and Squeeze

Comparable to crush and soak procedures for polyacrylamide gels, this method is easy and straightforward, but it suffers from poor yields.

Silica-Based Methods

Silica or glass milk strategies are fast and efficient because the same buffer can be used for dissolving the gel and capturing the nucleic acid. Problems may arise when agarose concentrations are very high (larger volumes of buffers are required, reducing DNA concentration), nucleic acid concentration is very low (recovery is poor), fragment size is too small or large (irreversible binding and shearing, respectively), or if agarose dissolution is incomplete. Finally, some silica resins will not bind nucleic acids in the presence of TBE. When in doubt use TAE buffers (Ausubel et al., 1998).

Low Melting Point Agarose (LMP Agarose)

LMP agarose melts between 50 and 65°C. Some applications tolerate the presence of LMP agarose (Feinberg and Vogelstein, 1984), but for those that don't, DNA can be precipitated directly or isolated by phenol treatment (*http://mycoplasmas.vm.iastate.edu/lab_site/methods/DNA/elutionagarose.html*). Another option is to digest the agarose with agarase. This DNA can either be used directly for some applications or be precipitated to remove small polysaccharides and concentrate the sample. Glass beads are another way to follow up on melting your agarose slice as mentioned above. The negative aspect of LMP agarose is that sample load and resolution power are lower than in standard agarose procedures.

What Are Your Options for Monitoring the Quality of Your DNA Preparation?

The limitations of assessing purity by $A_{260}:A_{280}$ ratio are described in Chapter 4 (spectrophotometer section). Nevertheless, $A_{260}:A_{280}$ ratios are useful as a first estimation of quality. For northerns and southerns, try a dot blot. Success of PCR reactions can be scouted out by amplification of housekeeping genes. If restriction fragments do not clone well, try purifying a control piece of DNA with the same method and religate.

BIBLIOGRAPHY

Amersham Pharmacia Biotech unpublished observations, Amersham Pharmacia Biotech Research and Development Department, 1996.

Ausubel, M., Brent, R., Kingston, R. E., Moore, D. D., Seidman, J. G., and Struhl, K. 1998. *Current Protocols in Molecular Biology*. Wiley, New York.

Azad, A. K., Coote, J. G., and Parton, R. 1992. An improved method for rapid purification of covalently closed circular plasmid DNA over a wide size range. *Lett. Appl. Microbiol.* 14:250–254.

Benson, S. A., and Spencer, A. 1984. A rapid procedure for isolation of DNA from agarose gels. *Biotech. Biofeedback.* 2:66–68.

Biek, D. B., and Cohen, S. N. 1986. Identification and characterization of recD, a gene affecting plasmid maintenance and recombination in *Escherichia coli*. *J. Bacteriol.* 167:594–603.

Birnboim, H. C., and Doly, J. 1979. A rapid alkaline extraction procedure for screening recombinant plasmid DNA. *Nucl. Acids Res.* 7:1513–1523.

Blattner, Th., Frederick, R., and Chuang, S. 1994. Ultrafast DNA recovery from agarose by centrifugation. *Biotech.* 17:634–636.

Bolivar, F., Rodriguez, R. L., Greene, P. J., Betlach, M. C., Heyneker, H. L., and Boyer, H. W. 1977. Construction and characterization of new cloning vehicles. II. A multipurpose cloning system. *Gene* 2:95–113.

Boom, R., Sol, C. J., Salimans, M. M., Jansen, C. L., Wertheim-van Dillen, P. M., and van der Noordaa, J. 1990. Rapid and simple method for purification of nucleic acids. *J. Clin. Microbiol.* 28:495–503.

Bostian, K. A., Lee, R. C., and Halvorson, H. O. 1979. Preparative fractionation of nucleic acids by agarose gel electrophoresis. *Anal. Biochem.* 95:174–182.

Britten, R. J., Graham, D. E., and Neufeld, B. R. 1974. Analysis of repeating DNA sequences by reassociation. *Meth. Enzymol.* 29:363–441.

DeFrancesco, L. 1999. Get the gel out of here. *Scientist* 13:21.

Dretzen, G., Bellard, M., Sassone-Corsi, P., and Chambon, P. 1981 A reliable method for the recovery of DNA fragments from agarose and acrylamide gels. *Anal. Biochem.* 112:295–298.

Ernst-Cabrera, K., and Wilchek, M. 1986. Silica containing primary hydroxyl groups for high-performance affinity chromatography. *Anal. Biochem.* 159: 267–272.

Evans, R. K., Xu, Z., Bohannon, K. E., Wang, B., Bruner, M. W., and Volkin, D. B. 2000. Evaluation of degradation pathways for plasmid DNA in pharmaceutical formulations via accelerated stability studies. *J. Pharm. Sci.* 89:76–87.

Farnert, A., Arez, A. P., Correia, A. T., Bjorkman, A., Snounou, G., and do Rosario, V. 1999. Sampling and storage of blood and the detection of malaria parasites by polymerase chain reaction. *Trans. R. Soc. Trop. Med. Hyg.* 93:50–53.

Feinberg, A. P., and Vogelstein, B. 1984. A technique for radiolabeling DNA restriction endonuclease fragments to high specific activity. *Addendum Anal. Biochem.* 137:266–267.

Feliciello, I., and Chinali, G. 1993. A modified alkaline lysis method for the preparation of highly purified plasmid DNA from *Escherichia coli*. *Anal. Biochem.* 212:394–401.

Fernley, H. N. 1971. In Boyer, P. D., ed., *The Enzymes*, vol. 4. Chapter 2, Mammalian Alkaline Phosphateses. Academic Press, NY. pp. 417–447.

Ferreira, G. N., Cabral, J. M., and Prazeres, D. M. 1998. Purification of supercoiled plasmid DNA using chromatographic processes. *J. Mol. Recognit.* 11:250–251.

Ferreira, G. N., Cabral, J. M., and Prazeres, D. M. 1999. Development of process flow sheets for the purification of supercoiled plasmids for gene therapy applications. *Biotechnol. Prog.* 15:725–731.

Ferreira, G. N., Monteiro, G. A., Prazeres, D. M., and Cabral, J. M. 2000. Downstream processing of plasmid DNA for gene therapy and DNA vaccine applications (Review). *Trends. Biotechnol.* 18:380–388.

Franken, J., and Luyten, B. J. 1976. Comparison of dieldrin, lindane, and DDT extractions from serum, and gas-liquid chromatography using glass capillary columns. *J. Assoc. Off. Anal. Chem.* 59:1279–1285.

Girvitz, S. C., Bacchetti, S., Rainbow, A. J., and Graham, F. L. 1980. A rapid and efficient procedure for the purification of DNA from agarose gels. *Anal. Biochem.* 106:492–496.

Gribanov, P. G., Shcherbakov, A. V., Perevozchikova, N. A., and Gusev, A. A. 1996. Use of aerosol A-300 amd GF/F (GF/C) filters for purifying fragments of DNA, plasmid DNA, and RNA. *Biokhimiia* 61:1064–1070.

Grimberg, J., Nawoschik, S., Belluscio, L., McKee, K., Turck, A., and Eisenberg, A. 1989. A simple and efficient non-organic procedure for the isolation of genomic DNA from blood. *Nucl. Acids Res.* 17:8390.

Goodman, N. C., Gulati, S. C., Redfield, R., and Spiegelman, S. 1973. Room-temperature chromatography of nucleic acids on hydroxylapatite columns in the presence of formamide. *Anal. Biochem.* 52:286–299.

Halfon, P., Khiri, H., Gerolami, V., Bourliere, M., Feryn, J. M., Reynier, P., Gauthier, A., and Cartouzou, G. 1996. Impact of various handling and storage conditions on quantitative detection of hepatitis C virus RNA. *J. Hepatol.* 25:307–311.

Harrison, P. R. 1971. Selective precipitation of ribonucleic acid from a mixture of total cellular nucleic acids extracted from cultured mammalian cells. *Biochem. J.* 121:27–31.

Hengen, P. N., 1994. *TIBS 19: Recovering DNA from Agarose Gels.*

Henrich, B., Lubitz, W., and Fuchs, E. 1982. Use of benzoylated-naphthoylated DEAE-cellulose to purify and concentrate DNA eluted from agarose gels. *J. Biochem. Biophys. Meth.* 6:149–157.

Hillen, W., Klein, R. D., and Wells, R. D. 1981. Preparation of milligram amounts of 21 deoxyribonucleic acid restriction fragments. *Biochem.* 20:3748–3756.

Hogrefe, C., and Friedrich, B. 1984. Isolation and characterization of megaplasmid DNA from lithoautotrophic bacteria. *Plasmid* 12:161–169.

Holmes, D. S., and Quigley, M. 1981. A rapid boiling method for the preparation of bacterial plasmids. *Anal. Biochem.* 114:193–197.

Huber, C. G. 1998. Micropellicular stationary phases for high-performance liquid chromatography of double-stranded DNA (Review). *J. Chromatogr. A.* 806: 3–30.

Humphreys, G. O., Willshaw, G. A., and Anderson, E. S. 1975. A simple method for the preparation of large quantities of pure plasmid DNA. *Biochim. Biophys. Acta* 383:457–463.

John, M. E. 1992. An efficient method for isolation of RNA and DNA from plants containing polyphenolics. *Nucleic Acids Res* 20:2381.

Johnson, S. R., Martin, D. H., Cammarata, C., and Morse, S. A. 1995. Alterations in sample preparation increase sensitivity of PCR assay for diagnosis of chancroid. *J. Clin. Microbiol.* 33:1036–1038.

Kahn, D. W., Butler, M. D., Cohen, D. L., Gordon, M., Kahn, J. W., and Winkler, M. E. 2000. Purification of plasmid DNA by tangential flow filtration. *Biotechnol. Bioeng.* 69:101–106.

Kahn, M., Kolter, R., Thomas, C., Figurski, D., Meyer, R., Remaut, E., and Helinski, D. R. 1979. Plasmid cloning vehicles derived from plasmids ColE1, F, R6K, and RK2. *Meth. Enzymol.* 68:268–280.

Kieser, T. 1984. Factors affecting the isolation of CCC DNA from *Streptomyces lividans* and *Escherichia coli*. *Plasmid*. 12:19–36.

Kim, C. S., Lee, C. H., Shin, J. S., Chung, Y. S., and Hyung, N. I. 1997. A simple and rapid method for isolation of high quality genomic DNA from fruit trees and conifers using PVP. *Nucleic Acids Res.* 25:1085–1086.

Kohler, T., Rost, A. K., and Remke, H. 1997. Calibration and storage of DNA competitors used for contamination-protected competitive PCR. *Biotech.* 23:722–726.

Kondo, T., Mukai, M., and Kondo, Y. 1991. Rapid isolation of plasmid DNA by LiCl-ethidium bromide treatment and gel filtration. *Anal. Biochem.* 198:30–35.

Krajden, M., Minor, J. M., Zhao, J., Rifkin, O., and Comanor, L. 1999. Assessment of hepatitis C virus RNA stability in serum by the Quantiplex branched DNA assay. *J. Clin. Virol.* 14:137–143.

Lang, D. 1969.Collapse of single DNA molecules in ethanol. *J. Mol. Biol.* 46:209.

Leonard, J. T., Grace, M. B., Buzard, G. S., Mullen, M. J., and Barbagallo, C. B. 1998. Preparation of PCR products for DNA sequencing. *Biotech.* 24:314–317.

Levy, M. S., O'Kennedy, R. D., Ayazi-Shamlou, P., and Dunnill, P. 2000a. Biochemical engineering approaches to the challenges of producing pure plasmid DNA. *Trends Biotechnol.* 18:296–305.

Levy, M. S., Collins, I. J., Tsai, J. T., Shamlou, P. A., Ward, J. M., and Dunnill, P. 2000b. Removal of contaminant nucleic acids by nitrocellulose filtration during pharmaceutical-grade plasmid DNA processing. *J. Biotechnol.* 76:197–205.

Li, Q., and Ownby, C. L. 1993. A rapid method for extraction of DNA From agarose gels using a syringe. Benchmark. *Biotech.* 15:976–978.

Liou, J. T., Shieh, B. H., Chen, S. W., and Li, C. 1999. An improved alkaline lysis method for minipreparation of plasmid DNA. *Prep. Biochem. Biotechnol.* 29:49–54.

Lis, J. T., and Schleif, R. 1975. Size fractionation of double-stranded DNA by precipitation with polyethylene glycol. *Nucl. Acids Res.* 2:383–389.

Lyddiatt, A., and O'Sullivan, D. A. 1998. Biochemical recovery and purification of gene therapy vectors (Review). *Curr Opin. Biotechnol.* 9:177–185.

Maitra, R., and Thakur, A. R. 1994. Multiple fragment ligation on glass surface: a novel approach. *Indian J. Biochem. Biophys.* 31:97–99.

Marmur, J. 1961. A procedure for the isolation of deoxyribonucleic acid from microorganisms. *J. Mol. Biol.* 3:208–218.

Martinson, H. G., and Wagenaar, E. B. 1974. Thermal elution chromatography and the resolution of nucleic acids on hydroxylapatite. *Anal. Biochem.* 61:144–154.

McCutchan, T. F., Gilham, P. T., and Soll, D. 1975. An improved method for the purification of tRNA by chromatography on dihydroxyboryl substituted cellulose. *Nucleic Acids Res.* 2:853–864.

Miller, D. L., Thomas, R. M., and Frazier, M. E. 1991. Ultrasonic cavitation indirectly induces single strand breaks in DNA of viable cells in vitro by the action of residual hydrogen peroxide. *Ultrasound Med. Biol.* 17:729–735.

Monteiro, G. A., Ferreira, G. N., Cabral, J. M., and Prazeres, D. M. 1999. Analysis and use of endogenous nuclease activities in Escherichia coli lysates during the primary isolation of plasmids for gene therapy. *Biotechnol. Bioeng.* 66:189–194.

Mukhopadhyay, M., and Mandal, N.C. 1983. A simple procedure for large-scale preparation of pure plasmid DNA free from chromosomal DNA from bacteria. *Anal. Biochem.* 133:265–270.
Muller, J., and Janz, S. 1993. Modulation of the H_2O_2-induced SOS response in *Escherichia coli* PQ300 by amino acids, metal chelators, antioxidants, and scavengers of reactive oxygen species. *Environ. Mol. Mutagen.* 22:157–163.
Munir, C. 1998. *Ultrafiltration and Microfiltration Handbook*. Technomic Publishing Lancaster, PA.
Narang, S. K., and Seawright, J. A. 1990. Hexane preserves the biological activity of isozymes and DNA. *J. Am. Mosquito Control Assoc.* 6:533–534.
Narayanan, S. 1996. Effects of anticoagulants used at blood specimen collection on clinical test results. *Rinsho Byori*, suppl 103:73–91.
Neudecker, F., and Grimm, S. 2000. High-throughput method for isolating plasmid DNA with reduced lipopolysaccharide content. *Biotech.* 28:107–109.
Norgard, M. V., Emigholz, K., and Monahan, J. J. 1979. Increased amplification of pBR322 plasmid deoxyribonucleic acid in Escherichia coli K-12 strains RR1 and chi1776 grown in the presence of high concentrations of nucleoside. *J. Bacteriol.* 138:270–272.
Nucleic Acid Purification Guide, 1996, Amersham Pharmacia Biotech.
O'Kennedy, R. D., Baldwin, C., and Keshavarz-Moore, E. 2000. Effects of growth medium selection on plasmid DNA production and initial processing steps. *J. Biotechnol.* 76:175–183.
O'Neill, M. A., Warrenfeltz, D., Kates, K., Pellerin, P., Doco, T., Darvill, A. G., and Albersheim, P. 1996. Rhamnogalacturonan-II, a pectic polysaccharide in the walls of growing plant cell, forms a dimer that is covalently cross-linked by a borate ester. In vitro conditions for the formation and hydrolysis of the dimer. *J. Biol. Chem.* 271:22923–22930.
Pham, T. T., Chillapagari, S., and Suarez, A. R. 1996. Preparation of pure plasmid or cosmid DNA using single-strand affinity matrix and gel-filtration spin columns. *Biotech.* 20:492–497.
Pich, U., and Schubert, I. 1993. Midiprep method for isolation of DNA from plants with a high content of polyphenolics. *Nucleic Acids Res.* 21:3328.
Radhakrishnan, I., and Patel, D. J. 1993. Solution structure of a purine.purine.pyrimidine DNA triplex containing G.GC and T.AT triples. *Structure.* 1:135–152.
Romanowski, G., Lorenz, M. G., and Wackernagel, W. 1991. Adsorption of plasmid DNA to mineral surfaces and protection against DNase I. *Appl. Environ. Microbiol.* 57:1057–1061.
Sambrook, J., Fritsch, E. F., and Maniatis, T. 1989. *Molecular Cloning: A Laboratory Manual*, 2nd ed. Cold Spring Laboratory, Cold Spring Harbor, NY.
Schratter, P., Krowczynska, A. M., and Leonard, J. T. 1993. Ultrafiltration in molecular biology. *Am. Biotechnol. Lab.* 11:16.
Schwarz, H., Whitton, J. L. 1992. A rapid, inexpensive method for eluting DNA from agarose without using chaotropic materials. *Biotech.* 13:205–206.
Sharova, E. G. 1977. Physico-chemical state of DNA after long-term storage. *Ukr. Biokhim. Zh.* 49:45–50.
Sinnett, D., Richer, C., and Baccichet, A. 1998. Isolation of stable bacterial artificial chromosome DNA using a modified alkaline lysis method. *Biotech.* 24:752–754.
Sippel, A. E. 1973. Purification and characterization of adenosine triphosphate: Ribonucleic acid adenyltransferase from *Escherichia coli*. *Eur. J. Biochem.* 37:31–40.

Skolimowski, I. M., Knight, R. C., and Edwards, D. I. 1983. Molecular basis of chloramphenicol and thiamphenicol toxicity to DNA in vitro. *J. Antimicrob. Chemother.* 12:535–542.

Smith, H. O. 1980. Recovery of DNA from gels. *Meth. Enzymol.* 65:371–380.

Stoker, N.G., Fairweather, N.F., and Spratt, B.G. 1982. Versatile low-copy-number plasmid vectors for cloning in Escherichia coli. *Gene* 18:335–341.

Strongin, A. Y., Kozlov, Y. I., Debabov, V. G., Arsatians, R. A., and Zlochevsky, M. L. 1977. A reliable technique for large-scale DNA separation. *Anal. Biochem.* 79:1–10.

Strzelecka, E., Spitkovsky, D., and Paponov, V. 1983. The effect of heparin on the porcine lymphocyte chromatin—I. The comparative study of DNA in different chromatin fractions. *Int. J. Biochem.* 15(3):293–296.

Studier, F. W., and Moffatt, B. A. 1986. Use of bacteriophage T7 RNA polymerase to direct selective high-level expression of cloned gencs. *J. Mol. Biol.* 189:113–130.

Summers, D. K., and Sherratt, D. J. 1984. Multimerization of high copy number plasmids causes instability: CoIE1 encodes a determinant essential for plasmid monomerization and stability. *Cell* 36:1097–1103.

Tabak, H. F., and Flavell, R. A. 1978. A method for the recovery of DNA from agarose gels. *Nucl. Acids Res.* 5:2321–2332.

Takahashi, R., Matsuo, S., Okuyama, T., and Sugiyama, T. 1995. Degradation of macromolecules during preservation of lyophilized pathological tissues. *Pathol. Res. Pract.* 191:420–426.

Visvikis, S., Schlenck, A., and Maurice, M. 1998. DNA extraction and stability for epidemiological studies. *Clin. Chem. Lab. Med.* 36:551–555.

Vogelstein, B., and Gillespie, D. 1979. Preparative and analytical purification of DNA from agarose. *Proc. Natl. Acad. Sci. U.S.A.* 76:615–619.

Wils, P., Escriou, V., Warnery, A., Lacroix, F., Lagneaux, D., Ollivier, M., Crouzet, J., Mayaux, J. F., and Scherman, D. 1997. Efficient purification of plasmid DNA for gene transfer using triple-helix affinity chromatography. *Gene Ther.* 4:3233–3330.

Woods, W. G. 1994. An introduction to boron: history, sources, uses, and chemistry. *Environ. Health. Perspect.* (suppl.) 102:5–11.

Yashima, E., Suehiro, N., Miyauchi, N., Akashi, M. 1993a. Affinity gel electrophoresis of nucleic acids. Specific base- and shape-selective separation of DNA and RNA on polyacrylamide-nucleobase conjugated gel. *J. Chromatogr. A.* 654:159–166.

Yashima, E., Suehiro, N., Miyauchi, N., and Akashi, M. 1993b. Affinity gel electrophoresis of nucleic acids. Nucleobase-selective separation of DNA and RNA on agarose-poly(9-vinyladenine) conjugated gel. *J. Chromatogr. A.* 654:151–158.

8

RNA Purification

Lori A. Martin, Tiffany J. Smith, Dawn Obermoeller,
Brian Bruner, Martin Kracklauer, and
Subramanian Dharmaraj

Selecting a Purification Strategy 198
Do Your Experiments Require Total RNA or mRNA? 198
Is It Possible to Predict the Total RNA Yield from a Certain Mass of Tissue or Number of Cells? 201
Is There Protein in Your RNA Preparation, and If So, Should You Be Concerned? 202
Is Your RNA Physically Intact? Does It Matter? 202
Which Total RNA Isolation Technique Is Most Appropriate for Your Research? 203
What Protocol Modifications Should Be Used for RNA Isolation from Difficult Tissues? 207
Is a One-Step or Two-Step mRNA-(poly(A) RNA)-Purification Strategy Most Appropriate for Your Situation? 209
How Many Rounds of Oligo(dT)–Cellulose Purification Are Required? 210
Which Oligo(dT)–Cellulose Format Is Most Appropriate? 210
Can Oligo(dT)–Cellulose Be Regenerated and Reused? ... 211
Can a Kit Designed to Isolate mRNA Directly from the Biological Sample Purify mRNA from Total RNA? ... 212
Maximizing the Yield and Quality of an RNA Preparation ... 212
What Constitutes "RNase-Free Technique"? 212

Molecular Biology Problem Solver, Edited by Alan S. Gerstein.
ISBN 0-471-37972-7

How Does DEPC Inhibit RNase? 213
How Are DEPC-Treated Solutions Prepared? Is More DEPC Better? 213
Should You Prepare Reagents with DEPC-Treated Water, or Should You Treat Your Pre-made Reagents with DEPC? 214
How Do You Minimize RNA Degradation during Sample Collection and Storage? 214
How Do You Minimize RNA Degradation during Sample Disruption? 215
Is There a Safe Place to Pause during an RNA Purification Procedure? 218
What Are the Options to Quantitate Dilute RNA Solutions? 218
What Are the Options for Storage of Purified RNA? 219
Troubleshooting 220
A Pellet of Precipitation RNA Is Not Seen at the End of the RNA Purification 220
A Pellet Was Generated, but the Spectrophotometer Reported a Lower Reading Than Expected, or Zero Absorbance 221
RNA Was Prepared in Large Quantity, but it Failed in a Downstream Reaction: RT PCR is an Example 221
My Total RNA Appeared as a Smear in an Ethidium Bromide-stained Denaturing Agarose Gel; 18S and 28S RNA Bands Were not Observed 222
Only a Fraction of the Original RNA Stored at –70°C Remained after Storage for Six Months 222
Bibliography 222

SELECTING A PURIFICATION STRATEGY

Do Your Experiments Require Total RNA or mRNA?

One of the first decisions that the researcher has to make when detecting or quantitating RNA is whether to isolate total RNA or poly(A)-selected RNA (also commonly referred to as mRNA). This choice is further complicated by the bewildering array of RNA isolation kits available in the marketplace. In addition the downstream application influences this choice. The following section is a short primer in helping make that decision.

From a purely application point of view, total RNA might suffice for most applications, and it is frequently the starting material for applications ranging from the detection of an mRNA species by Northern hybridization to quantitation of a message by

RT-PCR. The preference for total RNA reflects the challenge of purifying enough poly(A) RNA for the application (mRNA comprises <5% of cellular RNA), the potential loss of a particular message species during poly(A) purification, and the difficulty in quantitating small amounts of purified poly(A) RNA. If the data generated with total RNA do not meet your expectations, using poly(A) RNA instead might provide the sensitivity and specificity that your application requires. The pros and cons with either choice are discussed below. Your experimental data will provide the best guidance in deciding whether to use total or poly(A) RNA. Be flexible and open minded; there are many variables to consider when making this decision.

Two situations where using poly(A) RNA is essential are cDNA library construction, and preparation of labeled cDNA for hybridization to gene arrays. To avoid generating cDNA libraries with large numbers of ribosomal clones, and nonspecific labeled cDNA it is crucial to start with poly(A) RNA for these procedures.

The next section gives a brief description of the merits and demerits of using total RNA or poly(A) RNA in some of the most common RNA analysis techniques. Chapter 14, "Nucleic Acid Hybridization," discusses the nuances and quirks of these procedures in greater depth. For detailed RNA purification protocols, see Krieg (1996) Rapley and Manning (1998), and Farrel (1998).

Northern Hybridizations

Northern analysis is the only technique available that can determine the molecular weight of an mRNA species. It is also the least sensitive. Total RNA is most commonly used in this assay, but if you don't detect the desired signal, or if false positive signals from ribosomal RNA are a problem, switching to poly(A) RNA might be a good idea. Since only very small amounts of poly(A) RNA are present, make sure that it is feasible and practical to obtain enough starting cells or tissue. Theoretically you could use as much as 30 μg of poly(A) RNA in a Northern, which is the amount found in approximately 1 mg of total RNA. Will it be practical and feasible for you to sacrifice the cells or tissue required to get this much RNA? If not, use as much poly(A) RNA as is practical.

One drawback to using poly(A) RNA in Northern hybridizations is the absence of the ribosomal RNA bands, which are ordinarily used to gauge the quality and relative quantity of the RNA samples, as discussed later in this chapter. Fortunately there are other strategies besides switching to poly(A) RNA that can be used to increase the sensitivity of Northern hybridizations. You could alter the hybridization conditions of the DNA probe

(Anderson, 1999), or you could switch to using RNA probes in the hybridization, which are 3- to 5-fold more sensitive than DNA probes in typical hybridization buffers (Ambion Technical Bulletin 168, and references therein). Dramatic differences in the sensitivity of Northern blots can also be seen from using different hybridization buffers.

If you remain dissatisfied with the Northern data, and you are not interested in determining the size of the target, switching to a more sensitive technique such as nuclease protection or RT-PCR might help. Nuclease protection assays, which are 5- to 10-fold more sensitive than traditional membrane hybridizations, can accommodate 80 to 100 μg of nucleic acid in a single experiment. RT-PCR can detect extremely rare messages, for example, 400 copies of a message in a 1 μg sample as described by Sun et al. (1998). RT-PCR is currently the most sensitive of the RNA analysis techniques, enabling detection and quantitation of the rarest of targets. Quantitative approaches have become increasingly reliable with introduction of internal standards such as in competitive PCR strategies (Totzke et al., 1996; Riedy et al., 1995).

Dot/Slot Blots

In this procedure, RNA samples are directly applied to a membrane, either manually or under vacuum through a filtration manifold. Hybridization of probe to serial dilutions of sample can quickly generate quantitative data about the expression level of a target. Total RNA or poly(A) RNA can be used in this assay. Since the RNA is not size-fractionated on an agarose gel, a potential drawback to using total RNA in dot/slot blots is that signal of interest cannot be distinguished from cross-hybridization to rRNA. Switching to poly(A) RNA as the target source might alleviate this problem. However, it is crucial that relevant positive and negative controls are run with every dot/slot blot, whether the source of target nucleic acid is total RNA or poly(A) RNA.

Hybridization to Gene Arrays and Reverse Dot Blots

Gene arrays consist of cDNA clones (sometimes in the form of PCR products, sometimes as oligonucleotides) or the corresponding oligos spotted at high density on a nylon membrane, glass slide, or other solid support. By hybridizing labeled cDNA probes reverse transcribed from mRNA, the expression of potentially hundreds of genes can be simultaneously analyzed. This procedure requires that the labeled cDNA be present in excess of the target spotted on the array. This is difficult to achieve unless poly(A) RNA is used as template in the labeling reaction.

Ribonuclease Protection Assays

Either total RNA or poly(A) RNA can be used as starting material in nuclease protection assays. However, total RNA usually affords enough sensitivity to detect even rare messages, when the maximum amount (as much as 80 to 100 μg) is used in the assay. If the gene is expressed at extremely low levels, requiring weeklong exposure times for detection, a switch to poly(A) RNA might prove beneficial and may justify the added cost. Although very sensitive, nuclease protection assays do require laborious gel purification of the full-length probe to avoid getting confusing results.

RT-PCR

RT-PCR is the most sensitive method for detecting and quantitating mRNA. Theoretically, even very low-abundance messages can be detected with this technique. Total RNA is routinely used as the template for RT-PCR, (Frohman, 1990) but some cloning situations and rare messages require the use of poly(A) RNA (Amersham Pharmacia Biotech, 1995).

Note that one school of thought concerning RT-PCR considers it advisable to treat the sample RNA with DNase I, since no purification method produces RNA completely free of contaminating genomic DNA. RT-PCR is sensitive enough that even very small amounts of genomic DNA contamination can cause false positives. A second school of thought preaches avoidance of DNase I, as discussed in Chapter 11, "PCR."

cDNA Library Synthesis

As mentioned earlier, high-quality mRNA that is essentially free of ribosomal RNA is required for constructing cDNA libraries. Unacceptably high backgrounds of ribosomal RNA clones would be produced if total RNA were reverse transcribed to prepare cDNA.

Is It Possible to Predict the Total RNA Yield from a Certain Mass of Tissue or Number of Cells?

The data provided in this section are based on experimentation at Ambion, Inc. using a variety of samples and different purification products. The reader is cautioned that these are theoretical estimates, and yields can vary widely based on the type of tissue or cells used for the isolation, especially when dealing with difficult samples, as discussed later. The importance of rapid and complete tissue disruption, and homogenizing at subfreezing temperatures cannot be overemphasized. In addition, yields from very small amounts of starting material are subject to the law of diminishing returns. Thus, if the option is available, always choose more starting material rather than less. Samples can be pooled together, if possible, to maximize yields.

For example, 5 mg of tissue or 2.5×10^6 cells yields about 10 μg of total RNA, comprised of 8 μg rRNA, 0.3 μg mRNA, 1.7 μg tRNA, and other RNA. In comparison, 1 g of tissue or 5×10^8 cells yields about 2 mg of total RNA, comprised of 1.6 mg rRNA + 60 μg mRNA + 333 μg tRNA and other RNA.

Is There Protein in Your RNA Preparation, and If So, Should You Be Concerned?

Pure RNA has an $A_{260}:A_{280}$ absorbance ratio of 2.0. However, for most applications, a low $A_{260}:A_{280}$ ratio probably won't affect the results. Researchers at Ambion, Inc. have used total RNA with $A_{260:280}$ ratios ranging from 1.4 to 1.8 with good results in RNase protection assays, Northern analysis, in vitro translation experiments, and RT-PCR assays. If protein contamination is suspected to be causing problems, additional organic extractions with an equal volume of phenol/chloroform/isoamyl alcohol (25:24:1 mixture) may remove the contaminant. Residual phenol can also lower the $A_{260}:A_{280}$ ratio, and inhibit downstream enzymatic reactions. Chloroform/isoamyl alcohol (24:1) extraction will remove residual phenol. Chapter 4, "How to Properly Use And Maintain Laboratory Equipment," discusses other artifacts that raise and lower the $A_{260:280}$ ratio. Some tissues will consistently produce RNA with a lower $A_{260:280}$ ratio than others; the $A_{260:280}$ ratio for RNA isolated from liver and kidney tissue, for example, is rarely above 1.7.

Is Your RNA Physically Intact? Does It Matter?

The integrity of your RNA is best determined by electrophoresis on a formaldehyde agarose gel under denaturing conditions. The samples can be visualized by adding 10 μg/ml of Ethidium Bromide (EtBr) (final concentration) to the sample before loading on the gel. Compare your prep's 28S rRNA band (located at approximately 5 Kb in most mammalian cells) to the 18S rRNA band (located at approximately 2.0 Kb in most mammalian cells). In high-quality RNA the 28S band should be approximately twice the intensity of the 18S band (Figure 8.1).

The most sensitive test of RNA integrity is Northern analysis using a high molecular weight probe expressed at low levels in the tissues being analyzed. However, this method of quality control is very time-consuming and is not necessary in most cases.

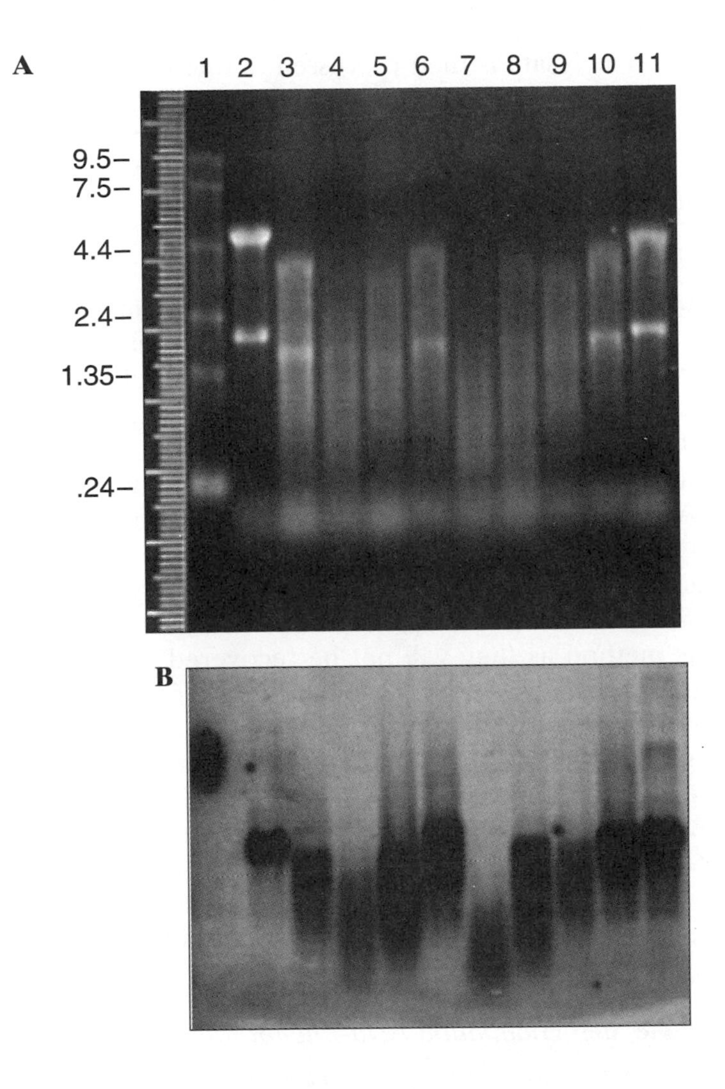

Figure 8.1 Assessing quality of RNA preparation via agarose gel electrophoresis (*A*) This gel shows total RNA samples (5 μg/lane) ranging from high-quality, intact RNA (lane 2) to almost totally degraded RNA (lane 7). Note that as the RNA is degraded, the 28S and 18S ribosomal bands become less distinct, the intensity of the ribosomal bands relative to the background staining in the lane is reduced, and there is a significant shift in their apparent size as compared to the size standards. (*B*) This is an autorad of the same gel after hybridization with a biotinylated GAPDH RNA probe followed by nonisotopic detection. The exposure is 10 minutes the day after the chemiluminescent substrate was applied. Note that the signal in lane 2, from intact RNA, is well localized with minimal smearing, whereas the signals from degraded RNA samples show progressively more smearing below the bands, or when the RNA is extremely degraded, no bands at all (lane 7). Reprinted by permission of Ambion, Inc.

Northern analysis is not tolerant of partially degraded RNA. If samples are even slightly degraded, the quality of the data is severely compromised. For example, even a single cleavage in 20% of the target molecules will decrease the signal on a Northern blot by 20%. Nuclease protection assays and RT-PCR analyses will tolerate partially degraded RNA without compromising the quantitative nature of the results.

Which Total RNA Isolation Technique Is Most Appropriate for Your Research?

There are three basic methods of isolating total RNA from cells and tissue samples. Most rely on a chaotropic agent such as guanidium or a detergent to break open the cells and simultaneously inactivate RNases. The lysate is then processed in one of several ways to purify the RNA away from protein, genomic DNA, and other cellular components. A brief description of each method along with the time and effort involved, the quality of RNA obtained, and the scalability of the procedures follow.

Guanidium-Cesium Chloride Method

Slow, laborious procedure, but RNA is squeaky clean; unsuitable for large sample numbers; little if any genomic DNA remains.

This method employs guanidium isothiocyanate to lyse cells and simultaneously inactivate ribonucleases rapidly. The cellular RNA is purified from the lysate via ultracentrifugation through a cesium chloride or cesium trifluoroacetate cushion. Since RNA is more dense than DNA and most proteins, it pellets at the bottom of the tube after 12 to 24 hours of centrifugation at ≥32,000 rpm.

This classic method yields the highest-quality RNA of any available technique. Small RNAs (e.g., 5S RNA and tRNAs) cannot be prepared by this method as they will not be recovered (Mehra, 1996). The original procedures were time-consuming, laborious, and required overnight centrifugation. The number and size of samples that could be processed simultaneously were limited by the number of spaces in the rotor. Commercial products have been developed to replace this lengthy centrifugation (Paladichuk, 1999) with easier, less time-consuming methods. However, if the goal were to isolate very high-quality RNA from a limited number of samples, this would be the method of choice (Glisin, Crkuenjakov and Byus, 1974).

Single- and Multiple Step Guanidium Acid-Phenol Method

Faster, fewer steps, prone to genomic DNA contamination, somewhat cumbersome if large sample numbers are to be processed.

The guanidium-acid phenol procedure has largely replaced the cesium cushion method because RNA can be isolated from a large number of samples in two to four hours (although somewhat cumbersome) without resorting to ultracentrifugation. RNA molecules of all sizes are purified, and the technique can be easily scaled up or down to process different sample sizes. The single-step method (Chomczynski and Sacchi, 1987) is based on the propensity of RNA molecules to remain dissolved in the aqueous phase in a solution containing 4 M guanidium thiocyanate, pH 4.0, in the presence of a phenol/chloroform organic phase. At this low pH, DNA molecules remain in the organic phase, whereas proteins and other cellular macromolecules are retained at the interphase.

It is not difficult to find researchers who swear by GITC—phenol procedures because good-quality RNA, free from genomic DNA contamination is quickly produced. However, a second camp of researchers avoid these same procedures because they often contain contaminating genomic DNA (Lewis, 1997; S. Herzer, personal communication). There is no single explanation for these polarized opinions, but the following should be considered.

Problems can occur in the procedure during the phenol/chloroform extraction step. The mixture must be spun with sufficient force to ensure adequate separation of the organic and aqueous layers; this will depend on the rotor as can be seen in Table 8.1. For best results the centrifuge brake should not be applied, nor should it be applied to gentler settings.

The interface between the aqueous and organic layers is another potential source of genomic contamination. To get high-purity RNA, avoid the white interface (can also appear cream colored or brownish) between the two layers; leave some of the aqueous layer with the organic layer. If RNA yield is crucial, you'll probably want as much of the aqueous layer as possible, again leaving the white interface. In either case you can repeat the organic extraction until no white interface is seen.

Residual salt from the precipitation step, appearing as a huge white pellet, can interfere with subsequent reactions. Excessive salt should be suspected when a very large white pellet is obtained from an RNA precipitation. Excess salt can be removed by washing the RNA pellet with 70% EtOH (ACS grade). To the RNA pellet, add about 0.3 ml of room temperature (or –20°C) 70% ethanol per 1.5 ml tube or approximately 2 to 3 ml per 15 to 40 ml tube. Vortex the tube for 30 seconds to several minutes to dislodge the pellet and wash it thoroughly. Recover the RNA with a low speed spin, (approximately 3000 × *g*; approximately 7500 rpm in a microcentrifuge, or approximately 5500 rpm in a SS34 rotor), for 5 to 10 minutes at room temperature or at 4°C.

Table 8.1 Spin Requirements for Phenol Chloroform Extractions

Volume Tube	Speed	Spin Time
1.5 ml	10,000 × *g*	5 minutes
2.0 ml	12,000 × *g*	5 minutes
15 ml	12,000× *g*	15 minutes
50 ml	12,000× *g*	15 minutes

Remove the ethanol carefully, as the pellets may not adhere tightly to the tubes. The tubes should then be respun briefly and the residual ethanol removed by aspiration with a drawn out Pasteur pipet. Repeat this wash if the pellet seems unusually large.

Non-Phenol-Based Methods

Very fast, clean RNA, can process large sample numbers, possible genomic contamination.

One major drawback to using the guanidium acid-phenol method is the handling and disposal of phenol, a very hazardous chemical. As a result phenol-free methods, based on the ability of glass fiber filters to bind nucleic acids in the presence of chaotropic salts like guanidium, have gained favor. As with the other methods, the cells are first lysed in a guanidium-based buffer. The lysate is then diluted with an organic solvent such as ethanol or isopropanol and applied to a glass fiber filter or resin. DNA and proteins are washed off, and the RNA is eluted at the end in an aqueous buffer.

This technique yields total RNA of the same quality as the phenol-based methods. DNA contamination can be higher with this method than with phenol-based methods (Ambion, Inc., unpublished observations). Since these are column-based protocols requiring no organic extractions, processing large sample numbers is fast and easy. This is also among the quickest methods for RNA isolation, usually completed in less than one hour.

The primary problem associated with this procedure is clogging of the glass fiber filter by thick lysates. This can be prevented by using a larger volume of lysis buffer initially. A second approach is to minimize the viscosity of the lysate by sonication (on ice, avoid power settings that generate frothing) or by drawing the lysate through an 18 gauge needle approximately 5 to 10 times. This step is more likely to be required for cells grown in culture than for lysates made from solid tissue. If you are working with a tissue that is known to be problematic (i.e., high in saccharides or fatty acids), an initial clarifying spin or extraction with an equal volume of chloroform can prevent filter-clogging problems. A reasonable starting condition for the clarifying spin is 8 minutes at 7650 × *g*. If a large centrifuge is not available, the lysate can be divided into microcentrifuge tubes and centrifuged at maximum speed for 5 to 10 minutes. Avoid initial clarifying spins on tissues rich in glycogen such as liver, or plants containing high molecular-weight carbohydrates. If you generate a clogged filter, remove the remainder of the lysate using a pipettor, place it on top of a fresh filter, and continue with the isolation protocol using both filters.

What Protocol Modifications Should Be Used for RNA Isolation from Difficult Tissues?

RNA isolation from some tissues requires protocol modifications to eliminate specific contaminants, or tissue treatments prior to the RNA isolation protocol. Fibrous tissues and tissue rich in protein, DNA and RNases, present unique challenges for total RNA isolation. In this section we address problems presented by difficult tissues and offer troubleshooting techniques to help overcome these problems. A separate section will discuss the homogenization needs of various sample types in greater detail.

Web sites that discuss similar issues are *http://www.nwfsc.noaa.gov/protocols/methods/RNAMethodsMenu.html* and *http://grimwade.biochem.unimelb.edu.au/sigtrans.html.*

Fibrous Tissue

Good yields and quality of total RNA from fibrous tissue such as heart and muscle are dependent on the complete disruption of the starting material when preparing homogenates. Due to low cell density and the polynucleate nature of muscle tissue, yields are typically low; hence it is critical to make the most of the tissue at hand. Pulverizing the frozen tissue into a powder while keeping the tissue completely frozen (use a chilled mortar and pestle) is the key to isolating intact total RNA. It is critical that there be no discernible lumps of tissue remaining after homogenization.

Lipid and Polysaccharide–Rich Tissue

Plant and brain tissues are typically rich in lipids, which makes it difficult to get clean separation of the RNA and the rest of the cellular debris. When using phenol-based methods to isolate total RNA, white flocculent material present throughout the aqueous phase is a classic indicator of this problem. This flocculate will not accumulate at the interface even after extended centrifugation. Chloroform:isoamyl alcohol (24:1) extraction of the lysate is probably the best way to partition the lipids away from the RNA. To minimize loss, back-extract the organic phase, and then clean up the recovered aqueous RNA by extraction with phenol:chloroform:isoamyl alcohol (25:24:1).

When isolating total RNA from plant tissue using a non-phenol-based method, polyvinylpyrrolidone-40 (PVP-40) can be added to the lysate to absorb polysaccharide and polyphenolic contaminants. When the lysate is centrifuged to remove cell debris, these contaminants will be pelleted with the PVP (Fang, Hammar, and Grumet, 1992; see also the chapter by Wilkins and Smart, "Isolation of RNA from Plant Tissue," in Krieg, 1996, for a list of references and protocols for removing these contaminants from plant RNA preps). Centrifugation on cesium trifluoroacetate has also been shown to separate carbohydrate complexes from RNA (Zarlenga and Gamble, 1987).

Nucleic Acid and Nuclease-Rich Tissue

Spleen and thymus are high in both nucleic acids and ribonuclease. Good homogenization is the key to isolating high-quality RNA from these tissues. Tissue samples should be completely pulverized on dry ice, under liquid nitrogen, to facilitate rapid homogenization in the lysis solution, which inhibits nucleases. Cancerous cells and cell lines also contain high amounts of DNA and RNA, which makes them unusually viscous, causing poor separation of the organic and aqueous phases and potentially clogging RNA-binding filters. Increasing the ratio of sample mass : volume of lysis buffer can help alleviate this problem in filter-based isolations. Multiple acid–phenol extractions can be done to ensure that most of the DNA is partitioned into the organic phase during acid-phenol-based isolation procedures. Two to three extractions are usually sufficient; one can easily tell if a lysate is viscous by attempting to pipet the solution and observing whether it sticks to the pipette tip. The DNA in the lysate can alternatively be sheared, either by vigorous and repeated aspiration through a small gauge needle (18 gauge) or by sonication (10 second sonication at 1/3 maximum power on ice, or until the viscosity is reduced).

Hard Tissue

Hard tissue, such as bone and tree bark, cannot be effectively disrupted using a Polytron™ or any other commonly available homogenizer. In this case heavy-duty tissue grinders that pulverize the material using mechanical force are needed. SPEX Certiprep, Metuchen, NJ, makes tissue-grinding mills that chill samples to liquid nitrogen temperatures and pulverize them by shuttling a steel piston back and forth inside a stationary grinding vial.

Bacteria and Yeast

Bacterial and yeast cells can prove quite refractory to isolating good-quality RNA due to the difficulty of lysing them. Another problem with bacteria is the short half-life of most bacterial messages. Lysis can be facilitated by resuspending cell pellets in TE and treating with lysozyme, subsequent to which the actual extraction steps are performed. A potenial drawback of using lytic enzymes is that they can introduce RNases. Use the highest-quality enzymes to reduce the likelihood of introducing contaminants. Yield and quality from phenol-based extraction protocols can also be improved by conducting the organic extractions at high temperatures (Lin et al., 1996).

Lysis of yeast cells is accomplished by vigorous vortexing in the presence of 0.4 to 0.5 mm glass beads. If using a non-phenol-based procedure for RNA isolation, the lysis can be monitored by looking for an increase in A_{260} readings. Yeast cells can also be treated with enzymes such as zymolase, lyticase, and glucolase to facilitate lysis (Ausubel et al., 1995).

Is a One-Step or Two-Step mRNA–(poly(A) RNA)–Purification Strategy Most Appropriate for Your Situation?

One-step procedures purify poly(A) RNA directly from the starting material. A two-step strategy first isolates total RNA, and then purifies poly(A) RNA from that.

Sample Number

One-step strategies involve fewer manipulations to recover poly(A) RNA. When comparing different one-step strategies, consider that two additional washing steps multiplied by 20 samples can consume significant time and materials, and arguably, faster purification strategies decrease the chance of degradation. Centifugation, magnetics, and other technologies sound appealing and fast, but the true speed of a technique is determined by the total manipulations in a procedure. High-throughput applications such as hybridization of gene arrays are usually best supported by one-step purification procedures.

Sample Mass

The percentage of poly(A) RNA recovery is similar between one- and two-step strategies. So, when experimental sample is limited, a one-step procedure is usually the more practical procedure.

Yield

Commercial one-step products are usually geared to purify small (1–5 μg) or large (25 μg) quantities of poly(A) RNA, and manufacturers can usually provide data generated from a variety of sample types. If you require more poly(A) RNA, a two-step procedure is usually more cost effective.

How Many Rounds of Oligo(dT)–Cellulose Purification Are Required?

One round of poly(A) RNA selection via oligo(dT)–cellulose typically removes 50 to 70% of the ribosomal RNA. One round of selection is adequate for most applications (i.e., Northern analysis and ribonuclease protection assays). A cDNA library generated from poly(A) RNA that is 50% pure is usually sufficient to identify most genes, but to generate cDNA libraries with minimal rRNA clones, two rounds of oligo(dT) selection will remove approximately 95% of the ribosomal RNA. Remember that 20 to 50% of the poly(A) RNA can be lost during each round of oligo(dT) selection, so multiple rounds of selection will decrease your mRNA yield. The use of labeled cDNA to screen gene arrays is severely compromised by the presence of rRNA-specific probes, so two rounds of poly(A) selection might be justified.

Which Oligo(dT)–Cellulose Format Is Most Appropriate?

Resins

Commercial resins are derivatized with oligo(dT) of various lengths at various loading capacities—mass of oligo(dT) per mass of cellulose. The linkage between the oligo(dT) and celluose is strong but not covalent; some nucleic acid will leave the resin during use. Oligo(dT) chains 20 to 50 nucleotides long, bound to cellulose at loading capacities of approximately 50 mg/ml, are commonly used in column and batch procedures. Some suppliers refer to this as Type 7 oligo(dT)-cellulose. The word "Type" refers to the nature of the cellulose. Type 77F cellulose is comprised of shorter strands than Type 7, and it does not provide good flow in a chromatography column. Type 77F does work very well in a batch mode, binding more mRNA than Type 7.

Column Chromatography

Oligo(dT)-cellulose can be scaled up or down using a variety of column sizes. Column dimension isn't crucial, but the frit or membrane that supports the oligo(dT)-cellulose is. The microscopic cellulose fibers can clog the frits and filter discs in a gravity chromatography column. Test the ability of several ml of buffer or water to flow through your column before adding your RNA sample. If your column becomes clogged during use, resuspend the packed resin with gentle mixing, and prepare a new column using a different frit, or do a batch purification on the rescued resin as described below. Some commercial products pack oligo(dT)-cellulose in a syringelike system so that the plunger can forcefully push through the matrix. The frits in these push-systems accommodate flow under pressure. Applying pressure to a clogged, standard oligo(dT)-cellulose chromatography column usually worsens matters. Occasionally air bubbles become trapped within the spaces of the frit. Gentle pressure or a very gentle vacuum applied to the exit port of the column can release these trapped bubbles and improve flow.

Batch Binding or Spin Columns

Batch binding consists of directly mixing the total RNA with oligo(dT)-cellulose in a centrifuge tube, and using a centrifuge to separate the celluose from the supernate in the wash and elution steps. Batch binding and washing of the matrix and spun columns circumvent the problems of slow flow rates, and clogged columns often experienced with gravity-driven chromatography. Scaling reactions up and down is convenient and economical, using the guidelines of 100 A_{260} units of total RNA per 0.5 g of oligo(dT)-cellulose. Increasing the incubation times for the poly(A) RNA hybridization to the oligo(dT)-cellulose can sometimes increase yields by 5 to 10%.

Tissues that lyse only with difficulty, and viscous lysates, can interfere with oligo(dT) binding by impeding the movement of oligo(dT)–coated particles. Additional lysis buffer, or repeated passage through a fine-gauge (21 gauge) needle with a syringe to shear the DNA and proteins, can reduce this viscosity. Lysates with excessive amounts of particulates should be cleared by centrifugation before attempting to select poly(A) RNA.

Can Oligo(dT)-Cellulose Be Regenerated and Reused?

Oligo(dT)-cellulose can theoretically be regenerated and reused, but the reader is strongly recommended not to do so. The hydroxide wash that regenerates the resin should destroy any lingering mRNA, but it is difficult to prove 100% destruction. Also the more a reagent is manipulated, the more likely it is to become contaminated with trace amounts of RNase. However some researchers still reuse oligo(dT)-cellulose until poor flow or reduced binding leads them to prepare fresh oligo(dT)-cellulose. Be especially wary of regenerated oligo(dT)-cellulose that appears pink or slimy.

If you must reuse oligo(dT)-cellulose, first wash it with 10 bed volumes of elution buffer followed by 10 bed volumes of 0.1 N NaOH. (One bed volume equals the volume of cellulose settled in the column.) The NaOH degrades any RNA remaining after elution. After the 0.1 N NaOH treatment, wash the oligo(dT)-

cellulose with 10 bed volumes of water followed by 10 bed volumes of absolute alcohol. If the regenerated oligo(dT)-cellulose is to be stored for longer than a couple of weeks, dry it under a vacuum and store it with desiccant at –20°C. For short-term storage, refrigerate at 4°C after the NaOH and water washes; desiccation isn't required.

If the oligo(dT)-cellulose is to be reused immediately after removing residual RNA with the NaOH wash, equilibrate the column in 10 bed volumes of elution buffer followed by 10 bed volumes of binding buffer. The column is now ready for sample application.

To use resin that has previously been regenerated and stored, resuspend the oligo(dT)-cellulose in elution buffer, pour into the column, and wash with 10 bed volumes of binding buffer.

Can a Kit Designed to Isolate mRNA Directly from the Biological Sample Purify mRNA from Total RNA?

One-step procedures that obtain mRNA from intact cells or tissue typically employ a denaturing solution to generate a lysate, which is directly added to the oligo(dT)-cellulose. Washing with specific concentrations of salt buffers ultimately separates poly(A) RNA from DNA and other RNA species.

Typically total RNA can be substituted into one-step procedures by skipping the homogenization steps, adjusting the salt concentration of the total RNA to 500 mM and adding this material to the oligo(dT)-cellulose. Consult the manufacturer of your product for their opinion on this approach, and verify the binding capacity of the oligo(dT)-cellulose for total RNA.

MAXIMIZING THE YIELD AND QUALITY OF AN RNA PREPARATION

What Constitutes "RNase-Free Technique"?

Fundamentals

RNase contamination is so prevalent, special attention must be given to the preparation of solutions. Solutions should be prepared in disposable, RNase-free plasticware or in RNase-free glassware prepared in the lab. Glassware can be made RNase-free by baking at 180°C for 8 hours to overnight, or by treating with a commercial RNase decontaminating solution. Alternatively, RNase can be removed by filling containers with 0.1% DEPC, incubating at 37°C for 2 hours, rinsing with sterile water and then either heating to 100°C for 15 minutes, or autoclaving for 15 minutes to eliminate RNase.

Electrophoresis apparatus used for RNA analysis can be made RNase-free by filling with a 3% hydrogen peroxide solution, incubating for 10 minutes at room temperature and rinsing with DEPC-treated water.

When preparing RNase-free solutions, wear gloves and change them often. Regardless of the method used to prepare RNase-free solutions, keep in mind that they can easily become contaminated after preparation. This occurs when solutions are open and used regularly, or when they are shared with others. It is wise to prepare small volumes of solutions and aliquot larger volumes into RNase-free containers. Solutions should be clearly labeled "RNase-free" to avoid contamination and should only be used with RNase-free pipettes and pipette tips. Also adhere to the maxim "when in doubt, throw it out."

How Does DEPC Inhibit RNase?

The most common method of preparing RNase-free solutions is diethylpyrocarbonate (DEPC) treatment. DEPC inactivates RNases by carboxyethylation of specific amino acid side chains in the protein (Brown, 1991). DEPC is a suspected carcinogen, and it should always be used with the proper precautions.

How Are DEPC-Treated Solutions Prepared? Is More DEPC Better?

Most protocols specify adding DEPC to solutions at a concentration 0.1%, followed by mixing and room temperature incubation for several hours to overnight. The container lid should be loosened for the extended incubation because a considerable amount of pressure can form during the reaction. Finally, the solution is autoclaved; this inactivates the residual DEPC by hydrolysis, and releases CO_2 and EtOH as by-products.

The EtOH by-product can combine with trace carboxylic acid contaminates in the vessel to form volatile esters, which impart a slightly fruity smell to the solution. This does not mean that trace DEPC remains in solution. DEPC has a half-life of 30 minutes in water, and at a DEPC concentration of 0.1%, solutions autoclaved for 15 minutes/liter can be assumed to be DEPC-free. Be aware that increasing the concentration of DEPC to 1% can inhibit more RNase but can also inhibit certain enzymatic reactions, so more is usually not better.

Should You Prepare Reagents with DEPC-Treated Water, or Should You Treat Your Pre-made Reagents with DEPC?

Some researchers prefer to DEPC-treat preprepared solutions, while others opt for preparing DEPC-treated water first and combining it with ultrapure RNase-free powdered reagents. It should be noted that many reagents commonly used in RNA studies contain primary amines, such as Tris, MOPS, HEPES, and PBS, and cannot be DEPC-treated because the amino group "sops up" the DEPC, making it unavailable to inactivate RNase. These solutions should be prepared with ultrapure reagents and DEPC-treated water. When preparing solutions in this manner, use RNase-free spatulas and magnetic stirrers, wear gloves and change them often. Spatulas and magnetic stirrers can be made RNase-free by soaking in 0.1% DEPC followed by autoclaving (as described above for containers) or by using a commercial RNase decontamination solution according to the manufacturer's directions. Either method of solution preparation is acceptable. Other options are commercially prepared RNase-free solutions available from several vendors, or recently-introduced alternatives to DEPC treatment.

How Do You Minimize RNA Degradation during Sample Collection and Storage?

RNase is present in all cells and tissues; hence they must be immediately inactivated when the source organism dies. Samples should be immediately frozen in liquid nitrogen, or immediately disrupted in a chaotropic solution (i.e., GITC). In some cases RNase activity can eventually be restored even in the presence of a chaotrope if the extract is not frozen (Amersham Pharmacia Biotech, unpublished observations). In other experiments homogenized tissue has been stored for at least one week at room temperature, or two months at 4°C without any loss of RNA in a lysis buffer (Ambion, Inc., unpublished observations). A commercial RNase inhibitor also exists that can prevent RNA degradation within mammalian tissue, cells, and some plant tissues stored above freezing temperature for long periods. However periodically sampling the integrity of RNA purified from frozen stock materials is recommended in light of reports of RNA degradation in samples frozen under protective conditions.

Mammalian Tissues and Cells

Tissues can be harvested and immediately immersed in liquid nitrogen. However, large pieces of tissue do not freeze instantaneously, allowing RNase to degrade RNA found in the interior of the sample. The smaller the tissue pieces, the faster it freezes. Once frozen, tissue should be immediately moved to a –70°C freezer, or stored on dry ice until it can be transferred to a freezer for long-term storage. In frozen tissue, RNA may be stable indefinitely, but periodic sampling for RNase degradation is recommended to avoid unpleasant surprises.

If the sample tissue is relatively soft (see the discussion of disruption methods below), and samples are few, they can be harvested directly into the lysis solution, immediately homogenized, and stored up to 12 months at –70°C without affecting RNA quality. Such lysates can be thawed on ice, an aliquot removed for processing, and refrozen. Firm or hard tissue requires more physical disruption as described below.

Mammalian cells are typically easy to homogenize. After a quick wash in culture media to remove debris, pipetting or vortexing in the presence of lysis solution will usually suffice. Cell lysates should be stored at –70°C. Alternatively, washed cell pellets can be quick-frozen by immersing the tube containing the pellet into liquid nitrogen. The tube can then be transferred to –80°C for long-term storage. The disadvantage to freezing cell pellets is that except for very small ones, they will have to be pulverized in liquid nitrogen for RNA isolation.

Bacteria and Yeast

Most gram-negative bacteria can be pelleted and frozen. Small samples (milliliters) of *E. coli* can be lysed and frozen as described above for mammalian cells; larger volumes (liters) will require enzymatic digestion or isolation procedures that incorporate lysis (e.g., an SDS lysis/isolation procedure). Some gram-positive bacteria and most yeast cells resist disruption and require more aggressive methods as described below.

How Do You Minimize RNA Degradation during Sample Disruption?

Fast and complete lysis of any sample is arguably the most critical element of RNA purification. When purifying RNA from a sample type for the first time, test your homogenization procedure for speed, efficiency, and ease of use in a small-scale experiment. A purification procedure involving 20 precious samples is the wrong time to discover the practical limits of an extraction procedure.

RNase inhibition provided by chaotropes and other reagents can be overwhelmed by adding too much starting material. Follow your procedure's recommendation. Scale up if necessary.

Monitor Disruption

Disruption can usually be monitored by close inspection of the lysate. Visible particulates should not be observed, except when disrupting materials containing hard, noncellular components, such as connective tissue or bone. Disruption of microorganisms, such as bacteria and yeast, can be monitored by spectrophotometry. The A_{260} reading should increase sharply as lysis begins and then level off when lysis is complete. Lysis can also be observed as clarification in the suspension or by an increase in viscosity.

Mammalian Tissues and Cells

Most animal tissues can be processed fresh (unfrozen). It is important to keep fresh tissue cold and to process it quickly (within 30 minutes) after dissection. If tissues are necrotic, the RNA can begin degrading in vivo. Ideally pre-dispense the lysis solution into the homogenizer, and then add the tissue and begin homogenizing. Samples should never be left sitting in lysis solution undisrupted.

Electronic rotor-stator homogenizers (e.g., Polytron) can effectively disrupt all but very hard or fibrous tissues. In addition, they do the job rapidly. If you have access to an electronic homogenizer, for most tissues, you should use it. If you can only use manual homogenizers, soft tissues can be thoroughly disrupted in a Dounce homogenizer, but firm tissues, however, especially connective tissues, will be homogenized more thoroughly in a ground glass homogenizer or TenBroeck homogenizer (available from Bellco, Vineland, NJ). Very hard tissues such as bone, teeth, and some hard tumors may require a milling device as described for yeast. A comparison of tissue disrupters is described in Johnson (1998). Enzymatic methods may also be used for specific eukaryotic tissues, such as collagenase to break down collagen prior to cell lysis.

Animal tissues and any type of relatively large cell pellets that have been frozen after collection must be disrupted by grinding in liquid nitrogen with a mortar and pestle. During this process it is important that the equipment and tissue remain at temperatures well below 0°C. The tissue should be dry and powdery after grinding. After grinding, thoroughly homogenize the sample in lysis solution using a manual or electronic homogenizer. Processing frozen tissue this way is cumbersome and time-consuming, but very effective.

Mammalian cells are normally easy to disrupt. Cells grown in suspension are collected by centrifugation, washed in cold 1× PBS, and resuspended in a lysis solution. Lysis is completed by immediate vortexing or vigorous pipetting of the solution. Rinse adherent cells in cold 1× PBS to remove culture medium. Then add lysis solution directly to the plate or flask, and scrape the cells into the solution. Finally, transfer the cells to a tube and vortex or pipette to completely homogenize the sample. Placing the flask or plate on ice while washing and lysing the cells will further protect the RNA from endogenous RNases released during the disruption process.

Plant Tissues

Soft, fresh plant tissues can often be disrupted by homogenization in lysis solution alone. Other plant tissues, like pine needles, can be frozen with liquid nitrogen, then ground dry. Some hard woody plant materials may require freezing and grinding in liquid nitrogen or milling. The diversity of plants and plant tissue make it impossible to give a single recommendation for techniques specific to your tissue. (See Croy, 1993, and Krieg, 1996, for guidance in preparing RNA from plant sources.)

Yeast and Fungi

Lysozyme and zymolase are frequently used with bacteria and yeast to dissolve cell walls, envelopes, coats, capsules, capsids, and other structures not easily sheared by mechanical methods (Ausubel et al., 1995). Sonication, homogenization, or vigorous vortexing in a lysis solution usually follows enzymatic treatment. Yeast can be extremely difficult to disrupt because their cell walls may form capsules or nearly indestructible spores. Bead mills that vigorously agitate a tube containing the sample, lysis buffer, and small beads will completely disrupt even these tough cells within a few minutes. Bead mills are available from Biospec Products, Inc., Bartlesville, OK, and Bio 101, Vista, CA. Alternatively, yeast cell walls can be lysed with hot phenol (Krieg, 1996) or digested with zymolase, glucalase, and/or lyticase to produce spheroplasts, which are readily lysed by vortexing in a lysis solution. Check that the enzyme you select is RNase-free.

To disrupt filamentous fungi, scrape the mycelial mat into a cold mortar, add liquid nitrogen, and grind to a fine powder with a pestle. The powder can then be thoroughly homogenized or sonicated in lysis solution to completely solubilize (Puyesky et al., 1997).

Bacteria

Bacteria, like plants, are extremely diverse; therefore it is difficult to make one recommendation for all bacteria. Bead milling will lyse most gram-positive and gram-negative bacteria, including mycobacteria (Cheung et al., 1994; Mangan et al., 1997; Kormanec and Farkasovshy, 1994). Briefly, glass beads and lysis solution are added to a bacterial cell pellet, and the mixture is milled for a few minutes. Some gram-negative bacteria can be lysed by sonication in lysis solution, but this approach is sufficient only for small cultures (milliliters), not large ones (liters).

Bacterial cell walls can be digested with lysozyme to form spheroplasts, which are then efficiently lysed with vigorous vortexing or sonication in sucrose/detergent lysis solution (Reddy and Gilman, 1998). Gram-positive bacteria usually require more rigorous digestion (increased incubation time and temperature, etc.) than gram-negative organisms (Krieg, 1996; Bashyam and Tyage, 1994).

Is There a Safe Place to Pause during an RNA Purification Procedure?

Ideally RNA should be purified without interruption, no matter which procedure is used. If a pause is unavoidable, stop when the RNA is precipitated or is in the presence of a chaotrope. For example, when using an organic isolation procedure, the RNA isolation can be stopped when the samples have been homogenized in a chaotrophic lysis solution. They can be stored for a few days at –20°C or –80°C without degradation.

What Are the Options to Quantitate Dilute RNA Solutions?

Spectrophotometry

The most common quantitative approach is to dilute a small volume of the RNA prep to meet the sample volume requirement of the cuvette. If the concentration of your RNA stock is low, the absorbance of the diluted RNA may fall outside the linear range of the spectrophotometer (see Chapter 4, "How to Properly Use and Maintain Laboratory Equipment").

Cuvettes are commercially available to accommodate sample volumes below 10 μl; some instruments can accept capillaries that hold less than 1 μl. If your spectrophotometer can tolerate these cuvette's minute sample windows, sample dilution might be unnecessary.

Dilute solutions can be concentrated by precipitation and microfiltration. Centrifugation-based RNase-free concentrators are available from Millipore corporation. (Bedford, MA), and glycogen enhances the precipitation of RNA from dilute solutions (Amersham Pharmacia Biotech, *MRNA Purification Kit Instruction Manual*, 1996). Adjust the NaCl concentration of 1.0 ml of an aqueous solution of RNA to 300 mM using a 3 M NaCl stock prepared in 10 mM Tris, 1 mM EDTA, pH 7.4. Add 10 μl of a 10 mg/ml glycogen solution (prepared in RNase-free water). Next, add 2.5 ml of ice-cold ethanol. Mix. Chill at –20°C for at least 2 hours, then centrifuge at 4°C for 10 minutes at 12,000 × g to recover the precipitated RNA. Be aware that since it is from a biological source, glycogen can contain protein (e.g., RNase) and nucleic acid (e.g., DNA) contaminants.

The riskiest option is to place your undiluted RNA prep into a cuvette. If this is your only option, carefully rinse the quartz cuvette with concentrated acid (check with your cuvette supplier to determine acid stability) followed by extensive rinsing in RNase-free water. Avoid hydrofluoric acid, which etches quartz and UV grade silica. Concentrated hydrochloric and nitric acid are tolerated by cuvettes of solid quartz or silica, but can damage cuvettes comprised of glued segments. A better option is to treat the cuvette with a commercial RNase decontamination solution.

Fluorometry

An alternative quantitation strategy is staining RNA with dyes such as Ribogreen®, SYBR®Green, and SYBR Gold (all available from Molecular Probes, Eugene, OR). Ribogreen is the most sensitive of these dyes for RNA; it is designed to be detected with a fluorometer for RNA quantitation in solution. With Ribogreen and a fluorometer, 1 to 10 ng/ml RNA can be detected. In contrast, both SYBR Green and SYBR Gold are designed to quantify RNA in a gel-based format, and they require the use of a densitometer or other gel documentation system that allows pixel values to be converted into numerical data. This method provides only rough approximations of the RNA loaded on a gel; it is valid for concentrations of 1 to 5 μg/lane. These dyes do not bind irreversibly to the RNA and do not have negative effects on downstream applications.

WHAT ARE THE OPTIONS FOR STORAGE OF PURIFIED RNA?

RNase activity and pH >8 will destroy RNA. For short-term storage of a few weeks or less, store your RNA in RNase-free Tris-EDTA or 1 mM EDTA at –20°C in aliquots. For long-term storage, RNA should be stored in aliquots at –80°C in TE, 1 mM EDTA, formamide, or as an ethanol/salt precipitation mixture.

TROUBLESHOOTING

A Pellet of Precipitated RNA Is Not Seen at the End of the RNA Purification.

The RNA Pellet Is There, but You Can't See It

- Pellets containing 0.5 to 2.0 μg of RNA should be visible but might not be as obvious as DNA pellets of the same mass. RNA pellets can range from clear to milky white in appearance. Pellets typically form near the bottom of the tube, but can also smear along the side depending on the rotor angle. Colored coprecipitants can help to visualize RNA pellets, but use them only if they are RNase-free. Marking the centrifuge tube to indicate the anticipated location of the pellet can help locate barely visible pellets.
- Remove the solution used to precipitate the RNA. This sometimes makes the pellet easier to see.
- Proceed as if a pellet is present, and quantitate the solution via a spectrophotomoter, fluorometer, or electrophoresis.

The RNA concentration was too low for precipitation by standard techniques

- The efficiency of RNA precipitation can be increased by adding 50 to 150 μg/ml glycogen or 10 to 20 μg/ml linear acrylamide to typical salt/ethanol precipitations. Glycogen does not appear to inhibit cDNA synthesis, Northern, or PCR reactions, but it may contain DNA, which could result in confusing RT-PCR results. Linear acrylamide is free of contaminating nucleic acids, but it is neurotoxic. Exercise great caution when handling RNA precipitated with acrylamide. Refer to manufacturers' Material Safety Data Sheets for more information on toxicity of linear acrylamide solutions.

The RNA pellet is truly absent

Sample Source Issues

Was the sample obtained from an unhealthy source? Did the tissue appear to be necrotic?

Was the sample quantity insufficient for the purification procedure?

Storage Issues

When originally isolated, was the sample allowed to linger at room temperature, or was it flash frozen immediately?

Was it stored in a frost-free freezer, hence subjected to thawing?

Was the pH of the stored preparation below 8.5?

Homogenization Issues

Was the sample immediately homogenized, or was it left intact for any period of time?

Was the extraction fast, thorough, and complete? Was the RNA too dilute to be effectively precipitated?

Was the Pellet Accidentally Discarded While Removing a Supernatant?

Nonsiliconized tubes decrease the likelihood of this happening.

A Pellet Was Generated, but the Spectrophotometer Reported a Lower Reading Than Expected, or Zero Absorbance

Refer to the troubleshooting example in Chapter 2, "Getting What You Need from a Supplier."

Did the RNA completely dissolve? Are visible pellet remnants (usually small white flecks) visible?

- Heat the RNA to 42°C, and vortex vigorously. Remove remaining debris by centrifugation. Overdried RNA pellets can be extremely difficult to resuspend; avoid drying with devices like a Speed Vac.

RNA Was Prepared in Large Quantity, but it Failed in a Downstream Reaction: RT PCR is an Example

Is the RNA at fault?

- Did the first strand cDNA synthesis reaction succeed, and the PCR reaction fail?
- Was the quality of the RNA evaluated?
- Was total RNA or poly(A)RNA used in the reaction? Using poly(A)RNA might work where total RNA failed.
- Was the poly(A)RNA purified once or twice on oligo(dT)cellulose. A second round will increase purity but will decrease yield up to 50%.

Is the RT-PCR reaction at fault?

- Did you test the positive control RNA and PCR primers?
- Did you test your gene specific PCR primers?

My Total RNA Appeared as a Smear in an Ethidum Bromide-stained Denaturing Agarose Gel; 18S and 28S RNA Bands Were not Observed

The RNA was degraded

Is it an electrophoresis artifact?

Did the RNA markers produce the correct banding pattern? If not, the buffers and loading dye could be the problem.

Could the gel be overloaded? 10 to 30 μg/lane of RNA is the maximum amount that should be loaded.

Only a Fraction of the Original RNA Stored at –70°C Remained after Storage for Six Months

The RNA is degraded.

Was the RNA stored as a wet ethanol precipitate or in formamide?

Was the RNA stored as aliquots?

Was the pH of the stored preparation <8.5?

Was the RNA frozen immediately after it was isolated?

Did you verify the calculations used to quantitate the RNA?

The RNA adsorbed to the walls of the storage container.

Is the RNA concentration <0.5 μg/ml, which increases the impact of loss due to adsorbtion?

Is the storage vessel siliconized, which decreases the risk of adsorbtion?

BIBLIOGRAPHY

Anderson, M. L. M. 1999. *Nucleic Acid Hybridization*. Springer, New York.

Amersham Pharmacia Biotech. *Mouse ScFv Module/Recombinant Phage Antibody System Instruction Manual*, Revision 4. 1995. Piscataway, NJ.

Ausubel, M., Brent, R., Kingston, R.E., Moore, D.D., Seidman, J.G. and Struhl, K. 1995. *Current Protocols in Molecular Biology*. Wiley, New York.

Bashyam, M., and Tyage, A. 1994. An efficient and high yielding method for isolation of RNA from Mycobacteria. *Biotech*. 17:834–836.

Brown, T. A., ed. 1991. *Molecular Biology Labfax*. Bios Scientific Publishers, Oxford, U.K.

Cheung, A. L., Eberhardt, K. J., and Fischetti, V. A. 1994. A method to isolate

RNA from gram-positive bacteria and mycobacteria. *Anal. Biochem.* 222:511–514.
Chomczynski, P., and Sacchi, N. 1987. Single-step method of RNA isolation by acid guanidinium thiocyanate-phenol-chloroform extraction. *Anal. Biochem.* 162:156–169.
Croy, R. D., ed. 1993. *Plant Molecular Biology*. Bios Scientific Publishers, Oxford, U.K.
Fang, G., Hammar, S., and Grumet, R. 1992. A quick and inexpensive method for removing polysaccharides from plant genomic DNA. *Biotech.* 13:52–54.
Farrell Jr., R. E. 1998. RNA Methodologies, 2nd ed. Academic Press, San Diego, CA.
Frohman, M. 1990. RACE: Rapid amplification of cDNA ends. In Innis, M. A., Gelfand, D. H., Sninsky, J., and White, T., eds., *PCR Protocols*. Academic Press, San Diego, CA, pp. 28–38.
Glisin, V., Crkvenjakov, R., and Byus, C. 1974. Ribonucleic acid isolated by cesium chloride centrifugation. *Biochem.* 13:2633–2637.
Herzer, S. 1999. Amersham Pharmacia Biotech, Piscataway, NJ, personal communication.
Johnson, B. 1998. Breaking up isn't hard to do: A cacophony of sonicators, cell bombs, and grinders. *The Scientist* 12:23.
Kormanec, J., and Farkasovshy, M. 1994. Isolation of total RNA from yeast and bacteria and detection of rRNA in northern blots. *Biotech.* 17:839–842.
Krieg, P. A. 1996. *A Laboratory Guide to RNA*, Wiley-Liss, New York.
Lewis, R. 1997. Kits take the trickiness out of RNA isolation: Purification. *Scientist* 11:16.
Lin, R., Kim, D., Castanotto, D., Westaway, S., and Rossi, J. 1996. RNA preparations from yeast cells. In Krieg, P. A., ed., *A Laboratory Guide to RNA*. Wiley-Liss, New York, pp. 43–50.
Mangan, J. A., Sole, K. M., Mitchison, D. A., and Butcher, P. D. 1997. An effective method of RNA extraction from bacteria refractory to disruption, including mycobacteria. *Nucl. Acids Res.* 25:675–676.
Mehra, M. 1996. RNA isolation from cells and tissues. In Krieg, P. A., ed., *A Laboratory Guide to RNA*. Wiley-Liss, New York, pp. 1–21.
Paladichuk, A. 1999. Isolating RNA: Pure and simple. *Scientist* 13:20.
Puyesky, M., Ponce-Noyale, P., Horwitz., B. S., and Herrera-Estrella, A. 1997. Glyceraldehyde-3-phosphate dehydrogenase expression in Trichoderma harzianum is repressed during conidiation and mycoparasitism. *Microbiol.* 143:3157–3164.
Rapley, R., and Manning, D. L. 1998. *Methods in Molecular Biology: RNA Isolation and Characterization Protocols*, vol. 86. Humana Press, Totowa, NJ.
Reddy, K., and Gilman, M. 1998. Preparation of Bacterial RNA. In Ausubel, F. M., Brent, R., Kingston, R. E., Moore, D. D., Seidman, J. G., Smith, J. A., and Struhl, K., eds., *Current Protocols in Molecular Biology*. Wiley, New York, pp. 4.4.4–4.4.6.
Riedy, M. C., Timm Jr., E. A., and Stewart, C. C. 1995. Quantitative RT-PCR for measuring gene expression. *Biotech.* 18:70–74, 76.
Sun, R., Ku J., Jayakar, H., Kuo, J. C., Brambilla, D., Herman, S., Rosenstraus, M., and Spadoro, J. 1998. Ultrasensitive reverse transcription–PCR assay for quantitation of human immunodeficiency virus Type 1 RNA in plasma. *J. Clin. Microbiol.* 36:2964–2969.
Technical Bulletin 168. *Increasing Sensitivity in Northern Analysis with RNA Probes*. Ambion Inc. Austin, TX. 1996.
Totzke, G., Sachinidis, A., Vetter, H., and Ko, Y. 1996. Competitive reverse tran-

scription/polymerase chain reaction for the quantification of p53 and mdm2 mRNA expression. *Mol. Cell. Probes* 10:427–433.
Wilkens, T., and Smart, L. 1996. Isolation of RNA from Plant Tissue. In Krieg, P. A., ed., *A Laboratory Guide to RNA*. Wiley-Liss, New York, pp. 21–42.
Zarlenga, D. S., and Gamble, H. R. 1987. Simultaneous isolation of preparative amounts of RNA and DNA from *Trichinella spiralis* by cesium trifluoroacetate isopycnic centrifugation. *Anal. Biochem.* 162:569–574.

9

Restriction Endonucleases

Derek Robinson, Paul R. Walsh, and Joseph A. Bonventre

Background Information 226
Which Restriction Enzymes Are Commercially Available? 226
Why Are Some Enzymes More Expensive Than Others? 227
What Can You Do to Reduce the Cost of Working with Restriction Enzymes? 228
If You Could Select among Several Restriction Enzymes for Your Application, What Criteria Should You Consider to Make the Most Appropriate Choice? 229
What Are the General Properties of Restriction Endonucleases? 232
What Insight Is Provided by a Restriction Enzyme's Quality Control Data? 233
How Stable Are Restriction Enzymes? 236
How Stable Are Diluted Restriction Enzymes? 236
Simple Digests 236
How Should You Set up a Simple Restriction Digest? 236
Is It Wise to Modify the Suggested Reaction Conditions? 237
Complex Restriction Digestions 239
How Can a Substrate Affect the Restriction Digest? 239
Should You Alter the Reaction Volume and DNA Concentration? 241
Double Digests: Simultaneous or Sequential? 242

Molecular Biology Problem Solver, Edited by Alan S. Gerstein.
ISBN 0-471-37972-7

Genomic Digests 244
When Preparing Genomic DNA for Southern Blotting, How Can You Determine If Complete Digestion Has Been Obtained? 244
What Are Your Options If You Must Create Additional Rare or Unique Restriction Sites? 247
Troubleshooting 255
What Can Cause a Simple Restriction Digest to Fail? 255
The Volume of Enzyme in the Vial Appears Very Low. Did Leakage Occur during Shipment? 259
The Enzyme Shipment Sat on the Shipping Dock for Two Days. Is It still Active? 259
Analyzing Transformation Failure and Other Multiple-Step Procedures Involving Restriction Enzymes 260
Bibliography 262

BACKGROUND INFORMATION

Molecular biologists routinely use restriction enzymes as key reagents for a variety of applications including genomic mapping, restriction fragment length polymorphism (RFLP) analysis, DNA sequencing, and a host of recombinant DNA methodologies. Few would argue that these enzymes are not indispensable tools for the variety of techniques used in the manipulation of DNA, but like many common tools that are easy to use, they are not always applied as efficiently and effectively as possible. This chapter focuses on the biochemical attributes and requirements of restriction enzymes and delivers strategies to optimize their use in simple and complex reactions.

Which Restriction Enzymes Are Commercially Available?

While as many as six to eight types of restriction endonucleases have been described in the literature, Class II restriction endonucleases are the best known, commercially available and the most useful. These enzymes recognize specific DNA sequences and cleave each DNA strand to generate termini with 5′ phosphate and 3′ hydroxyl groups. For the vast majority of enzymes characterized to date within this class, the recognition sequence is normally four to eight base pairs in length and palindromic. The point of cleavage is within the recognition sequence. A variation on this theme appears in the case of Class IIS restriction endonucleases.

These recognize nonpalindromic sequences, typically four to seven base pairs in length, and the point of cleavage may vary from within the recognition sequence up to 20 base pairs away (Szybalski et al., 1991).

To date, nearly 250 unique restriction specificities have been discovered (Roberts and Macelis, 2001). New prototype activities are continually being discovered. The REBASE database (*http://rebase.neb.com*) provides monthly updates detailing new recognition specificities as well as commercial availability.

These enzymes naturally occur in thousands of bacterial strains and presumably function as the cell's defense against bacteriophage DNA. Nomenclature for restriction enzymes is based on a convention using the first letter of the genus and the first two letters of the species name of the bacteria of origin. For example, *Sac*I and *Sac*II are derived from *Streptomyces achromogenes*. Of the bacterial strains screened for these enzymes to date, well over two thousand restriction endonucleases have been identified—each recognizing a sequence specificity defined by one of the prototype activities. Restriction enzymes isolated from distinct bacterial strains having the same recognition specificity are known as isoschizomers (e.g., *Sac*I and *Sst*I). Isoschizomers that cleave the same DNA sequence at a different position are known as neoschizomers (e.g., *Sma*I and *Xma*I).

Why Are Some Enzymes More Expensive Than Others?

The distribution of list prices for any given restriction enzyme can vary among commercial suppliers. This is due to many factors including the cost of production, quality assurance, packaging, import duties, and freight. For many commonly available enzymes produced from native overexpressors or recombinant sources, the cost of production is relatively low and is generally a minor factor in the final price. Recombinant enzymes (typically overexpressed in a well-characterized *E. coli* host strain) are often less expensive than their nonrecombinant counterparts due to high yields and the resulting efficiencies in production and purification. In contrast, those enzyme preparations resulting in very low yields are often difficult to purify, and they have significantly higher production costs. In general, these enzymes tend to be dramatically more expensive (per unit of activity) than those isolated from the more robust sources. As these enzymes may not be available at the same unit activity levels of the more common enzymes, they can be less forgiving in nonoptimal reaction conditions, and can be more problematic with initial use. The important point is that the relative price of a given restriction enzyme (or isoschizomer) may not be the best barometer of its performance in a specific application or procedure. The enzyme with the highest price does not necessarily guarantee optimal performance; nor does the one with the lowest price consistently translate into the best value.

Most commercial suppliers maintain a set of quality assurance standards that each product must pass in order to be approved for release. These standards are typically described in the supplier's product catalogs and detailed in the Certificate of Analysis. When planning to use an enzyme for the first time, it is important to review the corresponding quality control specifications and any usage notes regarding recommended conditions and applications.

What Can You Do to Reduce the Cost of Working with Restriction Enzymes?

Most common restriction enzymes are relatively inexpensive and often maintain full activity past the designated expiration date. Restriction enzymes of high purity are often stable for many years when stored at –20°C. In order to maximize the shelf life of less stable enzymes, many laboratories utilize insulated storage containers to mitigate the effects of freezer temperature fluctuations. Periodic summary titration of outdated enzymes for activity is another way to reduce costs for these reagents. For most applications, 1 μl is used to digest 250 ng to 1 μg of DNA. Enzymes supplied in higher concentrations may be diluted prior to the reaction in the appropriate storage buffer. A final dilution range of 2000 to 5000 Umits/ml is recommended. However, reducing the amount of enzyme added to the reaction may increase the risk of incomplete digestion with insignificant savings in cost. Dilution is a more practical option when using very expensive enzymes, when sample DNA concentration is below 250 ng per reaction, or when partial digestion is required. When planning for partial digestion, serial dilution (discussed below) is recommended. Most diluted enzymes should be stable for long periods of time when stored at –20°C. As a rule it is wise to estimate the amount of diluted enzyme required over the next week and prepare the dilution in the appropriate storage buffer, accordingly. For immediate use, most restriction enzymes can be diluted in the reaction buffer, kept on ice, and used for the day. Extending the reaction time to greater than one hour can often be used to save enzyme or ensure complete digestion.

If You Could Select among Several Restriction Enzymes for Your Application, What Criteria Should You Consider to Make the Most Appropriate Choice?

Each restriction endonuclease is a unique enzyme with individual characteristics, which are usually listed in suppliers' catalogs and package inserts. When using an unfamiliar enzyme, these data should be carefully reviewed. In addition some enzymes provide additional activities that may impact the immediate or downstream application.

Ease of Use

For many applications it is desirable and convenient to use 1 μl per reaction. Most suppliers offer standard enzyme concentrations ranging from 2000 to 20,000 units/ml (2–20 units/μl). In addition many suppliers also offer these enzymes in high concentration (often up to 100,000 units/ml), either as a standard product, or through special order. Enzymes sold at 10 to 20 units/μl are common and usually lend themselves for use in a wider variety of applications. When planning to use enzymes available only in lower concentrations (near 2000 units/ml), be sure to take the final glycerol concentration and reaction volume into account. By following the recommended conditions and maintaining the final glycerol concentration below 5%, you can easily avoid star activity.

Star Activity

When subjected to reaction conditions at the extremes of their operating range, restriction endonucleases are capable of cleaving sequences that are similar, but not identical, to their canonical recognition sequences. This altered specificity has been termed "star activity." Star sites are related to the recognition site, usually differing by one or more bases. The propensity for exhibiting star activity varies considerably among restriction endonucleases. For a given enzyme, star activity will be exhibited at the same relative level in each lot produced, whether isolated from a recombinant or a nonrecombinant source.

Star activity was first reported for *Eco*RI incubated in a low ionic strength high pH buffer (Polisky et al., 1975). Under these conditions, while this enzyme would cleave at its canonical site (G/AATTC), it also recognized and cleaved at N/AATTC. This reduced specificity should be a consideration when planning to use a restriction endonuclease in a nonoptimal buffer. It was also found that substituting Mn^{2+} for Mg^{2+} can result in star activity

(Hsu and Berg, 1978). Prolonged incubation time and high enzyme concentration as well as elevated levels of glycerol and other organic solvents tend to generate star activity (Malyguine, Vannier, and Yot, 1980). Maintaining the glycerol concentration to 5% or less is recommended. Since the enzyme is supplied in 50% glycerol, the enzyme added to a reaction should be no more than 10% of the final reaction volume.

When extra DNA fragments are observed, especially when working with an enzyme for the first time, star activity must be differentiated from partial digestion or contaminating specific endonucleases. First, check to make sure that the reaction conditions are well within the optimal range for the enzyme. Then, repeat the digest in parallel reactions, one with twice the activity and one with half the activity of the initial digest. Partial digestion is indicated as the cause when the number of bands is reduced to that expected after repeating the digestion with additional enzyme (or extending incubation time). If extra bands are still evident, contact the supplier's technical support resource for advice. Generally speaking, star activity and contaminating activities are more difficult to differentiate. Mapping and sequencing the respective cleavage sites is the best method to distinguish star activity from a partial digest or contaminant activity.

Site Preference

The rate of cleavage at each site within a given DNA substrate can vary (Thomas and Davis, 1975). Fragments containing a subset of sites that are cleaved more slowly than others can result in partial digests containing lighter bands visualized on an ethidium stained agarose gel. Certain enzymes such as *EcoR*II require an activator site to allow cleavage (Kruger et al., 1988). Substrates lacking the additional site will be cleaved very slowly. For certain enzymes (*Nae*I), adding oligonucleotides containing the site or adding another substrate containing multiple sites can improve cutting. In the case of *PaeR*7I, it has been shown that the surrounding sequence can have a profound effect on the cleavage rate (Gingeras and Brooks, 1983). In most cases this rate difference is taken in to account because the unit is defined at a point of complete digestion on a standard substrate DNA (e.g., lambda DNA) that contains multiple sites. Problems can arise when certain sites are far more resistant than others, or when highly resistant sites are encountered on substrates other than the standard substrate DNA. If a highly resistant site is present in a common cloning vector, then a warning should be noted on the data card or in the catalog.

Methylation

Methylation sensitivity can interfere with digestion and cloning steps. Many of the *E. coli* cloning strains express the genes for *EcoK*I methylase, dam methylase, or dcm methylase. The dam methylase recognizes GATC and methylates at the N6 position of adenine. *Mbo*I recognizes GATC (the same four base-pair sequence as dam methylase) and will only cleave DNA purified from *E. coli* strains lacking the dam methylase. *Dpn*I is one of only a few enzymes known to cleave methylated DNA preferentially, and it will only cleave DNA from dam^+ strains (Lacks and Greenberg, 1977). Another *E. coli* methylase, termed dcm, was found to block *Aat*I and *Stu*I (Song, Rueter, and Geiger, 1988). The dcm methylase recognizes CC(A/T)GG and methylates the second C at the C5 position.

The restriction enzyme recognition site doesn't have to span the entire methylation site to be blocked. Overlapping methylation sites can cause a problem. An example is the *Xba*I recognition site 5′ TCTAGA 3′. Although it lacks the GATC dam methylase target, if the preceding 5′ two bases are GA giving **GA**TCTAGA or the following 3′ bases are TC giving TCTAGA**TC**, then the dam methylase blocks *Xba*I from cutting. *E. coli* strains with deleted dam and dcm, like GM2163, are commercially available and should be used if the restriction site of interest is blocked by methylation. The first time a methylated plasmid is transformed into GM2163 the number of colonies will be low due to the important role played by dam during replication.

Methylation problems can also arise when working with mammalian or plant DNA. DNA from mammalian sources contain C5 methylation at CG sequences. Plant DNA often contains C5 methylation at CG and CNG sequences. Bacterial species contain a wide range of methylation contributed by their restriction modification systems (Nelson, Raschke, and McClelland, 1993). Information regarding known sensitivities to methylation can be found on data cards in catalog tables, by searching REBASE, and in the preceding review by Nelson.

Cloning problems can arise when working with DNA methylated at the C5 position. Most *E. coli* strains have an mcr restriction system that cleaves methylated DNA (Raleigh et al., 1988). A strain deficient in this system must be used when cloning DNA from mammalian and plant sources.

Substrate Effects

More on this discussion appears in the question below, *How Can a Substrate Affect the Restriction Digest?*

WHAT ARE THE GENERAL PROPERTIES OF RESTRICTION ENDONUCLEASES?

In general, commercial preparations of restriction endonucleases are purified and stored under conditions that ensure optimal reactivity and stability over time; namely –20°C. They are commonly supplied in a solution containing 50% glycerol, Tris buffer, EDTA, salt, and reducing agent. This solution will conveniently remain in liquid form at –20°C but will freeze at temperatures below –30°C. Those enzymes shipped on dry icc, or stored at –70°C, will have a white crystalline appearance; they revert to a clear solution as the temperature approaches –20°C. As a rule repeated freeze-thaw cycles are not recommended for enzyme solutions because of the possible adverse effects of shearing (more on the question, *How Stable are Restriction Enzymes?* appears below).

As a group (and by definition), Class II restriction endonucleases require magnesium (Mg^{2+}) as a cofactor in order to cleave DNA at their respective recognition sites. Most restriction enzymes are incubated at 37°C, but many require higher or lower (i.e., *Sma*I requires incubation at 25°C) temperatures. Percent activity tables of thermophilic enzymes incubated at 37°C can be found in some suppliers' catalogs. For most reactions, the pH optima is between 7 and 8 and the NaCl concentration between 50 and 100 mM. Concentrated reaction buffers for each enzyme are provided by suppliers. Typically each enzyme is profiled for optimal activity as a function of reaction temperature, pH (buffering systems), and salt concentration. Some enzymes are also evaluated in reactions containing additional components (BSA, detergents). Generally, these characteristics are documented in the published literature and referenced by suppliers.

Interestingly, a number of commonly used enzymes can display a broad range of stability and performance characteristics under fairly common reaction conditions. They may vary considerably in activity and may exhibit sensitivity to particular components. In an effort to minimize these undesirable effects, suppliers often adjust enzyme buffer components and concentrations to ensure optimal performance for the most common applications.

There is a wealth of information about the properties of these enzymes in most suppliers' catalogs, as well as on their Web sites. The documentation supplied with the restriction endonuclease should contain detailed information about the enzyme's properties and functional purity. It is important to read the Certificate of Analysis when using a restriction enzyme for the first time, as it may provide important information concerning particular substrate DNAs or alternative reaction conditions for a specific application.

What Insight Is Provided by a Restriction Enzyme's Quality Control Data?

Restriction enzymes are isolated from bacterial strains that contain a variety of other enzyme activities required for normal cell function. These additional activities include other nucleases, phosphatases, and polymerases as well as other DNA binding proteins that may inhibit restriction enzyme activity. In preparations where trace amounts of these activities remain, the end-structure of the resulting DNA fragments may be degraded, thus inhibiting subsequent ligation. Likewise plasmid substrates may be nicked, thus reducing transformation efficiencies.

Ideally the restriction enzyme preparation should be purified to homogeneity and free of any detectable activities that might interfere with digestion or inhibit subsequent reactions planned for the resulting DNA fragments. In order to provide researchers with a practical means to conveniently evaluate thc suitability of a given restriction enzyme preparation, suppliers include a Certificate of Analysis with each product, detailing the preparation's performance in a defined set of Quality Control Assays. In order to establish a standard reference for the amount of enzyme and substrate used in these assays, each supplier must first define the unit substrate and reaction conditions for each product.

Unit Definition

A unit of restriction endonuclease is defined as the amount of enzyme required to completely cleave 1 μg of substrate DNA suspended in 50 μl of the recommended reaction buffer in one hour at the recommended assay buffer and temperature. The DNA most often used is bacteriophage Lambda or another well-characterized substrate. Note that the unit definition is not based on classic enzyme kinetics. The enzyme molar concentration is in excess. A complete digest is determined by the visualized pattern of cleaved DNA fragments resolved by electrophoresis on an ethidium bromide-stained gel. Some restriction enzymes will behave differently when used outside the parameters of the unit definition. The number of sites (site density) or the particular type of DNA substrate may have an effect on "unit activity," but it is not always proportional (Fuchs and Blakesley, 1983).

Quality Control Assays—Maximum Units per Reaction

When using procedures requiring larger quantities of enzyme and/or extended reaction times, an appreciation of the quality control data can help determine a safe amount of enzyme for your application.

Overnight Assay

Increasing amounts of restriction endonuclease are incubated overnight (typically for 16 hours) in their recommended buffer with 1 μg of substrate DNA in a volume of 50 μl. The characteristic limit digest banding pattern produced by the enzyme in one hour is compared to the pattern produced from an excess of enzyme incubated overnight. A sharp, unaltered pattern under these conditions is an indication that the enzyme preparation is free of detectable levels of nonspecific endonucleases. The maximum number of units yielding an unaltered pattern is reported. Enzymes listing 100 units or more, a 1600-fold over digestion (100 units $\times$ 16 h), will not degrade DNA up to megabase size in mapping experiments and can be assumed to be virtually free of nonspecific endonuclease (Davis, T. and Robinson, D., unpublished observations).

Nicking Assay

Another sensitive test for contaminating endonucleases is a four hour incubation with a supercoiled plasmid that lacks a site for the enzyme being tested. The supercoil is very sensitive to nonspecific nicking by a single-stranded endonuclease, cleavage by a double-stranded endonuclease, or topoisomerase activity. If a single-stranded nick occurs, the supercoiled molecule, RFI, unwinds and assumes the circular form, RFII. If a double-stranded cleavage occurs, the circle will become linear. High levels of single-stranded nicking leads to linear DNA. All three forms of DNA have distinct electrophoretic mobilities on agarose gels. Enzymes converting 5% or less of the plasmid to relaxed form using 100 units of enzyme for four hours can be considered virtually free of nicking activity. High-salt buffers, especially at elevated temperature, can cause some conversion to relaxed form. A control reaction, including buffer and DNA but lacking enzyme, is incubated and run on the agarose gel for comparison.

Exonuclease Assay

Suppliers use a variety of assays to check for exonuclease activity. A general assay mixture contains a restriction endonuclease with 1 μg of a mixture of single- and double-stranded, ^{3}H-labeled *E. coli* DNA (200,000 cpm/μg) in a 50 μl reaction volume with the supplied buffer. Incubations (along with a background control containing no enzyme) are at the recommended temperature for four hours. Exonuclease contamination is indicated by the percent of the total labeled DNA in the reaction that has been rendered TCA-soluble. The limit of detectability of this assay is approximately 0.05%. Enzymes showing background levels of degradation with 100 units incubated for four hours can be considered virtually free of exonuclease.

Ligation/Recut Assay

Ligation and recutting is a direct determination of the integrity of the DNA fragment termini upon treatment with the restriction enzyme preparation. Ligation and recut of greater than 90% with a 10- to 20-fold excess of enzyme creating ends with overhangs or 80% for blunt ends indicate an enzyme virtually free of exonuclease or phosphatase specific for the overhang being tested. Alternative assays (i.e., end-labeling) are used to evaluate Type IIS restriction enzymes (e.g., *Fok*I, *Mbo*II). Since these enzymes cleave outside of their recognition sequence, the standard ligation assay would not determine a loss of terminal nucleotides due to exonuclease. The resulting ends could still ligate, and since their recognition sites remain intact, the enzyme would still be able to recut.

Blue-White Screening Assay

The β-galactosidase blue-white selection system is also applied to determine the integrity of the DNA ends produced after digestion with an excess of enzyme to test ligation efficiency. An intact gene gives rise to a blue colony; while an interrupted gene, which contains a deletion due to degraded DNA termini, gives rise to a white colony. Restriction enzymes tested using this assay should produce fewer than 3% white colonies.

The values given for the number of units added giving "virtually contaminant-free" preparations are somewhat arbitrary. They are useful, however, for determining maximum levels of enzyme to use in a reaction for most common applications. Enzymes with quality control results significantly below these values can still be used with confidence under simple assay conditions. As discussed later for complex restriction digestions, caution should be considered when extending reaction times and adding more than 1 to 2 μl of enzyme to 1 μg DNA in 50 μl.

How Stable Are Restriction Enzymes?

As a class, most restriction enzymes are stable proteins. Even during purification periods lasting two weeks, many enzymes lose no appreciable activity at 4°C. At the final stage of purification, the enzyme preparation is typically dialyzed into a 50% glycerol storage buffer and subsequently stored at –20°C. At this temperature the glycerol solution does not freeze. Most enzymes are stable for well over a 12-month period when properly stored. In one stability test of 170 restriction enzymes, activity was assessed after storage for 16 hours at room temperature. Of the enzymes tested, 122 (or 72%) exhibited no loss in activity (McMahon, M., and Krotee, S., unpublished observation). This point is important to note in case of freezer malfunction.

Even under optimal storage conditions, however, some enzymes may begin to lose noticeable activity within a six-month period. The supplier's expiration date, Certificate of Analysis, or catalog will provide more specific information regarding these enzymes. It is best to use these enzymes within a reasonable amount of time after they have been received. Some users employ a freezer box designed to maintain a constant temperature (for short periods at the bench) to store enzymes within the freezer. Alternatively, most enzymes can be stored at –70°C for extended periods. Repeated freeze–thaw cycles from –70°C to 0°C is not recommended. Each time the enzyme preparation solution is frozen, the buffer comes out of solution prior to freezing. As a result some enzymes may lose significant activity each time a freeze–thaw cycle is repeated. Often the extent of an enzyme's stability during storage at –20°C is buffer-related. Identical enzyme preparations obtained from two suppliers, when maintained in their respective storage buffers, may have significantly different shelf lives.

How Stable Are Diluted Restriction Enzymes?

For a discussion, refer above to the question *What Can You Do to Reduce the Cost of Working with Restriction Enzymes.*

SIMPLE DIGESTS

How Should You Set up a Simple Restriction Digest?

Reaction Conditions

Most restriction digests are designed either to linearize a cloning vector or to generate DNA fragments by cutting a given target DNA to completion at each of the corresponding restriction sites. To ensure success in any subsequent manipulations (i.e., ligation), the enzyme treatment must leave each of the resulting DNA termini elements intact.

To 1 μg of purified DNA in 50 μl of 1× reaction buffer, 1 μl of enzyme is added and the reaction is incubated for one hour at the recommended reaction temperature. In most instances the amount of DNA can be safely varied from about 250 ng to several micrograms and the volume can be varied between 20 μl and 100 μl. Suitable reaction times may be as little as 15 minutes or as long as 16 hours. Common DNA purification protocols, as well as commercially available kits, yield DNA that is suitable for most digestions. Most commonly used restriction enzymes are of high purity, inexpensive, and provided at concentrations of 5 to 20 units/μl. Using 1 to 2 μl will overcome any expected variability in DNA source, quantity, and purity. The length of incubation time may be decreased to save time or increased to ensure complete digestion of the last few tenths of a percent of substrate, as the reaction asymptotically approaches completion.

Control Reactions

Aside from the mere discipline of maintaining "good laboratory practice," the ultimate savings realized in time and effort by running a simple control reaction is often underestimated. Control reactions can often reveal the cause of a failed digest or point to the step within a series of reactions responsible for generating an unexpected result. For every experimental restriction enzyme reaction set performed, a control reaction (containing sample DNA, reaction buffer, and no restriction enzyme) should also be included and analyzed on the agarose gel. Degradation of DNA in the control reaction may indicate nuclease contamination in the DNA preparation or in the buffer. The control reaction products run alongside the sample reaction products on the agarose gel enables for a more accurate assessment of whether the reaction went to completion. Running the appropriate size markers is also recommended.

Is It Wise to Modify the Suggested Reaction Conditions?

Suppliers devote considerable effort in formulating specific enzyme preparations and the corresponding reaction buffers in order to ensure sufficient enzyme activity for most common applications. In addition suppliers often provide data (Activity Table) indicating the relative activity of each enzyme when incubated under standard reaction conditions for a variety of reaction buffers provided. This is a useful guide when planning multiple restriction enzyme digests. For enzymes with low activity in these standard buffers, specialized buffers are typically supplied. Restriction enzymes also have a broad range of activity in nonchloride salt buffers. Some suppliers also offer a potassium-acetate or potassium-glutamate single-buffer system that is formulated to be compatible with a significant subset of their enzymes. (McClelland et al., 1988; O Farrell, Kutter, and Nakanishe, 1980). The reaction buffers themselves are typically supplied as concentrated solutions, ranging from 2× to 10×, and should be properly mixed upon thawing prior to final dilution.

It is important to note that the reaction buffer supplied with a given enzyme is the same buffer in which all quality assurance assays are performed, and documented in the Certificate of Analysis provided. Consequently certain modifications to the recommended reaction conditions (i.e., adding components or changing reaction volume, temperature, or time of incubation) may produce unexpected results. Restriction enzymes can vary considerably in sensitivity to particular changes in their reaction parameters. While salt concentration may have a significant effect on activity, salt type (i.e., NaCl vs. KCl) is usually not critical. One exception would be in the case of *Sma*I, which has a strong preference for KCl. For most sensitive enzymes the Certificate of Analysis will detail any reaction modifications not recommended as well as any suggestions for alternative reaction conditions. In order to determine whether a given enzyme may be sensitive to an intended variation in reaction conditions, the Activity Table is also a useful reference. As a rule the most robust enzymes exhibit high relative activity across the range of buffers listed (*Pvu*II). Conversely, those enzymes showing a narrow range for high activity may require additional consideration prior to any change in reaction conditions (*Sal*I) and the technical resources provided by the supplier should be consulted.

All restriction enzymes, as do most other nucleases, require Mg^{2+} as a cofactor for the DNA cleavage reaction; most buffers for restriction enzymes contain 10 mM Mg^{2+}. To protect DNA preparations in storage buffer from any trace nucleases, EDTA (a Mg^{2+} chelator) is used, often stocked as a disodium salt solution. This is commonly used in various stop-dye solutions as well as electrophoresis buffer. DNA preparations with excessive concentrations of EDTA may inhibit restriction endonuclease cleavage, especially if the DNA solution represents a high proportion of the final reaction volume. Addition of Mg^{2+} will alleviate the inhibition.

A reducing agent, like dithiothreitol or β-mercaptoethanol, is a frequent buffer component even though it is not required for enzyme activity. However, as reaction buffers are typically diluted to their final reaction volume with distilled water, oxidation (i.e., from dissolved oxygen) could significantly reduce enzyme activity in the absence of sufficient reducing agent. BSA is frequently added as a stabilizing component to restriction enzyme preparations (Scopes, 1982). BSA increases the overall protein concentration and, by coating the hydrophobic surfaces of plastic vials, prevents possible denaturation. The activity level of many restriction enzymes in a reaction may be significantly enhanced if the final BSA concentration is around 100 μg/ml. Sometimes non-ionic detergents, like Triton ×-100 or Tween 20, are added as stabilizers for particular enzymes (*Eco*RI, *Not*I). A few restriction endonucleases, like *Bsg*I, have their activity significantly increased by the addition of *S*-adenosylmethionine (REBASE).

As most restriction enzymes are isolated from mesophilic bacteria, the vast majority exhibit excellent activity at 37°C in a near-neutral pH buffer. An increasing number of enzymes are being isolated from thermophilic bacteria, which display optimal activity within the range of 50°C to 75°C. As it happens, a good number of these enzymes also retain adequate activity at 37°C, and while this temperature may not be optimal for a particular enzyme, a supplier may list it as such for convenience in double-digest applications.

COMPLEX RESTRICTION DIGESTIONS

Complex reactions include double digests, reactions using nonoptimal buffers, reactions with DNA containing sites close to the ends, reactions with PCR products, and reactions involving multiple steps. In addition these include reactions with DNA concentrations that are significantly higher or lower than the recommended 1 μg/50 μl as well as simple reactions that simply didn't work the first time.

How Can a Substrate Affect the Restriction Digest?

PCR Products

Restriction endonucleases can often be used directly on PCR products in the PCR reaction mix. Suppliers often provide data indicating relative enzyme activity under these reaction conditions. Restriction endonuclease activity is influenced by the buffer used for PCR as well as the enzyme's ability to cleave in the presence of primers. The excess primers present in PCR reactions have been shown to inhibit *Sma*I and *Nde*I (Abrol and Chaudhary, 1993), but many restriction endonucleases can cleave in the presence of a 100-fold molar excess of primers. If your PCR products were not digested satisfactorily, eliminate the primers by gel purification, desalting column chromatography, membrane filtration or glass (Bhagwat, 1992).

Ends of Linear Fragments

Restriction endonucleases differ in their ability to cleave at recognition sites close to the end of a DNA fragment. Cleavage close to the end of a fragment is important when two restriction sites are close together in the cloning region of a plasmid and when cleaving near the ends of PCR products. Many restriction enzymes can cleave near a DNA end having one base pair in addition to a 1 to 4 single-base overhang produced by an initial cleavage; others require at least 3 base pairs in addition to an overhang (Moreira and Noren, 1995). When designing PCR primers containing restriction sites, adding eight random bases 5′ of the restriction site is recommended for complete digestion of the restriction sites.

Plasmids

Supercoiled plasmids often require more restriction endonuclease to achieve complete digestion than linear DNA. Manufacturers' catalogs often contain tables listing the number of units of restriction enzyme required to completely cleave 1 μg of commonly used supercoiled plasmids.

Inhibitors

Contaminants in the DNA preparation can inhibit restriction endonuclease activity. Residual SDS from alkaline lysis procedures can inhibit restriction endonucleases. High concentrations of NaCl, CsCl, other salts, or EDTA can inhibit restriction enzymes. Salt is concentrated when the DNA is alcohol precipitated. Washes containing 70% alcohol following the initial precipitation will solubilize some salt, but dialysis is preferred.

Protein contaminants in the DNA preparation can influence the restriction digests. Double strand specific exonucleases can co-purify with plasmid DNA when using column purification procedures (Robinson, D., and Kelley, K., unpublished observation). Phenol chloroform extraction followed by ethanol precipitation is an efficient method of removing proteins from DNA samples. The phenol and chloroform as well as the alcohol must be thoroughly removed to ensure restriction enzyme activity. Residual phenol and chloroform are removed by the alcohol precipitation and 70% alcohol wash steps. Alcohol is removed by desiccation. Dialysis can be used to remove residual alcohol that may be present from a DNA sample that was resuspended before the alcohol was completely removed. Alcohol can be introduced as a wash before elution when using diatomaceous earth as a resin for DNA purification. The resin must be thoroughly dried before DNA elution to remove the alcohol.

Core histones present on eukaryotic chromosomes can be difficult if not impossible to remove. Proteinase K followed by phenol chloroform extraction is often used in these preparations. Proteinase K is also used when preparing intact chromosomal DNA embedded in agarose for megabase mapping by pulse field gel electrophoresis (PFGE). Proteinase K must be inactivated using phenol chloroform or PMSF. Since the inhibition of proteinase K by a proteinase inhibitor such as PMSF is reversible, agarose blocks containing proteinase K should be extensively washed by changing the buffer multiple times. Most restriction enzymes are active in solutions containing PMSF.

Should You Alter the Reaction Volume and DNA Concentration?

Reaction Volume

A standard reaction volume to cleave 1 to 2 μg of DNA is 50 μl. Caution must be used when decreasing the reaction volume. Star activity tends to increase with decreasing reaction volume. The increase is most likely due to the higher glycerol concentration in the smaller volumes. Using 2 μl of *BamH*I containing 50% glycerol in a 10 μl reaction gives a final glycerol concentration of 10%. Increasing the reaction volume is not common unless more than 1 μg of DNA is being digested. Increasing the volume should be less problematic than decreasing the volume.

DNA Concentration

Varying the DNA concentration significantly from the standard (1 μg in 50 μl) can cause problems. Decreasing the amount of DNA or increasing the amount of overdigestion can increase star activity. An additional fourfold overdigestion occurs when 250 ngs are digested compared to 1 μg when using the same number of units of restriction enzyme. Low DNA concentrations near the K_m of a restriction enzyme could inhibit cleavage. The K_m for lambda DNA is 1000-fold less than 1 μg/50 μl (Fuchs &

Blakesley, 1983). Increasing the amount of DNA in 50 μl in most cases will not have a negative impact on the reaction. *Hind*III has been reported to work more efficiently on higher concentration DNA (Fuchs & Blakesley, 1983). Increasing the number of units or length of reaction will make up for the excess DNA. Care must be taken with the addition of extra enzyme, to keep the glycerol concentration to less than 5%. When digesting large quantities of DNA, using a concentrated enzyme is desirable. Inhibition may become a problem if the DNA has contaminants that influence enzyme activity. Salt and other contaminants in the DNA solution are more likely to be problematic if the DNA solution represents a large percentage of the final reaction mix.

Reaction Time

Extended digestion times can be used to increase the performance of a restriction enzyme, but the stability of the restriction enzyme in reaction should be checked by consulting the manufacturer's "survival in reaction" tables. BSA added to 100 μg/ml can increase survival. One should also consider that any trace contaminants in the preparation may continue to be active during an extended reaction. Often lower reaction temperatures can be used with unstable enzymes to increase performance when used for extended periods. One Unit of *Pme*I will digest 1 μg of DNA in two hours at 37°C but can digest 2 μg lambda in two hours at 25°C (Robinson, D., unpublished observation). When using *Pme*I for digesting agarose–embedded DNA, an incubation at 4°C overnight followed by one to two hours at 37°C is suggested.

Double Digests: Simultaneous or Sequential?

Simultaneous

The most convenient way to produce two different ends is to cut both at the same time in one reaction mix. Often the conditions for one enzyme or the other is not ideal. Manufacturers' buffer charts give the percent activity in buffers other than the one in which the enzyme is titered. If there is a buffer that indicates at least 50% activity for each enzyme, a coordinated double digest can be performed. Inexpensive, highly pure enzymes with no notes warning against star activity can be used in excess with confidence. A 10- to 20-fold excess of enzyme is recommended to increase the chances of success. Two microliters of a 10 unit/μl stock will give a 10-fold overdigest when used for one hour on 1 μg in a buffer giving 50% activity. If the enzyme is stable in reaction, then incubating for longer periods will increase the amount of overdigestion. Consult the manufacturer's stability information.

If the reaction produces extra fragments, possibly caused by star activity, reduce the reaction time or the amount of enzyme. If the reaction is incomplete, individually test each enzyme to determine it's ability to linearize the plasmid. A lack of cutting may indicate an inactive enzyme, absence of the expected site, or inhibitors in the template preparation. Test the enzyme on a second target as a control. If both enzymes are active, and the restriction sites are within several bases of each other, there may be a problem cutting close to the end of the fragment.

Sequential

Enzyme sets that are not compatible for double digests require sequential digestion. Always perform the first digest with the enzyme requiring the lower salt buffer. Either salt (or the corresponding 10× reaction buffer) may then be added to the reaction and the second enzyme can be used directly. To prevent the first enzyme from exhibiting star activity in the second buffer, it is wise to heat inactivate prior to addition of the second enzyme. Addition of BSA, reducing agents, or detergents has no adverse effects on restriction enzymes and may be safely added as required to the reaction.

If the pH requirements between the two enzymes differ by more than 0.5 pH units or the difference in salt requirement is critical (NaCl vs. KCl), alcohol precipitation between enzyme treatments is commonly performed. Alternatively, drop dialysis (see procedure D at the end of this chapter) is an option. A strategy that can often save a dialysis step would be to perform the first reaction in a 20 μl volume and then add 80 μl containing 10 μl of the higher salt buffer and enzyme to the initial reaction. The second reaction approximates the standard conditions for that enzyme.

Expensive enzymes should be optimized and used first in sequential reactions. When planning to use enzymes from different suppliers, first consider their optimal activity by looking at the NaCl or KCl requirements. Compare the buffer charts of both suppliers to determine if the enzyme is used in a standard or optimized buffer. Enzymes that are sold with optimized buffers should be used in those buffers when possible. If the same enzyme is sold by both suppliers, compare the two reaction buffers. Remember, the enzyme is titered in the buffer that is supplied. One supplier may choose to improve titer using a detergent and BSA, while the other may be using a different salt, pH, or enzyme concentration. In some cases a supplier may be categorizing an enzyme into a core buffer system by increasing the molar concentration of the enzyme. If used in an optimized buffer, this enzyme would titer at higher activity. If an enzyme from another supplier is used in this suboptimal core buffer, poor activity may result.

GENOMIC DIGESTS

When Preparing Genomic DNA for Southern Blotting, How Can You Determine if Complete Digestion Has Been Obtained?

Southern blotting involves the digestion of genomic DNA, gel electophoresis, blotting onto a membrane, and probing with a labeled oligonucleotide. The restriction pattern after gel electrophoresis is usually a smear, which may contain some distinguishable bands when visualized by ethidium bromide staining. It is often difficult to judge if the restriction digest has gone to completion or if degradation from star activity or nonspecific nuclease contamination is occurring. A twofold serial digest of genomic DNA enables a stable pattern, representing complete digestion, to be distinguished from an incomplete or degraded pattern.

Complete digestion is indicated when a similar smear of DNA appears in consecutive tubes of decreasing enzyme concentration within the serial digest. If the tubes with high enzyme concentration show smears that contain fragments smaller than those seen in tubes containing lesser enzyme, then it is likely that degradation is occurring. If the tube containing the most enzyme is the only sample demonstrating a complete digest, then the subsequent tubes (containing less enzyme) will demonstrate progressively larger fragments. A uniformly banded pattern will not occur in serial tubes unless the samples are all completely cut or completely uncut (Figure 9.1).

If the size of the smear does not change even at the greatest enzyme concentration, the digest may appear to have failed. A second possibility is that the fragments are too large to be resolved by standard agarose gel electrophoresis. Rare cutting enzymes may produce fragments greater than 50 kb, may not cleave a subset of sites due to methylation, or their recognition sequence might be underrepresented in the genome being studied. Pulse field gel electrophoresis must be used to resolve these fragments. Tables listing the average size expected from digestion of different species' DNA may be found in select suppliers' catalogs.

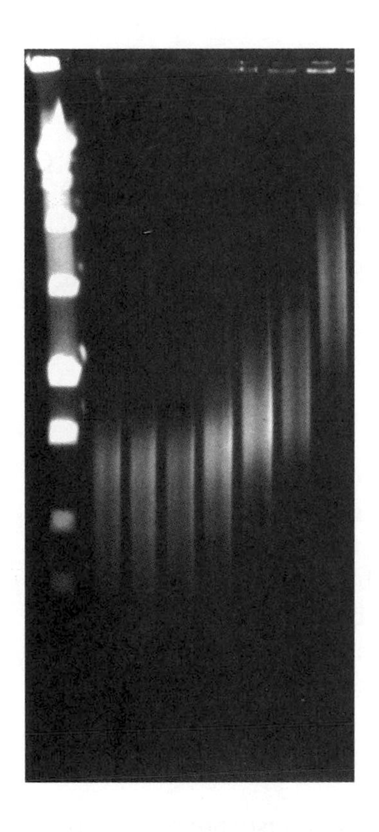

Figure 9.1 Testing for complete digestion of genomic DNA. Twofold serial digest using New England Biolabs *Avr*II of Promega genomic human DNA (cat. no. G304), 0.5 μg DNA in 50 μl NEB Buffer 2 for 1 hour at 37°C. *Avr*II added at 20 units and diluted to 10 units, etc., with reaction mix. The marker NEB Low Range PFG Marker (cat. no. N03050S). Complete digestion is indicated by lanes 2–4. Photo provided by Vesselin Miloushev and Suzanne Sweeney New England Biolabs. Reprinted by permission of New England Biolabs.

How Should You Prepare Genomic Digests for Pulsed Field Electrophoresis?

Pulse field electrophoresis techniques including CHEF, TAFE, and FIGE have made possible the resolution of DNA molecules up to several million base pairs in length (Birren et al., 1989; Carle, Frank, and Olson, 1986; Carle and Olson, 1984; Chu, Vollrath, and Davis, 1986; Lai et al., 1989; Stewart, Furst, and Avdalovic, 1988). The DNA used for pulsed field electrophoresis is trapped in agarose plugs in order to avoid double-stranded breaks due to shear forces. Protocol A has been used at New England Biolabs, Inc. for the preparation and subsequent restriction endonuclease digestion of *E. coli* and *S. aureus* DNA (Gardiner, Laas, and Patterson, 1986; Smith et al., 1986). This protocol may be modified as required for the cell type used.

Protocol A: Preparation of E. coli and S. aureus DNA

Cell Culture

1. Cells are grown under the appropriate conditions in 100 ml of media to an OD_{590} equal to 0.8 to 1.0. The chromosomes are then

aligned by adding 180 mg/ml chloramphenicol and incubating an additional hour.
2. The cells are spun down at 8000 rpm at 4°C for 15 minutes.
3. The cell pellet is resuspended in 6 ml of buffer A at 4°C. Alternatively 1.5 g of frozen cell paste may be slowly thawed in 20 ml of buffer A. Lysed cells from the thawing process are allowed to settle and the intact cells suspended in the supernatant are decanted and pelleted by centrifugation and washed once with 20 ml of buffer A. The pelleted cells are resuspended in 20 ml of buffer A.

DNA Preparation and Extraction

1. The suspended cells are warmed to 42°C and mixed with an equal volume of 1% low-melt agarose* in 1× TE at 42°C. For *S. aureus* cells, lysostaphin is added to a final concentration of 1.5 mg/ml. The agarose solution may be poured into insert molds. Alternatively, the agarose may be drawn up into the appropriate number of 1 ml disposable syringes that have the tips cut off.
2. The molds or syringes are allowed to cool at 4°C for 10 minutes. The agarose inserts are removed from the molds or extruded from the 1 ml syringes.
3. A 12 ml volume of the agarose inserts is suspended in 25 ml of buffer B (for *E. coli*), or 25 ml of buffer C (for *S. aureus*). Lysozyme (for *E. coli*) or Lysostaphin (for *S. aureus*) is added to a final concentration of 2 mg/ml. The solution is incubated for two hours at 37°C with gentle shaking. These solutions may also contain 20 μg/ml RNase I (DNase-free).
4. The agarose inserts are equilibrated with 25 ml buffer D for 15 minutes with gentle shaking. Replace with fresh buffer and repeat. Replace with 25 ml of buffer D containing 2 mg/ml proteinase K. This solution is incubated for 18 to 20 hours at 37°C with gentle shaking.
5. The inserts are again subjected to 15 minutes gentle shaking with 25 ml of buffer E. Replace with fresh buffer and repeat. Then incubate for 1 hour in buffer E, with 1 mM Phenylmethylsulfonyl fluoride (PMSF) to inactivate Proteinase K. As before, wash twice more with buffer E.
6. The inserts are washed twice with 25 ml of buffer F. The inserts are stored in buffer F at 4°C.

**Pulse field grade agarose should be used. The efficiency of the restriction enzyme digestion may vary with different lots of other low-temperature gelling agaroses.*

Digestion of Embedded DNA

Most restriction enzymes can be used to cleave DNA embedded in agarose, but the amount of time and enzyme required for complete digestion varies. Many enzymes have been tested for their ability to cleave embedded DNA (Robinson et al., 1991).

1. Agarose slices containing DNA (20 μl) are equilibrated in 1.0 ml of restriction enzyme buffer. The cylinders of agarose may be drawn back up into the 1 ml syringes in order to accurately dispense 20 μl of the agarose. The solution is gently shaken at room temperature for 15 minutes.
2. The 1 ml wash is decanted or aspirated from the agarose slice. The insert slice is submerged in 50 μl of restriction enzyme buffer. The appropriate number of units of the restriction enzyme with or without BSA is added to the reaction mixture and digested for a specific time and temperature as outlined by Robinson et al. (1991).
3. Following the enzyme digestion, the inserts may be treated to remove proteins using Proteinase K following the steps outlined above. Alternatively, the slices may be loaded directly onto the pulse field gel. Long-term storage of the endonuclease digested inserts is accomplished by aspirating the endonuclease reaction buffer out of the tube and submerging the insert in 100 ml of buffer E at 4°C. Insert slices that have been incubated at 50°C during the endonuclease digestion should be placed on ice for 5 minutes before handling the sample for loading or aspirating the buffer.

List of Buffers

Buffer A Cell suspension buffer: 10 mM Tris-HCl pH 7.2 and 100 mM EDTA.

Buffer B Lysozyme buffer: 10 mM Tris-HCl pH 7.2, 1 M NaCl, 100 mM EDTA, 0.2% sodium deoxycholate, and 0.5% *N*-lauryl-sarcosine, sodium salt.

Buffer C Lysostaphin buffer: 50 mM Tris-HCl, 100 mM NaCl, and 100 mM EDTA.

Buffer D Proteinase K buffer: 100 mM EDTA pH 8.0, 1% *N*-lauryl-sarcosine, sodium salt, and 0.2% sodium deoxycholate.

Buffer E Wash buffer: 20 mM Tris-HCl pH 8.0 and 200 mM EDTA.

Buffer G Storage buffer: 1 mM Tris-HCl pH 8.0 and 5 mM EDTA.

What Are Your Options If You Must Create Additional Rare or Unique Restriction Sites?

Cleavage at a single site in a genome may occur by chance using restriction endonucleases or intron endonucleases, but the number of enzymes with recongition sequences rare enough to generate megabase DNA fragments is relatively small. When no natural recognition site occurs in the genome, an appropriate sequence can be introduced genetically or in vitro via different multiple step reactions.

Genetic Introduction

Recognition sites have been introduced into *Salmonella typhimurium* and *Saccharomyces cerevisiae* genomes by site specific recombination or transposition (Hanish and McClelland, 1991; Thierry and Dujon, 1992; Wong and McClelland, 1992). Endogenous intron endonuclease recognition sites are found in many organisms. In cases where restriction enzymes and intron endonucleases cleave too frequently, it may be possible to use lambda terminase. The 100 bp lambda terminase recognition site does not occur naturally in eukaryotes. Single-site cleavage has been demonstrated using lambda terminase recognition sites introduced into the *E. coli* and *S. cerevisiae* genomes (Wang and Wu, 1993).

Multiple-Step Reactions

The remainder of this discussion reviews multiple-step procedures that have been used to generate megabase DNA fragments. Our intention is to provide a clear explanation of each procedure and highlight some of the complexities involved. Providing detailed protocols for each is beyond the scope of this chapter but can be found in the references cited.

Increasing the complexity of multiple-step reactions decreases the chances of success. Conditions needed for one step may not be compatible with the next. All of the steps must function well using agarose-embedded DNA as a substrate.

Altering Restriction Enzyme Specificity by DNA Methylation

DNA methylases can block restriction endonuclease cleavage at overlapping recognition sites, decreasing the number of cleavable restriction sites and increasing the average fragment size (Backman, 1980; Dobrista and Dobrista, 1980). Unique cleavage specificities can be created by using different methylase/restriction endonuclease combinations (Nelson, Christ, and Schildkraut, 1984; Nelson and Schildkraut, 1987). The following well-characterized, two-step reaction involves the restriction endonuclease *Not*I and a methylase (Gaido, Prostko, and Strobl, 1988; Qiang et al., 1990; Shukla et al., 1991).

The *Not*I recognition site

$$5' \ldots GC_{\wedge}GGCCGC \ldots 3'$$

$$3' \ldots CGCCGG_{\wedge}CG \ldots 5'$$

will not cleave when methylation at the following cytosine occurs in the *Not*I recognition site:

$$5' \ldots GCGGC^{m}CGC \ldots 3'$$

$$3' \ldots CGCCGGCG \ldots 5'$$

or

$$5' \ldots GCGGCCGC \ldots 3'$$

$$3' \ldots CG^{m}CCGGCG \ldots 5'$$

*Not*I sites that overlap the recognition site of the methylases M. *Fnu*DII, M. *Bep*I, or M. *Bsu*I can be modified as shown above. These methylases recognize the following sequence:

$$5' \ldots CGCG \ldots 3'$$

$$3' \ldots GCGC \ldots 5'$$

They methylate the first cytosine in the 5′ to 3′ direction:

$$5' \ldots {}^{m}CGCG \ldots 3'$$

$$3' \ldots GCG^{m}C \ldots 5'$$

Now the subset of *Not*I sites that are preceded by a C or followed by a G will be resistant to subsequent cleavage by *Not*I.

Resistant sites

$$5' \ldots \mathbf{C}GCGGCCGC \ldots 3'$$

$$3' \ldots \mathbf{G}CG^{m}CCGGCG \ldots 5'$$

or

$$5' \ldots GCGGC^{m}CGC\mathbf{G} \ldots 3'$$

$$3' \ldots CGCCGGCG\mathbf{C} \ldots 5'$$

which are sites flanked by any of the following combinations, will be cleaved by *Not*I:

$$5' \ldots \{A, G, T\}\, GC_{\wedge}GGCCGC\, \{A, C, T\} \ldots 3'$$

$$3' \ldots \{T, C, A\}\, CGCCGG_{\wedge}CG\, \{T, G, A\} \ldots 5'$$

This methylation reaction followed by *Not*I digestion statistically reduces the number of *Not*I sites by nearly half. The larger fragments produced may be more easily mapped using PFGE. A table of other potentially useful cross-protections for megabase mapping can be found in Nelson and McClelland (1992) and Qiang et al. (1990). A potential problem is that certain methylation sites may react slowly allowing partial cleavage events (Qiang et al., 1990).

DNA Adenine Methylase Generation of 8 to 12 Base-Pair Recognition Sites Recognized by DpnI

*Dpn*I is a unique restriction enzyme that recognizes and cleaves DNA that is methylated on both strands at the adenine in its recognition site (Lacks and Greenberg, 1975, 1977; Vovis, 1977).

*Dpn*I recognizes the following site:

5′ . . . G ^{m}A T C . . . 3′

3′ . . . C T ^{m}A G . . . 5′

The adenine methylases M. *Taq*I (McClelland, Kessler, and Bittner, 1984; McClelland, 1987), M. *Cla*I (McClelland, Kessler, and Bittner, 1984; McClelland, 1987; Weil and McClelland, 1989), M. *Mbo*II (McClelland, Nelson, and Cantor, 1985), and M. *Xba*I (Patel et al., 1990) have been used to generate a *Dpn*I recognition site with the apparent cleavage frequency of a 8 to 12 base-pair recognition sequence (Nelson and McClelland, 1992). The M. *Taq*I/*Dpn*I reaction is detailed below.

The M. *Taq*I recognition site

5′ . . . TCGA . . . 3′

3′ . . . AGCT . . . 5′

methylates the adenine on both strands of the above sequence to produce

5′ . . . T C G ^{m}A . . . 3′

3′.^{m}A G C T . . . 5′

Hemimethylated *Dpn*I sites (in bold below) will be generated when the sequence surrounding the site above is as follows:

5′ . . . T C **G ^{m}ATC** . . . 3′

3′ . . . ^{m}A G **C TAG** . . . 5′

or

5′ . . . **G A T C** G^{m}A . . . 3′

3′ . . . **C T^{m}A G** C T . . . 5′

The hemimethylated *Dpn*I site is cleaved at a rate 60× slower than the fully methylated site (Davis, Morgan, and Robinson, 1990). M. *Taq*I generates a fully methylated *Dpn*I site when two M. *Taq*I recognition sequences occur next to each other. The fully methylated *Dpn*I site is shown in bold below:

5′. . . . TC**G** ^{m}**A T C** G^{m}A . . . 3′

3′ . . . ^{m}AG**C T** ^{m}**A G** C T . . . 5′

The apparent recognition site of the M. *Taq*I/*Dpn*I reaction can be simply represented by the eight base pairs 5′ . . . TCGATCGA . . . 3′. The 10 base pair recognition site of the M. *Cla*I/*Dpn*I reaction can be represented by the sequence 5′ . . . ATCGATCGAT . . . . 3′. Notice that M. *Cla*I creates a *Dpn*I site by a slightly different overlap than demonstrated by the M. *Taq*I reaction. The M. *Cla*I/*Dpn*I reaction has been demonstrated on a bacterial and yeast genome (Waterbury et al., 1989; Weil and McClelland, 1989). The M. *Xba*I/*Dpn*I reaction can be represented by the 12 base-pair sequence 5′..TCTAGATCTAGA..3′. This reaction has been demonstrated on a bacterial genome (Hanish and McClelland, 1990).

We performed an extensive study of the M. *Taq*I/*Dpn*I reaction. The goal was to provide a mixture of the two enzymes that could be used in a single-step reaction cleaving the eight base-pairs 5′ . . . TCGATCGA . . . 3′. Several potential problems concerning M. *Taq*I were overcome. M. *Taq*I, a thermophile with a recommended assay temperature of 65°C, maintains greater than 50% of its activity at 50°C. This is the maximum working temperature for low-melt agarose. M. *Taq*I works well on DNA embedded in agarose. Trace *E. coli* Dam methylase contamination was removed from the recombinant M. *Taq*I by heat treatment at 65°C for 20 minutes. This is important because Dam methylase recognizes 5′ . . . GATC . . . 3′ and methylates the adenine creating *Dpn*I sites (Geier and Modrich, 1979). Two properties of the *Dpn*I make the reaction problematic. *Dpn*I does not function well on DNA embedded in agarose and hemimethylated sites are cleaved slowly (Davis, Morgan, and Robinson, 1990; Nelson and McClelland, 1992). A hemimethylated site generated at position 1129 on pBR322 could be completely cleaved with 60 units of *Dpn*I in one hour using the manufacturer's recommended conditions. Partial digestion products were observed with greater than 5 units of *Dpn*I.

As an alternative to agarose plugs, agarose microbeads (Koob and Szybalski, 1992) should be prepared and the DNA embedded as described. The reduced diffusion distance offered by the agarose microbead matrix provides the enzyme with more effective access to the embedded DNA substrate. *Dpn*I should be diffused into the microbeads by keeping the reaction mix on ice for at least four hours prior to the 37°C incubation. To ensure complete digestion, we suggest a range of *Dpn*I concentrations from 1 to 10 units. Incubation time should not exceed two hours with DpnI concentrations over 5 units.

Reducing the Number of Cleavable Sites via Blocking Agents Coupled with a Methylase Reaction—Achilles's Heel Cleavage

Three classes of blocking reactions have been developed. All three classes rely on the ability of a methylase to protect all but one or more selected DNA sites from digestion by a restriction endonuclease. We can summarize the methodology as follows:

- A restriction endonuclease/methylase recognition site is occupied by a blocking agent.
- The DNA is methylated, blocking subsequent cleavage at all unoccupied sites.
- The blocking agent and methylase are removed.
- Restriction enzyme is added. Cleavage occurs only at previously blocked sites.

1. Achilles' Heel Cleavage–DNA Binding Protein. A blocking reaction using DNA binding proteins followed by restriction enzyme cleavage is termed "Achilles' heel cleavage" (AC) (Koob, Grimes, and Szybalski, 1988a). Unwanted cleavage can occur if the blocking agent interacts with sites other than the one of interest, so blocking conditions should be optimized to minimize nonspecific interactions. These conditions must also allow the methylase to function properly. If the blocking agent doesn't stay bound to the site for the duration of the methylation reaction, the blocking site will be methylated, reducing the yield of the desired product. Finally, all steps must work well on DNA substrates embedded in agarose. The lac and lambda repressors were the first blocking reagents used in this type of reaction (Koob, Grimes, and Szybalski, 1988b); phage 434 repressor (Grimes, Koob, and Szybalski, 1990), and integration host factor (IHF) (Kur et al., 1992) have also been used. Single-site cleavage has been attained using the lac repressor site introduced into yeast and *Escherichia coli* genomes (Koob and Szybalski, 1990).

   Limitations to this strategy include the absence of natural binding protein sites and the low frequency of restriction/methylation sites. Binding protein sites have been engineered into the target DNA, and degenerate sites containing the required restriction/methylation sites have also been added (Grimes, Koob, and Szybalski, 1990). However, modifications in the recognition sequence of the binding protein can decrease the complex's half-life, allowing unwanted methylation at the AC site.

2. Achilles' Heel Cleavage–Triple Helix Formation. The second Achilles' cleavage reaction uses oligonucleotide-directed triple-helix formation as a sequence specific DNA binding protein blocking agent (Hanvey, Schimizu, and Wells, 1990; Maher, Wold, and Dervan, 1989). Pyrimidine oligonucleotides bind to homopurine sites in duplex DNA to form a stable triple-helix structure. The blocking reaction is followed by methylation, removal of the pyrimidine oligonucleotide and methylase, and cleavage by the restriction endonuclease. Single-site cleavage has been demonstrated on yeast chromosomes by blocking with a 24bp pyrimidine oligo, (Strobel and Dervan, 1991a, 1992) and on human chromosome 4 using a 16bp oligo (Strobel et al., 1991b). An advantage of this method over the DNA binding protein AC is the increase in frequency of sites. Insertion of the AC site into the genome is not required. Relatively short purine tracts can be targeted using sequence data. Degenerate probes can be used to screen for overlapping methylation/restriction endonuclease sites when suitable sequence data are not available (Strobel et al., 1991b).

   Reaction conditions for successful pyrimidine oligonucleotide AC are complex (Strobel and Dervan, 1992). Triple helix formation using spermine can inhibit certain methylases, or precipitate DNA in the low-salt reaction conditions required by some methylases. The narrow pH range for the protection reaction may not be compatible with conditions required for efficient methylation. Neutral or slightly acidic conditions promote highly stable triple helices but reduce sensitivity to single base mismatches (Moser and Dervan, 1987). Oligonucleotides that bind and protect mismatched sites allow nontarget restriction sites to remain unmethylated and subsequently cleaved. Increasing the pH from 7.2 to 7.8 can decrease the binding to similar sites (Strobel and Dervan, 1990). In higher pH reactions, the oligo does not stringently bind to the intended target, allowing some methylation to occur at the target site. The unwanted methylation reduces cleavage at the Achilles' site, lowering the yield of the desired DNA fragment.

3. Achilles' Heel Cleavage–RecA-Assisted Restriction Endonuclease. RecA-assisted restriction endonuclease (RARE) cleavage is the most versatile of Achilles' cleavage reaction discovered to date

(Ferrin and Camerini-Otero, 1991; Koob and Szybalski, 1992). In vitro studies indicate that in the presence of ATP, recA protein promotes the strand exchange of single-stranded DNA fragments with homologous duplex DNA. The three distinct steps in the reaction are (1) recA protein binds to the single-strand DNA, (2) the nucleoprotein filament binds the duplex DNA and searches for a homologous region, and (3) the strands are exchanged (Cox and Lehman, 1987; Radding, 1991). Stable triple-helix structures, termed "synaptic complexes," can be formed if the nonhydrolysable analog Adenosine 5′-(γ-Thio) triphosphate (ATPγS) is substituted for ATP (Honigberg et al., 1985). The nucleoprotein filament protects against methylation at a chosen site and is easily removed exposing the AC site. Any duplex DNA stretch containing a restriction endonuclease/methylase recognition site, 15 nucleotides (nt) or longer in length, can be targeted (Ferrin and Camerini-Otero, 1991). RARE cleavage has been used to generate single cuts in the *E. coli* genome by single-stranded oligonucleotides in the 30 nt range and on HeLa cell DNA with oligos in 60 nt range (Ferrin and Camerini-Otero, 1991). RecA-mediated Achilles' cleavage of yeast chromosomes using a 36-mer and 70-mer has been demonstrated (Koob and Szybalski, 1992). YACs (yeast artificial chromosomes) have been cleaved using nucleoprotein filaments in the 50 nt range (Gnirke et al., 1993).

Synaptic complex formation can also block cutting by a restriction endonuclease (Ferrin, 1995). Combined with the fact that many restriction enzymes are active in the buffer used to form these complexes, RARE can be applied to eliminate one of a pair of identical restriction sites in a cloning vector. Partial digestion has been applied to achieve a similar result, but this can fail if the desired site is cut at a comparatively slow rate.

The complexities of the recA-mediated Achilles' cleavage reaction include:

- A titration is required to find the exact ratio of recA to oligonucleotide (Ferrin and Camerini-Otero, 1991; Koob and Szybalski, 1992).
- Excess recA inhibits the methylation reaction.
- Complete hybridization of the oligonucleotide is required for stable triplex formation.
- The nucleoprotein complex diffuses slowly into agarose; microbeading is recommended when using this procedure.
- Nucleoprotein filaments produced with oligonucleotides less than 40 nt may not be stable for the length of time required

for diffusion into agarose microbeads (Koob and Szybalski, 1992).

- RecA DNA-binding requires Mg^{2+}.
- The methylases used must be free of contaminating nucleases.

TROUBLESHOOTING

What Can Cause a Simple Restriction Digestion to Fail?

Faulty Enzyme or Problem Template Preparation?

If the suspect enzyme fails to digest a second or control target, the titer of the enzyme activity should be measured by either a twofold serial or a volumetric titration as described below (procedures A and B).

If the titer assay indicates an active enzyme, and the enzyme cleaves a control template but not the experimental DNA, then an additional control digestion (procedure C) should be performed to test for an inhibitor in the template preparation. Often trans-acting inhibitors may be removed by the drop dialysis protocol (procedure D) detailed below. Spin columns may also be used to remove contaminants including primers, linkers, and nucleotides (Bhagwat, 1992). A linearized plasmid containing a single site may be used if cut and uncut samples are available as markers.

As a matter of course, restriction enzyme activity should be assayed by twofold serial titration if an enzyme has been stored for a period longer than a year, an enzyme shipment was delayed, or even if an enzyme was left on the bench overnight. This simple assay may be used to test enzymes under non-optimal conditions as well. Suppliers offer buffer charts that give an indication of an enzyme's expected activity in nonoptimal buffers, and this information may be useful when the sample DNA is in an alternative buffer due to a previous step or adapting digests so that the DNA samples will be optimized for subsequent steps.

Procedure A—Simple Twofold Serial Titer

Ideally the DNA should be the substrate on which the enzyme was titered by the supplier. Lambda phage DNA or adenovirus Type-2 DNA are common substrates used for enzyme titer. Any DNA that contains several sites that produce a distinguishable pattern may be applied.

1. For the following experiment, make a total of 200 μl of reaction mix. The reaction mix contains 1× reaction buffer, 1 μg DNA/50 μl reaction volume and BSA, if required. For this example, the enzyme is supplied with a vial of 10× reaction buffer and 10 mg/ml BSA. The final reaction mix requires 1× reaction buffer and 100 μg/ml BSA. Lambda DNA (commercially available at 500 μg/ml) is the substrate used to titer the enzyme.
   Add, in order:
   a. 170 μl of distilled water
   b. 20 μl of 10× buffer
   c. 2 μl of 10 mg/ml BSA
   d. 8 μl of 500 μg/ml Lambda DNA
2. Label six 1.5 ml microcentrifuge tubes (numbers 1–6). Pipette 50 μl of reaction mix into tube 1 and 25 μl of mix into the remaining tubes.
3. Add 1 μl of restriction endonuclease to the first tube containing 50 μl of reaction mix. With the pipette set at 25 μl, mix by gently pipetting several times.
4. From the 50 μl reaction mix/enzyme, transfer 25 μl to the second tube. This dilutes the enzyme concentration in half for each subsequent tube.
5. Repeat step 4 until the final tube is reached. The final tube has the most dilute enzyme, but indicates the highest titer. If the final tube, in the following series, shows a complete digestion, then the titer is at least 32,000 units/ml.
6. Cover each tube and incubate at the appropriate reaction temperature for one hour.
7. The reaction is stopped by adding at least 10 μl stop dye/50 μl reaction volume (50% 0.1 M EDTA, 50% glycerol, 0.05% bromophenol blue). The DNA fragments are resolved by agarose gel electrophoresis, stained with ethidium bromide, and visualized using ultraviolet light.
8. The titer is determined as follows:
   Tube 1 complete: titer ≥1000 units/ml
   Tube 2 complete: titer ≥2000 units/ml
   Tube 3 complete: titer ≥4000 units/ml
   Tube 4 complete: titer ≥8000 units/ml
   Tube 5 complete: titer ≥16,000 units/ml
   Tube 6 complete: titer ≥32,000 units/ml

The titer is based on the unit definition: 1 unit of restriction enzyme digests 1 μg DNA to completion in 1 hour. If the digestion pattern from tube 1 is complete, then 1 μl of the enzyme added contains at least 1 unit of activity. The concentration 1 unit/μl is the same as 1000 units/ml. With a dilution factor of 2, a complete digestion pattern from tube 2 indicates that the enzyme concentration is at least 2×1000 units/ml = 2000 units/ml. If tube 4 results in a complete digestion, and tube 5 results in a partial banding pattern, the final titer of the enzyme may be conservatively estimated as 8000 units/ml. Similarly a more precise serial dilution may be designed to evaluate the titer value between 8000 and 16,000 units/ml.

Procedure B—Volumetric Titration

The exact method will vary among enzyme manufacturers. You should contact your supplier for the exact method if this information is not found in their catalog.

While not as convenient as serial titration for most benchtop applications, most suppliers use volumetric titration to assay the activity of the restriction endonucleases. This method may yield more consistent results, especially when the enzyme stock is in high concentration. Most volumetric titers require initial dilution of the enzyme (often in 50% glycerol storage buffer) and the use of substantial amounts of substrate DNA/reaction mix. This method maintains constant enzyme addition to increasing amounts of reaction mix volume, while keeping the concentration of DNA substrate constant. The protocol may differ depending on the concentration and dilution of the enzyme. This method is recommended when evaluating an enzyme sample to be ordered in bulk amounts or for diagnostic applications where internal QC evaluation is required.

Procedure C—Testing for Inhibitors

In a single vial with 1× reaction buffer, add 1 μg each of the control and the experimental DNA. Add the restriction enzyme and incubate at the recommended temperature and time. If there is an inhibitor (often salt or EDTA), the mixed control substrate will not cut.

Procedure D—Drop Dialysis (Silhavy, Berman, and Enquist, 1984)

Many enzymes are adversely affected by a variety of contaminating materials in typical DNA preparations (minipreps, genomic and $CsCl_2$ preparations, etc.). The following drop dialysis method has been successfully used to remove inhibitory substances (e.g., SDS, EDTA, or excess salt) from substrates intended for subsequent DNA manipulations. It is particularly effective for assuring complete cleavage of DNA by sensitive restriction endonucleases, increasing the efficiency of ligation and preparation of templates for DNA sequencing.

1a. For purification of genomic DNA, miniprep DNA, or DNA used as a standard template for DNA sequencing: Phenol extract, chloroform extract, and then alcohol precipitate the DNA. Pellet the DNA in a microcentrifuge, pour off the supernatant, and rinse the pellet with 70% ethanol. Dry the pellet and resuspend it in 50 μl H_20. (Proceed to step 2.)
1b. For purification of templates for DNA sequencing of PCR products: Phenol extract and then chloroform extract the aqueous layer of the PCR reaction. Follow this with an alcohol precipitation. Pellet the DNA by microcentrifugation, pour off the supernatant, and rinse the pellet with 70% ethanol. Dry the pellet and resuspend it in 50 μl H_2O. Alternatively, purify the PCR product through an appropriate spin column, precipitate, and recover the DNA as described above. PCR products that are not a single band on an agarose gel should be gel-purified in low-melt agarose and then treated with β-agarase I or a purification column technology. When using β-agarase, treatment should be followed by extraction, precipitation, and recovery, as described above. When using a purification column, consult the manufacturer's recommendations for the particular column employed.
2. Pour 30 to 100 ml of dialysis buffer, usually double-distilled water or 1× TE (10 mM Tris-HCl, 1 mM EDTA, pH 8.0), into a petri plate or beaker.
3. Float a 25 mm diameter, Type VS Millipore membrane (cat. no. VSWP 02500, MF type, VS filter, mean pore size = 0.025 mm, Millipore, Inc.) shiny side up on the dialysis buffer. Allow the floating filter to wet completely (about 5 minutes) before proceeding. Make sure there are no air bubbles trapped under the filter.
4. Pipette a few microliters of the DNA droplet carefully onto the center of the filter. If the sample has too much phenol or chloroform, the drop will not remain in the center of the membrane, and the dialysis should be discontinued until the organics are further removed. In most cases this is performed by alcohol precipitation of the sample. If the test sample remains in the center of the membrane, pipette the remainder onto the membrane.

5. Cover the petri plate or beaker. Dialyze at room temperature. Be careful not to move the dish or beaker. Dialyze for at least one hour and no more than four hours.
6. Carefully retrieve the DNA droplet with a micropipette.

Note that step 4 may be tricky for those with shaky hands or poor hand-eye coordination. The filter has a tendency to move briskly around the surface as you touch it with the pipette tip. Practice with buffer droplets to master the technique before you try using a valuable sample.

Dialysis against distilled water is also recommended, especially if one is proceeding to another step where EDTA might be a problem.

The Volume of Enzyme in the Vial Appears Very Low. Did Leakage Occur during Shipment?

Some enzymes (some offered at high concentration) may be supplied in a very low volume and the vial may appear empty. During shipment, the enzyme may be dispersed over most of the interior surface of the vial or trapped just under the cap. Follow the steps below to ensure that the enzyme volume is correct. (Since the volume is very low, it is important to keep the entire vial under ice or as cold as possible by working quickly.)

1. Carefully check the exterior of the enzyme vial, noting any signs of glycerol leakage.
2. Add the enzyme's expected volume as water to an identical vial (for a counterbalance).
3. Briefly spin the enzyme vial in a microcentrifuge along with the counterbalance.
4. With both vials on ice, estimate the volume of the enzyme by comparison to that of the counterbalance.

The Enzyme Shipment Sat on the Shipping Dock for Two Days. Is It Still Active?

Restriction enzymes are shipped on dry ice or gel ice packs, depending on the supplier. When enzyme shipments arrive, there should still be a good amount of dry ice left; or if shipped with ice packs, these should still be cold, solid and not soft. For overnight shipments, most suppliers include sufficient thermal mass to maintain proper shipping temperature for at least 36 hours. If the shipment was delayed en route, misplaced, or left in receiving for one or more days, you should:

- Examine the contents, noting the integrity of the container.
- If contents are still cold (but questionable in terms of actual temperature), place a thermometer in the container, re-seal the lid, and note the temperature after 10 minutes.
- After collecting details regarding the shipment's ordering information, contact the supplier. Customer service should provide detailed information regarding the specific products in question and, if warranted, shipping details for a replacement order.

Generally, if the enzyme package is still cold to the touch, most enzymes should be completely active, even if the 10× buffers have recently thawed. Due to their salt content, the concentrated buffers would be liquid even at 0°C. If the enzyme is required for use immediately and no alternative source is available, the enzyme may be tested for activity by serial titration, as described above. Also bear in mind that many enzymes retain their activity after a 16 hour incubation at room temperature (McMahon, M., and Krotee, S., unpublished observation).

Analyzing Transformation Failures and Other Multiple-Step Procedures Involving Restriction Enzymes

A restriction digest is rarely the ultimate step of a research procedure, but instead an early (and essential) reaction within a multiple-step process, as in the case of a cloning experiment. Therefore, when troubleshooting restriction enzymes, and more so than other reagents, it is essential to objectively list *all* the feasible explanations for failure as noted in step 2 of the troubleshooting strategy discussed in Chapter 2, "Getting What You Need From A Supplier." The following discussion illustrates the importance of identifying and investigating all the possible causes of what appears to be a restriction enzyme failure.

If background levels are high after transformation, the enzyme activity should be checked. Alternatively, the vector may have ligated to itself. If the vector had symmetric ends, were the 5′ phosphates removed by dephosphorylation? Was the effectiveness of the dephosphorylation proved? Incomplete vector digestion might be caused by contaminants in the DNA preparation, incompatible buffer, insufficient restriction enzyme, or sites that are located adjacent to each other. If the vector had two different termini, was the success of both digestions verified by recircularization experiments?

Exonuclease contamination in the restriction enzyme or DNA preparation can prevent insert ligation, but ligation might proceed if the ends are blunted by the exonuclease. In this scenario the restriction site would be lost and the reading frame shifted. Phenol chloroform extraction followed by ethanol precipitation will remove exonuclease from DNA preparations. Check the restriction enzyme quality control data for exonuclease, ligation, and blue-white selection. Do not extend the digestion time if an exonuclease problem is suspected.

DNA preparations can contain contaminants that inhibit ligation as well as restriction endonuclease digestion, and the use of very dilute DNA solutions can amplify inhibition. Higher stock vector and insert concentrations are preferable because less of the final reaction volume comes from the DNA solution. If the DNA is stored in Tris-EDTA, the EDTA may inhibit the ligation or restriction digest. Using dilute DNA solutions gives less flexibility when choosing the molar ratio of insert to vector and final DNA concentration of the reaction; both parameters directly affect the quantity of desirable products produced in the ligation reaction.

Failed ligation can occur if the molar ratio of insert to vector is not sufficient. A molar ratio of 3:1 insert to vector should be used for asymmetric ligations and symmetric ligations with small inserts. Symmetric ligations with inserts greater than 800 bp should use 8 μg/ml insert to 1 μg/ml vector (Revie, Smith, and Yee, 1988). In general, the vector concentration should be kept at 1 μg/ml. Total DNA concentration should be kept to 6 μg/ml or less (Bercovich, Grinstein, and Zorzopulos, 1992). Blunt ends are treated as symmetric, and overnight ligation at 16°C is recommended. The addition of 7% PEG 8000 can also stimulate ligation. Single-base overhangs are more difficult to ligate than blunt ends; overnight ligation at 16°C using concentrated ligase is also suggested here. Even so, less than 20% ligation is seen for *Tth111*I under these conditions. Filling in the 5′ single-base overhang with Klenow resulting in a blunt end will increase ligation to about 40% (Robinson, D., unpublished observation).

Transformants containing only deletions indicate problems with ligation or dephosphorylation. Blunt end ligation of a PCR product made with unphosphorylated primers into a dephosphorylated vector will result in a failed ligation, although competent cells will take up some linear molecules. Cells can scavenge the antibiotic resistance gene used for selection, and the scavenged gene is normally found on a vector containing a deletion. The miniprep DNA from the transformants will often run smaller than the control linearized vector.

Faulty DNA ligase, a reaction buffer lacking ATP, and the addition of too much ligation mix to the competent cells can result in low colony count. An antibiotic in the plate that doesn't match the resistance gene within the vector or leaky expression of a toxic protein can kill competent cells, which could mimic a restriction enzyme failure. Cells can be tested by transformation using uncut vector. In addition, as restriction enzymes are excellent DNA binding proteins, they can remain bound to DNA termini and inhibit ligation. Active restriction enzyme can recleave ligated DNA. Often, after incubation, this effect may be minimized by either heating the reaction to 65°C or proceeding with an alternative purification step.

Failure at any one of the many steps of a cloning experiment can give the impression of a restriction enzyme failure. The same principle holds true for the many other applications that involve restriction enzymes.

BIBLIOGRAPHY

Abrol, S., and Chaudhary, V. K. 1993. Excess PCR primers inhibit cleavage by some restriction endonucleases. *Biotech.*, 15:630–632.

Backman, K. 1980. A cautionary note on the use of certain restriction endonucleases with methylated substrates. *Gene* 11:169–171.

Bercovich, J. A., Grinstein, S., and Zorzopulos, J. 1992. Effect of DNA concentration on recombinant plasmid recovery after blunt-end ligation. *Biotech.* 12:190–193.

Bhagwat, A. S. 1992. *Restriction Enzymes: Properties and Use.* Academic Press, San Diego, CA.

Birren, B. W., Lai, E., Hood, L., and Simon, M. I. 1989. Pulsed field gel electrophoresis techniques for separating 1- to 50-kilobase DNA fragments. *Anal. Biochem.* 177:282–285.

Carle, G. F., Frank, M., and Olson, M. V. 1986. Electrophoretic separations of large DNA molecules by periodic inversion of the electric field. *Science* 232:65–68.

Carle, G. F., and Olson, M. V. 1984. Separation of chromosomal DNA molecules from yeast by orthogonal-field-alternation gel electrophoresis. *Nucl. Acids Res.* 12:5647–5664.

Chu, G., Vollrath, D., and Davis, R. W. 1986. Separation of large DNA molecules by contour-clamped homogeneous electric fields. *Science* 234:1582–1585.

Cox, M. M., and Lehman, I. R. 1987. Enzymes of General Recombination. *An. Rev. Biochem.* 56:229–262.

Davis, T., and Robinson, D. New England Biolabs, unpublished observations.

Davis, T. B., Morgan, R. D., and Robinson, D. P. 1990. *Dpn*I cleaves Hemimethylated DNA. In *Human Genome II*, Official Program and Abstracts. San Deigo, CA, p. 26.

Dobrista, A. P., and Dobrista S. V. 1980. DNA protection with the DNA methylase M.*Bbv*I from *Bacillus brevis var.* GB against cleavage by the restriction endonucleases *Pst*I and *Pvu*II. *Gene* 10:105–112.

Ferrin L. J., and Camerini-Otero, R. D. 1991. Selective cleavage of human

DNA: *RecA*-assisted restriction endonuclease (RARE) cleavage. *Science* 254:1494–1497.

Ferrin, L. J. 1995. *Manipulating and Mapping DNA with RecA-Assisted Restriction Endonuclease (RARE) Cleavage*. Plenum Press, New York. pp. 21–30.

Fuchs, R., and Blakesley, R. 1983. Guide to the Use of Type II Restriction Endonucleases. In *Enzymes in Recombinant DNA*. Academic Press, San Diego, CA, pp. 3–38.

Gaido, M. L., Prostko, C. R., and Strobl J. S. 1988. Isolation and characterization of *Bsu*E methyltransferase, a CGCG specific DNA methyltransferase from *Bacillus subtilis. J. Biol. Chem.* 263:4832–4836.

Gardiner, K., Laas, W., and Patterson, D. 1986. Fractionation of large mammalian DNA restriction fragments using vertical pulsed-field gradient gel electrophoresis. *Somatic Cell Mol. Genet.* 12:185–195.

Geier, G. E., and Modrich, P. 1979. Recognition sequence of the *dam* methylase of *Escherichia coli* K12 and mode of cleavage of *Dpn* I endonuclease. *J. Biol. Chem.* 254:1408–1413.

Gingeras, T. R., and Brooks, J. E. 1983. Cloned restriction/modification system from *Pseudomas aerigomosa. Proc. Nat. Acad. Sci. U.S.A.* 80:402–406.

Gnirke, A., Huxley, C., Peterson, K., and Olson, M. V. 1993. Microinjection of intact 200- to 500-kb fragments of YAC DNA into mammalian cells. *Genomics* 15:659–667.

Grimes, E., Koob, M., and Szybalski, W. 1990. Achilles' heel cleavage: Creation of rare restriction sites in lambda phage genomes and evaluation of additional operators, repressors and restriction/modification systems. *Gene* 90:1–7.

Hanish, J., and McClelland, M. 1990. Methylase-limited partial *Not*I cleavage for physical mapping of genomic DNA. *Nucl. Acids Res.* 18:3287–3291.

Hanish, J., and McClelland, M. 1991. Enzymatic cleavage of a bacterial chromosome at a transposon-inserted rare site. *Nucl. Acids Res.* 19:829–832.

Hanvey, J. C., Schimizu, M., and Wells, R. D. 1990. Site specific inhibition of *Eco*RI restriction/modification enzymes by a DNA triple helix. *Nucl. Acids Res.* 18:157–161.

Honigberg, S. M., Gonda, D. K., Flory, J., and Radding C. M. 1985. The pairing activity of stable nucleoprotein filaments made from recA protein, single-stranded DNA, and adenosine 5′-(g-Thio) triphosphate. *J. Biol. Chem.* 260:11845–11851.

Hsu, M.-T., and Berg, P. 1978. Altering the specificity of restriction endonuclease: Effect of replacing Mg^{2+} with Mn^{2+}. *Biochem.* 17:131–138.

Koob, M., Grimes, E., and Szybalski, W. 1988a. Conferring new specificity upon restriction endonucleases by combining repressor-operator interaction and methylation. *Gene* 74:165–167.

Koob, M., Grimes, E., and Szybalski, W. 1988b. Conferring operator specificity on restriction endonucleases. *Science* 241:1084–1086.

Koob, M., and Szybalski, W. 1990. Cleaving yeast and *Eschrichia coli* genomes at a single site. *Science* 250:271–273.

Koob, M., and Szybalski, W. 1992. Preparing and using agarose microbeads. *Meth. Enzymol.* 216:13–20.

Koob, M., Burkiewicz, A., Kur, J., and Szybalski, W. 1992. RecA-AC: Single-site cleavage of plasmids and chromosomes at any predetermined restriction site. *Nucl. Acids Res.* 20:5831–5836.

Koob M. 1992. Conferring new cleavage specificities of restriction endonucleases. *Meth. Enzymol. U.S.A.* 216:321–329.

Kruger, D. H., Barcak, G. J., Reuter, M., and Smith, H. O. 1988. *EcoRII* can be activated to cleave refractory DNA recognition sites. *Nucl. Acids Res.* 16:3997–4008.

Kur, J., Koob, M., Burkiewicz, A., and Szybalski W. 1992. A novel method for converting common restriction enzymes into rare cutters: Integration host factor-mediated Achilles' cleavage (IHF-AC). *Gene* 110:1–7.

Lacks, S., and Greenberg, B. 1977. Complementary specificity of restriction endonucleases of *Diplococcus pneumoniae* with respect to DNA methylation. *J. Mol. Biol.* 114:153–168.

Lacks, S., and Greenberg, B. 1975. A deoxyribonuclease of *Diplococcus pneumoniae* specific for methylated DNA. *J. Biol. Chem.* 250:4060–4066.

Lai, E., Birrcn, B. W., Clark, S. M., Simon, M. I., and Hood, L. 1989. Pulsed field gel electrophoresis. *Biotech.* 7:34–42.

Maher, L. J., Wold, B., and Dervan, P. B. 1989. Inhibition of DNA binding proteins by oligonucleotide-directed triple helix formation. *Science* 245:725–730.

Malyguine, E., Vannier, P., and Yot, P. 1980. Alteration of the specificity of restriction endonucleases in the presence of organic solvents. *Gene* 8:163–177.

McClelland, M., Kessler, L. G., and Bittner, M. 1984. Site-specific cleavage of DNA at 8- and 10-base-pair sequences. *Proc. Nat. Acad. Sci. U.S.A.* 81:983–987.

McClelland, M., Nelson, M., and Cantor, C. 1985. Purification of *Mbo*II methylase (GAAGmA) from *Moraxella bovis*: Site specific cleavage of DNA at nine and ten base pair sequences. *Nucl. Acids Res.* 13:7171–7182.

McClelland, M. 1987. Site-specific cleavage of DNA at 8-, 9-, and 10-bp sequences. *Meth. Enzymol.* 155:22–33.

McClelland, M., Hanish, J., Nelson, M., and Patel, Y. 1988. KGB: A single buffer for all restriction endonucleases. *Nucl. Acids Res.* 16:364.

McMahon, M., and Krotee, S. New England Biolabs, private communication.

Moreira, R. F., and Noren, C. J. 1995. Minimum duplex requirements for restriction enzyme cleavage near the termini of linear DNA fragments. *Biotech.* 19:56–59.

Moser, H. E., and Dervan, P. B. 1987. Sequence-specific cleavage of double helical DNA by triple helix formation. *Science* 238:645–650.

Nelson, M., Christ, C., and Schildkraut, I. 1984. Alteration of apparent restriction endonuclease recognition specificities by DNA methylases. *Nucl. Acids Res.* 12:5165–5173.

Nelson, M., and Schildkraut, I. 1987. The use of DNA methylases to alter the apparent recognition specificities of restriction endonucleases. *Meth. Enzymol.* 155:1–48.

Nelson, M., and McClelland, M. 1992. Use of DNA methyltransferase/edonuclease enzyme combinations for megabase mapping of chromosomes. *Meth. Enzymol.* 216:279–303.

Nelson, M., Raschke, E., and McClelland, M. 1993. Effect of site-specific methylation of restriction endonucleases and DNA modification methyltransferases. *Nucl. Acids Res.* 21:3139–3154.

Nickoloff, J. A. 1992. Converting restriction sites by filling in 5′ extensions. *Biotech.* 12:512–514.

O'Farrell, P. H., Kutter, E., and Nakanishe, M. 1980. A restriction map of the bacteriophage T4 genome. *Mol. Gen. Genet.* 170:411–435.

Patel, Y., Van Cott, E., Wilson, G. G., and McClelland, M. 1990. Cleavage at the twelve-base-pair sequence 5′-TCTAGATCTAGA-3′ using M.*Xba*I (TCTAGm6A) methylation and DpnI (Gm6A/TC) cleavage. *Nucl. Acids Res.* 18:1603–1607.

Pingoud, A., and Jeltsch, A. 1997. Recognition and cleavage of DNA by type-II restriction endonucleases. *Eur. J. Biochem.* 246:1–22.

Polisky, B., Greene, P., Garfin, D. E., McCarthy, B. J., Goodman, H. M., and Boyer, H. W. 1975. Specificity of substrate recognition by the *Eco*RI restriction endonuclease. *Proc. Nat. Acad. Sci. U.S.A.* 72:3310–3314.

Qiang, B., McClelland, M., Poddar, S., Spokauskas, A., and Nelson, M. 1990. The apparent specificity of *Not*I (5′-GCGGCCGC-3′) is enhanced by M. *FnuD*II or M. *Bep*I methyltransferases (5′-mCGCG-3′): Cutting bacterial chromosomes into a few large pieces. *Gene* 88:101–105.

Radding, C. M. 1991. Helical interactions in homologous pairing and strand exchange driven by RecA protein. *J. Biol. Chem.* 266:5355–5358.

Raleigh, E. A., Murray, N. E., Revel, H., Blumenthal, R. M., Westaway, D., Reith, A. D., Rigby, P. W. J., Elhai, J., and Hanahan, D. 1988. McrA and McrB restriction phenotypes of some *E. coli* strains and implications for gene cloning. *Nucl. Acids Res.* 16:1563–1575.

Revie, D., Smith, D. W., and Yee, T. W. 1988. Kinetic analysis for optimization of DNA ligation reactions. *Nucl. Acids Res.* 16:10301–10321.

Roberts, R. J., and Macelis, D. 2001. REBASE-restriction enzymes and methylases. *Nucl. Acids Res.* 29:268–269.

Robinson, D., Akbari, T., McMahon, M., and Davis, T. 1991. Digestion of agarose-embedded DNA. *NEB Transcript* 3:8–9.

Ronbinson, D., and Kelly, D. New England Biolabs, unpublished observation.

Sambrook, J., Fritsch, E. F., and Maniatis, T. 1989. *Molecular Cloning: A Laboratory Manual*, 2nd ed. Cold Spring Harbor Press, Plainview, NY.

Scopes, R. 1982. *Protein Purification, Principles and Practice.* Springer, New York.

Shukla, H., Kobayashi, Y., Arenstorf, H., Yasukochi, Y., and Weissman, S. M. 1991. Purification of BsuE methyltransferase and its application in genomic mapping. *Nucleic Acids Res.* 19:4233–4239.

Silhavy, T. J., Berman, M. L., and Enquist, L. W. 1984. *Experiments with Gene Fusions.* Cold Spring Harbor Laboratory, Cold Spring Harbor, NY.

Smith, C. L., Warburton, P. E., Gaal, A., and Cantor, C. R. 1986. *Genetic Engineering.* Plenum Press, Newark, NJ.

Song, Y.-H., Rueter, T., and Geiger, R. 1988. DNA cleavage by *Aat*I and *Stu*I is sensitive to *Escherichia coli* dcm methylation. *Nucl. Acids Res.* 16:2718.

Stewart, G., Furst, A., and Avdalovic, N. 1988. Transverse alternating field electrophoresis (TAFE). *Biotech.* 6:68–73.

Strobel, S. A., and Dervan, P. B., 1990. Site-specific cleavage of a yeast chromosome by oligonucleotide-directed triple-helix formation. *Science* 252: 73–75.

Strobel, S. A., and Dervan P. B., 1991a. Single-site cleavage of yeast genomic DNA mediated by triple helix formation. *Nature* 350:172–174.

Strobel, S. A., Doucitte-Stamm, L. A., Riba, L., Housman D. E., and Dervan, P. B. 1991b. Site-specific cleavage of human chromosome 4 mediated by triple-helix formation. *Science* 254:1639–1642.

Strobel, S. A., and Dervan, P. B. 1992. Triple helix-mediated single-site enzymatic cleavage of megabase genomic DNA. *Meth. Enzymol.* 216:309–321.

Szybalski, W., Kim, S. C., Hasan, N., and Podhajska, A. J. 1991. Class-IIS restriction enzymes: A review. *Gene* 100:13–26.

Thierry, A., and Dujon, B. 1992. Nested chromosomal fragmentation in yeast using the meganuclease I-*Sce*I: A new method for physical mapping of eukaryotic genomes. *Nucl. Acids Res.* 20:5625–5631.

Thomas, M., and Davis, R. W. 1975. Studies on the cleavage of bacteriophage lambda DNA with *EcoRI* restriction endonuclease. *J. Mol. Biol.* 91:315–328.

Vovis, G. F., and Lacks, S. 1977. Conplementary action of restriction enzymes endo R.*Dpn*I and endo R.*Dpn*II on bacteriophage f1 DNA. *J. Mol. Biol.* 115:525–538.

Wang, Y., and Wu, R. 1993. A new method for specific cleavage of megabase-size chromosomal DNA by l-terminase. *Nucl. Acids Res.* 21:2143–2147.
Waterbury, P. G., Rehfuss, R. P., Carroll, W. T., Smardon, A. M., Faldasz, B. D., Huckaby, C. S., and Lane, M. J. 1989. Specific cleavage of the yeast genome at 5′-ATCGATCGAT-3′. *Nucl. Acids Res.* 17:9493.
Weil, M. W., and McClelland, M. 1989. Enzymatic cleavage of a bacterial genome at a 10-base-pair recognition site. *Proc. Nat. Acad. Sci. U.S.A.* 86:51–55.
Wong, K. K., and McClelland, M. 1992. Dissection of the *Salmonella typhimurium* genome by use of a Tn5 derivative carrying rare restriction sites. *J. Bacteriol.* 174:3807–3811.

10

Nucleotides, Oligonucleotides, and Polynucleotides

Alan S. Gerstein

Nucleotides 268
Nomenclature: De facto and Du jour 268
What Makes a Nucleotide Pure? 269
Are Solution Nucleotides Always More Pure Than Lyophilized Nucleotides? 269
Are Solution Nucleotides More Stable Than Lyophilized Nucleotides? 270
Does Your Application Require Extremely Pure Nucleotides? 272
How Can You Monitor Nucleotide Purity and Degradation? 272

The author would like to thank Anita Gradowski of Pierce Milwaukee for contributing such thorough and helpful information regarding the preparation of nucleotide solutions. Special thanks also to Cica Minetti and David Remeta of Rutgers University for discussing a method to calculate the extinction coefficient of an oligonucleotide. The contributions to this chapter by Howard Coyer and Thomas Tyre, also of Pierce Milwaukee, are too numerous to list.

Molecular Biology Problem Solver, Edited by Alan S. Gerstein.
ISBN 0-471-37972-7

How Should You Prepare, Quantitate, and Adjust the pH of Small and Large Volumes of Nucleotides? 273
What Is the Effect of Thermocycling on Nucleotide Stability? 275
Is There a Difference between Absorbance, A_{260}, and Optical Density? 275
Why Do A_{260} Unit Values for Single-Stranded DNA and Oligonucleotides Vary in the Research Literature? 278
Oligonucleotides 279
How Pure an Oligonucleotide Is Required for Your Application? 279
What Are the Options for Quantitating Oligonucleotides? 279
What Is the Storage Stability of Oligonucleotides? 280
Your Vial of Oligonucleotide Is Empty, or Is It? 281
Synthetic Polynucleotides 281
Is a Polynucleotide Identical to an Oligonucleotide? 281
How Are Polynucleotides Manufactured and How Might This Affect Your Research? 282
Would the World Be a Better Place If Polymer Length Never Varied? 284
Oligonucleotides Don't Suffer from Batch to Batch Size Variation. Why Not? 284
How Many Micrograms of Polynucleotide Are in Your Vial? 284
Is It Possible to Determine the Molecular Weight of a Polynucleotide? 285
What Are the Strategies for Preparing Polymer Solutions of Known Concentration? 285
Your Cuvette Has a 10 mm Path Length. What Absorbance Values Would Be Observed for the Same Solution If Your Cuvette Had a 5 mm Path Length? 286
Why Not Weigh out a Portion of the Polymer Instead of Dissolving the Entire Contents of the Vial? 287
Is a Phosphate Group Present at the 5′ End of a Synthetic Nucleic Acid Polymer? 287
What Are the Options for Preparing and Storing Solutions of Nucleic Acid Polymers? 287
Bibliography 288

NUCLEOTIDES

Nomenclature: De facto and Du jour

Lehninger (1975) provides a thorough discussion of proper nucleotide nomenclature and abbreviations. Unfortunately, commercial catalogs and occasionally the research literature introduce different notations. Some consider "NTP" a general term for deoxynucleotides, but the absence of the letter "d" indicates a ribonucleotide to others. Commercial literature also describes ribonucleotides as "RTP's." If the letter "d" is present, the name describes a deoxynucleotide. If "d" is absent, check the literature piece closely to avoid a common purchasing error. Dideoxynucleotides are generically referred to as "ddNTP's."

What Makes a Nucleotide Pure?

Using dATP as an example, what categories of impurities could be present? One potential contaminant is a nucleotide other than dATP, such as dCTP. A second class of impurity could be the mono-, di-, or tetraphosphate form of the deoxyadenosine nucleotide. Since most if not all commercial nucleotides are chemically synthesized from highly analyzed precursors, contamination with a nucleotide not based on deoxyadenosine is very unlikely. A third class of impurities is the non-UV-absorbing organic and inorganic salts accumulated during the synthesis and purification procedures.

While essentially all commercial nucleotides are chemically synthesized, the final products are not necessarily identical. Manufacturing processes vary; raw materials and intermediates of the nucleotide synthesis reactions are subjected to different purification strategies and processes. It is these intermediate steps, and the scrutiny of the products' final specifications, that allow manufacturers to legitimately claim that nucleotides are extremely pure.

A formal definition of *extremely pure* does not exist, but commercial preparations of such products typically contain greater than 99% of the desired nucleotide in the triphosphate form. Contaminating nucleotides are rarely detected in commercial preparations, even using exceedingly stringent high-performance chromatography procedures, but some contaminants escape HPLC detection. Freedom from non-UV-absorbing materials is typically judged by comparison of a measured molar extinction (A_m) coefficient to published extinction coefficients (ε)values. Nuclear magnetic resonance (NMR) may also be used to monitor for contaminants such as pyrophosphate.

Are Solution Nucleotides Always More Pure Than Lyophilized Nucleotides?

Nucleotides were first made commercially available as solvent-precipitated powders. The lyophilized and extremely pure solution forms appeared in the early 1980s. Some lyophilized preparations approach 98% purity or more but rarely match the >99% achieved by extremely pure solutions. Generally, solution nucleotides are purer than the lyophilized version, but unless supporting quality control data are provided, it should not be concluded that a solution nucleotide is extremely pure or even more pure than a lyophilized preparation.

Are Solution Nucleotides More Stable Than Lyophilized Nucleotides?

Peparations of deoxynucleoside triphosphates decompose into nucleoside di- and tetraphosphates via a disproportionation reaction. This reaction is concentration and temperature dependent. At temperatures above 4°C, lyophilized preparations of deoxynucleotides undergo disproportionation faster than nucleotides in solution. In contrast, the rate of degradation for both forms is less than 1% per year at –20°C and below (Table 10.1). Solutions of dideoxynucleotides and ribonucleotides are similarly stable for many months at temperatures of –20°C and below. Most, but not all, dideoxy- and ribonucleotides are stable for many months at 4°C.

Table 10.1 Storage Stability of Nucleotides

		% Triphosphate Form			
	Months	–70°C	–20°C	4°C	21°C
Powder					
dATP	54	99.44	99.14	97.47	93.93 (48 mo) 97.78 (3 mo)
dCTP	54	98.46	95.46	39.3 (33 mo)	39.45 (2.75)
dGTP	54	96.95	95.37	25.74 (27 mo)	34.4 (1.75)
dTTP	54	97.29	94.28	27.4 (30 mo)	39.45 (2.75 mo)
dUTP	N.A.	N.A.	N.A.	N.A.	N.A.
Solution (100 mM)					
dATP	54	99.2	98.75	95.3	91.8 (2 mo) 37.07 (39 mo)
dCTP	54	99.38	99.15	96.98	95.2 (2 mo) 21.25 (42 mo)

Table 10.1 (*Continued*)

	Months	% Triphosphate Form			
		−70°C	−20°C	4°C	21°C
Powder					
dGTP	54	99.63	98.83	95.47	90.5 (2 mo) 19.7 (42 mo)
dTTP	54	99.44	98.87	93.54	95.6 (2 mo) 0.07 (42)
dUTP	54	99.23	98.02	71.55	90.1 (1.2 mo) 40.13 (6 mo)
Solution (10 mM)					
dATP	15		99.68	99.59 (12 mo)	88.6 98.5 (2 mo)
dCTP	15		98.2	99.56 (12 mo)	86.11 98.85 (2 mo)
dGTP	15		98.6	99.51 (12 mo)	89.47 98.35 (2 mo)
dTTP	15		93.57	99.29 (12 mo)	81.05 98.86 (2 mo)
dUTP	15		93.8	99.45 (12 mo)	84.95 98.5 (2 mo)
Solution ddNTP (10 mM)					
ddATP	3		99.69	99.49	94.52
ddCTP	3		100	98.51	97.38
ddGTP	3		98.4	98.08	94.23
ddTTP	3		99.36	99.13	87.06
Solution ddNTP (5 mM)					
ddATP	3		99.77	98.12	68.56
	4		99.63	96.31	2
ddCTP	3		98.77	100	98.4
	4		99.27	99.46	93.72
ddGTP	3		95.61	98	96.67
	4		98.25	97.9	93.68
ddTTP	3		93.1	55.09	49.03
	4		94.25	63.23	3.6
RTP Solutions (100 mM)					
ATP	3		98.57	98.18	95.39
CTP	3		99.25	99.43	98.43
GTP	3		98.46	98.44	96.82
UTP	3		99.71	99.69	97.99

Source: Data based on chromatographic separation of nucleotide species via high performance chromatography on an Amersham Pharmacia Biotech FPLC® System.
Notes: Each sample, 0.2 μmoles (0.2 ml of a 1 mM solution) was injected onto a Mono Q® Ion Exchange column. Using the following buffers:

Buffer A, 5mM sodium phosphate, pH 7.0.
Buffer B, 5mM sodium phosphate, 1M NaCl, pH 7.0.

purification was achieved via a gradient of 5–35% NaCl over 15 minutes using a flow rate of 1 ml/min. Nucleotide peaks were detected at of 254 nm. (Data from Amersham Pharmacia Biotech, 1993a.)

Does Your Application Require Extremely Pure Nucleotides?

Only you can answer this question. Most applications have supporters and detractors for the use of extremely pure nucleotides.

How Can You Monitor Nucleotide Purity and Degradation?

Nucleotides produce very specific spectroscopic absorbance data. Absorbance ratios not within predicted ranges (Table 10.2) indicate a contaminated deoxy- or ribonucleotide, such as if dATP and dCTP were accidentally mixed together. This technique is adequate to quickly determine if a large contamination problem exists, but a high-performance liquid chromatography approach is required to detect minor levels of impurities.

The absorbance ratio will not indicate when the triphosphate form of a nucleotide breaks down into the di- and tetraphosphate forms. This form of degradation can be monitored most effectively

Table 10.2 Nucleotide Absorbtion Maxima

Nucleotide	Lambda Maximum (pH 7.0)	A_m (pH 7.0) molar extinction coefficient
2′-dATP	259 nm	15.2×10^3 [d]
2′-dCTP	280 nm[a]	13.1×10^3 [a,e]
2′-dGTP	253 nm	13.7×10^3 [f]
2′-dITP	249 nm	12.2×10^3 [b,h]
2′-dTTP	267 nm[b]	9.6×10^3 [g]
2′-dUTP	262 nm	10.2×10^3 [i]
c7-2′-ATP	270 nm	12.3×10^3 [j]
c7-2′-dGTP	257 nm	10.5×10^3 [c]
2′,3′-ddATP	259 nm	15.2×10^3 [d]
2′,3′-ddCTP	280 nm[a]	13.1×10^3 [a,e]
2′,3′-ddGTP	253 nm	13.7×10^3 [f]
2′,3′-ddTTP	267 nm	9.6×10^3 [g]
ATP	259 nm	15.4×10^3
CTP	280 nm[a]	13.0×10^3 [a]
GTP	252 nm	13.7×10^3
UTP	262 nm	10.2×10^3

Note: The spectral terms and definitions used are those recommended by the National Bureau of Standards Circular LCD 857, May 19, 1947.
[a] Spectral analysis done at pH 2.0.
[b] Spectral analysis done at pH 6.0.
[c] Value determined at Amersham Pharmacia Biotech.
[d] 2′-dAMP NRC reference spectral constants employed.
[e] 2′-dCMP NRC reference spectral constants employed.
[f] 2′-dGMP NRC reference spectral constants employed.
[g] 2′-dTMP NRC reference spectral constants employed.
[h] 2′-dIMP NRC reference spectral constants employed.
[i] 2′-dU NRC reference spectral constants employed.
[j] Leela and Kehne (1983).

by high-performance chromatography, but when such equipment is unavailable, thin layer chromatography can provide qualitative data (Table 10.3).

How Should You Prepare, Quantitate, and Adjust the pH of Small and Large Volumes of Nucleotides?

The following procedure can be used to prepare solutions of deoxynucleotides, ribonucleotides, and dideoxynucleotides provided that the different formula weights are taken into account.

A 100 mM solution of a solid nucleotide triphosphate is prepared by dissolving about 60 mg per ml in purified H_2O. The exact weight will depend on the formula weight, which will vary by nucleotide, supplier, and salt form. As solid nucleotide triphosphates are very unstable at room temperature, they should be stored frozen until immediately before preparing a solution.

Quantitation

Spectroscopy

The most accurate method of quantifying a solution is to measure the absorbance by UV spectrophotometry. A dilution should be made to obtain a sample within the linear range of the spectrophotometer. The sample should be analyzed at the specific λ_{max} for the nucleotide being used. The concentration can then be obtained by multiplying the UV absorbance reading by the dilution factor, and dividing by the characteristic A_m for that nucleotide. These data are provided in Table 10.2.

Table 10.3 TLC Conditions to Monitor dNTP Degradation

dNTP	R_f, Principal	R_f, Trace	Solvent System
dATP	0.25	0.35 (dADP)	A
dCTP	0.15	0.21 (dCDP)	A
dGTP	0.27	0.34 (dGDP)	B
dTTP	0.14	0.21 (dTDP)	A

Note: Solvent System A: Isobutyric acid/concentrated NH_4OH/water, 66/1/33; pH 3.7. Add 10 ml of concentrated NH_4OH to 329 ml of water and mix with 661 ml of isobutyric acid.
Solvent System B: Isobutyric acid/concentrated NH_4OH/water, 57/4/39; pH 4.3. Add 38 ml of concentrated NH_4OH to 385 ml of water and mix with 577 ml of isobutyric acid.
TLC Plates: Eastman chromagram sheets (#13181 silica gel and #13254 cellulose).

Weighing

One would think that the mass of an extremely pure nucleotide could be reliably determined on a laboratory balance. Not so, because during the manufacturing process, nucleotide preparations typically accumulate molecules of water (via hydration) and counter-ions (lithium or sodium, depending on the manufacturer), which signficantly contribute to the total molecular weight of the nucleotide preparation. Unless you consider the salt form and the presence of hydrates, you're adding less nucleotide to the solution than you think. The presence of salts and water also contribute to the molecular weights of oligo- and polynucleotides, which are also most reliably quantitated by spectroscopy.

pH Adjustment

The pH of a solution prepared by dissolving a nucleotide in water will vary, depending on the pH at which the nucleotide triphosphate was dried. An aqueous solution of nucleotide triphosphate prepared at Amersham Pharmacia Biotech will have a pH of approximately pH 4.5. The pH may be raised by addition of NaOH (0.1 N NaOH for small volumes, up to 5 N NaOH for larger volumes). Approximately 0.002 mmol NaOH per mg nucleotide triphosphate is required to raise the pH from 4.5 to neutral pH. If the pH needs to be lowered, addition of a H^+ cation exchanger to the nucleotide solution will lower the pH without adding a counter-ion. The amount of cation-exchanger resin per volume of 100 mM nucleotide solution varies greatly depending on the starting and ending pH. For very small volumes (<5 ml) of nucleotide solutions, a 50% slurry of SP Sephadex can be added dropwise. For larger volumes (>5 ml), solid cation exchanger can be added directly in approximately 0.2 cm^3 increments. The cation exchanger can be removed by filtration when the desired pH is obtained.

The triphosphate group gives the solution considerable buffering capacity. If an additional buffer is added, the pH should be checked to ensure that the buffer is adequate. The pH should be adjusted when the solution is at or near the final concentration. A significant change in the concentration will change the pH. An increase in concentration will lower the pH, and dilution will raise the pH, if no other buffer is present.

Similar results will be obtained for all of the nucleotide triphosphates. Monitor the pH of the solutions as a precaution; purines are particularly unstable under pH 4.5, and all will degrade at acid pH.

Example

To prepare a 10 mM solution from a 250 mg package of dGTP, the dGTP may be dissolved in about 40 ml of purified H_2O. The pH may then be adjusted from a pH of about 4.5 to the desired pH with 1 N NaOH, carefully added dropwise with stirring. About 0.5 ml of 1 N NaOH will be needed for this example. A dilution of 1:200 will give a reading in the linear range of most spectrophotometers. Spectroscopy should be performed at the nucleotide's absorbance maximum, which is 253 nm for dGTP. In this example an absorbance of about 0.700 is expected. The formula for determining the concentration is:

$$\frac{\text{Absorbance at } \lambda_{max} \times \text{ dilution factor}}{A_m} = \text{molar concentration}$$

Using the A_m for dGTP of 13,700, the concentration in this example is found to be

$$\frac{0.700 \times 200}{13,700} = 0.0102\,\text{M}, \quad \text{or} \quad 10.2\,\text{mM dGTP}$$

What Is the Effect of Thermocycling on Nucleotide Stability?

Properly stored, lyophilized and solution nucleotides are stable for years. The data in Table 10.4 (Amersham Pharmacia Biotech, 1993b) describe the destruction of nucleotides under common thermocycling conditions. Fortunately, due to the excess presence of nucleotides, thermal degradation does not typically impede a PCR reaction.

Is There a Difference between Absorbance, A_{260}, and Optical Density?

Readers are strongly urged to review Efiok (1993) for a thorough and clearly written discussion on the spectrophotometric quantitation of nucleotides and nucleic acids.

Absorbance (A)

Absorbance (A), also referred to as optical density (OD), is a unitless measure of the amount of light a solution traps, as measured on a spectrophotometer. The Beer-Lambert equation (Efiok, 1993) defines absorbance in terms of the concentration of the solution in moles per liter (C), the path length the light travels through the solution in centimeters (l), and the extinction coefficient in liter per moles times centimeters (E):

Table 10.4 Breakdown of Nucleotides under Thermocycling Conditions

	Nucleotides	% Purity of Triphosphate	
		0 PCR Cycles	25 PCR Cycles
Experiment 1	dATP	99.31	92.41
	dCTP	99.47	93.64
	dGTP	99.14	92.43
	dTTP	99.06	93.38
Experiment 2	dATP	99.56	94.17
	dCTP	99.80	95.36
	dGTP	99.78	94.02
	dTTP	99.60	94.17
Experiment 3	dATP	99.40	92.02
	dCTP	99.66	93.84
	dGTP	99.39	92.68
	dTTP	99.15	93.69
Experiment 4	dATP	99.44	92.77
	dCTP	99.59	93.89
	dGTP	99.43	92.88
	dTTP	99.19	93.65

Source: Data from Amerhsam Pharmacia Biotech (1993b).
Note: Each nucleotide was mixed with 10× PCR buffer from the GeneAmp® PCR Reagent Kit (Perking Elmer catalogue number N801-0055)to give a final nucleotide concentration of 0.2 mM in 1× PCR buffer. Noncycled control samples (0 cycles) were immediately assayed. Test samples were cycled for 25 rounds in a Perkin Elmer GeneAmp® PC System 9600 using the cycling program of 94°C for 10 seconds, 55°C for 10 seconds, and 72°C for 10 seconds. After cycling, the samples were stored on ice until assayed.

For analysis, samples were diluted to give a nucleotide concentration of 0.133 mM. The diluted samples were then assayed on FPLC® System using a MonoQ® column. The assay time for a sample was 10 minutes using a sodium chloride gradient (50–400 mM) in 20 mM Tris-HCl at pH 9.0. Nucleotide peaks were detect using a wavelength of 254 nm.

$$A = ClE$$

Since the units of C, l, and E all cancel, A is unitless.

Absorbance Unit

Also referred to as an optical density (OD) unit, an absorbance unit (AU) is the concentration of a material that gives an absorbance of one and therefore is also a unitless measure. Typically, when working with nucleic acids, we express the extinction coefficient in ml per mg times cm:

$$E = \frac{\text{ml}}{\text{mg} \times \text{cm}}$$

Using an extinction coefficient expressed in these terms, one A_{260} unit of double-stranded DNA has a concentration of DNA of 50 μg/ml.

For practical reasons, suppliers typically define the total volume of material to be one milliliter when selling their nucleic acids.

Note that from a supplier's perspective, an A_{260} unit specifies an amount of material and not a concentration. It is the amount of material in one milliliter that gives an absorbance of one. The A_{260} unit value provided by a supplier cannot be substituted into the Beer-Lambert equation to calculate concentration. If this substitution is done, the concentration will be off by a factor of 1000.

*Extinction Coefficient (*E*)*

Also known as absorption coefficient, absorptivity, and absorbency index, the proportionality constant E is a constant value inherent to a pure compound. E will not vary between different lots of a chemical. The units of E are typically ml/mg-cm or L/g-cm. It is experimentally measured by utilizing a method that is not affected by the presence of a contaminant. For example, the extinction coefficient of a nucleotide can be determined by measuring the amount of phosphorous present.

As in the Beer-Lambert equation, the concentration (C) of a solution in mg/ml or g/L = A/El.

Molar Extinction Coefficient (ε) versus A_m

The molar extinction coefficient (also referred to as molar absorbtivity) describes the absorbance of 1 ml of a 1 molar solution measured in a cuvette with a 1 cm path length. For practical reasons a manufacturer may measure a molar coefficient by weighing an amount of the solid material, mixing into a solution and measuring the absorbance of that solution. This way, a molar coefficient is calculated that is not a true molar extinction coefficient because it is affected by the presence of contaminants. To set this measured coefficient apart from a true molar extinction coefficient, companies use the symbol A_m. The A_m for a given chemical will vary from preparation to preparation depending on the presence of contaminants. Using nucleotides as an example, the number of sodium and water molecules present in the finished product can vary from lot to lot, causing the A_m values to also vary slightly between lots. The units of A_m are L/mol-cm.

*Suppose that you have 100 μl of a 5 mM solution of a nucleotide with a molar extinction coefficient of 10.4×10^3, how many A_{260} units do you have? Using the Beer-Lambert equation, the undi-

**Reprinted with minor changes, with permission, Amersham Pharmacia Biotech, 1990.*

luted 5 mM solution of this nucleotide will have an absorbance of 52. $A = 10.4 \times 10^3$ L/(mol × cm) × 0.005 M × 1 cm = 52. This measure of absorbance is a unitless measure of the opacity of the solution and is independent of the volume of the solution.

To calculate the A_{260} units present as a supplier would define an A_{260} unit, the volume of the solution must be taken into account. This is simply done by multiplying the volume of the solution in milliliters by the absorbance measurement. For the 100 μl of a solution with an absorbance of 52, the number of A_{260} units present is 5.2 units (i.e., 52 × 0.1 ml = 5.2 units).

Why Do A_{260} Unit Values for Single-Stranded DNA and Oligonucleotides Vary in the Research Literature?

The A_{260} unit values are generated by rearranging the Beer-Lambert equation as per Efiok (1993):

$$OD = ECL$$

$$\frac{C}{OD} = \frac{1}{E} = \frac{1}{\mathrm{AU}}$$

Substituting the value of $E_{1\mathrm{cm}}^{1\mathrm{mg/ml}}$ in Table 10.5 generates the conversion factors to A_{260} data into mg/ml of nucleic acid.

Manufacturer technical bulletins (Amersham Pharmacia Biotech, 2000) and protocol books (Ausubel et al., 1995; Sambrook, Fritsch, and Maniatis, 1989) frequently cite different values for single-stranded DNA and oligonucleotides. Since nucleotide sequence and length alter the value of an extinction coefficient, the variability amongst A_{260} conversion factors is likely caused by the use of different nucleic acid samples to calculate the extinction coefficient. In practice, this means that it probably does not matter which value you use for your work as long as you consistently use the same value for the same type of nucleic acid. However, consider the existence and impact of different conversion factors when attempting to reproduce the work of another researcher.

Table 10.5 Nucleic Acid

	$E_{1\mathrm{cm}}^{1\mathrm{mg/ml}}$ 1	$A_{260,}$ (μg/ml)
Double-stranded DNA	20	50
Single-stranded DNA or RNA (>100 nucleotides)	25	40
Single-stranded oligos (60–100 nucleotides)	30	33
Single-stranded oligos (<40 nucleotides)	40	25

Source: From Effiok (1993).

OLIGONUCLEOTIDES

How Pure an Oligonucleotide Is Required for Your Application?

During standard solid phase oligonucleotide (oligo) synthesis, nucleotides are coupled one at a time to a growing chain attached at its 3′ end to a solid support (unlike enzymatic DNA synthesis, chemical DNA synthesis occurs in the 3′ to 5′ direction). To prepare an oligonucleotide where the majority of the product is full length, a coupling efficiency of ≥98% at each nucleotide addition is required. At lower coupling efficiencies, the synthesis will yield a significant amount of oligos that are not full length (failure sequence).

Oligonucleotide impurities may consist of various forms of the desired sequence as well as impurities from the reagents used in synthesis. The ammonium hydroxide that detaches the oligonucleotide from the solid support of a DNA synthesizer and buffer salts carried over from a purificaton process can also be troublesome. Ammonium ions are inhibitory to T4 Polynucleotide kinase, so if the the oligo isn't properly de-salted, subsequent end-labeling reactions will fail.

Your application dictates the level of acceptable purity. The ammonium ions carried over from detaching the oligo from the solid support can completely inhibit end labeling but not other reactions. An oligo preparation that contains less than 50% full-length product will produce miserable sequencing results, but might function as a PCR primer. If your oligo functions reproducibly and verifiably generates data, it's sufficiently pure.

What Are the Options for Quantitating Oligonucleotides?

The concentration of oligonucleotides is most commonly approximated by applying the Beer-Lambert law and a conversion factor ranging from from 25 to 37 μg per A_{260} unit. This approach is inexact, but it is reliable for common molecular biology techniques as long as its limitations are considered. Computer software that predicts an extinction coefficient based on nucleotide sequence and nearest-neighbor analysis is also available. Such predictive software should be employed with caution, since it does not take into account a number of factors, such as the degree of base stacking and the presence of alternate structures commonly found among nucleic acids, that significantly influence the magnitude of the extinction coefficient.

If an exact extinction coefficient is required, a method that directly calculates the quantity of the nucleic acid is required. The phosphate analysis method of Griswold et al. (1951) is described below.

The method of Griswold et al. (1951) is based on a colorimetric assay (A_{820}) employing ANS (aminonaphtosulfonic acid) dissolved in a sulfite/bisulfite solution. The reaction requires the presence of molybdate prepared in 10 N sulfuric acid. A carefully prepared phosphate solution is utilized to obtain a standard curve by serial dilution (10–100 μM phosphate). DNA test solutions of known absorbance at 260 nm are digested with nuclease P1 and alkaline phosphatase. The phosphate released from the digestion is quantified by monitoring the blue color development at 820 nm following reaction with ANS solution in the presence of molybdate in acidic solution and incubation at 95°C for 10 minutes. The extinction coefficient is determined in accordance with the following equation:

$$E_{260} = \frac{A_{260\,\mathrm{nm}}}{\text{phosphate } (\mu\mathrm{M}) \times (n-1)}$$

where $A_{260\mathrm{nm}}$ is the original absorbance of the DNA solution, phosphate (μM) represents the value obtained in triplicate of the digested DNA solution extrapolated from the standard phosphate curve, and n is the number of bases comprising the oligonucleotide.

As with nucleotides, determining the amount of an oligo is best done by measuring the absorbance. If you prefer to measure the mass on a very accurate analytical balance, take into account the presence of contaminating salts and water.

What Is the Storage Stability of Oligonucleotides?

The fundamentals of safe DNA storage are discussed in Chapter 7, "DNA Purification," and RNA storage is discussed in Chapter 8, "RNA Purification." Lyophilized oligonucleotides are stable for months or years stored at –20°C and colder in frost-free or non-frost-free freezers. Solutions of DNA oligonucleotides are best stored at –20°C and below at neutral pH. Non-frost-free freezers are preferred to eliminate potential nicking due to freeze–thawing.

In one instance, which was not further investigated, approximately 10% of the phosphate groups were lost from the 5′ ends of phosphorylated oligo dT (approximately 15 nucleotides in length) after 12 months of storage at –20°C (Amersham Pharmacia Biotech, unpublished observations).

Your Vial of Oligonucleotide Is Empty, or Is It?

Lyophilization does not always produce a neat pellet at the bottom of the vial. The material might be dispersed throughout the inner walls of the vial in a very thin layer that is difficult to see. The best method to confirm the absence of the material is to dissolve the vial's contents by thoroughly pipetting the solvent on the vial's inner walls and measuring the absorbance at 260 nm.

SYNTHETIC POLYNUCLEOTIDES

Is a Polynucleotide Identical to an Oligonucleotide?

Manufacturers typically define polynucleotides as single- or double-stranded nucleic acid polymers whose length exceeds 100 nucleotides. Double-stranded polymers can be comprised solely of DNA or RNA, or DNA:RNA hybrids. As illustrated in Figure 10.1, a single preparation of a synthetic polynucleotide contains a highly disperse population of sizes. In comparison, oligonucleotides are almost always single-stranded molecules (RNA or DNA) shorter than 100 nucleotides and typically comprised of a nearly homogeneous population in length and sequence.

Polymer nomenclature is not universally accepted, but the major suppliers apply the following strategy:

- Poly dA—single-stranded DNA homopolymer containing deoxyadenosine monophosphate.
- Poly A—single-stranded RNA homopolymer comprise of adenosine monophosphate.
- Poly A · oligo $dT_{12\text{-}18}$—Double-stranded molecule, with one strand comprised of an RNA homopolymer of adenosine

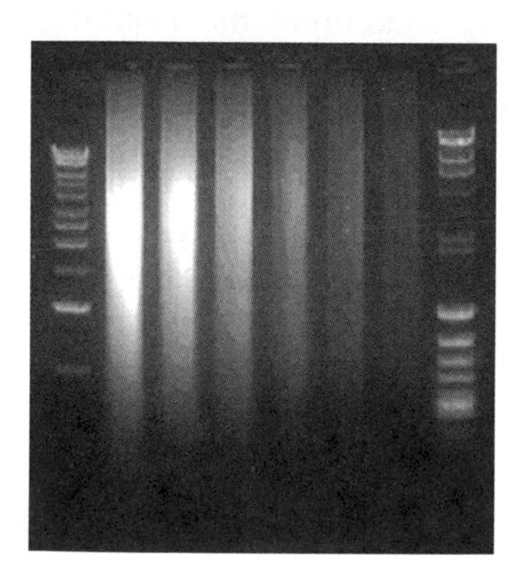

Figure 10.1 Lane 1–1 kb ladder; lane 2–7 poly (dI-dC) · (dI-dC); lane 2–2.0 μg; lane 3–1.5 μg; lane 4–1.0 μg; lane 5–0.5 μg; lane 6–0.25 μg; lane 7–0.125 μg; lane 8–Lambda HindIII/phi X174 Hinc II marker.

monophosphate; a mixture of DNA oligonucleotides 12 to 18 deoxythymidine monophosphates in length and randomly bound throughout the poly A strand.

- Poly dA-dT single-stranded DNA polymer comprised of alternating deoxyadenosine and deoxythymidine monophosphates.
- Poly dA·dT double-stranded DNA polymer containing deoxyadenosine monophosphate in one strand, and deoxythymidine monophosphate in the complementary strand.
- Poly (dA-dT)·(dA-dT) double-stranded DNA polymer comprised of alternating deoxyadenosine and deoxythymidine monophosphates in each strand.

Do double-stranded polynucleotides possess blunt or sticky ends? Yes to both, as explained below.

How Are Polynucleotides Manufactured and How Might This Affect Your Research?

The length of commercially produced polynucleotides varies from lot to lot. Polynucleotides are synthesized by polymerase replication of templates or by the addition of nucleotides to the 3′ ends of oligonucleotide primers by terminal transferase or poly A polymerase. These enzymatic reactions are difficult to regulate, so polymer size significantly varies between manufacturing runs. A second factor that affects the size of double-stranded polynucleotides is that these polymers are affected by annealing conditions. Double-stranded polymers may be produced by synthesizing each strand indpendently and then annealing the two independent strands. In reality, the annealing reaction consists of annealing two populations of strands, each with its own distribution of sizes. Depending on the actual composition of these two populations and the exact annealing conditions, the resulting population of the annealed double-stranded polymer may vary widely (see the discussion about structural uncertainty below for a related case).

Manufacturers apply analytical ultracentrifugation, gel electrophoresis, or chromatography to analyze polymer length. Commercial suppliers provide an average size of the polymer population, but they usually don't indicate the proportion of the different size polymers within a preparation. For example, two lots might have an average size of 500 bp; lot 1 might have a larger proportion of 800 bp polymers and lot 2 a larger proportion of polymers 300 bp in length. Will this affect your experiments? This question can be answered conclusively only at the lab bench, so it is a good idea to consider performing control experiments when using a new lot of polymer for the first time.

Structural Uncertainty

What is the basic structure of a double-stranded polymer? Is it blunt ended? Will it have overhangs? How long are the overhangs? There is no single answer to these questions due to the heterogeneous nature of the product and the impact of the exact conditions used for dissolving the polymer. The buffer composition, temperature of dissolution, and volume of buffer used will all affect the final structure of the dissolved polymer.

Heterogeneous Nature

If you add equimolar amounts of a disperse mixture of poly dA and a disperse mixture of poly dT, what are the odds that two strands bind perfectly complementary to form a blunt-ended molecule? What's the likelihood of generating the same overhang within the entire population of double-stranded molecules? Does one strand of poly dA always bind to one strand of poly dT, or do multiple strands interact to form concatamers? See Figure 10.2 for examples. Considering the heterogeneous population of the starting material, one should assume that a highly heterogeneous population of double-stranded polymers forms.

Buffer Composition

Double-stranded polynucleotides are usually supplied as lyophilized powders that may or may not contain buffer salts. The pH, salt concentration, and temperature of the final suspension affect the structure of the dissolved polymer. For example, at any specific temperature, the strands of poly dA·dT resuspended in water dissociate much more frequently than the same polymer dissolved in 100 mM sodium chloride. Heating a polymer solution to 85°C for 10 minutes followed by quick chilling on ice produces a different population of polymers compared to poly dA·dT dissolved in the same buffer at room temperature.

Consider these solution variations when attempting to reproduce your experiments and those cited in the literature.

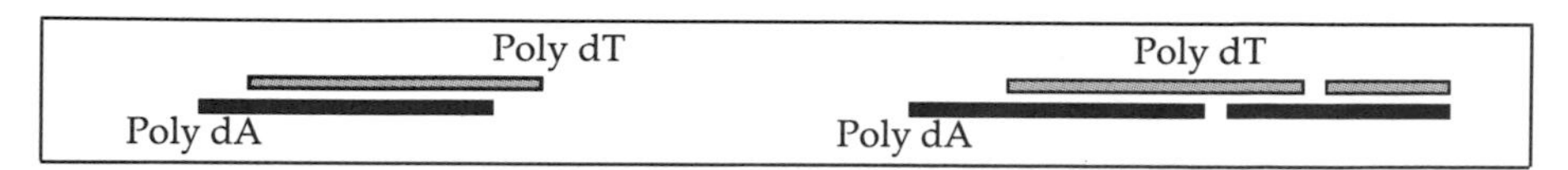

Figure 10.2 Variable products when annealing synthetic polynucleotides.

Would the World Be a Better Place If Polymer Length Never Varied?

Poly (dI-dC)·(dI-dC) is commonly applied to reduce nonspecific binding of proteins to DNA in band shift (gel retardation) experiments. The polymer's average size varies from hundreds of base pairs to several kilobase pairs. Two researchers from one laboratory used the same lot of poly (dI-dC)·(dI-dC) in experiments with different protein extracts. This one lot of poly (dI-dC)·(dI-dC) produced wonderful band shift results for the first scientist's protein extract, and miserable results for the second researcher's extract. Is this Nature's mystique or a lack of optimized band shift conditions?

Oligonucleotides Don't Suffer from Batch to Batch Size Variation. Why Not?

Oligonucleotides are almost always chemically synthesized on computer-controlled instruments, minimizing variation between batches. Different batches of the same oligonucleotide are identical in sequence and length provided that they are purified to homogeneity.

How Many Micrograms of Polynucleotide Are in Your Vial?

At least one manufacturer of polymers reports the absorbance units/mg specification for each lot of polymer. The data from three lots of poly (dI-dC)·(dI-dC) are listed below:

	Absorbance units/mg	μg/absorbance unit
Lot *A*	9.0	111
Lot *B*	13.7	73
Lot *C*	10.4	96

Why is there so much μg/unit variation among the three lots? How should you calculate the mass of material in different lots of this polymer? Should you use 50 μg/unit as you would for double-stranded DNA, or the μg/unit calculated above?

In the tradition of answering one question with another, ponder this. Why do manufacturers quantitate most of their polymer products in terms of absorbance units rather than micrograms? What are the possible explanations?

- It's easier to quantitate polymers on a spectrophotometer than to weigh them on a scale.

- DNA isn't the only material present in the polymer preparation.
- 100 units sounds more generous than 5 mg.

Despite multiple purification procedures that include extensive dialysis, other materials such as water and salts can accumulate in polynucleotide preparations. Since polynucleotides absorb light at 260 nm and the common contaminants do not, manufacturers package polymers based on absorbance units to guarantee that researchers get a consistent amount of nucleic acid.

So, if you choose to define experimental conditions using mass of polymer, use spectrophotometry and a conversion factor. Common conversion factors are 50 μg/absorbance unit (260 nm) for double-stranded DNA polynucleotides, 37 or 33 μg/absorbance unit for single-stranded DNA, and 40 μg/absorbance unit for single-stranded RNA. A conversion factor for synthetic RNA: DNA hybrids has not been defined. Some researchers apply 45 μg/absorbance unit, a compromise between the RNA (40 μg) and DNA (50 μg) values.

Be careful about weighing out an amount of polymer for use in an experiment, or quantitating polymers based on the absorbance units/mg reported within the package insert of a commercial product. Both approaches assume that the polymer is 100% pure and are likely to give higher variation in experimental conditions when changing lots of polymer from the same manufacturer or switching between manufacturers of a polymer.

Is It Possible to Determine the Molecular Weight of a Polynucleotide?

Once the average length of the polymer is known, a theoretical average molecular weight can be calculated based on the molecular weight of each strand or the molecular weight of nucleotide base pairs. Just remember that these calculations are based on the average lengths of disperse populations of polymers.

What Are the Strategies for Preparing Polymer Solutions of Known Concentration?

Suppose that your task was to prepare a 10 μM solution of poly dT. Theoretically you could prepare a solution that was 10 μM relative to the poly dT polymer (molarity calculations would be based on the average molecular weight reported on the manufacturer's certificate of analysis), or 10 μM relative to the deoxythymidine monophosphate (dT) nucleotide that comprises the polymer.

The preferred approach for preparing a polymer solution of a particular molar concentration is to express all concentrations in a concentration of bases or base pairs. The reason for this is that the best way to determine the amount of polymer present is by measuring absorbance. In addition, since the population of polymer molecules is so disperse, approximating the concentration of polymer based on strands of polymer may be misleading. Finally, this approach will maximize the reproducibility of your experiments between different lots of polymer and for those who try to reproduce your work.

10 μM of the dT Nucleotide

As described above, polymer solutions are best quantitated via a spectrophotometer. Before you go to the lab, grab some paper and perform a couple of quick calculations. First, using the molar extinction coefficient, calculate the absorbance of a 10 μM solution. The molar absorbtivity of poly dT is 8.5×10^3 L/mol-cm-base at 264 nm and pH 7.0. This means one mole of dT monomers in one liter will give an absorbance of 8500. Therefore a 10 μM solution (i.e., 0.000010 M) will have an absorbance of 0.085 (i.e., 8500×0.000010).

Next calculate the dilution required of 50 absorbance units to give the absorbance of a 10 μM solution (i.e., 0.085). If you have a vial with 50 absorbance units of polymer and you dissolved the entire 50 absorbance units in 1 ml of buffer, the spectrophotometer would hypothetically measure an absorbance close to 50. To obtain an absorbance of 0.085, the total dilution of the 50 absorbance units would be 588-fold (i.e., 50/0.085 = 588).

In the lab you would never dissolve the entire 50 absorbance units in 588 ml. First, this would limit you to using the polymer at concentrations of 10 μM or less. Second, the dilution may not work as you theoretically calculated. And finally, if the dilution did work as you expected, the solution would have an absorbance of less than 0.100 and therefore not be reliably measured by a spectrophotometer. In practice, you would prepare a stock solution of approximately 10 times the final desired concentration and then dilute to a range that can be measured by a spectrophotometer.

Your Cuvette Has a 10 mm Path Length. What Absorbance Values Would Be Observed for the Same Solution If Your Cuvette Had a 5 mm Path Length?

Half the path length, half the absorbance.

Why Not Weigh out a Portion of the Polymer Instead of Dissolving the Entire Contents of the Vial?

As discussed earlier, would you be weighing out DNA polymer or DNA polymer and salt? Also DNA polymers are very stable in solution when stored at –20°C or colder. (If you have a choice, store unopened vials of polymer at –20°C or colder; see below.) Aliquot your polymer stocks to avoid freeze–thaw nicking and contamination problems.

Is a Phosphate Group Present at the 5′ End of a Synthetic Nucleic Acid Polymer?

Synthetic DNA and RNA polymers are produced by adding nucleotides to the 3′ end of an oligonucleotide primer or by replicating a template by a nucleic acid polymerase. If the primer is phosphorylated, and if the mechanism of the DNA polymerase produces 5′ phosphorylated product, one could conclude that the polymer contains a 5′ phosphate. If your purpose is to end-label a polymer via T4 polynucleotide kinase, it's safest to assume that a phosphate is present, and either dephosphorylate the polymer or perform the kinase exchange reaction (Ausubel et al., 1995).

What Are the Options for Preparing and Storing Solutions of Nucleic Acid Polymers?

Synthetic polymers comprised of RNA and DNA are most stable (years) when stored as lyophilized powders at –20°C or –70°C. Polymer solutions are stable for several months or longer when prepared and stored as described below.

Double-Stranded Polymers

Concentrated Stock Solutions

To maintain principally the double-stranded form of synthetic DNA and DNA–RNA hybrids requires a minimum of 0.1 M NaCl, or lower concentrations of bivalent salts present in the solution (Amersham Pharmacia Biotech, unpublished observations). In the absence of salt, the two strands within a polymer can separate (breathe) throughout the length of the molecule. While its presence won't harm polymers during storage, salt could hypothetically interfere with future experiments. If this is a concern, polymers destined for use in double-stranded form can also be safely stored for months or years in neutral aqueous buffers (i.e., 50 mM Tris, 1 mM EDTA) at –20°C or –70°C, even though they will likely be in principally single-stranded form when heated to room temperature and above.

Preparing Solutions for Immediate Use

DNA alternating co-polymers such as poly (dI-dC)·poly (dI-dC) can be prepared in the salt buffers described above, heated to 60°–65°C, and slowly cooled (no ice) to room temperature to reanneal the strands. Duplexes of poly (dA)·poly (dT) require the salt buffers above, and should be heated to 40°C for 5 minutes, and slowly cooled to room temperature. Duplexes of poly (dI)·poly (dC) and RNA·DNA hybrids require salt buffers and heating to 50°C for 5 minutes, followed by slow cooling. Poly (dG)·poly (dC) can be difficult to dissolve. Even after heating to 100°C and intermittent vortexing, some polymer would not go into solution (A. Letai and J. Fresco, Princeton University, 1986, personal communication).

Single-Stranded Polymers

Single-stranded DNA and RNA polymers are stable in neutral aqueous buffers. Depurination will occur if DNA or RNA polymers are exposed to solutions at pH 4 or lower. In addition, for RNA polymers, pH of 8.5 or greater may cause cleavage of the polymer. Carefully choose your buffer strategy for RNA work, since the pH of some buffers (i.e., Tris) will increase with decreasing temperature.

If a single-stranded DNA polymer is difficult to dissolve in water or salt, heat the solution to 50°C. If heating interferes with your application, make the polymer solution alkaline, and after the polymer dissolves, carefully neutralize the solution (Amersham Pharmacia Biotech, unpublished observations).

BIBLIOGRAPHY

Amersham Pharmacia Biotech. 1993a. *Analects* 22(1):8.

Amersham Pharmacia Biotech. 1993b. *Analects* 22(3):8.

Amersham Pharmacia Biotech, 2000, Catalogue 2000. Amersham Pharmacia Biotech. 1990. Tech Digest Issue 13.

Amersham Pharmacia. 1990. *Biotech Tech Digest Issue* 10 (February); 13 (October).

Ausubel, F. M., Brent, R., Kingston, R. E., Moore, D. D., Seidman, J. G., and Struhl, K. 1995. *Current Protocols in Molecular Bology*. Wiley, New York.

Efiok, B. J. S., 1993. *Basic Calculations for Chemical and Biological Analyses*. AOAC International, Arlington, VA.

Griswold, B. L., Humoller, F. L., and McIntyre, A. R. 1951. Inorganic Phosphates and Phosphate Esters in Tissue Extracts. *Anal. Chem.* 23:192–194.

Leela, F., and Kehne, A. 1983. 2′-Desoxytubercudin-Synthese eines 2′-Desoxyadenosin-Isosteren durch Phasentransferglycosylierung. *Liebigs. Ann. Chem.*, 876–884.

Lehninger, A. L. 1975. *Biochemistry*. 2nd ed. Worth, New York.
Letai, A., and Fresco, J. 1986. Personal Communication. Princeton University.
Sambrook, J., Fritsch, E. F., and Maniatis, T. 1989. *Molecular Cloning: A Laboratory Manual*. Cold Spring Harbor, NY.

PCR

Kazuko Aoyagi

Introduction 292
Developing a PCR Strategy: The Project Stage 293
Assess Your Needs 293
Identify Any Weak Links in Your PCR Strategy 295
Manipulate the Reaction to Meet Your Needs 296
Developing a PCR Strategy: The Experimental Stage 296
What Are the Practical Criteria for Evaluating a DNA Polymerase for Use in PCR? 296
How Can Nucleotides and Primers Affect a PCR Reaction? 303
How Do the Components of a Typical PCR Reaction Buffer Affect the Reaction? 305
How Can You Minimize the Frequency of Template Contamination? 306
What Makes for Good Positive and Negative Amplification Controls? 308
What Makes for A Reliable Control for Gene Expression? 309
How Do the Different Cycling Parameters Affect a PCR Reaction? 309
Instrumentation: By What Criteria Could You Evaluate a Thermocycler? 309
How Can Sample Preparation Affect Your Results? 311

Molecular Biology Problem Solver, Edited by Alan S. Gerstein.
ISBN 0-471-37972-7

How Can You Distinguish between an Inhibitor Carried over with the Template and Modification of the DNA Template? 312
What Are the Steps to Good Primer Design? 312
Which Detection and Analysis Strategy Best Meets Your Needs? 315
Troubleshooting 315
RT-PCR 321
Summary 322
Bibliography 322
Appendix A: Preparation of Plasmid DNA for Use as PCR Controls in Multiple Experiments 327
Appendix B: Computer Software for Selecting Primers 327
Appendix C: BLAST Searches 328
Appendix D: Useful Web Sites 328

INTRODUCTION

The principle of the polymerase chain reaction (PCR) was first reported in 1971 (Kleppe et al., 1971), but it was only after the discovery of the thermostable Taq DNA polymerase (Saiki et al., 1988; Lawyer et al., 1989) that this technology became easy to use. Initially the thermal cycling was handled manually by transferring samples to be amplified from one water bath to another with the addition of fresh enzyme per cycle after the denaturation step (Saiki et al., 1986; Mullis et al., 1986). Today, 30 years later, we are fortunate to have thermal cyclers, along with enzymes and other reagents dedicated for various PCR applications. The advances in PCR technology and the number of annual publications using PCR in some area of the research has grown tremendously from a single-digit number to 1.6×10^4 in 1999 (Medline search). The popularity of the PCR method lies in its simplicity, which permits even a lay person without a molecular biology degree to run a reaction with minimum training.

However, this easy "entry" can also act as a "trap" to encounter common problems with this technology. The purpose of this chapter is to help you select and optimize the most appropriate PCR strategy, to avoid problems, and to help you think your way out of problems that do arise. While your research topic may be unique, the solutions to most PCR problems are less so. Employing one or a combination of methods mentioned in this chapter could solve problems. I encourage readers to spend time in setting up the lab, choosing the appropriate protocol, optimizing the conditions and analysis method *before* running the first PCR reaction. In the long run, you will save time and resources.

This chapter provides practical guidelines and references to in-depth information. Other useful information is added in the Appendix to help you navigate through various tools available in today's market.

DEVELOPING A PCR STRATEGY: THE PROJECT STAGE

Assess Your Needs

First ask yourself what outcome you need to achieve to feel successful with your experiment (Table 11.1). What kind of information do you need to get? Is it qualitative or quantitative? Are you setting up a routine analysis to run for the next two years, or is this for the manuscript you need to send to the editor in a hurry in order for your paper to get accepted? Your priorities will help you choose the method that best fits your needs.

Table 11.2a shows an example of a list for a researcher who needs to develop a PCR method where approximately 48 genes will be studied for relative gene expression in response to various drug treatments to be given over a three-year period. In contrast, Table 11.2b shows a list of a scientist who wishes to clone a gene with two different mRNA forms generated by alternative splicing

Table 11.1 Priority Check List

Objectives	High/Medium/Low
Quantitative	
Sensitivity	
Fidelity	
High-throughput	
Reproducibility	
Cost-sensitive	
Long PCR product	
Limited available starting material	
Short template size	
Gel based	
Simple method	
Nonradioactivity involved	
Automated	
Long-term project	
DNA PCR	
RNA PCR	
Multiple samples	
Multiplex	

Table 11.2a Priority List: Researcher 1

Objectives	High/Medium/Low
Quantitative	H
Sensitivity	H
Fidelity	M
High-throughput	M
Reproducibility	H
Cost-sensitive	M
Long PCR product	L
Limited available starting material	M
Short template size	H
Gel based	L
Simple method	H
Nonradioactivity involved	H
Automated	H
Long-term project	H
DNA PCR	L
RNA PCR	H
Multiple samples	H
Multiplex	H

Table 11.2b Priority List: Researcher 2

Objectives	High/Medium/Low
Quantitative	M
Sensitivity	M
Fidelity	H
High-throughput	L
Reproducibility	H
Cost-sensitive	H
Long PCR product	L
Limited available starting material	M
Short template size	M
Gel based	H
Simple method	L
Nonradioactivity involved	L
Automated	L
Long-term project	L
DNA PCR	L
RNA PCR	H
Multiple samples	L
Multiplex	L

mechanisms. The purpose of Researcher 2 is to amplify the cDNA and to demonstrate the size difference by separating the two forms by gel electrophoresis. The data are needed for a manuscript due in two months. You can see the differences between the priorities and needs of the two researchers.

After setting clear objectives of what your PCR reaction must accomplish, check that you have the adequate resources. This includes not only budget but also head count, skill level, time, equipment, sequence information, sample supply, and other issues. If time is most critical, then you may require a colleague's assistance or a new instrument to do the project as quickly as possible. In a similar token, if the sample is difficult to obtain in abundance, the choice of PCR that minimizes the sample requirement becomes more important.

Selecting one PCR strategy that optimally satisfies every research need is unlikely. At this early planning stage, a compromise between competing needs will likely be required. Remember that after all the planning is complete, the final PCR strategy still has to evolve at the lab bench.

Identify Any Weak Links in Your PCR Strategy

There are many parameters that affect the outcome of a PCR reaction. Some examples are as follows:

- PCR reaction chemistry (enzyme, nucleotide, sample, primer, buffer, additives).
- PCR instrument type (ramp time, well-to-well homogeneity, capacity to handle many samples).
- Thermal cycling conditions (two-step, three-step, cycle segment length—i.e., denaturation, annealing, and extension—ramp time, etc.).
- Sample collection, preparation, and storage (DNA, RNA, microdissected tissue, cells, and archived material).
- PCR primer design.
- Detection method (simultaneous detection, post PCR detection).
- Analysis method (statistical analysis).

Like the weakest link in a chain, your final result will be limited by the parameter that is least optimum. For example, suppose that you're studying the tissue-specific regulation of two mRNA forms. Regardless of the time spent optimizing the PCR reaction, instrument type, and everything to near-perfection, the use of agarose gel electrophoresis may not allow you to reach the conclusion that there are two different mRNA forms if their molecular weights are similar. You might require a separation technique with greater resolving power.

Suppose that your objective requires quantitative PCR. RNA from 30 samples is collected and RT-PCR is performed. The PCR reaction is run in duplicate and repeated twice on two different days. One-step RT-PCR is done using the same RNA samples, and PCR products are analyzed by polyacrylamide gel electrophoresis (PAGE). For some unknown reason, the second experiment shows different quantitative data. Which data are correct? Without a sufficient number of samples to calculate standard deviation, one cannot make any quantitative analysis. For quantitative PCR, the sample size has to be large enough and the standard curve must show that PCR was linear within the range one is examining. To do this, serial dilution of a positive control must be run simultaneously, and the test samples have to fall within this range of amplification. Minimums of three to four samples are required for reliable statistical analysis of the data. It is also a good idea to generate enough cDNA to run multiple experiments to reduce error due to differences in the cDNA synthesis step. The positive control must also be properly stored so that loss or damage of DNA does not generate false negative results.

High-tech, automated PCR synthesis and detection systems are useless if the sample preparation destroys the mRNA, co-purifies PCR inhibitors, or the PCR primer design amplifies genomic DNA. Your results will only be as good as the weakest parameter in your PCR strategy.

Manipulate the Reaction to Meet Your Needs

Table 11.3 describes positive and negative effectors of the PCR reaction. These data can help you plan your experiment or modify your strategy if your results aren't satisfactory.

DEVELOPING A PCR STRATEGY: THE EXPERIMENTAL STAGE

What Are the Practical Criteria for Evaluating a DNA Polymerase for Use in PCR?

An appreciation of what your research objective requires from a PCR product should be central to your selection of a thermostable DNA polymerase. Were you planning to identify a rare mutation in a heterogeneous population as in allelic polymorphisms (Frohman, Dush, and Martin, 1988)? As the copy number gets smaller (less than 10), the need for high-fidelity enzyme or enzyme mixes increases, as discussed below. In contrast, if you're screening a batch of transgenic mice for the presence or absence of a marker gene via Southern hybridization, enzyme fidelity might not be as crucial. Most applications do not require high

Table 11.3 Positive and Negative Effectors of a PCR Reaction

To Enhance This Parameter	Manipulate One or More of These Components	
Fidelity and specificity	Enzyme	Select an enzyme with potent 3′–5′ Exonuclase activity.
	Primer design	Include mismatches at 3′ end, which can help discriminate against homologous sequences such as pseudogenes. Enzyme selection can enhance this effect. With Taq polymerase, relative amplification efficiencies with 3′-terminal mismatches is greater for A:G and C:C than for other nucleotide pairs (Kwok et al., 1995). Use longer primers (refer to section "What Are the Steps to Good Primer Design?". Primers less than 15 nucleotides do not give enough specificity from a statistical point of view.
	PCR cycling condition	Increase annealing temperature. Reduce cycle segment time (denaturation, annealing, etc.). Lower cycling number.
	Reaction chemistry	Decrease $[Mg^{2+}]$. Apply a hot start strategy (Erlich, Gelfand, and Sninsky, 1991). Check that concentration and pH of dNTP solution(s) is correct. Decrease primer concentration.
	Template	Confrim that template is intact, not nicked, and free of contaminants and inhibitors. Confirm the presence of sufficient starting copy number.
	Method of analysis	Minimize contamination and handling errors; use an automated analysis system. Use sufficient sample number to enable reliable statistical analysis. Check for erroneous manipulation (pipetting errors, etc.).
	Clean lab practice	Use a positive displacement pipette. Use a separate room to set up experiments. Wear gloves. Use UNG and dUTP (Longo, Berninger, and Hartley, 1990).
	Cycler	Check that the temperature profile is consistent at every position in the heating block. Decrease ramp time. Check for tight fit between reaction vessels and heating block.
Efficiency of doubling/cycle	Reaction	Increase concentration of dNTPs and enzymes. Use minimal concentrations of DMSO, DMF, formamide, SDS, gelatin, glycerol (see Table 11.7).

Table 11.3 (*Continued*)

To Enhance This Parameter	Manipulate One or More of These Components	
	Template	Confirm that template is unnicked, free of contaminants and inhibitors. Use a smaller size template DNA (get more molecules per pg of input template, and less complexity for primer annealing). For example, PCR product vs. genomic DNA. Decrease amplicon size.
	Enzymes	Taq > Pfu, >>Stoffel fragment.
	Cycling	Decrease cycling time or use a shuttle profile (Cha et al., 1992). Decrease the size of the reaction tube. Check for tight fit between reaction vessels and heating block.
	Cycler	Decrease ramp time.
	Primer design	Use forward and reverse primers that have similar length and GC content. Confirm that primers do not form primer-dimer or hairpin structure.
Reproducibility	Sample	Ensure that template is clean and intact. Confirm presence of sufficient starting template and sufficient sample number for statistical analysis.
	Reagents	Use the same lots of primer and buffers between experiments. Store enzyme in small aliquots. Investigate for presence of contaminating template and inhibitors to PCR reaction.
	Controls	Include positive and negative controls with all experiments.
	Cycling	Use a hot-start strategy (Kellogg et al., 1994). Use the same cycler between experiments.
Quantitative	Template	Confirm the quantity of the template. Confirm template preparation is clean. Investigate for presence of contaminating template and inhibitors to PCR reaction.
	Experimental design	Include triplicate or quadruplicate samples. Use a statistically sufficient number of samples. Prepare a standard curve to demonstrate the range over which PCR product yield provides a reliable measure of the template input. Robust: Confirm that chemistry, primer design, tubes, thermal cycler, and other factors are optimized.
	Analysis	Confirm the analytical method's accuracy/resoluton. Is it accurate during the exponential phase of PCR?

Table 11.3 (*Continued*)

To Enhance This Parameter	Manipulate One or More of These Components	
		Use appropriate controls. Repeat experiments when data are outside of standard deviation limits. Minimize the manipulations from start to finish.
	Cycler	Check that the temperature profile is consistent at every position in the heating block.
	Control	Confirm that controls have similar sequence profile and amplification efficiency. Confirm that PCR was linear by producing a standard curve.
	Analysis	Use an automated system to reduce handling steps.
	Detection	Check the detection strategy's senitivity and ability to measure yield in the exponential phase of PCR. Confirm that the technique has high sensitivity and magnitude over a wide dynamic range.
High-throughput	Instrument	Select a system that handles microtiter plates and multiple sample simultaneously.
	Reaction	Use a hot-start PCR strategy (D'Aquilla et al., 1991; Chous et al., 1992; Kellogg et al., 1994). Use a master PCR reagent mix. Use aliquots taken from the same lot of material; don't mix aliquots from different lots.
	Sample preparation	Use of robotics. Storage of sample as cDNA or ethanol precipitate, rather than RNA in solution.
	Cycling	Use one cycling strategy for all samples. Decrease the cycling time.
	Analysis	Use an automated system.
	Detection	Use an automated detection system to monitor the exponential phase.
Sensitivity	Detection	Monitor specific PCR product formation by hybridization via nucleic acid probe. Use fluorescent intercalating dye (Wittwer et al., 1997).
	Reaction	Use a nested PCR strategy (Simmonds et al., 1990). Note: Sensitivity is gained at the expense of quantitation. Use a hot-start PCR strategy. Use UNG and dUTP to prevent carryover.
	Analysis	Use a real time PCR strategy that detects low levels of amplicon missed by gel

Table 11.3 (*Continued*)

To Enhance This Parameter		Manipulate One or More of These Components
		electrophoresis. When hybridization probes are used, primer-dimer formation will not mask the authentic product, even after 40 cycles. This is not true for SYBR® Green or Amplifluor.
		Nested PCR or extra manipulation may be needed for other non-real-time PCR based techniques. "Hot" nested PCR is one such example that elegantly combines the qualities of nested PCR with the high resolution of PAGE (Jackson, Hayden, and Quirke, 1991).
	Control	Include positive and negative controls; when the target is not detected, one can conclude that target was below 100 copies, etc., which makes the data more meaningful than just saying it was not detected.
	Lab setup	Clean lab.
		No contamination.
	Experimental design	Check primer design. If amplifying related genes is a concern, design the primer to create mismatches at the 3′ end using the most heterogeneous sequence region.

fidelity, but one needs to be aware when high fidelity has to be considered. During planning, one should also consider the many ways a PCR reaction can be manipulated to achieve a given end, as discussed throughout this chapter.

The data in Table 11.4 are provided to highlight the biochemical properties of common PCR-related enzymes and help you develop a selection strategy. For a comprehensive comparison of thermostable DNA polymerases, see Perler, Kumar, and Kong (1996), Innis et al. (1999), and Hogrefe (2000). However, biochemical data and logic can't always predict the most appropriate enzyme for PCR; experimentation might still be required to determine which enzyme works best. Abu Al-Soud and Radstrom (1998) demonstrate that contaminants inhibitory to PCR vary with the sample source, and that experimentation is required to determine which thermostable DNA polymerase will produce successful PCR. A second illustration of the difficulty in predicting success based on enzymatic properties are the Archae DNA polymerases, which have not become premiere PCR enzymes despite their extreme thermostability and good proofreading activity.

Table 11.4 Selected Properties of Common Thermostable DNA Polymerases

Enzyme	Proofreading (3′–5′ exonuclease)	5′–3′ Exonuclease	Heat Stability (min before 50% activity remains)	Processivity (dNTP/ binding)	Extension Rate (dNTP/ s/mol)
Taq DNA polymerase	Absent	Present	9 at 97.5°C (40–60 at 95°C. depending on protein concentration[a,b]	50–60	60–150
Stoffel fragment	Absent	Absent	21 at 97.5°C	5–10	130
Tth DNA polymerase	Absent	Present			25
rTth XL	Trace	Present		30–40	
AmpliTaq CS	Absent	Absent		50–60	
UlTma DNA polymerase	Present (low)	Absent	50 at 97.5°C		
Pfu DNA polymerase (native and recombinant)	Present	Absent	1140 at 95°C[a]	10[a]	60
Pfu DNA polymerase (exo-form)	Absent	Absent	1140 at 95°C[a]	11[a]	
(*Pyrococcus species* GB-D)	Present	Absent	1380 at 95°C[c]		>80
(aka Deep Vent®)			480 at 100°C		
Tli Pol (aka Vent®)	Present	Absent	402 at 95°C[c] 108 at 100°C	7	67
Herculase enhanced DNA polymerase	Present	Present[a]			
Tbr DNA polymerase (Dynazyme™)[e]	Absent	Present	150 at 96°C		
Platinum *Pfx*[f]	Present	Absent	720 at 95°C 180 at 100°C	100–200	100–300
Platinum *Taq*[f]	Absent	Present	96 at 95°C	50–60	60–150
Advantaq Polymerase[g]	Absent	Absent	40 at 95°C		40
Tac Pol	Present	Absent	30 at 75°C		
Mth Pol	Present	Absent[d]	12 at 75°C		
ThermalAce™ *Pyolobus fumarius*[h]	Present	Absent		5-fold greater than *Taq* DNA Polymerase	
Hot *Tub (T. flaius)*[i]	Present	Absent	Similar to Taq		

Source: Unless otherwise noted, all data from Perler, Kumar, and Kong (1996).
[a] Data provided by H. Hogrefe, Stratagene, Inc.
[b] New England Biolabs Catalog, 2000.
[c] Z. Kelman (JBC 274:28751); present according to Perler.
[d] Data provided by D. Titus, MJ Resesarch, Inc.
[e] Data provide by D. Hoekzema, Life Technologies Inc.
[f] Data provided by J. Ambroziak, Clonetech Laboratories Inc.
[g] Data provided by Invitrogen, Inc.
[h] Lawyer et al. (1993). PCR Methods & application pp. 275–286.
[i] Data provided by Amersham Pharmacia Biotech, Inc.

Fidelity

Fidelity could be defined as an enzyme's ability to insert the proper nucleotide and eliminate those entered in error. As thoroughly reviewed by Kunkel (1992), fidelity is not a simple matter; there are several steps during the polymerization of DNA where mistakes can be made and corrected. Still most practical discussions of fidelity focus on the proofreading function provided by an enzyme's 3′–5′ exonuclease activity. Cline, Braman, and Hogrefe (1996) compared the fidelity of several thermostable DNA polymerases side by side, taking care to optimize the conditions for each enzyme. They observed the following fidelity rates (mutation frequency/bp/duplication), in order: *Pfu* (1.3×10^{-6}) > *Deep Vent* (2.7×10^{-6}) > *Vent* (2.8×10^{-6}) > *Taq* (8.0×10^{-6}) > exo^{-} *Pfu* and *UlTma* ($\sim 5 \times 10^{-5}$). These and similar data should be viewed in relative rather than absolute terms, because assay methods affect the absolute number of detected misincorporations (André et al., 1997), and sample source can affect the performance of enzymes differentially and unpredictably (Abu Al-Soud and Radstrom, 1998).

Proofreading activity can also reduce PCR yield, especially in reactions that generate long PCR products. The greater time required to extend the fragment increases the chance of primer degradation by the 3′–5′ exonuclease activity (de Noronha and Mullins, 1992 and Skerra, 1992). The problem of reduced yield can be corrected by including an enzyme with strong proofreading activity into a PCR reaction with a polymerase that lacks a strong proofreading activity (Barnes, 1994; Cline, Braman, and Hogrefe, 1996; MJ Research Inc. Application Bulletin, 2000).

Heat Stability

Is a higher reaction temperature always helpful and necessary? No. For most DNA-based PCR, the consensus is that hot-start PCR increases both sensitivity and yield by preventing nonspecific PCR product formation (Faloona et al., 1990). Higher temperatures can melt secondary structures, but there are limitations to the use of heat. Very high denaturation temperatures can also damage DNA, through depurination and subsequent fragmentation, especially during long PCR reactions (Cheng et al., 1994). It can also increase hydrolysis of RNA in one step RT-PCR in the presence of magnesium ions (Brown, 1974). In order to reduce heat-induced damage, incorporation of additives such as DMSO is used (see later section on additives).

Choosing an enzyme with specialized activities will not produce the desired results unless the appropriate conditions are applied. For example, UlTma™ DNA polymerase has a pH optimum for polymerase activity of 8.3 and exonuclease activity at pH 9.3 (Bost et al., 1994). Likewise, presence of metal ions can favor one activity over the other for many polymerases.

Long PCR

Additives such as single-stranded binding protein (Rapley, 1994), T4 gene 32 protein (Schwarz, Hansen-Hagge, and Bartram, 1990), and proprietary commercial products may increase the production efficiency of long PCR. However, fidelity was also shown to be crucial to the replication of large products via PCR (Barnes, 1994). By supplementing PCR reactions containing *Taq* DNA polymerase (which lacks proofreading activity) with proofreading-rich *Pfu* DNA polymerase, Barnes generated fragments up to 35 kb. Bear in mind that proofreading activity can potentially reduce yield, especially with large PCR products. As discussed above, this problem can be avoided by utilizing a combination of polymerases that possess and lack strong proofreading activity.

The availability of specialized, designer enzymes are an attractive strategy that shouldn't be ignored. However, selecting the right enzyme(s) is one step among many, and can't guarantee the desired result. One near-term example is the importance of enzyme concentration. The concentration of polymerase applied to a PCR reaction ranges from one to four units per 100 μL. Greater concentrations can increase formation of nonspecific PCR products. The importance of optimizing other parameters, such as buffer component, primer design, and cycling conditions is shown in Table 11.3 and Table 11.5.

How Can Nucleotides and Primers Affect a PCR Reaction?

Nucleotide Concentration

The standard concentration of each nucleotide in the final reaction is approximately 200 μM, which is sufficient to synthesize 12.5 μg of DNA when half of the dNTPs are incorporated. Adding more nucleotide is unnecessary and detrimental. Too much nucleotide reduces specificity by increasing the error rate of the polymerase and also chelates magnesium, changing the effective optimal magnesium concentration (Gelfand, 1989; Coen, 1995).

Primer Concentration

The standard primer concentration is 100 to 900 nM; too much primer can increase the formation of nonspecific products. It is

Table 11.5 Optimizing $MgCl_2$ Concentration for PCR

Component	Final Concentration	Per Reaction	1 mM	2 mM	3 mM	4 mM	6 mM	8 mM	10 mM
10× PCR buffer	1×	10 μl	40.0	40.0	40.0	40.0	40.0	40.0	40.0
50 μM forward primer	0.5 μM	1 μl	4.0	4.0	4.0	4.0	4.0	4.0	4.0
50 μM reverse primer	0.5 μM	1 μl	4.0	4.0	4.0	4.0	4.0	4.0	4.0
Template DNA	Optimum	10 μl	40.0	40.0	40.0	40.0	40.0	40.0	40.0
100 mM $MgCl_2$	Various	Various	4.0	8.0	12.0	16.0	24.0	32.0	40
25 mM dNTP mix	0.2 mM	0.8 μl	3.2	3.2	3.2	3.2	3.2	3.2	3.2
Taq polymerase 5 U/μl	2.5 U	0.5 μl	2.0	2.0	2.0	2.0	2.0	2.0	2.0
H_2O		To 100 ul	To 400 μl	To 400 μl	To 400 μl	To 400 μl	To 400 μl	To 400 μl	To 400 μl

especially important to adjust the primer concentration when the target sequence is rare or the template amount is low. Less primer is needed in these cases; too much primer will generate primer-dimers or smearing of the product visualized by agarose gel electrophoresis. For most applications it is practical to apply the standard concentrations cited above and to focus effort on optimizing other critical parameters. For real-time PCR multiplex applications, it is recommended that a primer matrix study be performed (Table 11.6a,b) to ensure the limiting primer concentration for an endogenous control. This way the target gene amplification is not compromised by competition for reagents in the same reaction tube (well). This recommendation applies to all housekeeping genes regardless of the abundance level (i.e., needed not only for rRNA but also for less abundant genes, e.g., glyceraldehyde 3-phosphate dehydrogenase, cyclophilin, and hypoxanthine-guanine phosphoribosyl-transferase).

The range of final concentration for forward and reverse primers is 100 to 900 nM in the matrix below. Perform an initial series of experiments to find the rough range of an optimum primer concentration. Follow with a second series of experiments to fine-tune the primer concentration range. In the following example, the final results suggest a forward primer concentration

Table 11.6a Primer Matrix Study

	Primer concentration (nM)	Forward Primer 100	300	600	900
Reverse Primer	100	++	+	–	–––
	300	+	+	–	–––
	600	–	––	––	–––
	900	––	––	–––	–––

Table 11.6b Primer Matrix Study: Final Primer Optimization Matrix

	Primer concentration (nM)	Forward Primer 100	120	140	160	180	200
Reverse Primer	100	+	+	+	++	++	++
	120	+	+	+	++	++	++
	140	+	+	++	++	++	+++
	160	+	+	+	+	+	++
	180	–	–	–	–	–	–
	200	–––	––	–	–	–	–

of 200 nM and reverse primer at 140 nM. Both the specificity and the yield can be scored for excellent (+++), good (++), fair (+), and similarly for poor (–), very bad (–––) based on no signal, smear, and low yield.

Nucleotide Quality

The benefits of using extremely pure solution nucleotides as compared to standard lyophilized nucleotides include proper pH and absence of nuclease. A nucleotide solution at too low or high a pH can shift the overall pH of the reaction buffer and decrease yield, as can unequal quantities of the four nucleotides. The proper quantitation and pH adjustment of nucleotide solutions is discussed in Chapter 10, "Nucleotides, Oligonucleotides, and Polynucleotides."

How do the Components of a Typical PCR Reaction Buffer Affect the Reaction?

The buffer impacts the amplification by maintaining pH range, minimizing effect of inhibitors, protecting enzymes from premature loss of activity, stabilizing template, and more. Because polymerases have a narrow optimum pH range, a slight shift of pH, as little as 0.5 to 1 can reduce the yield of the PCR products. Because

Tris buffer changes its pH with temperature, it is not an ideal buffer for Taq polymerase. Table 11.7 summarizes the effects of several common additives on Taq polymerase. Their impact and optimum concentrations might differ for other enzymes, but the data regarding Taq polymerase is a starting point. Consult the manufacturers of other enzymes for more details.

Magnesium

The concentration of $MgCl_2$ affects enzyme specificity and reaction yield. In general, lower concentrations of Mg^{2+} leads to specific amplification and the higher concentration encourages nonspecific amplification. The effective concentration of Mg^{2+} is dependent on the dNTP concentration as well as the template DNA concentration and primer concentration. The strategy illustrated in Table 11.5 can be used to optimize Mg^{2+} concentration as well as other additives described below.

Additives and Contaminants

Detergent, gelatin, and other components are often included to reduce the negative effect of contaminants (Gelfand, 1992) (Table 11.7). Tween eliminates the effects of SDS, which can be carried over from sample preparation. Detergent can also stabilize the activity of some enzymes, such as Taq polymerase. When the amount of template is very small, nuclease can degrade the precious DNA, but the presence of "carrier" DNA can prevent this. Gelatin helps prevent the template DNA from getting adsorbed to the surface of the reaction tube and also stabilizes polymerase activity.

The mechanisms behind the effects of some additives and contaminants are unclear. Less than 1% DMSO may affect the T_m of primers, the thermal activity of Taq polymerase and/or the degree of product strand separation. Higher DMSO concentration (10–20%) inhibits Taq polymerase activity from 50% to 90%. Ethanol does not affect activity up to concentrations of 10%.

How Can You Minimize the Frequency of Template Contamination?

Since the power of amplification is so great, the fear of getting a false positive is common (Dieffenbach and Dveksler, 1995). Here is a list of general PCR practices to minimize cross-contamination.

- Wear a clean lab coat and gloves when preparing samples for PCR.
- Have separate areas for sample preparation, PCR reaction setup, PCR amplification, and analysis of PCR products.

Table 11.7 Effects of Additives on *Taq* DNA Polymerase

Chemicals	Mode of Action	Amount for Enhancement or Inhibition
Ethanol		Slight enhancement at 10%
Urea	Lower target T_m for annealing.	Slight enhancement at 1–1.5 M, but inhibition at greater than 2 M
DMSO	Lower target T_m for annealing.	Enhancement at 1–10% (v/v) (*www.alkami.com*) 12–15% (v/v) (Baskaran et al., 1996)[a]
DMF	Lower target T_m for annealing.	Inhibition at 10% or greater
Formamide	Lower target T_m for annealing. Increase specificity and yield by changing T_m of primer-template hybridization and lower heat destruction of enzyme.	Enhancement at 1.25–10% (v/v); Inhibition at 15% or greater
SDS	Prevent aggregation of enzyme.	Inhibition at 0.01% or greater
Glycerol	Enhance specificity by changing T_m. Extends Taq polymerase resistance to heat damage.	Enhancement at 5–20% (v/v) (*www.alkami.com*)
Perfect match polymerase enhancer (PMPE) (Stratagene Inc.)		Approximately 1%
Ficoll 400 (Wittwer and Garling, 1991)		The optimal amount must be empirically determined.
Gelatin		100 μg/ml or 0.01–0.1% (w/v).
Tween 20/NP40		0.1–0.05% (v/v) Tween 20 0.05% (v/v) NP40
T4 Gene 32 protein (Schwarz et al., 1990)	Increase specificity and yield by changing T_m of primer-template hybridization.	0.05–0.1 nmole/amplification reaction (note: original publication incorrectly states 0.5 –1.0 nmole)
Triton X-100	Prevents enzyme from aggregating.	0.01 % (v/v)
Bovine Serum Albumin (BSA)	Neutralizes many factors found in tissue samples which can inhibit PCR.	10–100 μg/ml
Betaine		0.5–2.0 M (Roche Molecular Biochemicals Web site) (1.8–2.5 M) (Baskaran et al., 1996)[a]
Tetramethyl ammonium chloride (TMAC)		10–100 μm
PEG 6000		5–15% (w/v)
Spermidine	Reduces nonspecific reaction between polymerase and template DNA.	

Other references: For Taq DNA polymerase, Gelfand (1992, pp. 6–16); for the polymerase chain reaction, Coen (1995).

[a] Baskaran et al. (1996) claims that combination of DMSO (5–10%) and betaine (1.1–1.4 M) produces best results.

- Open PCR tube containing amplification products carefully, preferably in a room other than where the PCR reactions take place. Spin tubes briefly before opening a lid.
- Use screw cap microfuge tubes for templates and positive controls to control microaerosolization when opening tubes.
- Use a positive-displacement pipette or aerosol-resistant pipette tips.
- Discard pipette tips in a sealed container to prevent airborne contamination.
- Periodically clean lab benches and equipment with 10% bleach solution.
- Prevent contamination by using uracil-N-glycosylase (UNG) which acts on single- and double-stranded dU-containing DNA and destroys the PCR products (Longo, Berninger, and Hartley, 1990).
- Aliquot reagents, sterile water, primers, and other material into tubes to reduce the risk of contamination.
- When possible, avoid using plasmid DNA as a control. The DNA can contaminate the lab like a virus if not handled carefully. A safer control is a sample containing the target at high or low levels. Another method involves a synthetic oligonucleotide template that contains the sequence complimentary to primer binding region plus part of the sequence being amplified by the forward and reverse primers designed just for the initial testing of primers. They have major internal sequence deletions; thus they only serve to validate the primers. They are not amplified simultaneously with the test samples. If you must use plasmid DNA as a control, refer to the Appendix A for preparation of a plasmid DNA control solution that can be stored over a long period of time.

What Makes for Good Positive and Negative Amplification Controls?

The inclusion of reliable positive and negative controls in all your experiments will save time and eliminate headaches. Examples follow:

- Positive controls: Samples containing the target sequence at high copy number.
- Negative controls: One primer only, no Mg^{2+}, no enzyme, sample known to lack the target sequence, no RT step for RT-PCR.

Unfortunately, the above controls can also fail. Most often the failure originates in the preparation of the positive and negative controls. Plasmid DNA is unstable at low concentrations during storage, especially in plain water or TE (10 Tris, 1 mM EDTA, pH 7). At dilute concentration, DNA can be lost by adsorption to the inner wall of a tube or be degraded by nuclease activity. A good way to store plasmid DNA (or control cDNA or genomic DNA) is in TE with 20 μg/ml glycogen (molecular biology grade, nuclease free) in small aliquots in a –20°C freezer. Repeated freeze–thawing of control DNA should be avoided. The water used for any aspect of a PCR reaction should also be nuclease free, and stored in small volumes. Don't use a bottle of water that's been sitting in the lab for months. Microorganisms are too easily introduced.

What Makes for a Reliable Control for Gene Expression?

Good endogenous controls are constituitively expressed and change minimally while the target gene expression may vary greatly. Poor controls change their expression levels during the treatment, thus masking the target gene expression fluctuation. Bonini and Hofmann (1991) and Spanakis (1993) provide examples where inappropriate controls prevented the detection of biologically significant changes in gene expression. Some popular endogenous controls such as β-actin and glyceraldehyde dehydrogenase (GAPDH) are well known for having pseudogenes, and related genes, adding complexity to interpretation of results (Multimer et al., 1998; Raff et al., 1997). rRNA (28S, 18S, 5.8S, etc.) seems to be more constant in its level than other mRNA type housekeeping genes such as β-actin. Without a housekeeping gene that stays relatively constant (nothing really stays absolutely constant), a subtle change in gene expression will go undetected in the noise, and incorrect conclusions will result. The true level of a control should be monitored rather than taken for granted.

How Do the Different Cycling Parameters Affect a PCR Reaction?

The objective of the information in Table 11.8 is to provide guidelines to help you fine-tune a reaction based on your experimental observations. The data refer to Taq polymerase, but the trends hold true for most thermostable DNA polymerases.

Instrumentation: By What Criteria Could You Evaluate a Thermocycler?

Since the discovery of thermostable Taq DNA polymerase, numerous instrument companies have developed PCR cyclers, not

Table 11.8 Effect of Cycling Parameters on PCR

Stages of PCR	Standard Segment Time and Temperature	Below Optimum Duration	Above Optimum Duration
Initial denaturation	1–3 min 94°C (95°C for higher (55–60%) GC content)	Lower yield or no products Some genomic DNA needs more time, while PCR products or plasmid DNA need less time	Lower yield from premature loss of enzyme activity
Denaturation during cycling	5–20 s	Lower yield	Lower yield
Primer annealing	5–20 s 45–60°C Higher temperatures for more specific annealing	Lower yield	Nonspecific product formation
Primer extension	10–20 s 70–75°C	Lower yield	Lower yield Increased error rate
Cycle number	25–40	Lower yield	Nonspecific product formation
Final extension	1–2 min 70–75°C	Incomplete double-stranded DNA	Nonspecific product formation

only for amplification but for detection and analysis as well. A review of your current and anticipated needs will help you select the most appropriate machine within your budget.

Temperature Regulation

Consistent, predictable ramp times (the time required to transition from one temperature to the next) are crucial to achieve the desired PCR results. The time required to reach the 55°C annealing temperature from the 94°C denaturation temperature can vary one minute or more, depending on the cycler design. The consistency of the heating or cooling profile of samples can also vary with the instrument and introduce errors. If your goal is to run both tubes and plates, make sure that the tube fits the well snuggly, as ill-fit tubes do not transfer heat well.

Programming Capability

If you run different cycling parameters, the capacity to link preexisting programs rather than repeatedly installing old programs will save significant time. The ability to store many programs is also useful if you run many programs routinely or share a cycler with multiple users.

Minimum Manipulations

If your objective requires high-throughput analysis, it is recommended to use a cycler that combines amplification and analysis without further manipulation, such as gel electrophoresis or blotting. These postamplification processes require pipetting, opening and closing of reaction tubes, and so forth, which greatly increase the chance of contamination of other samples throughout the lab as the product contains enormous copies of the target sequence.

Reaction Vessels

Will your planned and unforeseen research require reactions in 0.2 ml, 0.5 ml tubes, or multiwell dishes? The ability to accommodate multiple sample formats usually pays off in the long run.

How Can Sample Preparation Affect Your Results?

Sample preparation can make the difference between good yield and no amplification. The purpose of sample preparation is to eliminate PCR inhibitors as well as to provide the DNA sequence available for PCR reaction. Compounds that inhibit PCR may co-purify with the DNA template and make PCR impossible (Reiss et al., 1995; Yedidag et al., 1996). Inhibitors do not have to be diffusible. Sometimes crosslinking of protein to DNA via carbohydrate groups can cause inhibition (Poinar et al., 1998). Addition of adjuncts such as bovine serum albumin (BSA) or T4 gene 32 protein can sometimes reverse the inhibition (Kreader, 1996). However, it is easiest to remove these inhibitors during the sample preparation than to figure how to reduce the degree of inhibition later. The qualities of good sample preparation follow:

- *Intact*: Undegraded and unnicked. DNA might appear intact immediately after isolation, but repeated use can result in nuclease-mediated degradation. This may result from incomplete removal of nucleases during the initial sample preparation or contamination of the sample during repeated usage; RNA requires a storage pH below 8.0 and special care to avoid RNase contamination.
- *Fixed*: DNA isolated from paraffin-embedded tissue sections and archived fixed tissues may pose problems due to nicking of DNA during tissue preparation. (Note: Human genome haploid equivalent is approximately 3 billion base pairs. Given that the distance between base pair is about 3.4A°, each human cell contains about 2 meters of DNA! A typical DNA isolation method shears genomic DNA in the process.)

- *Inhibitor-free*: Heparin, porpholin, SDS (<0.01%), sarkosyl, heme (Alkane et al., 1994), EDTA, sodium citrate, humic acid (Zhou et al., 1996), phenol, chloroform, xylene cyanol (Alkami PCR manual), and some heavy metals can inhibit PCR.
- *Clean*: $A_{260:280}$ ratio of 1.8 to 2.0; Free of protein and carbohydrate. (See Chapter 4, "How To Properly Use and Maintain Laboratory Equipment," for situations where $A_{260:280}$ ratios prove unreliable.)
- *RNA*: Free of DNA.

How Can You Distinguish between an Inhibitor Carried over with the Template and Modification of the DNA Template?

If it is diffusable inhibition of a thermostable DNA polymerase, adding the sample in smaller quantity lessens the effect whereas the effect worsens with more sample. If the problem is caused by template modification, dilution will have no effect. Compounds such as *N*-phenacylthiazolium bromide (PTB) may eliminate inhibition (Poinar et al., 1998) caused by agents crosslinking to the template. PCR inhibitors can be detected by performing reactions in the presence of commercially available exogenous internal positive controls, which can be added to your PCR reaction without hampering the amplification of your target.

What Are the Steps to Good Primer Design?

Step 1. Consider the Objectives

What must the PCR accomplish? What pressures does this put on the primers?

- Must you identify few or many targets? The identification of several targets requires numerous primers, increasing the difficulty of avoiding 3′ overlaps.
- Must you clone the full-length coding region of a gene? For long PCR, you may use the nearest-neighbor algorithm for selection of T_m (Rychlik et al., 1990).
- Must you generate quantitative data? PCR efficiency becomes more critical, as does avoiding primer-dimers.
- Must you design primers without knowing the exact sequence of the specific species based on information from another species (i.e., design primers for the rat gene X using mouse or human gene sequence for gene X)? If so, aligning as many sequences of gene X from as many organisms as you can collect in order to select the most conserved region for primer design increases the likelihood of success.

- Must you avoid amplifying pseudogenes? What is known about pseudogenes to your target? A preliminary review of the research literature can save you time and headaches. Unfortunately, there are more pseudogenes than are reported. One quick way to search for pseudogene amplification with your selected primer pairs is to do a BLAST search (see Appendix C). However, the only sure way to avoid pseudogenes is to design primers across exon–exon junctions and test for them at the bench by amplifying genomic DNA. Processed pseudogenes do not have introns, so they can be amplified when the PCR primer extend over the two exon junctions.
- Are you searching for a single nucleotide polymorphism (SNP)? SNP primer design requires specialized strategies (Kwok et al., 1995; Wu et al., 1991).
- Must you design a small amplicon to increase detection of the gene in samples where the chance of amplifying a long sequence is unlikely (i.e., paraffin embedded sections, forensic samples, and partially degraded samples)?

<u>Step 2.</u> Apply the Sequence Analysis Programs to Develop Candidate Primers

These programs are described in Appendix B.

<u>Step 3.</u> Apply Good Primer Design

Refer to the generally accepted elements of good primer design (Dieffenbach and Dveksler, 1995). The new nearest-neighbor model based on DNA thermodynamics data for PCR primer design is also recommended (SantaLucia, 1998).

- The optimum length of primers for use with Taq DNA polymerase is between 18 and 28 bases for specificity (This number may vary with enzymes with greater heat stability.) The longer primer gives more specificity but tends to anneal with lower efficiency and results in a significant decrease in yield. A good pair of primers has melting temperature (T_m) 55°C to 60°C. Shorter primers (less than 15 nucleotide long) anneal very efficiently, but they may not give sufficient specificity. Longer primers may be useful when distinguishing multiple gene forms sharing a high degree of sequence homology. The probability of finding a match using a set of 20 nucleotide long primers is $(\frac{1}{4})^{(20+20)} = 9 \times 10^{-26}$ (Cha and Thilly, 1995). It is likely that this set of primers will amplify another gene in the mammalian genome (3×10^9 bp per haploid genome).

- GC content should be between 40% and 60%. The T_m of both primers should be similar to each other and similar to the primer-binding sites at the ends of the fragment to be amplified to achieve an optimal annealing temperature and amplification.
- 3′-end complementarity between primers and self-complementarity within primers must be avoided because it may increase primer-dimer formation and reduce PCR efficiency. This is more problematic when you have a low number of target gene copies.
- Avoid runs of G/C, especially guanidine.
- When performing RT-PCR, design primers to go across exon–exon junctions to avoid amplifying genomic DNA. Since the use of DNase has a negative effect on RNA, it is better to avoid genomic DNA amplification by primer design (Huang, Fasco, and Kaminsky, 1996).
- Include controls lacking RT unless you have shown that this set of primers does not amplify genomic DNA.
- After designing the primers, search for specificity using BLAST (Basic Local Alignment Search Tool), a set of similarity search programs designed to explore all of the available sequence databases regardless of whether the query is protein or DNA (Appendix C). This is especially important for those genes with many pseudogenes and related genes. If you are executing RT-PCR, this is essential. Even a trace amount of genomic DNA left in the RNA sample preparation can give sufficient amplification of those genes. Often these PCR products are indistinguishable by gel electrophoresis and make data interpretation difficult.
- You may add exogenous sequence to the 5′ end of primers for cloning and other purposes.
- When sequence information is ambiguous, substitute deoxyinosine for the unknown nucleotide, and place the ambiguous sequence on the 5′ end. Design and test different primers to determine which works best. Inosine is naturally found in some tRNA. It base pairs with A, C, and U in the translation process (Martin et al., 1985; Kwok et al., 1995).
- Before testing the primers with your test sample, measure the quantity of your primers, and then test with the positive controls. You cannot assume that all primers have the correct sequence. Inefficient desalting, incorrect labeling, and other quality control problems can ruin a primer's performance.

Step 4. Develop and Apply a Primer Testing Strategy

If your goal is to study many genes, then you may want to consider setting a standard thermal cycling condition to run all your PCR reactions, even though they won't produce the optimal PCR results. If your goal is to study a few genes, then the design is more straightforward.

This discussion about primer design is relevant to basic PCR. The bibliography provides references for primer design relevant to multiplex or nested PCR applications.

Which Detection and Analysis Strategy Best Meets Your Needs?

It is crucial that your strategy be consistent with your purpose. If you require quantitative data, a hybridization-based strategy is not ideal. Probe labeling, membrane transfer, and hybridization conditions can introduce variability. If resolution is crucial to your study, PAGE rather than agarose electrophoresis might be required. Because the potential for variability usually increases with the number of manipulations required to generate the data, real-time PCR usually provides greater reproducibility along with time savings.

Table 11.9 compares commonly applied detection methods.

TROUBLESHOOTING

Even the most thorough, insightful planning cannot guarantee success, and PCR can generate results indicative of complete failure or a reaction in need of optimization. The troubleshooting section is organized to reflect the fact that any given PCR problem can have several underlying explanations. The optimization of cycling conditions, primer concentration, and other parameters discussed throughout the chapter can also help resolve a problem.

No Product

Template

Is the target sequence absent?

Amplify housekeeping gene or some gene you know is present as a control; perform standard curve assay with plasmid or amplicon to estimate the dynamic range of detection. This range also indicates the lower limit of detection.

Table 11.9 Comparison of Different Detection Strategies

	Method	Indicates Size	Quantitative	High-throughput	Specificity	Reproducibility
Post PCR	Agarose gel electrophoresis					
	Intact PCR product	Yes	Poor	Poor	Poor	Poor
	Restriction enzyme analysis	Yes	Fair	Poor	Fair	Fair, if the PCR reaction falls in the exponentional phase
	Blots/hybridization	Yes	Poor	Poor	Fair	Poor
	Scanning/densitometer	Yes	Fair	Poor	Fair	Poor; too many variables
	Nested PCR	Yes	Fair	Poor	Good	Fair
	PAGE electrophoresis (sequencing gel)	Yes	NA	Fair	Excellent	Poor
Real time	Automated detection and analysis systems	No	Excellent	Excellent	Excellent	Excellent

Enzyme

Is the enzyme inactive?

Did positive controls work?

Primer

Is the primer poorly designed?

Utilize several different amplicon locations to design the primers to increase your chance of success.

Cycling Parameters

Was there insufficient amplification?

Take a portion of the PCR products and amplify further or repeat with a larger quantity of starting material or test with nested PCR.

Lower or raise annealing temperature (See Tables 11.3 and 11.8).

Buffer

Were one or more buffer components faulty?

Include a positive control such as a commercially tested endogenous control, or a pretested set of reagents

Mg^{2+} concentration is not optimum?

Raise or lower the concentration as per Table 11.5.

Other

Was the detection method sufficiently sensitive?

Prepare a standard curve with a positive control to determine the detection limit.

Smear on the Gel

Template

Was the template copy number too large?
Was the template degraded?

Enzyme

Was too much enzyme and/or too much template included?

Primer

Is the primer design following the design guideline? Is the concentration of primer too low or too high? Do primers lack specificity?

Cycling Parameters

Too many amplification cycles?
Is the annealing temperature too low?

Buffer

Is the Mg^{2+} concentration optimal?

Lower the concentration as per Table 11.5.

Other

Was the appropriate electrophoresis buffer and/or gel concentration used?

Wrong Product

Template

Is the template copy number too large or is the template DNA degraded?

Test a negative control sample to determine if data represent an artifact.

Enzyme

Use a hot-start strategy (Ehrlich, Gelfand, and Sninsky, 1991) to increase specificity.

Primer

Inappropriate primer design?

Apply nested PCR, sequencing, restriction analysis or hybridization to troubleshoot.

Perform a BLAST search to assess possibility of amplifying a different gene (Appendix C).

Cycling Parameters

Too many amplification cycles?
Annealing temperature too low?

Buffer

Is the Mg^{2+} concentration optimal?

Lower the concentration as per Table 11.5.

Other

Was the appropriate electrophoresis buffer and/or gel concentration used?

Faint Band of the Correct Size/Low Yield

Template

Poor quality DNA or RNA?

Check the sample for degradation, inhibitor, or contamination.

Enzyme

Use a hot-start strategy to increase specificity.

Primer

Examine primer design for unmatched T_m of the forward and reverse primers, runs of pyrimidine and purine, or other unfavorable sequence; if a primer-dimer band (lower molecular weight) is visible, a hot-start strategy might increase the yield of the desired product.

Cycling Parameters

Insufficient amplification cycles?

Continue amplification with fresh reagents

Annealing temperature not optimum?

Increase/decrease for more yields.

Buffer

Nonoptimal Mg^{2+} concentration?

Increase concentration as per Table 11.5.

Positive Control Generated Product, but Your Sample Did Not

Template

The sample did not contain the target sequence at detectable level.

Pipetting problem? DNA sample never added to the reaction? It is always a good idea to do two to four reactions to exclude such a possibility.

Enzyme

Low specificity and yield?

Use modified form of Taq DNA polymerase such as TaqGold™ to increase both specificity and yield. This enzyme is inactive until thermal activation to provide a hot-start for increased specificity. At the same time this enzyme is time-released, providing more enzyme in the later cycles when more enzyme increases yield. Decreased mispriming also increases the amount of the desired PCR products (Abramson, 1999).

Primer

Primer design not optimal if primer-dimer is formed?

Redesign.

Cycling Parameters

Insufficient amplification cycles?

Continue amplification with fresh reagents.
If the yield of the positive control is also low, optimize annealing and denaturation temperature and duration of each hold time to increase yield.

Buffer

Mg^{2+} concentration is not optimal?

Increase concentration as per Table 11.5.

Other

Presence of PCR inhibitors?

Test an exogenous IPC (internal positive control) for troubleshooting, or do mixing experiment to test if addition of your sample inhibits the positive control.
Is an inhibitor crosslinked to the DNA template?
Try adding adjunct such as PTB

The troubleshooting discussion above further illustrates how appropriate controls can simplify or eliminate much of the troubleshooting effort. Prevention is the key.

Misincorporations of Nucleotides

Template

Too much single-stranded DNA sample due to insufficient extension time or not having enough quantity of one of the primers?

Enzyme

Too many units of DNA polymerase present?

Primer

Nonoptimal T_m causes pre-PCR annealing to secondary, unintended sites?

Check sequence for hot spots for mispriming.

Cycling parameters

Ramp time too long?
Annealing temperature too low?

Buffer

Mg^{2+} concentration too low?

Check dNTP and template concentration.
Adjust as per Table 11.5.

RT-PCR

Despite the increased interest in RT-PCR, this technique can be more challenging than DNA PCR in many ways. Here are some parameters to keep in mind:

- Isolation and purification of RNA requires greater care.
- Design of primers spanning a large intron may be necessary to avoid amplifying contaminating genomic DNA.
- DNase treatment of RNA preparation may affect different genes differentially for the subsequent PCR (Huang, Fasco, and Kaminsky, 1996); thus use it only as the last resort. Residual DNase I can reduce the yield of PCR products.
- The most frequently used reverse transcriptases are MuLV, *rTth* DNA polymerase, and SuperScript™ (Life Technologies). For RNA with excessive secondary structure or high GC content, apply *rTth* DNA polymerase. Its greater heat stability allows for higher reaction temperatures using a gene-specific reverse primer,

which increases specificity of the RT-PCR reaction. However, these conditions may increase hydrolysis of RNA.

- The choice of primers for the cDNA synthesis includes random primers (nonamers and hexamers), oligo dT and gene-specific primers. For cloning full-length gene, use oligo dT. Use random hexamers for multiplex or when the test sample may not be of good quality (i.e., clinical samples), where full-length mRNA is difficult to obtain (i.e., paraffin-embedded tissue), and where the position of the amplicon is distant from the poly (A) tail. The latter case is especially important when RNA secondary structure prevents full-length synthesis of the first-strand cDNA via the relatively low temperature (37–42°C) RT reaction. Therefore your choice of primers for RT depends on the relative distance between the priming site, the amplicon location and the gene structure. You may want to avoid oligo dT if the following conditions apply to your gene:

  Presence of long 3′-untranslated region (UTR) (>1 Kb) or the length of it is unknown.

  The amplicon site is at the 5′ end of a long transcript.

  The amplicon site is at the 5′ end of a GC-rich gene.

SUMMARY

This chapter has discussed basic PCR technology issues. The complexity of more advanced techniques such as allele-specific amplification, long PCR, RACE, DICE, competitive RT-PCR, touchdown, multiplex PCR, nested PCR, QPCR, and in situ PCR could not be covered in this review.

The intellectual and biochemical strategies discussed within this chapter were not designed to answer every question related to PCR, but to provide a foundation to help you, better ask and answer questions that you will encounter. Combined with the resources provided within this chapter, the author hopes this chapter provides you with new insight to evaluate and meet your PCR needs.

BIBLIOGRAPHY

Abramson, R. 1999. Thermostable DNA polymerases: An update. In Innis, M. A., Gelfand, D. H., and Sninsky, J. J., eds., *PCR Applications: Protocols for Functional Genomics*. Academic Press, San Diego, CA, pp. 39–57.

Abu Al-Soud, W., and Radstrom, P. 1998. Capacity of nine thermostable DNA polymerases to mediate DNA amplification .n the presence of PCR-inhibiting samples. *Appl. Environ. Microbiol.* 64:3748–3753.

Akane, A., Matsubara, K., Nakamura, H., Takahashi, S., and Kimura, K. 1994. Identification of the heme compound copurified with deoxyribonucleic acid

(DNA) from bloodstains, a major inhibitor of polymerase chain reaction (PCR) amplification. *J. Forensic Sci.* 39:362–372.

Altschul, S. F., Madden, T. L., Schaffer, A. A., Zhang J., Zhang, Z., Miller, W., and Lipman, D. J. 1997. Gapped BLAST and PSI-BLAST: A new generation of protein database search programs. *Nucl. Acids Res.* 25:3389–3402.

André, P., Kim, A., Khrapko, K., and Thilly, W. G. 1997. Fidelity and mutational spectrum of Pfu DNA polymerase on a human mitochondrial DNA sequence. *Genome Res.* 7:843–852.

Barnes, W. M. 1994. PCR amplification of up to 35-kb DNA with high fidelity and high yield from lambda bacteriophage templates. *Proc. Nat. Acad. Sci. USA.* 91:2216 –2220.

Baskaran, N., Kandpal, R. P., Bhargava, A. K., Glynn, M. W., Bale, A., and Weissman, S. M. 1996. Uniform amplification of a mixture of deoxyribonucleic acids with varying GC content. *Genome Res.* 6:633–638.

Bonini, J. A., and Hofmann, C. 1991. A rapid, accurate, nonradioactive method for quantitating RNA on agarose gels. *Biotech.* 11:708–710.

Bost, D. A., Stoffel, S., Landre, P., Lawyer, F. C., Akers, J., Abramson, R. D., and Gelfand, D. H. 1994. Enzymatic characterization of Thermotoga maritima DNA polymerase and a truncated form, UlTma DNA Polymerase. *Fed. Am. Soc. Exp. Biol. J.* 8:A1395.

Brown, D. M. 1974. Chemical reactions of polynucleotides and nucelic acids. In Tso, P. O. P., ed., *Basic Principles in Nucleic Acids Chemistry*. Academic Press, New York, pp. 43–44.

Cha, R. S., Zarbl, H., Keohavong, P., and Thilly, W. G. 1992. Mismatch amplification mutation assay (MAMA): Amplification to the C-H-ras gene. *PCR Meth. Appl.* 2:14–20.

Cha, R. S., and Thilly, W. G. 1995. Specificity, efficiency, and fidelity of PCR. In Dieffenbach, C. W., and Dveksler, G. S., eds., *PCR Primer: A Laboratory Manual*. Cold Spring Harbor Laboratory Press, New York, pp. 37–62.

Cheng, C., Fockler, S., Barnes, W. M., and Higuchi, R. 1994. Effective amplification of long targets from cloned inserts and human genomic DNA. *Proc. Nat. Acad. Sci. USA* 91:5695–5699.

Chou, Q., Russel, M., Birch, D. E., Raymond, J., and Block, W. 1992. Prevention of pre-PCR mispriming and primer dimerization improves low copy number amplifications. *Nucl. Acids Res.* 20:1717–1732.

Cline, J., Braman, J. C., and Hogrefe, H. H. 1996. PCR fidelity of *Pfu* DNA polymerase and other thermostable DNA polymerases. *Nuc. Acids Res.* 24: 3546–3551.

Coen, D. M. 1995. The polymerase chain reaction. In *Current Protocols in Molecular Biology*. Wiley, New York, ch. 15.

Compton, T. 1990. Degenerate primers for DNA amplification. In Innis, M. A., Gelfand, D. H., Sninsky, J. J., and White, T. J., eds., *PCR Protocols*. Academic Press, New York, pp. 39–45.

de Noronha, C. M., and Mullins, J. I. 1992. Inhibition of Vent-polymerase amplimer degradation in polymerase chain reaction by 3′ terminal phosphorothionate linkages. *PCR Methods Appl.* 2:131–136.

Dieffenbach, C. W., and Dveksler, G. S., eds. 1995. *PCR Primer: A Laboratory Manual.* Cold Spring Harbor Laboratory Press, New York.

D'Aquilla, R. T., Bechtel, L. J., Videler J. A., Eron J. J., Gorezyca, P., and Kaplan J. C. 1991. Maximizing sensitivity and specificity of PCR by pre-amplification heating. *Nucl. Acids Res.* 19:3749.

Erlich, H. A., Gelfand, D., and Sninsky. J. J. 1991. Recent advances in the polymerase chain reaction. *Science* 252:1643–1651.

Faloona, F., Weiss, S., Ferre, F., and Mullis, K. 1990. Direct detection of HIV sequences in blood. High gain polymerase chain reaction. Paper presented at the 6th Int. Conf. on AIDS, San Francisco. Abstract 1019.
Ford, O. S., and Rose, E. A. 1995. Long-PCR: Long-distance PCR. In Dieffenbach, C. W., and Dveksler, G. S, eds., *PCR Primer: A Laboratory Manual.* Cold Spring Harbor Laboratory Press, New York, pp. 63–77.
Frohman, M. A., Dush, M. K., and Martin, G. R. 1988. Rapid production of full-length cDNA from rare transcripts: Amplification using a single gene specific oligonucleotide primer. *Proc. Nat. Acad. Sci. USA* 85:8998–9002.
Fuqua, S. A. W., Fitzgerald, S. D., and McGuire, W. L. 1990. A simple polymerase chain reaction method for detection and cloning of low-abundance transcripts. *Biotech.* 9:206–211.
Gelfand, D. H. 1992. The design and optimization of the PCR. In Erlich, H. A., ed., *PCR Technology*. Oxford University Press, New York, pp. 7–16.
Gelfand, D. 1992. Taq DNA polymerase. In Erlich, H. A., ed., *PCR Technology*. Oxford University Press, New York, pp. 17–22.
Gelfand, D. H. 1989. Taq DNA polymerase. In Erlich, H. A., ed., *PCR Technology: Principles and Applications for DNA Amplification*. Stockton Press, New York, pp. 17–22.
Gelfand, D. H., and White, T. J. 1990. Thermostable DNA polymerases. In Innis, M. A., Gelfand, D. H., Sninsky, J. J., and White, T. J., eds., *PCR Protocols*. Academic Press, San Diego, CA, pp. 129–141.
Gelmini, S., Orlando, C., Sestini, R., Vona, G., Pinzoni, P., Ruocco, L., and Pazzagli, M. 1997. Quantitative polymerase chain reaction-based homogeneous assay with fluorogenic probes to measure C-Erb-2 oncogene amplification. *Clin. Chem.* 43:752–758.
Hogrefe, H. 2001. *Methods In Enzymology*. Academic Press, New York.
Huang, Z., Fasco, M. J., and Kaminsky, L. S. 1996. Optimization of DNase I removal of contaminating DNA from RNA for use in quantitative RNA-PCR. *Biotech.* 20:1012–1020.
Innis, M. A., and Gelfand, D. H. 1990. Optimization of PCR. In Innis, M. A., Gelfand, D. H., Sninsky, J. J., and White, T. J., eds., *PCR Protocols*. Academic Press, New York, pp. 3–12.
Innis, M. A., Gelfand, D. H., Sninsky, J. J., and White, T. J., Eds., 1999. *PCR Application*. Academic Press, New York.
Jackson, D., Hayden, J. D., and Quirke, P. 1991. Extraction of nucleic acid from fresh and archival material. In McPherson, M. J., Quirke, P., and Taylor, G. R., eds., *PCR: A Practical Approach.* IRL Press, Oxford, England.
Kellog, D. E., Rybalkin, I., Chen, S., Mukhamedova, N., Vlasik, T., Siebert, P. D., and Chenchik, A. 1994. Taq Start antibody: "Hot Start" PCR Facilitated by a neutralizing monoclonal antibody directed against Taq DNA polymerase. *Biotech.* 16:1134–1137.
Kleppe, K., Ohtsuka, E., Kleppe, R., Molineuox, I., and Khorana, H. G. 1971. Studies on polyunucleotides: XCVI. Repair replication of short synthetic DNA's as catalyzed by DNA polymerases. *J. Mol. Biol.* 56:341–361.
Krawetz, S. A., Pon, R. T., and Dixon, G. H. 1989. Increased efficiency of the Taq polymerase catalyzed polymerase chain reaction. *Nucl. Acids Res.* 17:819.
Kreader, C. A. 1996. Relief of amplification inhibition in PCR with bovine serum albumin or T4 gene 32 protein. *Appl. Environ. Microbiol.* 62:1102–1106.
Kunkel, T. A. 1992. DNA replication fidelity. *J. Biol. Chem.* 67:18251–18254.
Kwok, S., Chang, S.-Y., Sninsky, J. J., and Wang, A. 1995. Design and use of mismatched and degenerate primers. In Dieffenbach, C. W., and Dveksler, G. S., eds., *PCR Primer: A Laboratory Manual.* Cold Spring Harbor Laboratory Press, New York, pp. 143–155.

Kwok, S., Kellogg, D. E., McKinney, N. Spasic, D., Goda, L., and Sninsky, J. J. 1990. Effects of primer-template mismatches on the polymerase chain reaction: Human immunodeficiency virus type 1 model studies. *Nucl. Acids Res.* 18:999–1005.

Lawyer, F. C., Stoffel, S., Saiki, R. K., Chang, S. Y., Landre, P. A., Abramson, R. D., and Gelfand, D. H. 1993. High-level expression, purification, and enzymatic characterization of full-length *thermus aquaticus* DNA polymerase and a truncated form deficient in 5′ to 3′ exonuclease activity. *PCR Meth. Appl.* 2:275–287.

Lawyer, F. C., Stoffel, S., Saiki, R. K., Myambo, K., Drummond, R. and Gelfand, D. H. 1989. Isolation, characterization, and expression in Escherichia coli of the DNA polymerase gene from *Thermus aquaticus*. *J. Biol. Chem.* 264:6427–6437.

Ling, L. L., Keohawong, P., Dias, C., and Thilly, W. G. 1991. Optimization of the polymerization chain reaction with regard to fidelity: Modified Ti, Taq, and VenT DNA polymerases. *PCR Meth. Appl.* 1:63–69.

Longo, N., Berninger, N. S., and Hartley, J. L. 1990. Use of Uracil DNA glycosylase to control carry-over contamination in polymerase chain reactions. *Gene* 93:125–128.

MJ Research Application Bulletin, 2000. MJ Research (Waltham, MA). The brochure is available at http://www.mjr.com/html/consumables/finnzymes/dynazyme_ext.pdf.

Martin, F. H., Castro, M. M., Aboulela, F., and Tinoco Jr., I. 1985. Base pairing involving deoxyinosine: Implication for probe design: *Nucl. Acids Res.* 13:8927–8938.

Myers, T. W., and Sigua, C. L. 1995. Amplification of RNA: High temperature reverse transcription and DNA amplification with Thermusthermophillus DNA polymerase. In Innis, M. A., Gelfand, D. H., and Sninsky, J. J., eds., *PCR Strategies*. Academic Press, San Diego, CA, pp. 58–68.

Mullis, K. B., Faloona, F. A., Scharf, S. J., Saiki, R. K., Horn, G. T., and Erlich, H. A. 1986. Specific enzymatic amplification of DNA sequences in vitro: The polymerase chain reaction. *Cold Spring Harbor Symp. Quant. Biol.* 51:263–273.

Mullis, K. B., and Faloona, F. 1987. Specific synthethis of DNA in vitro via a polymerase catalyzed chain reaction. *Meth. Enzymol.* 155:335–350.

Multimer, H., Deacon, N., Crowe, S., and Sonza, S. 1998. Pitfalls of processed pseudogenes in RT-PCR. *Biotech.* 24:585–588.

Perler, F. B., Kumar, S., and Kong, H. 1996. Thermostable DNA polymerases. *Adv. Protein Chem.* 48:377–435.

Poinar, H. N., Hofreiter, M., Spaulding, W. G., Martin, P. S., Stankiewicz, B. A., Bland, H., Evershed, R. P., Possnert, G., and Paabo, S. 1998. Molecular coproscopy: dung and diet of the extinct ground sloth Nothrotheriops shastensis. *Science* 281:402–406.

Raff, T., vander Giet, M., Endemann, D., Wiederholt, T., and Paul, M. 1997. Design and testing of *β*-actin primers for RT-PCR that do not co-amplify processed pseudogenes. *Biotech.* 23:456–460.

Rapley, R. 1994. Enhancing PCR amplification and sequencing using DNA-binding proteins. *Mol. Biotechnol.* 2:295–298.

Reiss, R. A., Schwert, D. P., and Ashworth, A. C. 1995. Field preservation of coleoptera for molecular genetic analysis. *Environ. Entomol.* 24:716–719.

Rybicki, E. P., and Hughes, F. L. 1990. Detection and typing of maize streak virus and other distantly related geminiviruses of grasses by polymerase chain reaction amplification of a conserved viral sequence. *J. Gen. Virol.* 71:2519–2526.

Rychlik, W., Spencer, W. J., and Rhoads, R. E. 1990. Optimization of the annealing temperature for DNA amplification *in vitro*. *Nucl. Acids Res.* 18:6409–6412.

Saiki, R. K. 1992. The design and optimization of the PCR. In Erlich, H. A., ed., *PCR Technology*. Oxford University Press, New York, pp. 7–16.

Saiki, R. K., Bugawan, T. C., Horn, R. T., Mullis, K. B., and Erlich, H. A. 1986. Analysis of enzymatically amplified β-globin and HLA-DQDNA with allele-specific oligonucleotide probes. *Nature* 324:163–166.

Saiki, R. K., Gelfand, D. H., Stoffel, S., Scharf, S. J., Higuchi, R., Horn, G. T., Mullis, K. B., and Erlich, H. A. 1988. Primer-directed enzymatic amplification of DNA with a thermostable DNA polymerase. *Science* 239:487–491.

Saiki, R. S., Scharf, S., Faloona, F., Mullis, K. B., Horn, G. T., Erlich, H. A., and Arnheim, N. 1985. Enzymatic amplification of beta-globin genomic sequences and restriction site analysis for diagnosis of sickle cell anemia. *Science* 230:1350–1354.

SantaLucia Jr., J. 1998. A unified view of polymer, dumbbell, and oligonucleotide DNA nearest-neighbor thermodynamics. *Proc. Nat. Acad. Sci. USA* 95:1460–1465.

SantaLucia Jr., J., Allawi, H. T., and Senevitratne, P. A. 1996. Improved nearest-neighbor parameters for predicting DNA duplex stability. *Biochem.* 35:3555–3562.

Sarkar, G., Kapeiner, S., and Sommer, S. S. 1990. Formamide can drastically increase the specificity of PCR. *Nucl. Acids Res.* 18:7465.

Schwarz, K., Hansen-Hagge, T., and Bartram, C. 1990. Improved yields of long PCR products using gene 32 protein. *Nucl. Acids Res.* 18:1079.

Simmonds, P., Balfe, P., Peutherer, J. F., Ludlam, C. A., Bishop, J. O., and Brown, A. J. L. 1990. Human immunodeficiency virus-infected individuals contain prouirus in small numbers of peripheral mononuclear cells and at low copy numbers. *J. Virol.* 64:804.

Skerra, A. 1992. Phosphorothionate primers improve the amplification of DNA sequences by DNA polymerases with proofreading activity. *Nucl. Acids Res.* 20:3551–3554.

Smith, K. T., Long, C. M., Bowman, B., and Manos, M. M. 1990. Using cosolvents to enhance PCR amplification. *Amplifications* 9:16–17.

Spanakis, E. 1993. Problems related to the interpretation of autoradiographic data on gene expression using common constitutive transcripts as controls. *Nucl. Acids Res.* 21:3809–3819.

Thweatt, R., Goldstein, S., and Reis, R. J. S. 1990. A universal primer mixture for sequence determination at the 3′ ends of cDNAs. *Anal. Biochem.* 190:314–316.

Wittwer, C. T., and Garling, D. J. 1991. Rapid cycle DNA amplification: Time and temperature optimization. *Biotech.* 10:76–83.

Wittwer, C. T., Ririe, K. M., Andrew, R. V., David, D. A., Gundy, R. A., and Balis, U. J. 1997. The Lightcycler^{TM}: A microvolume multisample fluorimeter with rapid temperature control. *Biotech.* 2:171–181.

Wu, D. Y., Ugozzoli, L., Pal, B. K., Qian, J., and Wallace, R. B. 1991. The effect of temperature and oligonucleotide primer length on the specificity and efficiency of amplification by the polymerase chain reaction. *DNA Cell Biol.* 10:233–238.

Yap, E. P. H., and McGee, J. O'D. 1991. Short PCR product yields improved by lower denaturation temperatures. *Nucl. Acids Res.* 19:1713.

Yedidag, E. N., Koffron, A. J., Mueller, K. H., Kaplan, D. B., Fryer, J. P., Stuart, F. P., and Abecassis, M. 1996. Acyclovir triphosphate inhibits the diagnostic polymerase chain reaction for cytomegalovirus. *Transplantation* 62:238–242.

Zimmermann, K., Pischinger, K., and Mannhalter, J. W. 1994. Nested primer PCR detection limits of HIV-1 in a background of increasing numbers of lysed cells. *Biotech.* 17:18–20.

APPENDIX A

PREPARATION OF PLASMID DNA FOR USE AS PCR CONTROLS IN MULTIPLE EXPERIMENTS

Have you ever failed to amplify a section of a plasmid that previously produced the desired PCR product? Your problem is not unique. Often plasmid DNA at low concentration of DNA is degraded by nuclease or adsorbs to the wall of the plastic tube during storage and handling. This protocol produces plasmid DNA that is stable for months and years if stored at –20°C and generates reproducible standard curves.

The addition of glycogen (20 μg/ml final concentration) in 10 mM Tris, 1 mM EDT (pH8.0) buffer can protect DNA from degradation by nuclease as well as loss from adsorption to the tube. After making serial ten-fold dilutions (100 μl of DNA in 900 μl of TE), aliquot the solution in 100 μl or less volume and store at –20°C. For preparation of TE buffer, use fresh nuclease-free water. "Sterile" water sitting on the lab bench for one week or more may contain contaminants as well as nucleases.

APPENDIX B

COMPUTER SOFTWARE FOR SELECTING PRIMERS*

- Primer v. 1.4 (DOS)
- PINCERS (Macintosh)
- Oligonucleotide Selection Program (Macintosh, DOS, Digital VAX/VMS, SUN SPARC-based workstations)
- Right Primer: Primer Design Utility
- Gene Runner 3.0
- Oligo 5.0 (DOS)
- Oligo 4.0
- DNASIS 2.0 (Windows)
- MacDNASIS (Macintosh)
- GeneWorks (Macintosh)
- Lasergene (DOS, Windows, Macintosh)
- Eugene™ (DOS)
- GeneJockey (Macintosh)
- Wisconsin Sequence Analysis Package (Digital VAX/VMS, IBM RS6000, Sun SPARC-based workstations, Silicon Graphics Workstation)
- MacVector (Macintosh)
- PRIMER PRIMER (Macintosh, DOS, PowerMac)
- DesignerPCR (Macintosh)
- Vector NT1 (Windows, Macintosh)
- Primer Designer (Macintosh, DOS)
- Primer Express™
- HYTHER (PC-Windows, UNIX and Web-based platforms) available for license at *http://jsll.chem.wayne.edu/Hyther/hythermenu.html.*

**Data from Dieffenbach and Dveksler (1995).*

APPENDIX C

BLAST SEARCHES

There are many genes that share local sequence homology with a primer 18 to 30 nucleotide long. For example, the beta-actin primer shares 100% homology with pseudogenes, gamma-actin, and related genes. It is therefore misleading to use this primer to estimate the level of beta-actin gene expression. Some of the PCR products will be derived from these genes, but you have no way to tell how much came from the true beta-actin gene. Pseudogenes are not translated into protein and have no biological significance, so your RT-PCR result may not relate to immunological data or biochemical assays. In some cases it will amplify other genes not even related to the one you are investigating. For this reason it helps to know the BLAST search information before ordering primers.

The BLAST programs (*http://www.ncbi.nlm.nih.gov/BLAST*) have been designed for speed, with a minimal sacrifice of sensitivity to distant sequence relationships. The scores assigned in a BLAST search have a well-defined statistical interpretation, making real matches easier to distinguish from random background hits. BLAST uses a heuristic algorithm that seeks local as opposed to global alignments, and it is therefore able to detect relationships among sequences that share only isolated regions of similarity (Altschul et al., 1997). For a better understanding of BLAST, refer to the BLAST instructional course, which explains the basics of the BLAST algorithm.

APPENDIX D

USEFUL WEB SITES

Topics	Content and URL
Basic PCR information	Weizmann Institute of Science Genome and Bioinformatics site *http://bioinformatics.weizmann.ac.il/mb/bioguide/pcr/contents.html*
Collection of PCR protocols	Standard PCR protocols *http://www.protocol-online.net/molbio/PCR/standard_pcr.htm*
Optimization of PCR	Primer design and reaction optimisation. E. Rybicki, Department of Microbiology, University of Cape Town. In *Molecular Biology Techniques Manual: Third Molecular Biology Techniques Manual*, V. E. Coyne et al., eds. *http://www.uct.ac.za/microbiology/pcroptim.htm*
Standard PCR guideline	PCR Primer: Strategies to improve results provided by G. Afseth of Perkin Elmer at Northwestern University (1997) *http://www.biotechlab.nwu.edu/pe/index.html*
Molecular biology methods	Current Protocols in Molecular Biology *http://www.wiley.com/cp/cpmb/mb0317.htm*
	Elsevier Trends Journals Technical Tips online *http://research.bmn.com/tto*
	Molecular biology reagents and procedures. Dartmouth University *http://www.dartmouth.edu/artsci/bio/ambros/protocols*

APPENDIX D (*Continued*)

PCR protocols and online manual	Alkami Quick Guide™ for PCR. A laboratory reference for the polymerase chain reaction. 1999 *http://www.alkami.com/qguide/idxguide.htm*
PCR protocol	Roche Molecular Biochemicals PCR protocol *http://206.53.227.20/prod_inf/manuals/pcr_man/index.htm*
Links to many sources of basic PCR and other molecular biology information	ExPASy (Expert Protein Analysis System) proteomics server of the Swiss Institute of Bioinformatics (SIB) Molecular Biology Server *http://www.expasy.ch/*
Multiplex PCR	Multiplex PCR: Critical parameters and step-by step protocol. O. Hehegariu et al., *Biotech.* 23(1997):504–511 *http://info.med.yale.edu/genetics/ward/tavi/bt/BT(23)504.pdf*
Various PCR topics, including multiplex PCR	Tavi's PCR site (Octavian Henegariu) on variety of topics *http://info.med.yale.edu/genetics/ward/tavi/PCR.html*
Primers	Primers! Web site *http://www.alkami.com/cntprmr.cgi?url=http://www.wil liamstone.com/primers/javascript/*
	Hyther *http://jsll.chem.wayne.edu/Hyther/hythermenu.html*
PCR chat room	Protocol online (discussion) *http://www.protocol-online.net/discussion/index.htm*
Real-time PCR	References for TaqMan real-time assay *http://www.appliedbiosystems.com/ab/about/pcr/sds/ taqrefs.html*
Gene quantitation	References on absolute and relative gene quantitation by PE Biosystmes *http://www.appliedbiosystems.com/ab/about/pcr/sds/ taqrefs.html#rev*
Gene search and validation BLAST	BLAST (National Center for Biotechnology Information (NCBI) using the Basic Local Alignment Search Tool (BLAST) family of programs *http://www.ncbi.nlm.nih.gov/blast/blast.cgi?Jform=0*

Caution: The dynamic nature of the Web allows us to provide more up-to-date information. However, there are major challenges associated with information available on the Web. Some of the major challenges are as follows: (1) Since it is easier for anyone to publish on the Web, its content may not be evaluated nor accurate. (2) The URL address as well as its content may change or even disappear without notice, thus quickly invalidating any list of "useful" sites. All of the Web sites given in this section were selected to give the reader sources of information only and by no means recommended as "valid" source. It is up to the users to determine what is useful. The author highly recommends that readers use their own judgment before adapting any information given.

12

Electrophoresis

Martha L. Booz

Chemical Safety 334
What Is the Safest Approach to Working with Acrylamide? 334
What Are the Symptoms of Acrylamide Poisoning? 335
What Is the Medical Response to Accidental Acrylamide Exposure? 335
How Can You Dispose of Excess, Unusable Acrylamide? .. 335
What Is the Shelf Life of Acrylamide and Acrylamide Solutions? 336
Electrical Safety 336
What Are the Requirements for a Safe Work Area? 336
What Are the Requirements for Safe Equipment in Good Working Order? 337
Polyacrylamide (PAGE) Gels—Before Selecting a Gel: Getting the Best Results for Your Purpose 337
What Is the Mechanism of Acrylamide Polymerization? ... 338
What Other Crosslinkers Are Available, and When Should They Be Used? 338
How Do You Control Pore Size? 339
How Do You Calculate %T and %C? 341

I am grateful to Bruce Goodrich for the figure on degassing acrylamide, to Fiona Leung for the data regarding the molecular weight vs. relative mobility curve, and to Lee Olech and Dave Garfin for fruitful discussions about many of the questions in this chapter.

Molecular Biology Problem Solver, Edited by Alan S. Gerstein.
ISBN 0-471-37972-7

Why Should You Overlay the Gel? What Should You Use for an Overlay? 341
Regarding Reproducible Polymerization, What Practices Will Ensure That Your Bands Run the Same Way Every Time? 341
What Catalyst Concentration Should You Use? 343
What Is the Importance of Reagent Purity on Protein Electrophoresis and Staining? 343
Which Gel Should You Use? SDS-PAGE, Native PAGE or Isoelectric Focusing? 345
Will Your SDS Gel Accurately Indicate the Molecular Weight of Your Proteins? 345
Should You Use a Straight % Gel or a Gradient Gel? 345
What Issues Are Relevant for Isoelectric Focusing? 346
How Can You Resolve Proteins between Approximately 300 and 1000 kDa? 347
What Issues Are Critical for Successful Native PAGE? 348
Sample Solubility 348
Location of Band of Interest 348
How Can You Be Sure That Your Proteins Have Sufficient Negative Charge to Migrate Well into a Native PAGE Gel? 348
Buffer Systems for Native PAGE 349
What Can Go Wrong with the Performance of a Discontinuous Buffer System? 349
What Buffer System Should You Use for Peptide Electrophoresis? 350
Power Issues 350
Constant Current or Constant Voltage—When and Why? 351
Why Are Nucleic Acids Almost Always Separated via Constant Voltage? 352
Why Are Sequencing Gels Electrophoresed under Constant Power? 352
Should You Run Two Sequencing Cells off the Same Power Supply under Constant Power? 352
Improving Resolution and Clarity of Protein Gels 353
How Can You Generate Reproducible Gels with Perfect Bands Every Time? 353
Sample Preparation—Problems with Protein Samples 353
What Procedures and Strategies Should Be Used to Optimize Protein Sample Preparation? 353
Is the Problem Caused by Sample Preparation or by the Electrophoresis? 354

Is the Problem Caused by the Sample or the Sample Buffer? 354
How Do You Choose a Detergent for IEF or Native PAGE? 354
What Other Additives Can Be Used to Enhance Protein Solubility? 355
Agarose Electrophoresis 355
What Is Agarose? 355
What Is Electroendosmosis ($-M_r$ or EEO)? 355
Are Double-Stranded Markers Appropriate for Sizing Large Single-Stranded (Not Oligonucleotide) DNA? 356
What Causes Nucleic Acids to Migrate at Unexpected Migration Rates? 356
What Causes Commercial Preparations of Nucleic Acid Markers to Smear? 356
What Causes Fuzzy Bands? 357
Elution of Nucleic Acids and Proteins from Gels 357
Detection 357
What Should You Consider before Selecting a Stain? 357
Will the Choice of Stain Affect a Downstream Application? 359
Is Special Equipment Needed to View the Stain? 361
How Much Time Is Required for the Various Stains? 361
What If You Need to Quantify Your Stained Protein? 361
What Causes High Background Staining? 362
Will the Presence of Stain on Western-Blotted Proteins Interfere with Subsequent Hybridization or Antibody Detection Reactions? 363
Does Ethidium Bromide Interfere with the Common Enzymatic Manipulation of Nucleic Acids? 363
Standardizing Your Gels 363
What Factors Should Be Considered before Selecting a Molecular Weight Marker? 363
Are Double-Stranded Markers Appropriate for Sizing Large (Not Oligonucleotide) Single-Stranded DNA? If Not, Which Markers Are Recommended? 364
Can a Pre-stained Standard Be Applied to Determine the Molecular Weight of an Unknown Protein? 364
How Do You Determine Molecular Weight on a Western Blot? 365
What Are the Options for Determining pI and Molecular Weight on a 2-D Gel? 365

How Do You Measure the pH Gradient of a Tube IEF Gel or an IPG Gel? 366
Troubleshooting 368
What Is This Band Going All the Way across a Silver-Stained Gel, between Approximately 55 and 65 kDa? 368
How Can You Stop the Buffer Leaking from the Upper Chamber of a Vertical Slab Cell? 368
Bibliography 368
Appendix A: Procedure for Degassing Acrylamide Gel Solutions 371

Dangerously high voltage and acrylamide, a neurotoxin and suspected carcinogen, are inescapable elements of electrophoresis. Proper personal protection and good laboratory practice will minimize the risk of harming yourself or your colleagues.

CHEMICAL SAFETY

What Is the Safest Approach to Working with Acrylamide?

Unpolymerized, monomeric acrylamide is a neurotoxin in any form. *Bis*-acrylamide is equally dangerous. Protect yourself by wearing gloves, a lab coat, and safety glasses, and never pipet acrylamide solutions by mouth.

Acrylamide powders should be weighed and solutions prepared in a ventilated hood. Acrylamide can be detected in the air above a beaker of acrylamide solution and throughout the laboratory. Values in the single-digit ppm range are detected above a 10% solution at room temperature (Figure 12.1). The detection method involves passing air samples through an acrylamide-binding column, and analyzing the eluant via HPLC (Dow Chemical Company, 1988). The MSDS for acrylamide gives the OSHA permissible exposure limit for acrylamide as $0.3\,mg/m^3$ for personal exposure in an industrial setting.

The use of pre-cast gels and pre-mixed acrylamide solutions can reduce exposure to acrylamide and *bis*-acrylamide. Even after polymerization, a small fraction of the acrylamide remains in the neurotoxic monomeric form. Wear gloves when handling a polymerized gel.

If you need to cast your own gels, we suggest you use pre-mixed acrylamide solutions, which are also available from many vendors. The pre-mixed solutions avoid the weighing and mixing steps, and generally have a long storage life.

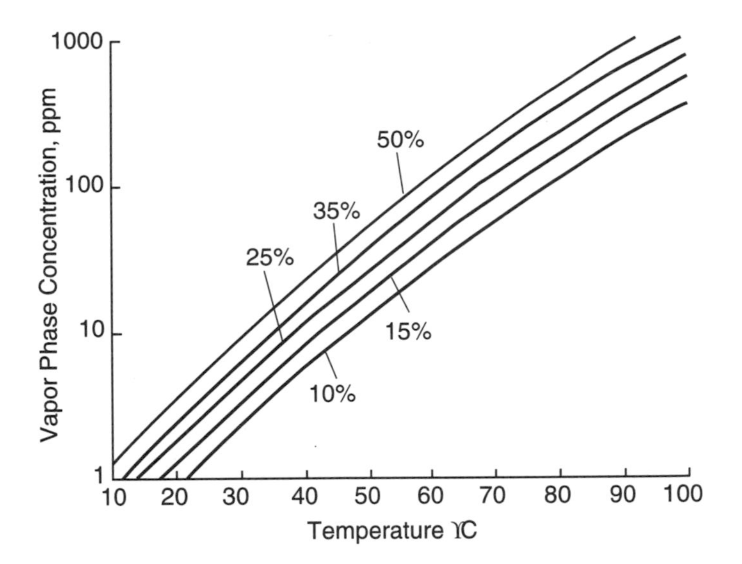

Figure 12.1 Vapor phase concentrations of acrylamide-water solutions (10–50% acrylamide). Cytec Industries Inc., 1995. Reprinted by permission of Cytec Inc.

What Are the Symptoms of Acrylamide Poisoning?

The initial symptoms of acrylamide poisoning on the skin are peeling of the skin at the point of contact, followed by tingling and numbness in the exposed area. If exposure by any means (touch, ingestion, inhalation) continues, muscular weakness, difficulty maintaining balance, and clumsiness in walking and in the use of the hands may develop. A large, acute exposure can produce confusion, disorientation, slurred speech and ataxia (severe loss of balance). Muscular weakness and numbness in the extremities may also follow. Anyone exposed to any form of acrylamide should be immediately examined by a medical doctor (Bio-Rad Laboratories, MSDS, 2000).

What Is the Medical Response to Accidental Acrylamide Exposure?

On your skin: Wash the affected skin several times with soap for at least 15 minutes under running water.

In your mouth: Rinse your mouth immediately with water and seek medical attention immediately.

Swallowed or inhaled: If swallowed, do not induce vomiting. Seek medical attention immediately. If breathed in, get to fresh air, and seek medical attention immediately (Bio-Rad Laboratories, MSDS, 2000).

How Can You Dispose of Excess, Unusable Acrylamide?

Check with your institutional or local county environmental regulators for the disposal requirements in your area. The safest way to dispose of a small amount of liquid acrylamide is to polymerize it in the hood in a closed plastic bag set into a beaker surrounded by a very large, tightly fastened plastic bag, to prevent spattering as the acrylamide polymerizes.

If you have more than 100 ml to dispose of, contact your local environmental safety officers to determine your recommended procedure. Acrylamide solutions emit significant heat during polymerization, and polymerization of large volumes of acrylamide can be explosive due to rapid heat buildup (Dow Chemical Company, 1988; Cytec Industries, 1995; Bio-Rad Laboratories, 2000).

Acrylamide and *bis*-acrylamide powders must be disposed of as solid hazardous waste. Consult your local environmental safety office.

What Is the Shelf Life of Acrylamide and Acrylamide Solutions?

Commercially prepared acrylamide solutions are stable for as long as one year, unopened, and for six months after opening. The high purity of the solution components and careful monitoring throughout the manufacturing process provides extended shelf life. The lifetime of homemade solutions similarly depends on the purity of the acrylamide and *bis*-acrylamide, the cleanliness of the laboratory dishes, and the purity of the water used to make the solutions.

Solid acrylamide breaks down with time due to oxidation and UV light, producing acrylic acid and ammonia. Acrylic acid in a gel can cause fuzzy bands, or fuzzy spots in the case of 2-D gels, streaking and smearing, and poor resolution (Allen and Budowle, 1994). Acrylamide decomposition occurs more quickly in solution, and it can be accelerated by any impurities within the water (Allen and Budowle, 1994). Thus acrylamide powder should be stored airtight at room temperature, and acrylamide solutions should be stored at 4°C, both in the dark.

Production facilities must establish standards and measures to determine the effective lifetime of unpolymerized acrylamide solutions.

ELECTRICAL SAFETY

What Are the Requirements for a Safe Work Area?

The voltages used in electrophoresis can be dangerous, and fires have occurred due to problems with electrophoresis cells. The following precautions should be observed to prevent accidents and fires.

- There should be no puddles of liquid on the horizontal surfaces of the electrophoresis cell.
- The area around the power supply and cell should be dry.
- The area for at least 6 inches around the power supply and cell should be bare of clutter and other equipment. Clear space means any fire or accident can be more easily controlled.

What Are the Requirements for Safe Equipment in Good Working Order?

The wires connecting the cell to the power supply must be in good condition, not worn or cracked, and the banana plugs and jacks must be in good condition, not corroded or worn. Broken or worn wires can cause rapid changes in resistance, slow electrophoresis or a halt in the run. All cables and connectors must be inspected regularly for breaks and wear.

The banana plugs on the ends of the wires should be removed from the power supply at the end of the run by pulling them straight out. Grasp the plug, not the wire. If pulled at an angle, the solder joint attaching the banana plugs to the wires can loosen and cause the loss of the electrical circuit. On the cell core, electrode banana posts with flattened baskets do not make good contact with the banana jack in the cell lid, and should be replaced. The banana jacks (female part) in the cell lid should be inspected regularly to make sure there is no corrosion.

Before starting an electrophoresis run, dry any liquid on the horizontal surfaces of the cell, especially near the banana plugs and jacks. Any liquid on the horizontal surfaces of the cell can arc during the run, damaging the cell and stopping the electrophoresis.

POLYACRYLAMIDE (PAGE) GELS—BEFORE SELECTING A GEL: GETTING THE BEST RESULTS FOR YOUR PURPOSE

Before choosing which gel to use, it is important to consider several questions, all of which can help you choose the gel that will give you the best results for your purpose. The next paragraphs provide information on how to select a gel percentage or pore size, when to use SDS-PAGE and when native PAGE, what buffer system to use, which crosslinker to use, and degree of resolution needed.

What Is the Mechanism of Acrylamide Polymerization?

Most protocols use acrylamide and the crosslinker *bis*-acrylamide (bis) for the gel matrix. TEMED (N,N,N′,N′-tetramethylethylenediamine) and ammonium persulfate are used to catalyze the polymerization of the acrylamide and bis. TEMED, a base, interacts with ammonium persulfate at neutral to basic pH to produce free radicals. The free radical form of ammonium persulfate initiates the polymerization reaction via the addition of a vinyl group (Figure 12.2). At an acidic pH, other catalysts must be used, as described in Andrews (1986), Hames and Rickwood (1981), and Caglio and Righetti (1993).

What Other Crosslinkers Are Available, and When Should They Be Used?

Bis-acrylamide is the only crosslinker in common use today. There are others available, for specialty applications. DHEBA (N,N′-dihydroxyethylene-bis-acrylamide) and DATD (N,N′-diallyltartardiamide) were both used historically with tube gels and radioactive samples (before slab gels came into common use). The tube gels were cut into thin discs, the disks were dissolved with periodic acid, and the radioactivity in the disks was counted in a scintillation counter. Of course the periodic acid destroyed some amino acids, so these crosslinkers are not useful for Edman sequencing or mass spectrometry.

Another crosslinker, BAC (*bis*-acrylylcystamine) can be dissolved by beta-mercaptoethanol. It is useful for nucleic acid electrophoresis (Hansen, 1981). However, proteins containing disulfide bonds do not separate on a BAC gel. The subunits with the sulfhydryl moiety bind to the gel matrix close to the origin of the gel, and separation does not occur, so BAC is not recommended for preparative protein electrophoresis, though it is useful for proteins which do not contain any sulfhydryl bonds.

One other crosslinker, piperazine diacrylamide (PDA), can replace *bis*-acrylamide in isoelectric focusing (classical tube gel or flatbed gel) experiments. PDA imparts greater mechanical strength to a polyacrylamide gel, and this is desired at the low acrylamide concentrations used in isoelectric focusing (IEF gels). Some proteomics researchers use PDA to crosslink the 2nd dimension SDS-PAGE slab gels as well, because of the increased mechanical strength, and because the background of a silver stained gel is much better when PDA is used (Hochstrasser, 1988). For further information on these crosslinkers, see Allen and Budowle, 1994.

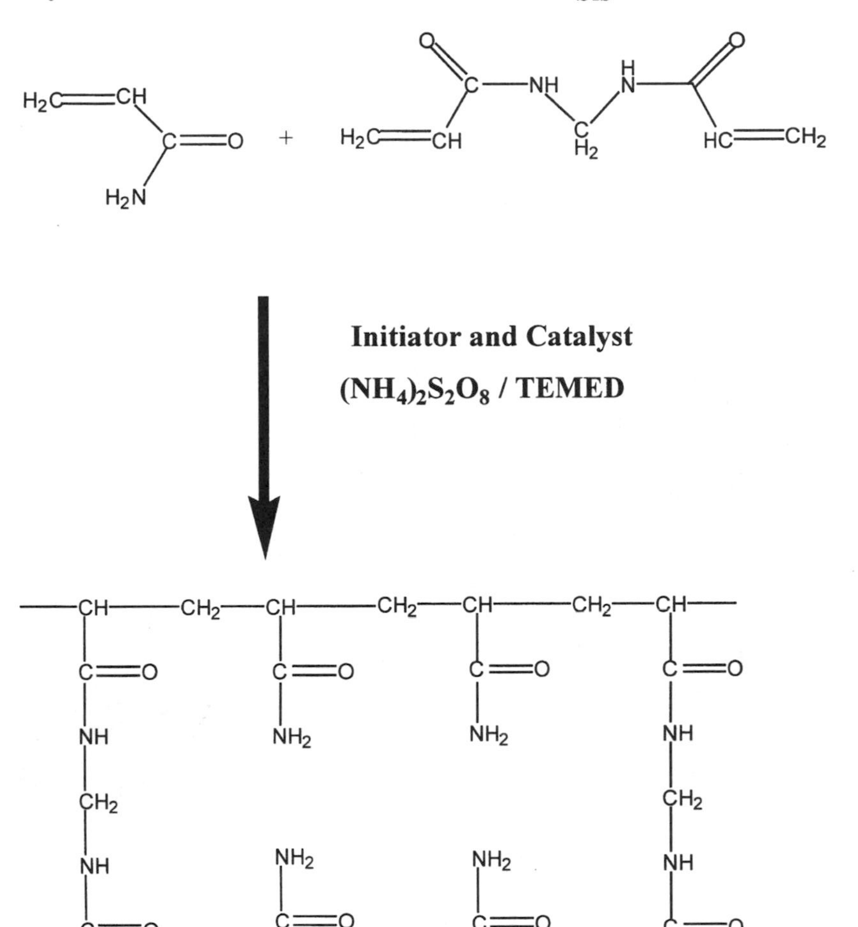

Figure 12.2 Polymerization of acrylamide. Reproduced with permission from Bio-Rad Laboratories.

How Do You Control Pore Size?

Pore size is most efficiently and predictably regulated by manipulating the concentration of acrylamide in the gel. Pore size will change with the amount of crosslinker, but the effect is minimal and less predictable (Figure 12.3). Note the greater impact of acrylamide concentration on pore size, especially at the levels of crosslinker usually present in gels (2.7–5%).

Practical experience with various ratios of acrylamide : bis have shown that it is best to change pore size by changing the acry-

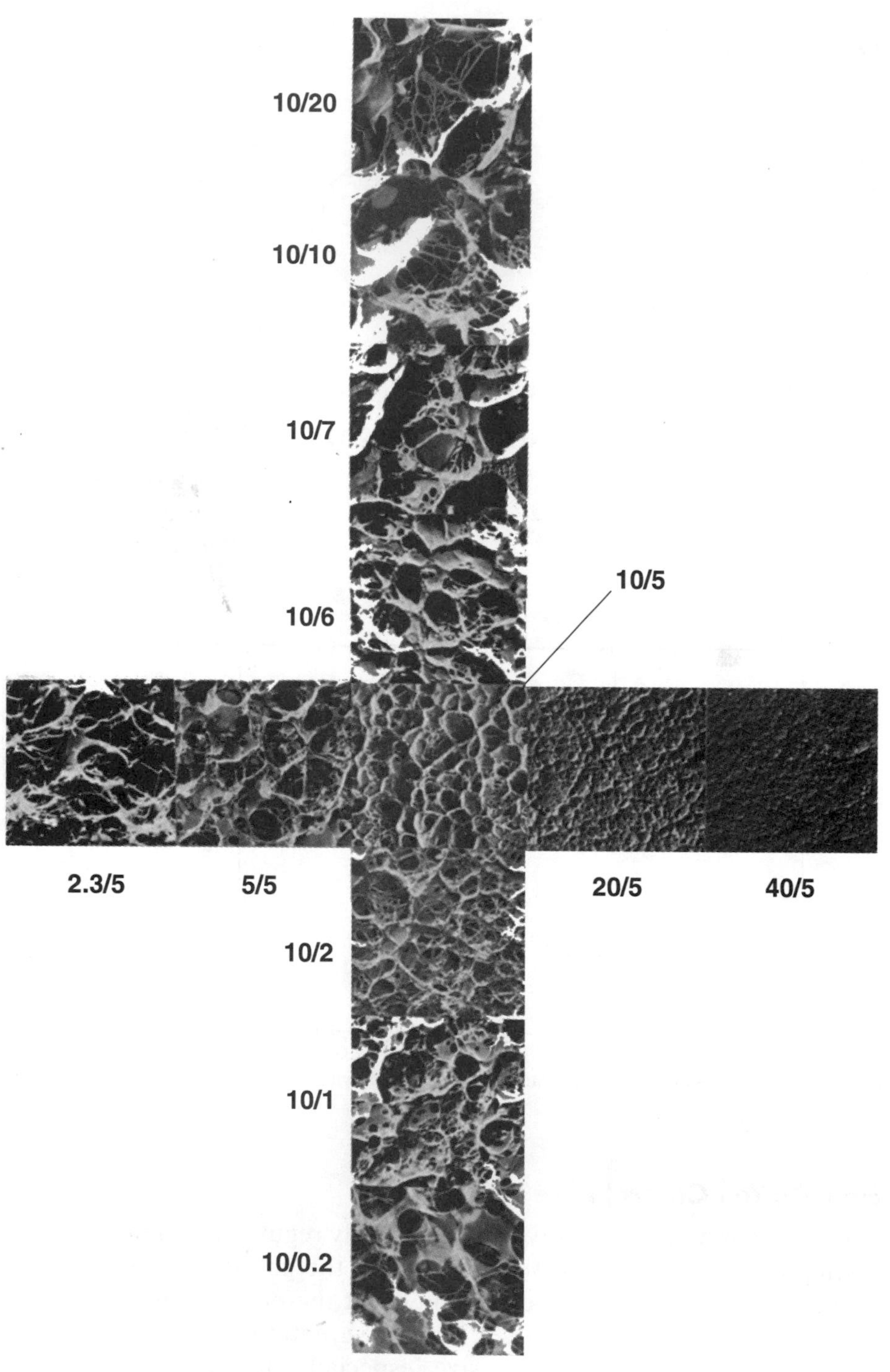

Figure 12.3 Electron micrograph of polyacrylamide gels of various $\%T$, showing the change in pore size with the change in $\%T$ and $\%C$. *From* Rüechel, Steere, and Erbe (1978, Fig. 3, p. 569). Reprinted from Journal of Chromatography, volume 166, Ruechel, R., Steere, R., and Erbe, E. Transmission-electron Microscopic Observations of Freeze-etched Polyacrylamide gels. pp. 563–575. 1978. With permission from Elsevier Science.

lamide concentration. A 19:1 ratio of acrylamide to bis (5% *C*; see below for calculation of *C*) is used in low concentration gels, such as IEF gels, and sequencing gels, to impart greater mechanical strength to the gel. A 29:1 ratio (3.4% *C*) is used for concentrations of acrylamide from 8% to 12%, and a 37.5:1 ratio (2.67% *C*) is used for concentrations of acrylamide above 12% to provide flexibility to the gel. SDS-PAGE and native gels are usually run at 10% to 12%. For comparison, a 12% acrylamide gel with a 5% crosslinker concentration will be brittle and will tear easily.

How Do You Calculate %*T* and %*C*?

Percent T is $\%T$ = (g acrylamide + g *bis*-acrylamide)/100 ml water × 100.

Percent C is $\%C$ = (g *bis*-acrylamide)/(g acrylamide + g *bis*-acrylamide) × 100.

Note that $\%C$ is not the grams *bis*-acrylamide/100 ml water, but rather the percentage of crosslinker as a function of the total weight of acrylamide and *bis*-acrylamide used.

Why Should You Overlay the Gel? What Should You Use for an Overlay?

An overlay is essential for adequate resolution. If you do not overlay, the bands will have the shape of a meniscus. Two closely spaced bands will overlap; the middle of the top band will extend down between the front and back of the bottom band. Overlaying the gel during polymerization will prevent this problem.

Common overlays are best quality water, the buffer used in the gel at a 1× dilution, and water-saturated t-butanol. The choice is a matter of personal preference. Many researchers prefer the alcohol overlay because it will not mix with the gel solution. However, alcohol will turn acrylic plastic (Perspex) from clear to white, and it is difficult to pipet without spills.

Regarding Reproducible Polymerization, What Practices Will Ensure That Your Bands Run the Same Way Every Time?

Reproducible polymerization is one of the most important ways to ensure that your samples migrate as sharp, thin bands to the same location in the gel every time. Attention to polymerization will also help keep the background of your stained gels low. Acrylamide polymerization is affected by the amount of oxygen gas dissolved in the solution, the concentrations and condition of the catalysts, the temperatures and pH of the stock solutions, and the purity of the gel components. The following paragraphs discuss how to ensure reproducible polymerization and therefore reproducible, excellent gels.

Eliminate Dissolved Oxygen

Oxygen quenches the free radicals generated by TEMED and APS, thus inhibiting the polymerization reaction. Dissolved oxygen must be eliminated via degassing with a bench vacuum or better (20–23 inches of mercury or better) for at least 15 to 30 minutes with stirring (see Appendix A). To achieve reproducible polymerization and consistent pore size, allow the gel solutions, which should be stored in the cold to inhibit breakdown, to come to room temperature before casting a gel. Note that cold gel solutions contain more dissolved oxygen, and low temperature directly inhibits the polymerization reaction. If the temperature during polymerization is not controlled, the pore size will vary from day to day.

Symptoms of Problems with Catalyst Potency

The best indicator of a problem catalyst is poor polymerization of the gel. If you're confident that you have good quality chemicals and water, and have degassed your solutions to remove oxygen, and still the sides of the wells do not polymerize around the teeth of the comb, a degraded catalyst is the likely explanation.

Separation of the gel from the spacers also indicates poor polymerization; the dye front will migrate in the shape of a frown. A third symptom of poor polymerization is *schlieren* in the body of the gel. *Schlieren* are swirls, changes in the refractive index of the gel, where polymerization has been very slow or has not occurred. The gel has no structure at the location of the *schlieren*. It breaks apart in pieces at the *schlieren* lines, when removed from the cassette. *Schlieren* can also be caused by inadequate mixing of the gel solution before pouring it into the gel cassette.

It is difficult to predict the potency of TEMED unless you know its history of use. TEMED is very hygroscopic and will degrade within six months of purchase if it becomes contaminated with water. Therefore store TEMED in a desiccator at room temperature if you use it frequently, or at 4°C if you use it less than once a week. Cold TEMED must be warmed to room temperature before the bottle is opened to prevent condensation from contaminating the TEMED liquid.

Determine the potency of APS by watching it dissolve, or by listening to it. Weigh out 0.1 g of APS in a small weigh boat, and then place the weigh boat with the APS onto a dark surface. Add 1 ml of highest purity water directly to the weigh boat, to make a 10% solution. If the APS is potent, you will see tiny bubbles fizzing off the surface of the APS crystals. No fizzing is observed with deteriorated APS. Or put 0.1 g of APS in a 1.5 ml Eppendorf tube, and add 1 ml of water. Cap it and listen for the fizzing. If you do not hear little crackling noises, like fizzing, it has lost its potency and should be replaced.

Stored solutions of TEMED and APS may polymerize gels, but if you want to minimize the chance of failure and maximize reproducibility, especially with protein gels, prepare APS fresh every day, store TEMED dry at room temperature in a desiccator, and degas your solutions before polymerization.

Temperature

The temperature of polymerization should be 20 to 22°C. If your lab is below 20°C, or if the temperature varies more than five degrees from day to day, reproducibility problems may arise. Note that cold delays polymerization, heat speeds it, and the reaction itself is exothermic.

What Catalyst Concentration Should You Use?

The appropriate catalyst concentration depends on what gel % you are polymerizing. Please refer to Table 12.1.

Note that these catalyst concentrations are for protein PAGE gels only. Sequencing gels are polymerized differently. The final concentrations of catalysts for a 6 %T sequencing gel, which allow the solution to be introduced into the gel sandwich before polymerization starts, are TEMED, 0.1% (v/v), and APS, 0.025% (w/v).

What Is the Importance of Reagent Purity on Protein Electrophoresis and Staining?

Reagent purity is extremely important for reproducible results. If the reagents and water you use are very pure, then the polymerization and electrophoresis will be controllable and reproducible from day to day. Any problems you have can be ascribed to the sample and its preparation. The following discussion goes into various reagent purity problems and their resolution.

Table 12.1 Gel Percentage vs. Catalyst Concentration

Gel % Concentration	APS Concentration (w/v)	TEMED (v/v)
4–7%	0.05%	0.1%
8–14%	0.05%	0.05%
⩾15%	0.05%	0.025%

Water

The common contaminants of water are metal ions, especially sodium and calcium, the halide ions, especially chloride, and various organic impurities (Chapter 3 discusses water impurities in greater depth.) Each kind of impurity has a different effect; we will not attempt to enumerate all these effects here. Copper ions inhibit acrylamide polymerization, but copper metal and other metals initiate polymerization. Ions can cause ionic interactions between the macromolecules in your sample, perhaps causing aggregation of certain proteins, with band smearing the result. The organic contaminants can also cause loss of resolution. The effects on staining the samples in the gel are also significant, as impurities in the water can bind the stain, causing bad background. A detailed discussions about preventing background in a stained gel is provided below. The principle here is that impurities in the water cause problems, and the purest water available should be used for electrophoresis to help prevent these problems.

Bacteria in your water purifier can also cause artifacts, such as vertical pinpoint streaks in your gel or on blots stained for total protein. Bacteria migrating up the hose from the sink to the filter cartridges is a common cause of contamination. Note that bacteria can grow in dishwater left to sit in the sink, so be careful where you place the end of the hose that carries water from the water purifier.

Another possible source of contamination in your water is the maintenance department in your institution, especially if your water purifier lacks a charcoal filter for removing organic contaminants. The maintenance department may add organic amine compounds to the distilled water system at your institution to keep scale off the walls of the pipes providing distilled water to your lab. This is commonly done every six months or so. Such contaminants will cause background problems in your stained gels, among other artifacts. The water used to prepare solutions for electrophoresis and staining procedures should be charcoal column-purified and deionized.

Reagents

Impure reagents—from gel components to buffer salts, stains, and dyes—can create problems similar to impure water. Gels will not be reproducible, resolution may be poor, and background staining may be substantial. For reproducible results and good resolution, always use the purest components available, electrophoresis grade.

WHICH GEL SHOULD YOU USE? SDS-PAGE, NATIVE PAGE OR ISOELECTRIC FOCUSING?

The strategy you choose depends on your goal, of course. If you want to determine the molecular weight of your protein, use SDS-PAGE. If you want to measure the isoelectric point of your protein, choose isoelectric focusing (IEF). For proteomics work, use 2-D electrophoresis (IEF followed by SDS-PAGE). Native PAGE is used to assay enzyme activity, or other biological activity, for example, during a purification procedure. Each kind of protein PAGE has issues to consider, and these issues are addressed in the next section. Improving gel resolution is addressed in a separate section below.

Will Your SDS Gel Accurately Indicate the Molecular Weight of Your Proteins?

Estimation of the molecular weight of the protein of interest, accurate to within 2000 to 5000 daltons, requires the protein band(s) to run within the middle two-thirds of the gel. This is illustrated in the graph of the log of the molecular weight of a set of standard proteins vs. the relative mobility of each one (Figure 12.4). Note that the proteins with a relative mobility below 0.3 or above 0.7 fall off the linear portion of the curve. Thus the most accurate molecular weight values are obtained when the relative mobility of the protein of interest is between 0.3 and 0.7. This means that if your protein doesn't enter the gel very well, you must change the gel %T before you can get a good molecular weight value. The sample may require a different (better) solubilization procedure also. (See comments on sample preparation, below.)

Should You Use a Straight % Gel or a Gradient Gel?

If you want to resolve proteins that are within a few thousand daltons of each other in molecular weight, then use a straight percent gel (the same concentration of acrylamide throughout the gel). To get baseline resolution for such proteins, that is, to get clear, unstained space between bands, you may need to use a

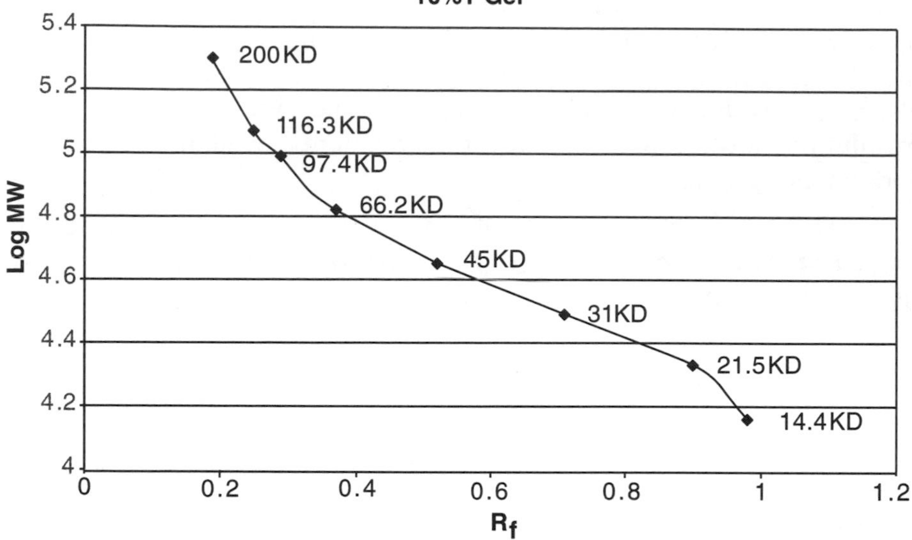

Figure 12.4 Log of the molecular weight (in daltons) of a protein versus the relative mobility. Reproduced with permission from Bio-Rad Laboratories.

longer gel. Mini gels have 6 to 8 cm resolution space. Large gels have 12 to 20 cm space. The closer the bands are in molecular weight, the longer the gel must be.

A gradient gel is used to resolve a larger molecular weight range than a straight percent gel. A 10% gel resolves proteins from 15 to 100 kDa, while a 4% to 20% gradient gel resolves proteins from 6 to 300 kDa, although the restriction about good molecular weight determination discussed above still holds. Accurate molecular weights can be determined with gradient gels (Podulso and Rodbard, 1980).

What Issues Are Relevant for Isoelectric Focusing?

Isoelectric focusing (IEF) measures the isoelectric point, or pI, of a protein. The main problem for IEF is sample solubility, seen as streaking or in-lane background on the stained IEF gel, or horizontal streaking on a 2-D gel. Sample solubilization should be optimized for each new sample; searching the scientific literature to identify protocols used for similar samples is a good starting point. Information on sample preparation is included below in the discussion about improving resolution.

At present there are two kinds of IEF gels in use: gels formed with carrier ampholytes, and gels formed with acrylamido buffers, known as IPG gels (immobilized pH gradient gels).

The two kinds of gels suffer from problems specific to each kind of gel. For gels formed with carrier ampholytes, the main problem is cathodic drift, the movement of the pH gradient off the basic part of the IEF gel with time. With cathodic drift, the pH gradient gradually drifts off the basic side of the gel, forming a plateau in the center of the pH gradient. Cathodic drift occurs after long focusing times. The drift is controlled by determining the optimum time of focusing in volt-hours, and then always, reproducibly, focusing your gels for the determined number of volt-hours. The optimum time of focusing is determined by performing a time course, setting up identical gels, and then taking them down one by one as time passes, and determining from the results when the proteins have reached the optimum resolution. Gels formed with carrier ampholytes are also limited in the amount of protein that can be focused, since with an overloaded gel, the gradient will deform before all the protein has moved to its pI.

Cathodic drift is completely avoided by the use of IPG gels for isoelectric focusing. The pH gradient is cast into the polyacrylamide gel, which is supported by a plastic backing. There is no cathodic drift because the pH gradient is fixed during the gel-casting step, rather than formed during the first part of the electrophoresis, as with carrier ampholyte gels.

There are major additional advantages to IPG gels: they are much more reproducible than carrier ampholyte gels, and they can focus much more protein than carrier ampholyte gels, up to 5 mg or more, because the fixed pH gradient cannot be overbuffered as above, and because electrophoresis can be carried out at much higher voltage potentials (up to 10,000 volts) and for much longer volt-hours (up to 100,000 volt-hours for 17–18 cm IPG gels). Proteins isolated using 2-D electrophoresis can be sequenced or analyzed by mass sprectrometry, and thus identified. The problems with IPG strips are still being identified. One problem for 2-D electrophoresis seems to be the loss of some hydrophobic (membrane) proteins during transfer of the proteins from the IPG strip to the SDS-PAGE gel (Adessi et al., 1997; Molloy, 2000). Very low and very high molecular weight proteins may also be problematic, as well as basic proteins. Procedures to avoid these problems must be worked out for each sample.

How Can You Resolve Proteins between Approximately 300 and 1000 kDa?

We suggest you use a composite gel for very large proteins. Composite gels are made of 1% acrylamide and 1% low melt agarose. The agarose makes the acrylamide strong enough to handle, and the acrylamide makes the pores in the agarose gel small enough to resolve proteins above about 300 kDa. Composite gels are tricky to pour, as the gel cassette must be warmed to about 40°C, and the gel mixture must be cooled to just above the agarose gelling point before pouring. The mixture must be introduced into the gel casette within a few seconds of adding the catalysts, as acrylamide polymerization takes place within one or two minutes at elevated temperatures. Andrews (1986) has a general procedure for composite gels.

Another option for very large proteins is the use of PAGE with some additive that may enlarge the pore size and thus permit the separation of very large proteins. We have not tested this option, and thus have no recommendations, but Righetti et al. (1992) have used PEG with a standard 5% *T* gel to form much larger pores than normal.

WHAT ISSUES ARE CRITICAL FOR SUCCESSFUL NATIVE PAGE?

Sample Solubility

Native PAGE is performed under conditions that don't denature proteins or reduce their sulfhydryl groups. Solubilizing samples for native PAGE is especially challenging because most nondenaturing detergents do not solubilize complex samples well, and the unsolubilized proteins stick on the gel origin and bleed in, causing in-lane background.

Location of Band of Interest

Sample proteins move in a native gel as a function of their charge as well as their mass and conformation, and because of this, the location of the protein band of interest may be difficult to determine. For instance, in some buffer systems, BSA, at 64 kDa, will move in front of soybean trypsin inhibitor, at 17 kDa (Garfin, 2000). The easiest way to detect the protein of interest is to determine its location by Western blotting. Alternatively, the protein's location can be monitored by enzyme activity or bioassay, which usually requires elution from the gel. Elution is discussed below.

How Can You Be Sure That Your Proteins Have Sufficient Negative Charge to Migrate Well into a Native PAGE Gel?

To determine this, it is useful to have some idea of the pI of the protein of interest. The pH of the buffer should be at least 2 pH units more basic than the pI of the protein of interest. An alternative is to use an acidic buffer system, and reverse the polarity of the electrodes. This works well for very basic proteins.

Buffer Systems for Native PAGE

Buffer systems for native PAGE are either continuous or discontinuous. Discontinuous buffer systems focus the protein bands into thin fine lines in the stacking gel, and these systems are preferred because they provide superior resolution and sample volumes can be larger and more dilute. In a discontinuous buffer system, the buffers in the separating gel and stacking gel, and the upper and lower tank buffers, may all be different in concentration, molecular species, and pH. The reader should initially try the standard Laemmli SDS-PAGE buffer system without the SDS and reducing agent. That buffer system is relatively basic, so most proteins will be negatively charged and run toward the anode. If this is not successful for your protein, consult Chrambach and Jovin (1983), who have published a set of discontinuous buffer systems covering the whole range of pH, for additional discontinuous buffer systems.

Continuous buffer systems have the same buffer throughout the gel, sample and running buffer. Continuous buffer systems can be found in McLellan (1982). Continuous buffer systems are easier to use. For protein gels, the choice between continuous and discontinuous buffer systems is usually made on the basis of what works, and the pI of the protein(s) of interest.

Nucleic acid gels, both PAGE and agarose gels, use the same buffer in all parts of the system: in the gel, in the sample and in the running buffer (urea, which is uncharged, may be omitted from the running buffer). The pH, type of buffer, and buffer concentration are the same throughout the system in most methods of nucleic acid electrophoresis. This makes the gels easy to pour and to run.

The disadvantage of a continuous buffer system is that the samples must be low volume, because the bands in such a system will be as tall or thick as the height of the sample in the well, in a vertical and horizontal slab gel. This is true of both protein or nucleic acid samples.

WHAT CAN GO WRONG WITH THE PERFORMANCE OF A DISCONTINUOUS BUFFER SYSTEM?

In protein electrophoresis, the Laemmli buffer system used for SDS-PAGE has four different buffers, all different in pH, composition, and concentration. Of course, the main voltage potential across the whole gel drives the proteins into and through the gel. However, the differences in buffer pH and concentration set up small voltage potentials within the cell voltage potential. These small voltage potentials form across areas in a lane where the number of ions is lower than elsewhere in the lane, causing the mobility of the macromolecules to increase or decrease, depending on the voltage potential in that specific location in the lane. This is the basis of the "stacking condition" (Hames and Rickwood, 1981).

If the discontinuous buffer protocol is not carried out properly, the small voltage potentials can occur in the wrong places, causing the protein bands to spread out sideways into the next lane, or causing the lane to narrow into a vertical streak of unresolved protein. Thus it is important to make up the buffers for a discontinuous buffer system properly. For instance, in the Laemmli buffer system, the resolving gel buffer is TRIS, pH 8.8 (some authors use pH 8.9). TRIS base is dissolved, and pH'd to the correct value with 6N HCl. If the pH is made too low, and base is added to correct the error, then the total ionic strength of the separating gel buffer will be too high, and the lanes in the gel will narrow. Or, if the pH is too high (not enough HCl), the bands will broaden and smear. (A TRIS-based separating gel buffer takes about 30 minutes to pH correctly. It is best to proceed slowly so that the buffer is made correctly.)

WHAT BUFFER SYSTEM SHOULD YOU USE FOR PEPTIDE ELECTROPHORESIS?

The most favored buffer system currently is that described by Schägger and von Jagow (1987). This discontinuous buffer system uses much higher concentrations of buffer salts, but the ratios of the salts are balanced. So the movement of the small proteins (peptides) is slowed, and they are separated behind the dye front. The results with this buffer system are excellent, and it has been widely used for several years for peptides and proteins up to 100 kDa.

POWER ISSUES

Macromolecules move through a polyacrylamide or agarose gel because they carry a charge at the pH of the buffer used in the system, and the voltage potential put across the cell by the power supply drives them through the gel. This is the effect of the main voltage potential, set by the power supply.

Constant Current or Constant Voltage—When and Why?

The choice of constant current or constant voltage depends on the buffer system, and especially on the size of the gel. Historically constant voltage was used because constant current power supplies were not available. However, currently available programmable power supplies, with constant voltage, constant current, or constant power options, permit any power protocol to be used as needed.

Generally speaking, constant current provides better resolution because the heat in the cell can be controlled more precisely (The higher the current, the higher the heat, and the poorer is the resolution, due to diffusion of the bands.) However, constant current runs will take longer than constant voltage runs (Table 12.2).

Table 12.2 Use of Power Supply Parameters

Procedure	Size of cell or inter–electrode distance	Buffer System	Power Parameter
SDS-PAGE	Mini cell: gel 6–8 cm long	Discontinuous	Constant voltage used routinely; better resolution with constant current
SDS-PAGE	Large cell: gel 16–20 cm long	Discontinuous	Constant current required; use of constant voltage degrades resolution significantly in the bottom $\frac{1}{3}$ of the gel
Native PAGE	Large or mini cell	Discontinuous	Constant current required; use of constant voltage degrades resolution significantly in the bottom $\frac{1}{3}$ of the gel
Native PAGE	Large or mini cell	Continuous	Constant voltage (no advantage to constant current; cooling recommended for good resolution)

Note: Recommended power conditions can vary among manufacturers.

Why Are Nucleic Acids Almost Always Separated via Constant Voltage?

Nucleic acids are usually separated with a continuous buffer system (the same buffer everywhere). Under these conditions, the runs take the same time with constant voltage as with any other parameter held constant, and the resolution is not improved using another parameter as constant. This is usually true for both agarose and polyacrylamide gel electrophoresis.

The use of continuous buffers in nucleic acid electrophoresis makes the gels easy to pour and to run. As with protein separation, small sample sizes must be utilized within continuous buffer systems, particularly when using vertical systems, to prevent bands from overlapping.

Why Are Sequencing Gels Electrophoresed under Constant Power?

Sequencing gels are run under constant voltage or constant power, at a temperature between 50 and 55°C. If constant voltage is used, then the voltage must be changed during the run, after the desired temperature is reached. If constant power is used, the power can be set, and the voltage and current will adjust as the run proceeds, maintaining the elevated temperature required for good band resolution. Elevated temperature and the urea in the sequencing gel maintain the DNA in a denatured condition.

Should You Run Two Sequencing Cells off the Same Power Supply under Constant Power?

If the power supply can draw enough current (power) to accommodate two sequencing cells, one might conclude that two sequencing gels could be run off the same power supply. *Don't do this!* If something happened to one cell, for instance, if the buffer level fell below the level of the gel so that the circuit in that cell was interrupted, then the other cell would carry the power needed for two. The buffer in the second cell would boil away, and the cell would likely catch fire. In practice, it is very difficult to get each cell to carry exactly the same current load through the entire run. When the current loads differ, a vicious cycle/runaway condition can arise, where one cell requires more current to maintain the voltage, causing the power supply to increase its output, but the second cell, because of its lower resistance, receives the additional power. It just isn't safe to run two sequencing cells on one power supply under constant power.

It is acceptable to run two sequencing cells under constant voltage from the same power supply, as long as the power supply can provide the needed current. It is urgently recommended that you remain in the room while the run is proceeding, in case a problem occurs.

IMPROVING RESOLUTION AND CLARITY OF PROTEIN GELS

How Can You Generate Reproducible Gels with Perfect Bands Every Time?

High-quality, reproducible results are generated by using pure, electrophoresis grade chemicals and electrophoresis grade water, by preparing solutions the same way every time and with exact measurement of volumes, by correctly polymerizing your gels the same way every time as discussed above, and by preparing the samples so that they enter the gel completely, without contaminating components that can degrade the resolution. The most important factors for good band resolution and clarity are correct sample preparation and the amount of protein loaded onto the gel, and they are discussed in greater detail below. Finally, the detection procedure must be followed carefully, with attention to detail and elapsed time.

SAMPLE PREPARATION—PROBLEMS WITH PROTEIN SAMPLES

Some samples require exceptional patience and work to determine an optimal preparation protocol. Beyond what follows, a literature search for procedures that worked for proteins similar to yours is recommended.

What Procedures and Strategies Should Be Used to Optimize Protein Sample Preparation?

Consider the cellular location of your protein of interest, and attempt to eliminate contaminating materials at the earliest stages of the purification. If it is a nuclear binding protein, first isolate the nuclei from your sample, usually with differential centrifugation, and then isolate the proteins from the nuclei. If it is a mitochondrial protein, use differential centrifugation to isolate mitochondria (spin the cell lysate at $3000 \times g$ to remove nuclei, then at $10,000 \times g$ to bring down mitochondria). If the protein is membrane bound, use a step gradient of sucrose or other centrifugation medium to isolate the specific membrane of interest. For soluble proteins, spin the cell lysate at $100,000 \times g$ to remove all cellular membranes and use the supernatant. Note that nucleic acids are very sticky; they can cause proteins to aggregate together with a loss of electrophoretic resolution. If you have smearing in your sample, add 1 μg/ml of DNase and RNase to remove the nucleic acids.

Is the Problem Caused by Sample Preparation or by the Electrophoresis?

If a nonprestained standard runs well in a gel, producing sharply defined, well-shaped bands, then any problems in the sample lanes lie in sample preparation or in the sample buffer. For this reason we urge you to run a standard on every gel.

Is the Problem Caused by the Sample or the Sample Buffer?

For lyophilized standards, make fresh standard buffer. Sometimes it is difficult to determine whether the problem is in the sample or the sample buffer. Run the standard both with and without the sample buffer to determine this. It is best to prepare the sample buffer without reducing agent—dithiothreitol (DTT), beta-mercaptoethanol (BME), or dithioerythritol (DTE)—freeze it into aliquots, and add the reducing agent to the aliquot before use. All these reducing agents evaporate readily from aqueous solution. Adding the reducing agent fresh for each use means the reducing agent will always be fresh and in full strength.

Buffer components may separate out during freezing, especially urea, glycerol, and detergents. Aliquots of sample buffer must be mixed thoroughly after thawing, to make sure the buffer is a homogeneous solution.

How Do You Choose a Detergent for IEF or Native PAGE?

Triton X-100 is often used to keep proteins soluble during IEF or native PAGE, but it may solubilize only 70% of the protein in a cell (Ames and Nikaido, 1976). SDS is the best solubilizer, but it cannot be used for IEF because it imparts a negative charge to the proteins. During the IEF, it is stripped off the proteins by the voltage potential, and the formerly soluble proteins precipitate in the IEF gel, resulting in a broad smear. Of course, SDS cannot be used in native PAGE because it denatures proteins very effectively. Some authors state that SDS may be used in combination with other detergents at 0.1% or less. It may help solubilize some proteins when used this way (Molloy, 2000). However, this is not recommended, as the protein loads must remain low, and other problems may result (Molloy, 2000).

Many non-ionic or zwitterionic detergents can be used for IEF or native PAGE to keep proteins soluble. CHAPS (3-[(3-cholamidopropyl)dimethylammonio]-1-propanesulfonate) is most often used, as it is a very good solubilizer, and is nondenaturing. It should be used from 0.1% up to 4.0%. Another very effective solubilizer is SB 3-10 (decyldimethylammoniopropane-sulfonate), but it is denaturing (Rabilloud et al., 1997). Other detergents, designed especially for IEF on IPG gels, have recently been designed and used successfully (Chevallet et al., 1998; Molloy, 2000). The minimum detergent concentration for effective solubilization must be determined for each sample (Rabilloud et al., 1999). Again, to learn what detergent might be effective for your sample, we suggest a literature search.

What Other Additives Can Be Used to Enhance Protein Solubility?

Some proteins are very difficult to solublize for electrophoresis. Urea can be used, from 2 to 8 M or 9.5 M. Thiourea can be used at up to 2 M; it greatly enhances solubility but cannot be used at higher concentration. This is because above 2 M, the urea, thiourea, or detergent may precipitate out (Molloy, 2000). The total urea concentration (urea + thiourea) cannot be above approximately 7.0 M if thiourea is used with a bis gel due to these solubility constraints.

AGAROSE ELECTROPHORESIS

What Is Agarose?

Agarose, an extract of seaweed, is a polymer of galactose. The polymer is 1,3-linked (beta)-D-galactopyranose and 1,4-linked 3,6-anhydro-(alpha)-L-galactopyranose. The primary applications are electrophoresis of nucleic acids, electrophoresis of very large proteins, and immunoelectrophoresis.

What Is Electroendosmosis ($-M_r$ or EEO)?

$-M_r$ is a measure of the amount of electroendosmosis that occurs during electrophoresis with a particular grade of agarose. Electroendoosmosis is the mass movement of water toward the cathode, against the movement of the macromolecules, which is usually toward the anode. High $-M_r$ means high electroendosmosis. The mass flow of water toward the cathode is caused by fixed negative charges in the agarose gel (sulfate and carboxyl groups on the agarose). Depending on the application, electroendosmo-

Table 12.3 Agarose Preparations of Different $-M_r$ Values

Application	Kind of Agarose
Chromosome separation	Pulsed field grade or chromosomal grade; each kind of agarose—molecular biology grade, pulsed field grade, or chromosomal grade—will result in different run times in a pulsed field run, depending on the size of the chromosomes.
Size separation and recovery of DNA or RNA	Low-melt agarose melts at 65°C, and nucleic acids can be recovered with a syringe filter above gelling temperature (35°C).
Isoelectric focusing of proteins	Zero $-M_r$ agarose
Immunoelectrophoresis of proteins (for a review of the many kinds of immunoelectrophoresis, see Axelsen et al., 1973)	Standard low $-M_r$ agarose

sis causes loss of resolution, or it can enable certain kinds of separations to occur, for instance, during counterimmunoelectrophoresis. Applications for agarose preparations of different $-M_r$ values are shown in Table 12.3.

Are Double-Stranded Markers Appropriate for Sizing Large Single-Stranded (Not Oligonucleotide) DNA?

A full discussion is given below under "Standardizing Your Gels."

What Causes Nucleic Acids to Migrate at Unexpected Migration Rates?

Supercoiled DNA is so twisted about itself that it has a smaller Stoke's radius (hydrated radius), and moves faster than some smaller DNA fragments. If supercoiled DNA is nicked, it will unwind or start to unwind during the electrophoresis, and become entangled in the agarose. As this occurs, the DNA slows down its migration, and produces unpredictable migration rates.

What Causes Commercial Preparations of Nucleic Acid Markers to Smear?

There are several reasons why nucleic acid markers smear:

1. Too much marker was added to the lane.
2. The markers were electrophoresed too fast (too hot).
3. The markers were contaminated with DNase.
4. The higher molecular weight markers were sheared by rough pipeting.

What Causes Fuzzy Bands?

The sample might have been degraded by endogenous DNase or that present in the enzymes or reagents used in sample preparation. You may see, "beards" or tails on the bands. For pulsed field samples (in agarose blocks), wash the gel blocks longer and at higher temperatures.

The gel may be running too hot, or the buffer may have been used up, causing high currents that overheat the gel. Turn the voltage down, and remake your buffers, paying careful attention to the dilution and mixing of the stock solution.

Samples loaded too high in the well (overloading) can also produce fuzzy results. DNA near the surface of the gel will run faster than the DNA remaining in solution within the well. The bands will run as inclined planes (\) rather than vertically (|). If the bands are viewed or imaged from directly above they will appear fuzzy. When viewed from a slight angle, the bands will appear normal. The sample should not fill the entire well. Rather, it should occupy half or less of the well. Also the samples should be level and parallel to the surface of the gel in the wells.

Poor-quality agarose can also contribute to a fuzzy appearance. Molecular biology grade or good-quality agarose will prevent this.

Bio-Rad technical support has had a report of a contamination in the user's water that was breaking down the DNA. When the water used for the preparation of the gel and buffers was autoclaved, the problem was eliminated.

ELUTION OF NUCLEIC ACIDS AND PROTEINS FROM GELS

Table 12.4 summarizes the features, benefits and limitations of different elution strategies. DNA purification and elution is also discussed in Chapter 7.

DETECTION

What Should You Consider before Selecting a Stain?

There are several factors to consider before selecting a stain, primary among them the sensitivity needed. Tables 12.5 and 12.6

Table 12.4 Comparison of Elution Strategies

Medium or Macromolecule	Feature	Benefit	Limitation
Agarose			
Nucleic acids	Freeze and Squeeze—Cut out the band of interest from the gel, put it in an Eppendorf microtube, and freeze it. This destroys the structure of the agarose gel. Then cut off the bottom of the Eppendorf tube, put the microtube into a slightly larger tube, and spin it down. The liquid containing the band of interest will be in the larger tube, and the agarose will remain in the smaller tube.	Easy and fast	Such kits don't work with oligos or very large nucleic acids
	Electroelution	None—not recommended	Recoveries low; nucleic acids bind to dialysis membrane
Oligonucleotides	Freeze and squeeze kits	Easy and fast	Not good below 30 bp, which don't electrophorese in an agarose gel
Proteins	Freeze and squeeze kits	Easy and fast	Buffer systems not worked out for very large proteins
Polyacrylamide			
Nucleic Acids	BAC crosslinkers	Excellent recoveries	Require subsequent column to separate nucleic acids from decrosslinked polyacrylamide
Oligonucleotides	Crush gels in an equal volume of elution buffer; let sit overnight	Easy to do, requires no equipment	Best reovery no more than 50%
Proteins	Electroelution	Excellent recoveries	Some proteins bind to dialysis membrane
	Crush gel piece in an equal volume of elution buffer, let sit overnight	Relatively easy to do, requires no equipment	Best recovery no more than 50%

(continued)

Table 12.4 (*Continued*)

Medium or Macromolecule	Feature	Benefit	Limitation
	BAC, DADT, DHEBA crosslinkers (significant amounts of acrylamide remain in polymerized gels with DADT and DHEBA being a safety issue). Not recommended.	Good recoveries possible with certain proteins, depending on subsequent application	Require subsequent column to separate protein from decrosslinked polyacrylamide; periodate oxidizes sulfhydryl containing amino acid sidechains; polypeptides with sulfhydryl groups bind to BAC-crosslinked matrix
	Preparative Electrophoresis	Excellent recoveries	May require fraction collector, peristaltic pumps, chillers, other accessories
Peptides	Electroelution	Excellent recoveries possible, depending on nature and size of peptide.	Time and power conditions must be optimized for especially small peptides to prevent their being driven into the dialysis membrane

provide a general guide to stain sensitivity, and mention other considerations.

Will the Choice of Stain Affect a Downstream Application?

This is an important question. Colloidal Coomassie and Sypro® Ruby can be used on 2-D gels when mass spectrometry (mass spec) is the detection procedure. Certain silver stains can also be used to stain samples for mass spec analysis because of improvements in the sensitivity of mass spectrometers. Sypro Red covers three orders of magnitude, Coomassie covers two, and silver stains provide coverage over one magnitude. Not all silver stains give good mass spectrometry results and those which are used are not as good as Coomassie or Sypro Ruby (Bio-Rad Laboratories, R&D).

For amino acid sequencing, the gel is usually blotted to PVDF, stained for the protein of interest, and then sequenced. Immunodetection or other more sensitive methods can be used, but usually the sequencing requires at least 1 μg of protein. For

Table 12.5 Common Protein Stains

Stain	Application	Sensitivity	Benefits/Limitations
Coomassie brilliant blue R-250 (with MeOH/HOAc)	SDS-PAGE	1 μg protein per band	Easy, traditional stain; low sensitivity, high disposal costs
Coomassie brilliant blue G-250 (colloidal, low or no MeOH)	SDS-PAGE, 2-D, native PAGE, IEF	100 ng per band	Much better sensitivity, easy disposal; long staining times for best results
Silver stain	SDS-PAGE, 2-D, native PAGE, IEF	10 ng per band	Excellent sensitivity, tricky to perform, requires excellent quality water
Copper stain (requires SDS to work)	SDS-PAGE only	10–100 ng per band	Fast and easy, good before blotting
Zinc stain (requires SDS to work)	SDS-PAGE only	10–100 ng per band	Fast and easy, good before blotting
Sypro Orange (requires SDS to work)	SDS-PAGE 2-D	10 ng per band	Published sensitivities may be difficult to attain; SDS concentration critical
Sypro Ruby	SDS-PAGE 2-D	10 ng per band	Easy to use, expensive, stain of choice for 2-D and subsequent mass spectrometry and quantitative analysis

Table 12.6 Common Nucleic Acid Stains

Stain	Application	Sensitivity
Ethidium bromide	Sub-cell gels. Note that this stain is carcinogenic and is viewed only on a UV light box. Good safety practices are mandatory with this stain. Disposal is also an issue.	1–10 ng
Silver stain	PAGE gels, agarose gels with certain silver stains. Disposal is an issue.	1–10 ng
Stains all	Stains various cell components with different colors.	100 ng–1 μg

this reason we suggest that you stain your blot with Coomassie. This does not interfere with sequencing. Note that if you want to blot your gel after staining, only reversible stains such as copper stain and zinc stain can be used with good success. If you stain your gel with Coomassie or silver, the proteins are fixed in the gel and are very difficult to transfer to a membrane. Only copper or zinc stains are recommended before blotting a gel for immune detection.

Is Special Equipment Needed to View the Stain?

A light box is helpful for viewing the colored stains—Coomassie, silver, copper, and zinc—on gels. Digitizing the stained image from the gel is the best way to store the data for silver-stained gels, as they darken when dried. Fluorescent stains require at least a UV light box, and may require a fluorescent imager or other specialized scanner, depending on the excitation and emission wavelengths of the chosen stain.

How Much Time Is Required for the Various Stains?

The speed of staining is quite variable depending on the quality of water, the temperature, and how closely the staining steps are timed. Gels stained with Coomassie can be left in stain from 30 minutes to overnight, but longer staining times will require much longer destaining times, and more changes of destain solution. Colloidal Coomassie may require several days in the stain for optimum sensitivity and uniformity of staining. Silver stain must be timed carefully for best results. There are many silver staining protocols; most can be completed in 1.5 to 4 hours. Both copper and zinc staining require only 5 to 10 minutes. The fluorescent stains have various time requirements, usually from a few minutes to an hour at most. It is recommended that the protocols for fluorescent staining be followed carefully for best results.

What If You Need to Quantify Your Stained Protein?

The amino acid composition of the protein of interest will affect stain performance. No general rules are available, but some proteins stain better with Coomassie, for instance, and others stain better with silver. Both of these stains are adequate for relative quantitation of your protein (i.e., "The treated band is 2× denser than the untreated sample."). It is useful to consult the literature for information on the staining characteristics of your protein of interest.

If you must obtain the absolute amount of your protein, the best standard to use is the protein of interest itself. If the protein of interest is not available in purified form to run in a separate lane in a known amount, then bovine gamma globulin gives a better standard curve than bovine serum albumin with Coomassie brilliant blue R-250 or G-250. BSA is stained much more densely with Coomassie than other proteins at the same concentration, restricting its use as a standard. We do not recommend any silver stain for quantitation, unless you are sure your protein of interest responds the same way to silver as the protein chosen as the standard.

Note also that most silver stains provide only one absorbance unit of linearity, whereas Coomassie will provide 2 to 2.5 absorbance units of linearity. Sypro Ruby is linear over 3 absorbance units. These generalizations may or may not apply to your protein of interest; the amount of linearity of a stain on a particular protein must be assessed anew for each protein.

What Causes High Background Staining?

Impure Reagents and Contaminants from Earlier Procedures

The effect of chemical impurities was discussed above. If the SDS within the PAGE gel is contaminated with C10, C14, or C16 forms of the detergent, Coomassie brilliant blue and silver may stain the background of the gel. These and other detergents, urea, carrier ampholytes, and other gel components may also be stained. They should be removed by fixation before the stain is applied.

Certain buffer and gel components can also contribute to background staining, which can be prevented if a gel is fixed before staining. Which fixative to use depends on the gel type and the stain. When using Coomassie (or colloidal Coomassie), SDS-PAGE gels should be fixed in the same solution used to prepare the stain. The several osmotic potentials that exist between the fixing solution and the buffers within the gel cause the TRIS, glycine, and SDS to leave the gel, making for a much cleaner background.

IEF gels should be fixed in 10% trichloroacetic acid, 40% MeOH, and if possible, 2.5% sulfosalicylic acid, since the latter helps remove carrier ampholytes. Immobilized pH gradient gels, IPG gels, are not usually stained with silver, but they can be stained with colloidal Coomassie. It is sometimes useful to stain the IPG strips as an aid in diagnosis of problems with the 2-D slab gels.

Will the Presence of Stain on Western-Blotted Proteins Interfere with Subsequent Hybridization or Antibody Detection Reactions?

Proteins can be detected on a blot after staining the blot with a general protein stain such as Coomassie or colloidal gold, but the interference with subsequent immunodetection will be high (Frank Witzman, 1999). The interference can be 50% or more, but this may not matter if the protein of interest is in high abundance.

Proteins which have been stained in the gel will not transfer out of the gel properly, and it is unlikely that an immuno detection procedure will be successful. It is usual to run duplicate gels or run duplicate lanes on the same gel and cut the gel in half, if you want to both stain and blot the protein of interest.

Does Ethidium Bromide Interfere with the Common Enzymatic Manipulation of Nucleic Acids?

Ethidium bromide does not usually interfere with the activities of most common DNA modifying enzymes. However, ethidium bromide has been shown to interfere with restriction endonucleases (Soslau and Pirollo, 1983; Parker et al., 1977).

STANDARDIZING YOUR GELS

What Factors Should Be Considered before Selecting a Molecular Weight Marker?

Ask yourself whether you need exact or approximate molecular weight values. If you need exact values, you must use a standard that will form thin tight bands at the same location from batch to batch. Most pre-stained standards do not form such thin, tight bands, and are good for only "ball park" molecular weight values and assessing transfer efficiencies.

You might also ask whether you will run native or denatured gels. Denatured gels, usually SDS-PAGE gels, provide exact molecular weights because of the elimination of the charge on the protein as a factor in the electrophoresis. (Negatively charged SDS coats the proteins, hiding the native charge on the proteins, and providing a constant charge to mass ratio.)

Native gels provide results which reflect the charge, size and shape of the proteins. It is not acceptable to measure molecular weight by native electrophoresis, because more than one parameter is measured during this technique. Some companies sell "molecular weight standards" for native gels, but these standards have no scientific validity. Molecular weights can be determined for native gels by means of a Fergusson plot (Andrews, 1986). Proteins can be used to measure whether the electrophoresis is reproducible, and can provide information on the relative separation of various bands from each other. However, because more than one parameter influences the movement of the proteins in the gel, they cannot be used to measure molecular weight.

Another factor that affects the migration rate in any kind of gel is the protein's amount and type of posttranslational modification. Proteins with significant glycosylation will run more slowly than their total molecular weight might suggest (Podulso, 1981). It is also possible to use gradient gels for molecular weight determination (Lambin and Fine, 1979; Podulso and Rodbard, 1980).

Are Double-Stranded Markers Appropriate for Sizing Large (Not Oligonucleotide) Single-Stranded DNA? If Not, Which Markers Are Recommended?

Double-stranded DNA size markers are not appropriate for sizing large single-stranded DNAs. Most labs with need of such markers obtain single-stranded DNA (usually phage DNA), calibrate it for size by sequencing it, and use that as a single-stranded DNA marker. Since the mobility of many single-stranded nucleic acids is variable, it is recommended to cross-calibrate with a second single-stranded source (e.g., a different phage).

Can a Pre-stained Standard Be Applied to Determine the Molecular Weight of an Unknown Protein?

Pre-stained protein standards usually run as broad, fuzzy bands, making them useful for approximate, but not exact, molecular weight determinations. Thus they can be used to report only approximate molecular weights (within 10,000 daltons of the molecular weight as determined by an unstained standard). The molecular weight values of most pre-stained standards vary from batch to batch because the conjugation reaction between marker protein and dye marker is not perfectly reproducible.

Some vendors now offer pre-stained recombinant proteins of known, reproducible molecular weights. The bands in these protein standards form thin, tight bands, and they can be used for accurate molecular weight determination.

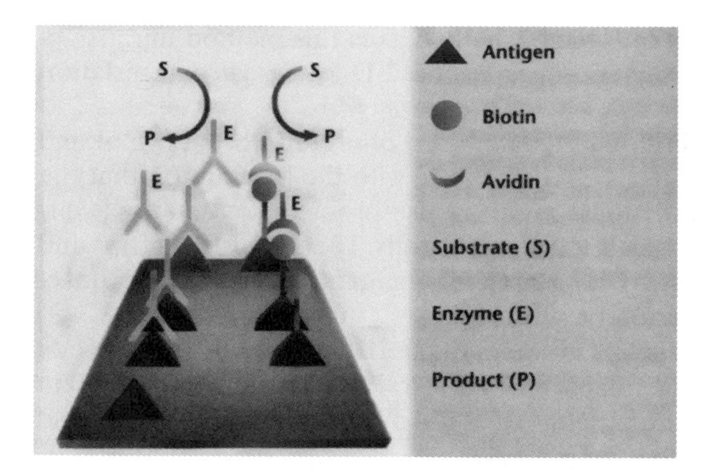

Figure 12.5 Use of biotinylated protein standards to calculate molecular weight on Western blots. Permission to use this Figure has been granted by Bio-Rad Laboratories, Inc.

How Do You Determine Molecular Weight on a Western Blot?

Use biotin-labeled molecular weight markers, and detect them with streptavidin-conjugated horseradish peroxidase or alkaline phosphatase. The streptavidin conjugate that will detect the markers is added to the solution containing the labeled secondary antibody (e.g., horseradish peroxidase or alkaline phosphatase) that will subsequently react with the sample proteins (Figure 12.5). These markers will provide precise molecular weight values.

The pre-stained recombinant proteins of known, reproducible molecular weights discussed above can also determine the molecular weights of proteins on a blot.

Some researchers will cut off the molecular weight standard lane from the blot and stain it with Coomassie or Amido Black, and then realign the stained standards with the rest of the blot once it has been processed. The problem with this approach is that the nitrocellulose can slightly shrink or swell, causing the bands to misalign. Other researchers simply feel uncomfortable about the prospect of perfectly aligning the segments after cutting, so this is not recommended.

What Are the Options for Determining pI and Molecular Weight on a 2-D Gel?

There are several ways to do this:

1. Add proteins of known (denatured) pI and MW to your sample and electrophorese the standards within the same gel. The added proteins are often difficult to detect within the

2-D spot pattern, which usually makes this method unsatisfactory. It may be appropriate for 2-D of *in vitro* translation products.

2. Use a 2-D standard comprised of proteins of known pI and MW, and run it on a separate gel, with the assumption that the gels will run identically. This is also problematic, since it is difficult to get the gels to run identically. The use of IPG strips and pre-cast slab gels helps, but drying artifacts may cause unacceptable variation between gels.

3. Measure the pH gradient of the IEF gel with a pH electrode (see below and Chapter 4, "How to Properly Use and Maintain Laboratory Equipment,") and use a MW standard in the second dimension to determine MW.

4. Carbamylate a protein of known (denatured) pI, and add it to the sample (Tollaksen, 1981). A protein with a MW not seen in the sample should be used. The carbamylated protein will run as a series of spots starting with the spot of known pI. Each spot to the acidic side will be 0.1 pH unit more acidic than the one to the basic side. Carbamylated proteins are also commercially available.

5. If you are electrophoresing a well-characterized sample, such as *E. coli* or mouse liver, compare your pI and MW data to online databases such as those available at *http://www.expasy.ch/*. This is the preferred option if your sample is present in such a database. If such a database is not available for your sample, you should use 2 of the above methods.

How Do You Measure the pH Gradient of a Tube IEF Gel or an IPG Gel?

Several methods are presented here. None are very satisfactory, as there are problems with them all.

To document the pH gradient, measure the migration distance for several proteins of known pI, and create a standard curve by plotting the pI value of your marker against the R_f value. You will need to normalize your standard proteins so that you can compare gels.

Several commercial products, comprised of colored proteins of known pI, are available for native IEF. However, these standards cannot be used for 2-D gels, since native pI values differ from the pI value of the same protein under denaturing conditions. The native pI value is based on the surface charge and conformational effects of the protein. In 2-D gels all amino acid side chains are exposed and affect the migration of the protein in denaturing conditions, thus altering the pI.

A second approach is to directly measure the pH throughout the length of the gel (this works only with carrier ampholyte tube gels). Slice the gel into 1, 5, or 10mm sections, and put the pieces into numbered tubes. Next, add 1.0ml of 50mM KCl to each tube, place them inside a vacuum dessicator without dessicant, and draw a vacuum on the tubes. Incubate overnight at room temperature, and measure the pH of the ampholyte solution, starting from the acidic end, after 24 hours. Incubation for 24 hours is recommended to ensure that equilibrium of the ampholyte concentration in the gel piece and the liquid has occurred. The potassium chloride and vacuum are required to prevent atmospheric CO_2 from affecting the pH of the solutions. The potassium chloride also helps the pH electrode work more easily in solutions with low concentrations of ampholytes. The problem with this procedure is that it is difficult to cut the gel into exact, reproducibly sized sections.

As decribed in Chapter 4, "How to Properly Use and Maintain Laboratory Equipment," electrodes are available that can directly measure the pH of a gel. There are two kinds: flat-bottomed electrodes, suitable for a flat strip gel, and microelectrodes, which must be inserted into the (tube) gel. Flat-bottomed electrodes usually have the reference electrode to the side, as a little piece of glass sticking out. The reference electrode must be parallel with the main electrode, at the same pH in use. The microelectrode has the reference electrode in a circular shape around the main electrode. Both types require some getting used to, but provide good results when used carefully and in a reproducible manner.

Veteran proteomics researchers identify proteins in their samples by comparison of their spot patterns to those in Web-based 2-D databases, and choose known proteins to sequence and measure by mass spectrometry. Once those proteins have been compared and identified for sure, they can be used as internal pI and MW standards. Usually constituitive proteins that do not vary in concentration are used. (Wilkins et al., 1997) Most 2-D data analysis software packages can establish a pH gradient once spots of known pI are specified.

Some groups report the use of pH paper to get a very rough idea of the pH gradient (personal communication from Bio-Rad customers), but this is not recommended because it lacks precision.

In the case of IPG strips, you may assume that if you have a pH 3 to 10 gel, that you can measure the length of the gel from end to end, and divide it up into pH units. This is valid only for a rough idea of the pI of a protein of interest. Manufacturers' specifications for the length of the gels ranges from ±5 to ±2 mm, and the pH gradient on the gel may also vary enough to change the location of a pH on the gel.

TROUBLESHOOTING

What Is This Band Going All the Way across a Silver-Stained Gel, between Approximately 55 and 65 kDa?

The band most likely contains skin keratin, originating from fingers, flakes of skin, or hair dander (dandruff) within the gel solutions or running buffer. This band, which may be quite broad, is usually detected only with more sensitive staining methods, such as silver. There is usually only one band and the molecular weight varies depending on the type of skin keratin. Ochs (1983) demonstrates conclusively that this band is due to skin keratin contamination.

How Can You Stop the Buffer Leaking from the Upper Chamber of a Vertical Slab Cell?

The upper chamber should be set up on a dry paper towel before the run with the upper buffer in it, and let stand for up to 10 minutes to determine if there are any leaks from the upper chamber. In some cells the leaks can be stopped by filling up the lower chamber to the same height as the liquid in the upper chamber. This eliminates the hydrostatic head causing the leak, and the run can proceed successfully. Otherwise, make sure the cell is assembled correctly, and if the problem persists, contact the cell's manufacturer.

BIBLIOGRAPHY

Adessi, C., Miege, C., Albrieux, C., and Rabilloud, T. 1997. Two-dimensional electrophoresis of membrane proteins: A current challenge for immobilized pH gradients *Electrophoresis* 18:127–135.

Albaugh, G. P., Chandra, G. R., Bottino, P. J. 1987. Transfer of proteins from plastic-backed isoelectric focusing gels to nitrocellulose paper. *Electrophoresis* 8:140–143.

Allen, R. C., and Budowle, B. 1994. *Gel Electrophoresis of Proteins and Nucleic Acids: Selected Techniques.* Walter de Gruyter, New York.

Allen, R. C., Saravis, C. A., and Maurer, H. R. 1984. *Gel Electrophoresis of Proteins and Isoelectric Focusing: Selected Techniques.* Walter de Gruyter, New York.

Ames, G. F. L., and Nikaido, K. 1976. Two-dimensional gel electrophoresis of membrane proteins. *Biochem.* 15:616–623.

Anderson, B. L., Berry, R. W., and Telser, A. 1983. A sodium dodecyl sulfate-polyacrylamide gel electrophoresis system that separates peptides and proteins in the molecular weight range of 2500 to 90,000. *Anal. Biochem.* 132:365–375.

Andrews, A. T. 1986. *Electrophoresis, Theory, Techniques and Biochemical and Clinical Applications*, 2nd ed. Monographs on Physical Biochemistry. Clarendon Press, Oxford, U.K.

Axelsen, N. H., Krilll, J., and Weeks, B., eds. 1973. A manual of quantitative immunoelectrophoresis. *Scand. J. Immunol.* suppl. 1, 2.

Bio-Rad Laboratories, 2000 Acrylamide Material Safety Data Sheet (MSDS). Document number 161–01000 MSDS CAS Number 79-06-01.

Caglio, S., and Righetti, P. G. 1993. On the pH dependence of polymerizaion efficiency, as investigated by capillary zone electrophresis. *Electrophoresis* 14:554–558.

Chevallet, M., Santoni, V., Poinas, A., Rouquie, D., Fuchs, A., Keiffer, S., Rossignol, M., Lunardi J., Garin, J., and Rabilloud, T. 1998. New zwitterionic detergents improve the analysis of membrane proteins by two-dimensional electrophroesis. *Electrophresis* 19:1901–1909.

Chiari, M., Chiesa, C., Righetti, P. G., Corti, M., Jain T., and Shorr R. 1990. Kinetics of cysteine oxidation in immoibilized pH gradient gels. *J. Chrom.* 499:699–711.

Chrambach, A., and Jovin, T. M. 1983. Selected buffer systems for moving boundary electrohporesis of gels at various pH values, presented in a simplified manner. *Electrophoresis* 4:190–204.

Cytec Industries. 1995. *Acrylamide Aqueous Solution, Handling and Storage Procedures*. Self-published booklet. West Paterson, New Jersey. p. 3.

Dow Chemical. 1988. Aqueous acrylamide monomer, safe handling and storage guide, health, environmental, and toxicological information, specifications, physical properties, and analytical methods. Unpulished binder. Midland, MI.

Garfin, D. Personal communication. 2000. Bio-Rad Laboratories Research and Development Dept., Hercules, CA.

Gianazza, E., Rabilloud, T., Quaglia, L., Caccia, P., Astrua-Testori, S., Osio, L., Grazioli, G., and Righetti, P. G. 1987. Additives for immobilized ph gradient two-dimensional separation of particulate material: Comparison between commerical and new synthetic detergents. *Anal. Biochem.* 165:247–257.

Görg, A., Postel, W., Günther, S., and Weser, J. 1985. Improved horizontal two-dimensional electrophoresis with hybrid isoelectroic focusing in immobilized pH gradients in the first dimension and laying-on transfer to the second dimension. *Electrophoresis* 6:599–604.

Granier, F. 1988. Extraction of plant proteins for two-dimensional electrophoresis. *Electrophoresis* 9:712–718.

Hames, B. D., and Rickwood, D., eds. 1981. *Gel Electrophoresis of Proteins: A Practical Approach*. IRL Press, Washington, DC.

Hansen, J. N. 1984. Personal communication.

Hansen, J. N. 1981. Use of solubilizable acrylamide disulfide gels for isolation of DNA fragments suitable for sequence analysis. *Anal. Biochem.* 116:146–151.

Hansen, J. N., Pheiffer, B. H., and Boehnert, J. A. 1980. Chemical and electrophoretic properties of solubilizable disulfide gels. *Anal. Biochem.* 105:192–201.

Herbert, B. R., Molloy, M. P., Gooley, A. A., Walsh, B. J., Bryson, W. G., and Williams, K. L. 1998. Improved protein solubility in two-dimensional electrophoresis using tributyl phosphine as reducing agent. *Electrophoresis* 19:845–851.

Herbert, B. 1999. Advances in protein solubilisation for two-dimensional electrophoresis. *Electrophoresis* 20:660–663.

Hjelmeland, L. M., Nebert, D. W., and Chrambach, A. 1978. Electrophoresis and electrofocusing of native membrane proteins. In Catsumpoolas, N., ed., *Electrophoresis '78*, Elsevier North-Holland, New York.

Hjelmeland, L. M., Nebert, D. W., and Osborne Jr., J. C. 1983. Sulfobetaine derivatives of bile acids: Nondenaturing surfactants for membrane biochemistry. *Anal. Biochem.* 130:72–82.

Hochstrasser, D. F., Patchornik, A., and Merril, C. R. 1988. Development of polyacrylamide gels that improve the separation of proteins and their detection by silver staining. *Anal. Biochem.* 173:412–423.

Kusukawa, N., Ostrovsky, M. V., and Garner, M. M. 1999. Effect of gelation conditions on the gel structure and resolving power of agarose-based DNA sequencing gels. *Electrophoresis* 20:1455–1461.

Kyte, J., and Rodriguezz, H. 1983. A discontinuous electrophoretic system for separating peptides on polyacrylamide gels. *Anal. Biochem.* 133:515–522.

Lambin, P., and Fine, J. M. 1979. Molecular weight estimation of proteins by electrophoresis in linear polyacrylamide gradient gels in the absence of denaturing agents. *Anal. Biochem.* 98:160–168.

McLellan, T. 1982. Electrophoresis buffers for polyacrylamide gels at various pH. *Anal. Biochem.* 126:94–99.

Molloy, M. 2000. Two-dimensional electrophoresis of membrane proteins on immobilized pH gradients. *Anal. Biochem.* 280:1–10.

Ochs, D. 1983. Protein contaminants of sodium dodecyl sulfate-polyacrylamide gels. *Anal. Biochem.* 135:470–474.

Parker, R. C., Watson, R. M., and Vinograd, J. 1977. Mapping of closed circular DNAs by cleavage with restriction endonucleases and calibration by agarose gel electrophoresis. *Proc. Natl. Acad. Sci. USA* 74:851–855.

Poduslo, J. F. 1981. Glycoprotein molecular-weight estimation using sodium dodecyl sulfate-pore gradient electrophoresis: Comparison of TRIS-glycine and TRIS-borate-EDTA buffer systems. *Anal. Biochem.* 114:131–139.

Podulso, J. F., and Rodbard, D. 1980. Molecular weight estimation using sodium dodecyl sulfate-pore gradient electrophoresis. *Anal. Biochem.* 101:394–406.

Rabilloud, T. 1996. Solubilization of proteins for electrophoresic analyses. *Electrophoresis* 17:813–829.

Rabilloud, T. 1998. Use of Thiourea to increase the solubility of membrane proteins in two-dimensional electrophoresis. *Electrophoresis* 19:758–760.

Rabilloud, T., Valette, C., and Lawrence, J. J. 1994. Sample application by in-gel rehydration improves the resolution of two-dimensional electrophoresis with immobilized pH gradients in the first dimension. *Electrophoresis* 15:1552–1558.

Rabilloud, T., Adessi, C., Giraudel, A., and Lunardi, J. 1997. Improvement of the solubilization of proteins in two-dimensional electrophoresis with immobilized pH gradients. *Electrophoresis* 18:307–316.

Rabilloud, T., Bilsnick, T., Heller, M., Luche, S., Aebersold, R., Lunardi, J., and Braun-Breton, C. 1999. Analysis of membrane proteins by two-dimensional electrophoresis: Comparison of the protein extracted from normal or *Plasmodium falciparum*-infected erythrocyte ghosts. *Electrophoresis* 20:3603–3610.

Righetti, P. G., Chiari, M., Casale, E., and Chiesa, C. 1989. Oxidation of alkaline immobiline buffers for isoelectric focusing in immobilized pH gradients. *Appl. Theoret. Electrophoresis* 1:115–121.

Righetti, P. G., Caglio, S., Saracchi, M., and Quaroni, S. 1992. "Laterally aggregated" polyacrylamide gels for electrophoresis. *Electrophoresis* 13:387–395.

Rüchel, R., Steere, R. L., and Erbe, E. F. 1978. Transmission-electron microscopic observations of freeze-etched polyacrylamide gels, *J. Chromatog.* 166:563–575.

Soslau, G., and Pirollo, K. 1983. Selective inhibition of restriction endonuclease cleavage by DNA intercalators. *Biochem Biophys. Res. Commun.* 115:484–489.
Schägger, H., and von Jagow, G. 1987. Tricine-sodium dodecyl sulfate-polyacrylamide gel electrophoresis for the separation of proteins in the range from 1 to 100 kDa. *Anal. Biochem.* 166:368–379.
Tollaksen, S. L., Edwards, J. J., and Anderson, N. G. 1981. The use of carbamylated charge standards for testing batches of ampholytes used in two-dimensional electrophoresis. *Electrophoresis* 2:155–160.
Wilkins, M. R., Williams, K. L., Appel, R. D., and Hochstrasser, D. F. (eds.). 1997. *Proteome research: New frontiers in fuctional genomics. Principles and practice.* Springer Verlag, Berlin.
Witzman, F. 1999. Personal communication.

APPENDIX A

PROCEDURE FOR DEGASSING ACRYLAMIDE GEL SOLUTIONS

Degas your acrylamide solution in a side-arm vacuum flask with a cork that is wider han the flask opening for 15 minutes with gentle stirring (Figure 12.6). Use at least a bench vacuum to degas (20–23 inches of mercury in most buildings); a water aspirator on the sink is not strong enough (at most 12–16 inches of mercury). A vacuum pump (>25 inches of mercury) is best. When the solution bubbles up and threatens to overflow into the side arm, release the vacuum by quickly removing the cork from the top of the flask. Then replace the cork, swirl the solution, and continue the procedure. The solution will bubble up four or five times, and then most of the air will be removed. Continue degassing for 15 minutes total. The degassing is a convenient time to weigh out 0.1 g of APS in a small weigh-boat and to test its potency as described in the text.

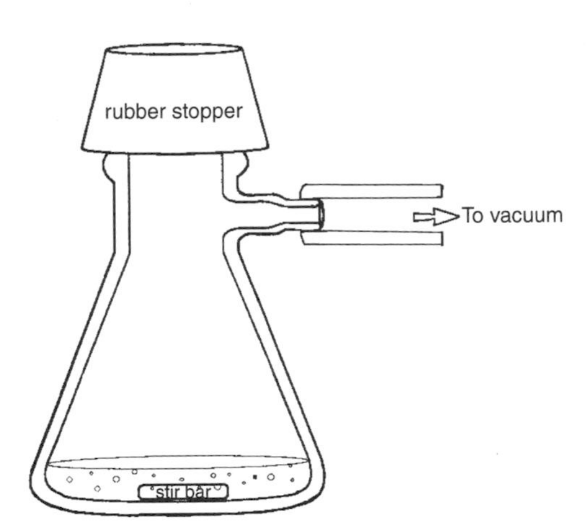

Figure 12.6 Vacuum flask strategy to eliminate dissolved oxygen from acrylamide solutions. Reproduced with permission from Bio-Rad Laboratories.

13

Western Blotting

Peter Riis

Physical Properties of Proteins 374
What Do You Know about Your Protein? 374
What Other Physical Properties Make Your Protein Unusual? 374
Choosing a Detection Strategy: Overview of Detection Systems 375
What Are the Criteria for Selecting a Detection Method? 377
What Are the Keys to Obtaining High-Quality Results? ... 379
Which Transfer Membrane Is Most Appropriate to Your Needs? 379
Blocking 380
Which Blocking Agent Best Meets Your Needs? 381
Washing 382
What Composition of Wash Buffer Should You Use? 382
What Are Common Blot Size, Format, and Handling Techniques? 382
The Primary Antibody 383
Are All Antibodies Suitable for Blotting? 383
How Should Antibodies Be Handled and Stored? 384
Secondary Reagents 384
How Important Is Species Specificity in Secondary Reagents? 385
Why Are Some Secondary Antibodies Offered as $F(ab')_2$ Fragments? 385
Amplification 387

Molecular Biology Problem Solver, Edited by Alan S. Gerstein.
ISBN 0-471-37972-7

Stripping and Reprobing 388
Will the Stripping Procedure Affect the Target Protein? ... 388
Can the Same Stripping Protocols Be Used for Membranes from Different Manufacturers? 389
Is It Always Necessary to Strip a Blot before Reprobing? 389
Troubleshooting 389
Setting Up a New Method 396
Bibliography 397

PHYSICAL PROPERTIES OF PROTEINS

What Do You Know about Your Protein?

In order to make informed choices among the bewildering range of available transfer and detection methods, it is best to have as clear an idea as possible of your own particular requirements. In large part these choices will depend on the nature of your target protein. Even limited knowledge can be used to advantage.

How abundant is your protein? It isn't necessary to answer the question in rigorously quantitative terms: an educated guess is sufficient. Are your samples easy to obtain and plentiful, or limited and precious? Is the sample likely to be rich in target protein (e.g., if the protein is overexpressed) or poor in target (perhaps a cytokine)? Obviously low protein concentration or severely limited sample size would require a more sensitive detection method.

What is the molecular weight of your target protein? Low MW proteins (12 kDa or less) are retained less efficiently than higher molecular weight proteins. Membranes with a pore size of 0.1 or 0.2 micron are recommended for transfer of these smaller proteins, and PVDF will tend to retain more low MW protein than nitrocellulose. The ultimate lower limit for transfer is somewhere around 5 kDa, although this depends largely on the protein's shape and charge.

The transfer of high molecular weight proteins (more than 100 kDa) can benefit from the addition of up to 0.1% SDS to the transfer buffer (Lissilour and Godinot, 1990). Transfer time can also be increased to ensure efficient transfer of high molecular weight proteins.

What Other Physical Properties Make Your Protein Unusual?

In cases where proteins are highly basic (where the pI of the protein is higher than the pH of the transfer buffer) the protein will not be carried toward the anode, since transfer takes place on the basis of charge. In these cases it is best to include SDS in the transfer buffer. Alternatively, the transfer sandwich can be assembled with membranes on both sides of the gel.

CHOOSING A DETECTION STRATEGY: OVERVIEW OF DETECTION SYSTEMS

Detection systems range from the simplest colorimetric systems for use on the benchtop to complex instrument-based systems (Table 13.1). The simplest is radioactive detection: a secondary reagent is labeled with a radioactive isotope, usually the low-energy gamma-emitter iodine-125. After the blot is incubated with the primary antibody, the labeled secondary reagent (usually Protein A, but it can be a secondary antibody) is applied, the blot

Table 13.1 Comparison of Detection Methods

Method	Features	Limitations	Sensitivity
Radioactive	Can quantitate through film densitometry; can strip and reprobe blots; no enzymatic development step	Use of radioactive material can be difficult and expensive; requires licensing and radiation safety training	1 pg
Colorimetric	Easy to perform; hard copy results directly on blot; minimal requirements for facilities and equipment	Relatively insensitive	200 pg
Chemiluminescent	Highly sensitive; can quantitate using film densitometry; can strip and reprobe	Requires careful optimization	1 pg (luminol) 0.1 pg (acridan)
Fluorescent	Good linear range for quantitation; data stored digitally	Equipment expensive; stringent membrane requirements; stripping and reprobing possible but difficult	1 pg

washed and exposed to film for hours or days. Radioactive blots can more quickly be detected using storage phosphor plates instead of film; the plates are read on a specialized scanning instrument. Detailed discussions about the features and benefits of detection by film and scanners are included in Chapter 14, Nucleic Acid Hybridization.

Enzymatic reactions are used in a number of different systems to indicate the presence of bound antibody. The simplest type of enzymatic detection is chromogenic. Here the secondary reagent is conjugated to an enzyme, either horseradish peroxidase (HRP) or alkaline phosphatase (AP). After incubation with the secondary reagent and washing, the blot is incubated with a substrate. The enzyme catalyzes a reaction in which the substrate is converted to a colored precipitate directly on the membrane, essentially coloring the band on which the primary antibody has bound. While not as sensitive as other methods, colorimetric detection is fast and simple, and requires no special facilities.

Chemiluminescent detection combines characteristics of both radioactive and chromogenic detection. Again, an enzyme label is used (commonly HRP, but there are systems for use with AP as well), but in this case the reaction produces light rather than a colored product as a result of reaction. The light is usually captured on X-ray film, just like a radioactive blot. Specialized imaging equipment for chemiluminescent blots is also available. Chemiluminescent detection is very sensitive, and the blots are easily stripped for subsequent reprobing.

There are significant differences in the various available chemiluminescent detection systems. The most widely used are the luminol-based HRP systems. These typically emit usable signals for an hour or two. There are also newer, higher-sensitivity HRP-based systems that emit light for more than 24 hours; however, these substrates are more expensive and require even more careful optimization than the luminol-based systems. AP-based chemiluminescent systems are also available. They are not widely used in Western blotting, but they are highly sensitive and also emit light for extended periods. Those systems producing extended light output have the advantage that several exposures can be taken from the same blot.

With the availability of fluorescence-scanning instruments, new methods for detection have come into use. It may seem at first glance that a secondary antibody could simply be coupled to a fluorescent molecule and the detection performed directly. Although this is possible, this method is not sufficiently sensitive for most purposes. The approach usually taken uses an enzyme-coupled secondary reagent (in this case usually AP) and a substrate that produces an insoluble, fluorescent product. The enzymatic reaction results in amplification of the signal, giving much better sensitivity than a fluorescently tagged secondary reagent. The blot is read on a fluorescent scanner and recorded as a digitized image.

What Are the Criteria for Selecting a Detection Method?

Sensitivity

There is a natural tendency to choose the most sensitive method available. High-sensitivity systems are desirable for detection of low-abundance proteins, but they are also desirable in cases where primary antibody is expensive or in limited supply, since these systems allow antibodies to be used at high dilutions. On the other hand, low-sensitivity systems, especially chromogenic systems, are easy to work with, require less exacting optimization, and tend to be less prone to background problems. Sensitivity overkill can be more trouble than it is worth.

What can you conclude from commercial sensitivity data? It can be difficult to compare the claims of sensitivity made by commercial suppliers. Although there is nothing wrong with the way these values are established, comparison between different systems can be difficult because the values depend on the exact conditions under which the determination was made. The primary antibody has a large effect on the overall sensitivity of any system, so comparisons between systems using different primary antibodies are less meaningful than they may seem at first glance. In order to compare two different detection systems, the target protein, the primary antibody, and, where possible, the secondary reagent should be the same. Such direct comparisons are hard to come by. Also sensitivity claims are usually made with purified protein rather than with crude lysate. For these reasons commercial sensitivity claims should be considered approximate, and it may be unrealistic to expect to attain the same level of sensitivity in your own system as that quoted by the manufacturer.

Signal Duration

Will your research situation require extended signal output in order to prepare several exposures from the same blot?

Ability to Quantitate

Film-based systems (chemiluminescent and radioactive) as well as fluorescence-scanning methods, allow quantitation of target proteins. Results on film are quantified by densitometry, while the digital raw data from fluorescence scanners (and storage-phosphor scanners for radioactive detection) is inherently quantitative. The linear range of film-based systems (limited by the response of the film) is a little better than one order of magnitude, while the manufacturers of fluorescent scanners claim something closer to two orders of magnitude.

There are several cautions to bear in mind when considering protein blot quantitation. Standards (known amounts of purified target protein—not to be confused with molecular weight standards) must be run on every blot, since even with the most consistent technique there can be blot-to-blot variation. The standard should be loaded on the gel, electrophoresed, and transferred in exactly the same way your samples are.

The determination of quantity can only be made within the range of standards on the blot: extrapolation beyond the actual standard values is not valid. This together with the limited linear range means that several dilutions of the unknown sample usually must be run on the same blot. Given all the lanes of standards and sample dilutions, the amount of quantitative data that can be extracted from a single blot is somewhat limited. Protein blot quantitation can be useful, but it is not a substitute for techniques such as ELISA or RIA.

Antibody Requirements

Typically the choice of available primary antibodies is not as wide as that of the other elements of the detection system. Primary antibodies can be obtained from commercial suppliers, non-profit repositories, and even other researchers. Tracking down a primary antibody can be time-consuming, but publications such as Linscott's Directory (Linscott, 1999, and *http://www.linscottsdirectory.com/index2.html*), the "Antibody Resource Page" (*http://www.antibodyresource.com*), the Usenet newsgroup "Methods and Reagents" (*bionet.molbio.methds-reagnts*), and Stefan Dubel's recombinant antibody page (*www.mgen.uni-heidelberg.de/SD/SDscFvSite.html*) and *www.antibody.com* can help.

If no antibodies against your target protein exist, your only options are to raise the antibody yourself or to have someone else do it. Companies such as Berkeley Antibody Company, Genosys, Rockland, and Zymed (among many others) can do this kind of work on a contract basis. Whichever route you choose, this is a time-consuming and potentially expensive undertaking.

Ability to Strip and Reprobe

Radioactive and chemiluminescent systems are ideally suited to stripping and reprobing. Other systems (chemifluorescent and chromogenic) leave insoluble precipitates over the bands of interest; these precipitates can be removed only with the use of solvents, which is an unpleasant extra step and can be hard on blots. Not all targets survive this treatment. (See below for important cautions regarding stripping.)

Equipment and Facility Requirements

Radioactivity can be used only after fulfilling stringent training and licensing requirements. Radioactive methods, like chemiluminescent methods, also require darkroom facilities (unless storage phosphor equipment is available). Fluorescent methods require specialized scanning equipment. Chromogenic methods do not require any specialized facilities or equipment.

What Are the Keys to Obtaining High-Quality Results?

Careful choice of materials, an understanding of the questions your experiments are intended to answer, and an appreciation of the fact that every new system requires optimization are all necessary for obtaining good results. Optimization takes time, but it will pay off in the final result. It is also important to develop consistency in technique from day to day, and to keep detailed and accurate records. Consistency and good record-keeping will make it much easier to isolate the source of any problem that may come up later.

Which Transfer Membrane Is Most Appropriate to Your Needs?

The same considerations go into the choice of membrane that go into the choice of any other component of your detection strategy. What is the molecular weight of your protein? What detection method will you use, and does this method have special membrane requirements? Do you intend to strip and reprobe your blots? (See Table 13.2.)

Nitrocellulose wets easily and gives clean backgrounds. Unfortunately, it is physically fragile (liable to tear and crack), especially when dry. This fragility makes nitrocellulose undesirable for use in stripping and reprobing. The problem of physical fragility has been overcome with the introduction of supported nitrocellulose, which has surfaces of nitrocellulose over a core or "web" of physically stronger material. The added physical strength comes at the cost of slightly higher background.

Table 13.2 Characteristics of Transfer Membranes

Membrane	Characteristics
Nitrocellulose	Low background. Easy to block. Physically fragile.
Supported nitrocellulose	Binding properties similar to nitrocellulose. Higher background than pure nitrocellulose. Physically strong.
PVDF	High protein binding capacity. Physically strong. Highly hydrophobic: requires methanol pre-wetting and dries easily. Good for stripping and reprobing.

PVDF (polyvinylidene difluoride) membranes are physically stronger and have higher protein-binding capacity than nitrocellulose. However, they are highly hydrophobic: so much so that they need to be pre-wetted with methanol before they can be equilibrated with aqueous buffer. When handling PVDF, you should take special care to ensure the membrane does not dry out, since uneven blocking, antibody incubation, washing, or detection can result. If the membrane does dry out, it should be re-equilibrated in methanol and then in aqueous buffer. The high affinity of PVDF for protein gives efficient transfer and high detection efficiency, but it can make background control more difficult. PVDF is the membrane of choice for stripping and reprobing.

Transfer membranes are available in several pore sizes. The standard pore size, suitable for most applications, is 0.45 micron. Membranes are also commonly available in 0.2 and even 0.1 micron pore size: these smaller pore sizes are suitable for transfer of lower molecular weight proteins, below about 12 kDa. Transfer efficiency is not good with membranes with a pore size of less than 0.1 micron.

BLOCKING

All transfer membranes have a high affinity for protein. The purpose of blocking is simply to prevent indiscriminate binding of the detection antibodies by saturating all the remaining binding capacity of the membrane with some irrelevant protein. (For a detailed discussion, see Amersham, n.d., from which much of the following is drawn.)

Which Blocking Agent Best Meets Your Needs?

The protein most commonly used for the purpose is nonfat dry milk, often referred to as "blotto," used at 0.5% in PBS containing 0.1% Tween-20. Any grocery-store brand of nonfat dry milk can be used.

Gelatin is isolated from a number of species, but fish skin gelatin is usually considered the best for Western blotting. Fish gelatin is usually used at a concentration of 2%. It is an effective blocker, and will not gel at this concentration at 4°C.

Bovine serum albumin (BSA) is available in a wide range of grades. Usually a blotting or immunological grade of BSA is appropriate. It is less expensive than fish skin gelatin, and can be used at 2%.

Normal serum (fetal calf or horse) is used sometimes, at a concentration of 10%. It can be an effective blocking agent, but is quite expensive. Since serum contains immunoglobulins, it is not compatible with Protein A and some secondary antibodies.

Casein can be used at 1%, but it is very difficult to get dry casein into solution. Casein and casein hydrolysate are the basis of some commercial blocking agents.

Different primary antibodies work better with different blocking agents: nonfat dry milk is usually a good first choice, but when setting up a new method, it is a good idea to evaluate different blockers.

It has been claimed that some blocking agents, nonfat dry milk in particular, can hide or "mask" certain antigens. Of course, there must be no component of the blocking agent that the primary or secondary antibodies can specifically react with.

Some researchers include a second blocking step prior to secondary antibody incubation. However, if the initial blocking is sufficient and reagent dilutions are optimal, this should not be necessary.

A more specific kind of blocking may be needed when avidin or streptavidin is used as a detection reagent and the sample contains biotin-bearing proteins. Because of this "endogenous biotin" the avidin or streptavidin will pick up these undesired proteins directly. If you suspect this may be a problem, a control reaction can be run with no primary antibody but with the avidin or streptavidin detection. The presence of bands in this control reaction will indicate that the avidin or streptavidin is binding to the endogenous biotin.

The remedy for such a situation is to treat the blot prior to antibody incubation first with avidin (to bind all the endogenous biotin) and then with free biotin (to block all remaining free binding sites on the added avidin). The free biotin is washed away, and antibody detection can proceed (Lydan and O'Day, 1991).

WASHING

Thorough washing is critical to obtaining clean blots, so washing times and solution volumes should always be generous. It is important to realize that protein binding and antibody interactions do not all occur at the surface but rather throughout the entire thickness of the membrane. For this reason, thorough soaking and equilibration of the membrane is critical at every step.

Washing should always be performed at room temperature and with thorough agitation. The exact volume of wash buffer will depend on the container used for washing, but the depth of the solution should be about 1 cm. When protocols call for changing wash solution, this should not be ignored. The higher the sensitivity of the detection method, the more important is scrupulous washing technique.

What Composition of Wash Buffer Should You Use?

Standard wash buffer simply consists of PBS or TBS with added detergent: Tween-20 is routinely used at 0.1%, although Tween concentrations can be raised to as high as 0.3% to help reduce background. Concentrations higher than this tend to disrupt antibody binding. Triton, NP-40 and SDS should not be used, as they may strip off bound antibodies or target proteins.

Another method sometimes used to increase the effectiveness of washing is increasing the concentration of salt in the wash solution. High salt reduces charge-mediated effects, which tend to be less specific, and favors hydrophobic interactions, which are more specific. The usual upper limit for NaCl concentration in wash buffers is 500 mM. (Standard PBS and TBS contain 130 mM NaCl.)

What Are Common Blot Size, Format, and Handling Techniques?

Small blots, or larger blots cut into strips for analysis with several different antibodies, can be incubated in large centrifuge tubes or specialized strip-incubation trays. Larger blots should be incubated in trays. Centrifuge tubes are convenient and allow small reagent volumes to be used. Even with trays, there only needs to be sufficient blocking or antibody solution to submerge the blot completely and allow free flow of the solution. Be generous, however, with volumes of stripping and washing solutions.

Incubations and washes should be performed with constant agitation. For tubes, a tube-roller or tilting platform can be used. For trays, an orbital platform shaker with adjustable speed is ideal. Antibody incubations are typically carried out for 30 minutes to 1 hour at room temperature; however, they can also be carried out at 4°C overnight. Overnight incubation allows lower antibody concentrations to be used and in some cases results in increased sensitivity. It is important that antibody concentrations be optimized under the same incubation conditions that will be used in the final procedure.

Membranes should never be handled with fingers. A forceps is best, but powder-free gloves can also be used. There is some evidence that residual powder from powdered gloves can mask chemiluminescent signals (Amersham Pharmacia Biotech, 1998).

Blots can be stored directly after transfer in buffer at 4°C overnight. Alternatively, the blocking step can be allowed to go overnight at 4°C without agitation. Blots should not be stored wet for longer than two days, as bacterial growth may occur.

After transfer or after stripping, blots can be air-dried and stored in airtight containers at 4°C. Do not air-dry blots without stripping them first if you intend to reprobe: dried-on antibody is almost impossible to strip.

THE PRIMARY ANTIBODY

Are All Antibodies Suitable for Blotting?

Successful blotting depends largely on the quality of the primary antibody. Not all primary antibodies that react with a target protein in solution will react with that same protein once it is bound to a membrane. During electrophoresis and transfer, proteins become denatured and reduced. This change in the target protein may render it nonreactive with some antibodies, particularly monoclonals. Before starting out, you should make sure that the primary antibody you intend to use is suitable for blotting. This information can be obtained from the originator or suppler of the antibody, or it can be determined by running control blots.

Polyclonal antibodies can be used simply as diluted raw sera, but in many cases (especially with low titer sera) the use of an Ig fraction can reduce background. Affinity purification is ideal, though not always feasible. Ammonium sulfate purification can also provide sufficient purity.

The same purification requirements hold for monoclonal antibodies, but given the small quantities available, especially when obtained from commercial sources, purification is not always practical. You should know the isotype of your primary antibody so you can choose an appropriate secondary reagent. IgMs are often considered less desirable as primary antibodies because they are more difficult to purify and require more specialized secondary reagents.

How Should Antibodies Be Handled and Stored?

Antisera and monoclonal antibodies should be divided into small aliquots, flash-frozen by plunging in a dry ice/ethanol or liquid nitrogen bath, and stored at –70°C. Under these conditions they are stable for years. Once thawed, aliquots should not be frozen and thawed again, but stored at 4°C. Sera and purified monoclonals are stable at 4°C (sometimes for as long as a year), but ascites fluids can contain proteases, so storage at 4°C is not recommended. Repeated freeze–thawing can cause aggregation of antibodies and loss of reactivity. Sodium azide may be used as a preservative at 0.02%.

Antibodies should always be diluted in buffer containing carrier protein. The actual antibody concentration in working solutions is so low that without added carrier, much of the antibody would be lost to adsorption to the walls of containers. Using 0.1% BSA is sufficient. Nonfat dry milk is not recommended, since it is not as clean as laboratory grade albumin and is prone to bacterial growth.

SECONDARY REAGENTS

A wide variety of secondary reagents can be used to detect primary antibodies. Besides secondary antibodies, there are the immunoglobulin-binding proteins Protein A and Protein G, as well as avidin and streptavidin. Some considerations apply to all secondary reagents. In general, secondary reagents are less stable than primary antibodies, since not just antibody binding activity but reporter activity must be retained. In fact stability of the reporter group is the main determinant in secondary antibody stability. Iodinated conjugates are stable for weeks, while enzyme conjugates typically are stable for months. These reagents usually should not be frozen, as repeated freeze–thaw cycles can result in aggregation or loss of reporter activity. Several labs, however, have reported good results in flash-freezing enzyme conjugates and storing them in single-use aliquots at –70°C.

How Important Is Species Specificity in Secondary Reagents?

The species in which a secondary antibody is raised is not usually important—goats and donkeys are often used because it is possible to obtain large amounts of serum from these animals. "Goat anti-rabbit" is simply an antibody raised against rabbit Ig, produced by immunizing a goat.

A good secondary antibody for blotting should be affinity purified: for example, a raw goat anti-rabbit antiserum is run over a column containing immobilized rabbit Ig. Everything in the serum that doesn't bind to rabbit Ig washes through the column and is discarded. Everything that does bind is then dissociated, eluted, and collected. This affinity-purified secondary antibody will have much less protein than the raw serum: the irrelevant proteins would only contribute to background without increasing the signal.

A further purification step is often performed to ensure species specificity. Cross-adsorption, as the process is known, is in some ways the mirror image of affinity purification. Anti-rabbit Ig is run through a column containing, for example, mouse Ig. Everything that washes through the column without binding is collected, thus removing any antibodies that react with mouse Ig. This process can be repeated with a number of columns containing Ig from different species, ensuring that the resulting antibody will only react with the Ig of a single species. Depending on the nature of your study, this species specificity may or may not be important. If there is not likely to be Ig from other species present in your sample, it is unnecessary. Furthermore no cross-adsorbed secondary reagent is completely species specific: there is enough homology between species that even a cross-adsorbed antibody will pick up a "foreign" Ig if enough of it is present. It is impossible to attain 100% species, class, or isotype specificity in secondary reagents, since there will always be some small degree of homology between the wanted and unwanted target.

Why Are Some Secondary Antibodies Offered as $F(ab')_2$ Fragments?

In blotting, there is usually no advantage to the use of these reagents. The only rare case in which an $F(ab')_2$ fragment would be advantageous would be one in which samples contained Fc

Table 13.3 Reactivity of Protein A and Protein G

Immunoglobulin	Protein A	Protein G
Mouse IgG1	+/–	++
Mouse IgG2a	++	++
Mouse IgG2b	++	++
Mouse IgG3	++	+++
Mouse IgA	–	?
Mouse IgM	+/–	?
Rat IgG1	+/–	?
Rat IgG2a	+/–	+++
Rat IgG2b	+/–	++
Rat IgG2c	+	++
Rat IgM	+/–	?
Goat Ig	+/–	+++
Sheep Ig	–	++
Rabbit Ig	+++	+++
Horse Ig	–	+++

Source: Adapted, with permission, from data provided by Amersham Pharmacia Biotech.
Note:
+++ Strong binding
++ Acceptable binding
+ Weak binding
– No binding
? No data

receptors (as do some bacteria and lymphocytes): the use of $F(ab')_2$ fragments would prevent the background binding of antibodies to these receptors through the Fc portion.

Protein A and Protein G

Protein A and Protein G are bacterial proteins that bind specifically to immunoglobulins from a variety of species. Table 13.3 lists some common immunoglobulins and their reactivity. Why use Protein A and Protein G rather than a secondary antibody? A species-specific secondary antibody will usually give stronger signal and better specificity than Protein A or G. The advantage of Protein A or G is versatility: the same secondary reagent can be used with a variety of primary antibodies. This is especially important for radioactive detection, since a stock of several different secondary antibodies would have to be constantly replenished because of radioactive decay.

Avidin and Streptavidin

Avidin, isolated from egg white, and streptavidin, a bacterial protein, bind biotin with extremely high affinity and specificity. Primary antibodies can be covalently conjugated to biotin, used on a blot, then detected with avidin or streptavidin. A wide range of avidin and streptavidin conjugates is commercially available. Since any avidin or streptavidin conjugate can be used with any biotinylated reagent, avidin and streptavidin are close to being universal detection reagents.

Some primary antibodies are available in biotinylated form, and there are also kits and reagents available for performing biotinylation in the lab. Coupling is usually accomplished through an *N*-hydroxy-succinimidyl ester, an amine-reactive functional group (Haugland and You, 1998). Ideally antibodies to be labeled by this chemistry should be free of carrier protein, since all proteins in the solution will react. Subsequent purification by column or dialysis is necessary, which means that you need to start with a large enough amount of protein to ensure a reasonable recovery.

Avidin and streptavidin can be used interchangeably. However, streptavidin is not charged at neutral pH and not glycosylated. It therefore tends to yield slightly lower backgrounds and better specificity than avidin.

One very useful application of biotin/streptavidin detection is in the determination of molecular weights. Biotinylated molecular weight markers are commercially available, and they can be run on gels and transferred just like normal molecular weight markers. The blot is treated as usual through primary antibody incubation and washing, but when the secondary antibody incubation is performed, labeled streptavidin is added to the solution so that incubation with secondary antibody (to bind the primary antibody) and streptavidin (to bind the biotinylated markers) take place simultaneously. The streptavidin should be labeled with the same reporter group as the secondary antibody. In this way both the molecular weight markers and the band of interest will show on the blot, without having to separate the blot into different pieces. Determination of molecular weight by electrophoresis is, however, always approximate.

AMPLIFICATION

Several strategies have been used to increase the signal on Western blots by increasing the amount of reporter group that binds to a given amount of target protein. If one primary antibody bound to its target protein results in the binding of, say, 50 HRP molecules rather than 2 or 3, this will clearly result in increased signal.

This approach is often taken through the use of the biotin-streptavidin system. The simplest way to accomplish this would be a three layer system: primary antibody-biotinylated secondary antibody-streptavidin reporter. The idea is that the binding of the second and third layer takes place on something better than a one-to-one basis; the additional layer multiplies this effect.

The same concept can be carried further through the use of special reporter groups: for example, multimeric complexes of enzyme. Such complexes are commercially available. The guiding idea is to bind as much reporter group as possible to a single primary antibody molecule.

Before chemiluminescent detection systems became widely available, this approach was about the only one used for obtaining very high sensitivity. The amplification methods can still be helpful in boosting the sensitivity of chromogenic detection systems. They can also be used with chemiluminescent systems, but here, the increase in sensitivity may not be balanced out by the higher background: with three layers the optimization becomes much more complex and demanding.

STRIPPING AND REPROBING

It is often an advantage to be able to perform detection on the same blot with more than one antibody. This can be done by dissociating or stripping antibodies off the blot after detection is complete so that the blot can be probed with a new set of antibodies.

Stripping is only feasible in cases where the detection system leaves no precipitate on the blot: colorimetric and chemifluorescent methods are not really suitable. (It is actually possible to strip such blots after treatment with organic solvents to dissolve the precipitate, but this is not recommended since membranes vary in their resistance to solvents and subsequent redetection is often not successful.) An alternative in cases where stripping is not practical is to run duplicate sets of lanes on the same gel and then to cut up the blot after transfer: the different portions of the blot can then be probed with different antibodies.

Will the Stripping Procedure Affect the Target Protein?

While stripping can be very useful, there are limitations to the technique. Treatment harsh enough to dissociate antibodies can be harsh enough to damage or dissociate target proteins. Loss of some target protein in each stripping cycle is inevitable. Sometimes a single treatment can result in complete loss of target protein (or at least its immunoreactivity). Even in favorable cases, 25% or more of the target can be lost in one stripping cycle. For this reason it is a good practice to probe for the least abundant target protein first, and then to move on to increasingly abundant proteins where more target loss can be tolerated.

The most common stripping technique uses 2% SDS and 100 mM 2-mercaptoethanol (2-ME) or dithiothreitol (DTT) and heating with agitation at 50 to 65°C, preferably in a fume hood (Amersham Pharmacia Biotech, 1998). This method is effective but can result in pronounced target loss. Another method is incubation at room temperature with glycine buffer at pH 2. This is more gentle but may not be as effective. With either method, thorough washing is necessary afterward. Reblocking is also necessary, as the stripping treatment tends to remove the blocking agent.

The effectiveness of stripping can be verified by repeating the secondary antibody incubation and detection steps (i.e., with no primary antibody). This should be done at least at the outset to confirm that the chosen stripping method is effective.

Can the Same Stripping Protocols Be Used for Membranes from Different Manufacturers?

In most cases the same protocols can be used with membranes of the same kind from different manufacturers. Unless there is something unique about a particular membrane, standard protocols can be followed.

Is It Always Necessary to Strip a Blot before Reprobing?

There are some situations in which blots can be redetected without first stripping. When peroxidase is used as a reporter group in chemiluminescent blots, the blot can be treated with dilute hydrogen peroxide (30 minutes in 15% H_2O_2 in PBS, followed by thorough washing). The radicals formed in the peroxidase reaction will irreversibly inactivate the enzyme. The blot can then be washed and carried through subsequent redetection with another primary antibody. This method, however, is only suitable in cases in which two different, non-cross-reacting secondary reagents are used. Otherwise, the secondary reagent used in the second detection cycle will pick up both the original and the new primary antibodies.

TROUBLESHOOTING

It is important to develop rational troubleshooting strategies (see Table 13.4). Problems are inevitable, so taking a systematic approach to troubleshooting will, in the long run, save time,

Table 13.4 Western Blotting Troubleshooting Logic Tree

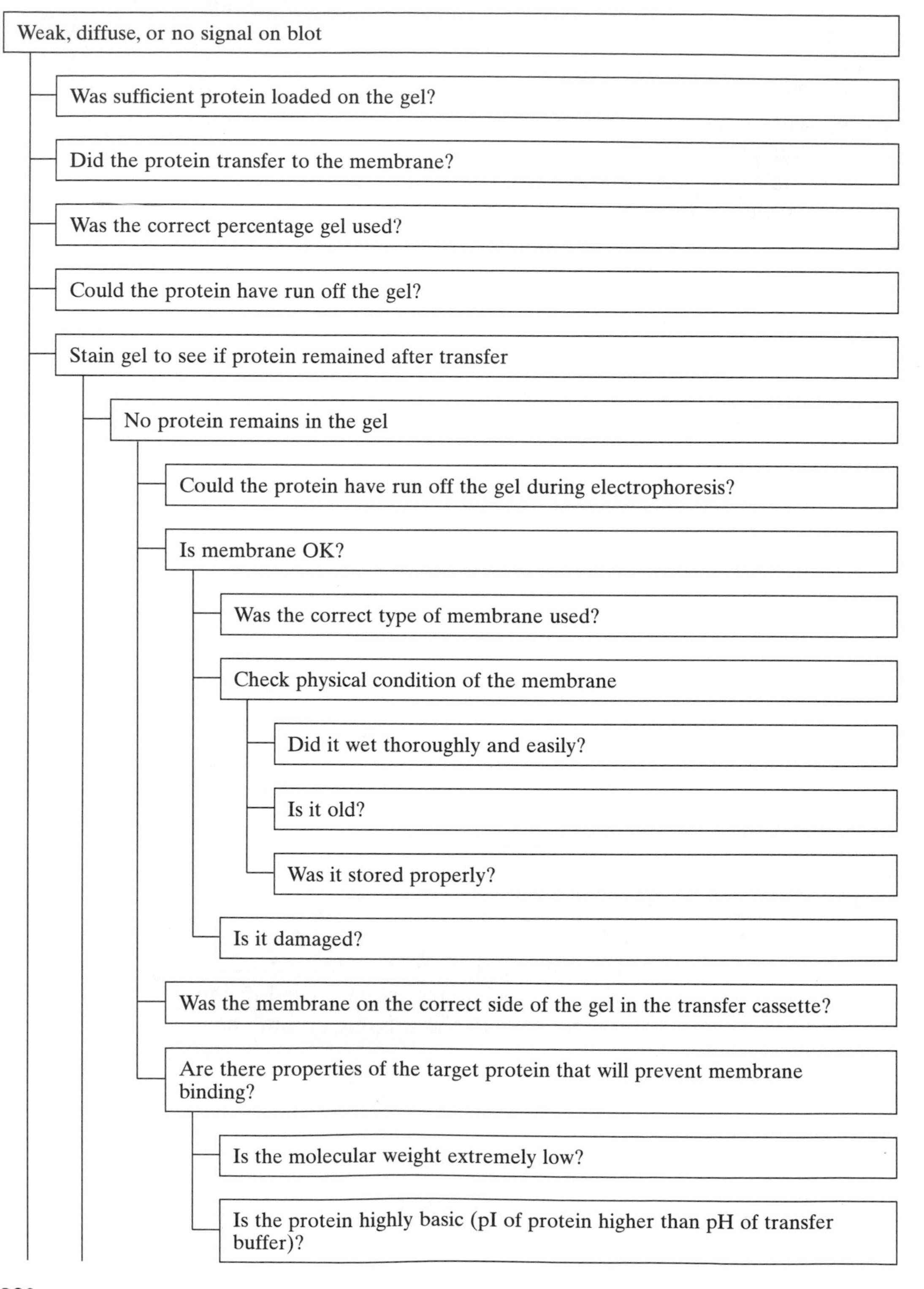

- Weak, diffuse, or no signal on blot
  - Was sufficient protein loaded on the gel?
  - Did the protein transfer to the membrane?
  - Was the correct percentage gel used?
  - Could the protein have run off the gel?
  - Stain gel to see if protein remained after transfer
    - No protein remains in the gel
      - Could the protein have run off the gel during electrophoresis?
      - Is membrane OK?
        - Was the correct type of membrane used?
        - Check physical condition of the membrane
          - Did it wet thoroughly and easily?
          - Is it old?
          - Was it stored properly?
        - Is it damaged?
      - Was the membrane on the correct side of the gel in the transfer cassette?
      - Are there properties of the target protein that will prevent membrane binding?
        - Is the molecular weight extremely low?
        - Is the protein highly basic (pI of protein higher than pH of transfer buffer)?

Table 13.4 (*Continued*)

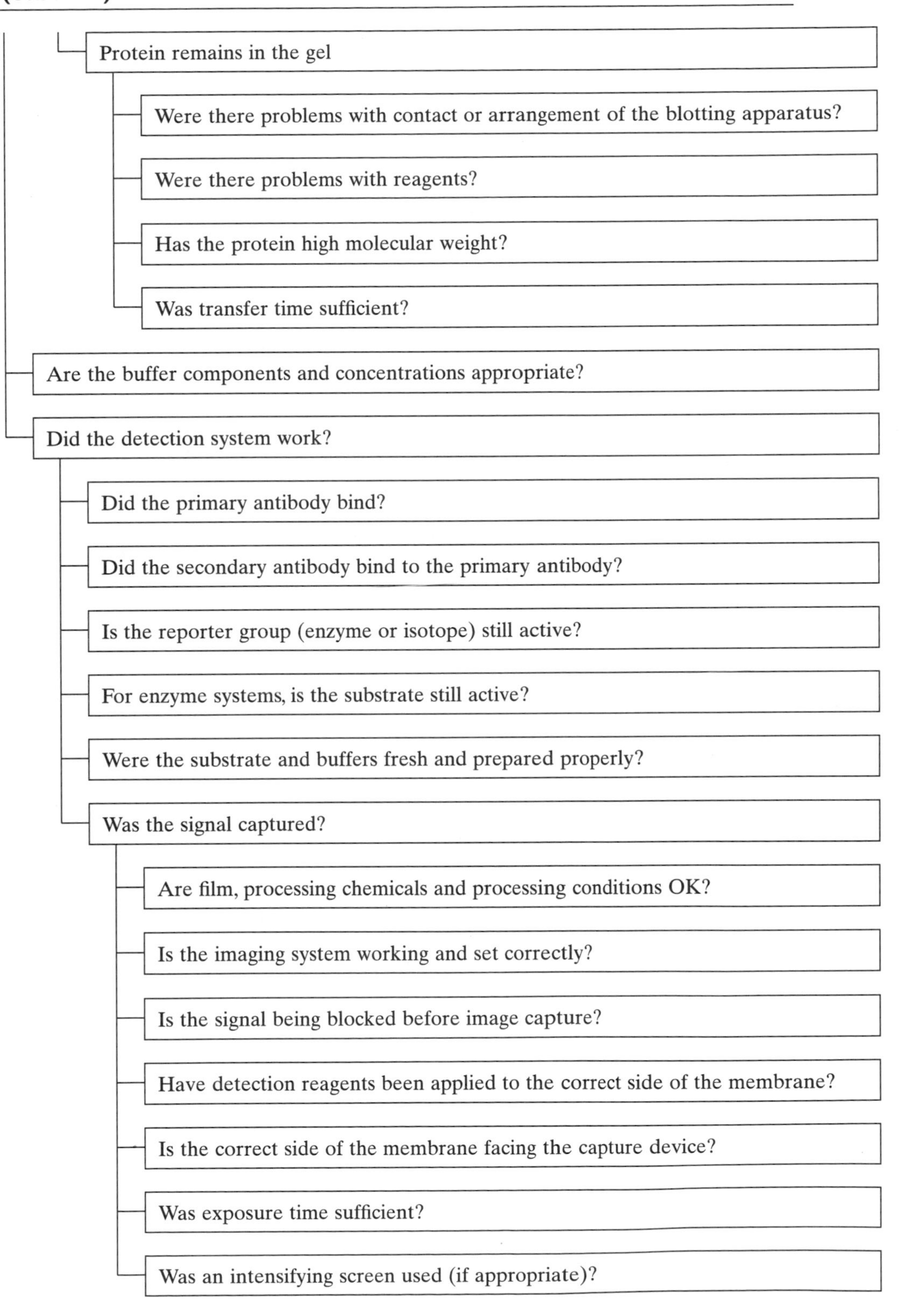

Table 13.4 (*Continued*)

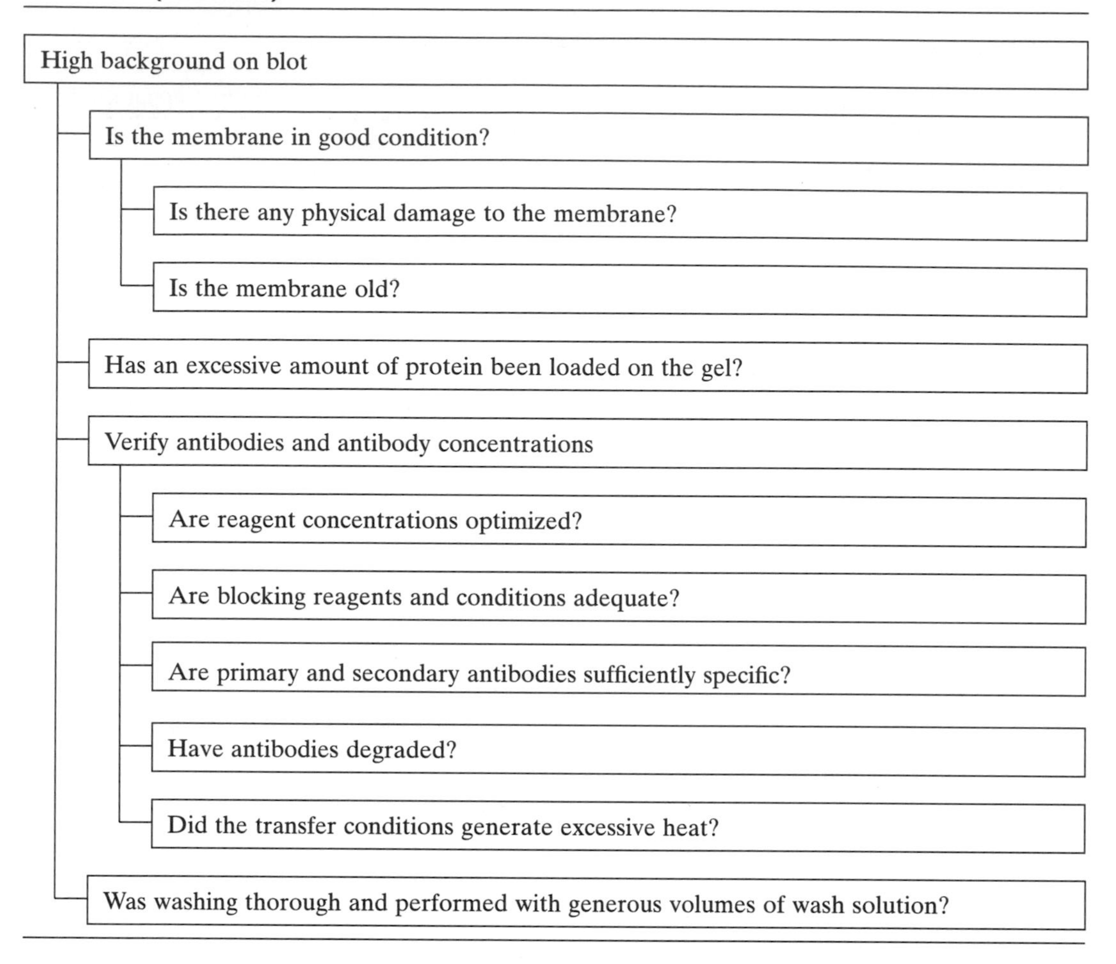

- High background on blot
  - Is the membrane in good condition?
    - Is there any physical damage to the membrane?
    - Is the membrane old?
  - Has an excessive amount of protein been loaded on the gel?
  - Verify antibodies and antibody concentrations
    - Are reagent concentrations optimized?
    - Are blocking reagents and conditions adequate?
    - Are primary and secondary antibodies sufficiently specific?
    - Have antibodies degraded?
    - Did the transfer conditions generate excessive heat?
  - Was washing thorough and performed with generous volumes of wash solution?

energy, and reagents. Examples of common and unusual problems are illustrated in Figures 13.1–13.6.

The guiding principle is to break the system into its component parts, and test each step in isolation. This ideal is not possible in every case. Rather, those components that can be isolated should be. Once validated, they can be used to test the other components.

Consider the case of weak or no signal. The first step would be to review your system overall and make sure there are no reagent incompatibilities. Certain detection reagents are incompatible with common buffers and buffer additives. Sodium azide is a powerful peroxidase inhibitor. Although it is often used as a buffer preservative, peroxidase conjugates must not be diluted in azide-containing buffer, nor should wash buffers containing azide be used with peroxidase conjugates. The presence of azide in con-

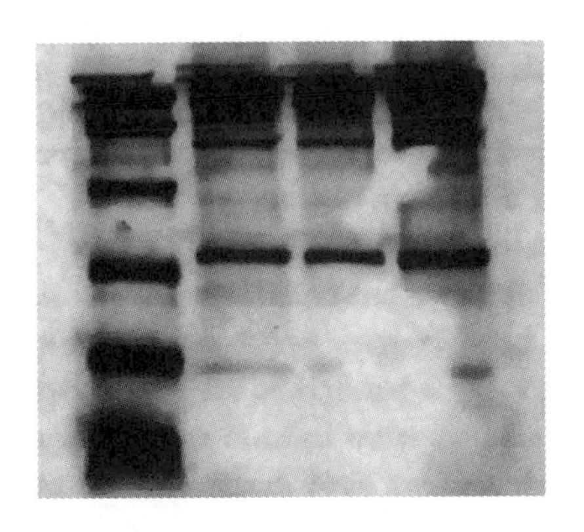

Figure 13.1 Western blot of fluorescein labeled Brome Mosaic Viral proteins prepared using a rabbit reticulocyte in vitro translation system, detected using an anti-fluorescein peroxidase conjugate and ECL. This effect is caused by poor contact between the polyacrylamide gel and the membrane in the electroblotting apparatus. Ensure that all fiber pads are of sufficient thickness; with use these pads will flatten. Periodically they must be replaced. Published by kind permission of Amersham Pharmacia Biotech UK Limited.

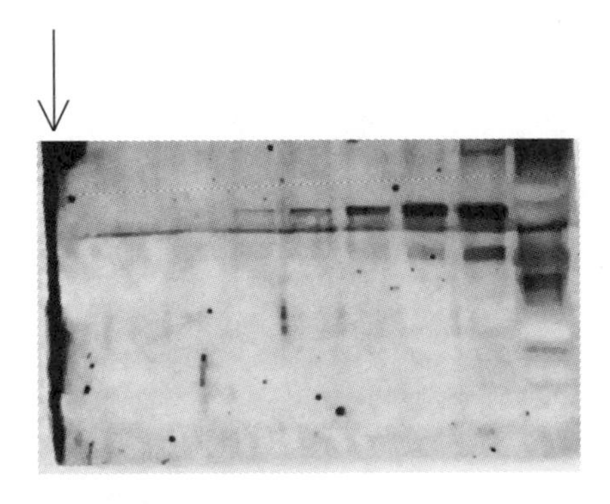

Figure 13.2 Rat brain homogenate Western blot immunodetected using an anti-transferrin antibody and ECL. This effect is caused by damage at the cut edge of the membrane resulting in a high level of nonspecific binding of the antibodies used during the immunodetection procedure. Membranes should be prepared using a clean sharp cutting edge, for example, a razor blade or scalpel. Published by kind permission of Amersham Pharmacia Biotech UK Limited.

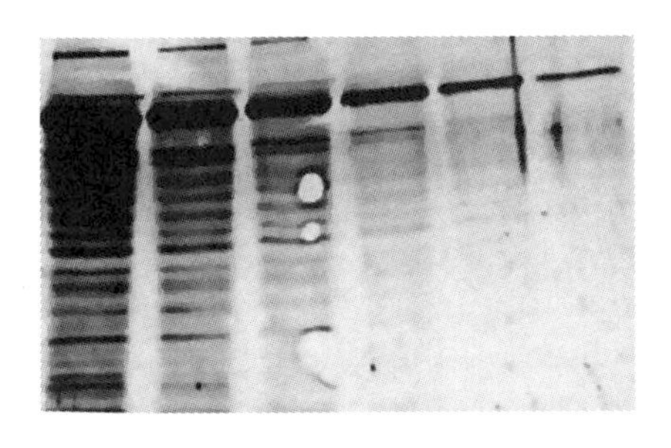

Figure 13.3 K562 cell lysate Western blot immunodetected using an anti-transferrin antibody and ECL. Air bubbles trapped between the gel and the membrane prevent transfer of the proteins, so no signal is produced. Air bubbles should be removed by rolling a clean pipette or glass rod over the surface of the polyacrylamide gel/membrane before assembling the electroblotting apparatus. Published by kind permission of Amersham Pharmacia Biotech UK Limited.

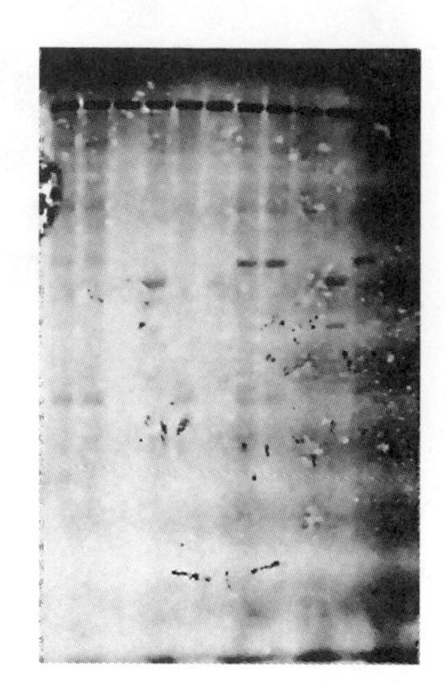

Figure 13.4 Western blot of fluorescein labelled Brome Mosaic Viral proteins prepared using a rabbit reticulocyte in vitro translation system, detected using an anti-fluorescein-peroxidase conjugate and ECL. This effect is caused by using dirty fiber pads in the electroblotting apparatus. The fiber pads should be cleaned after each use by soaking in Decon™ and rinsing thorougly in distilled water. Periodically the fiber pads must be replaced. Published by kind permission of Amersham Pharmacia Biotech UK Limited.

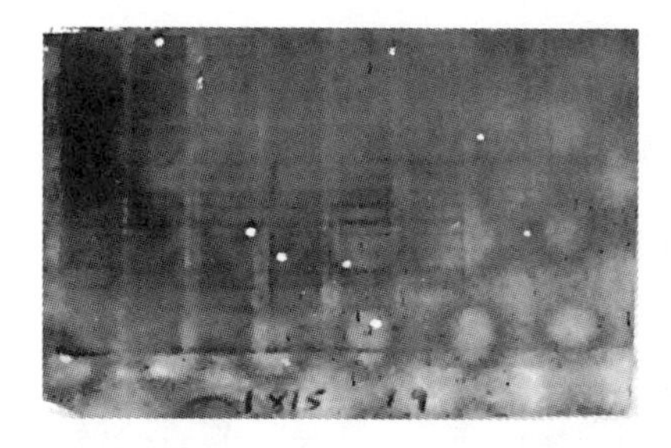

Figure 13.5 Rat brain homogenate Western blot stained with AuroDye Forte, a total protein stain. This effect is caused by fiber pads that are too thick for the electroblotting apparatus. Published by kind permission of Amersham Pharmacia Biotech UK Limited.

Figure 13.6 Rat brain homogenate Western blot detection of β-tubulin with the ECL Western blotting system. This effect is caused by too strong a dilution of secondary antibody. Antibodies and streptavidin conjugates should be titrated for optimum results. Published by kind permission of Amersham Pharmacia Biotech UK Limited.

centrated stocks of primary antibodies is not a problem, however, because the azide will be diluted and washed away before the HRP conjugate is applied.

Alkaline phosphatase should not be used with phosphate buffers. Use TRIS instead. The presence of phosphate will inhibit the phosphatase reaction.

Avidin and streptavidin should not be diluted in buffers containing nonfat milk. Nonfat milk contains free biotin, which will bind to avidin or streptavidin with high affinity, preventing binding with your biotinylated antibody (Hoffman and Jump, 1989).

If there are no problems with the choice of reagents, the next step is to demonstrate that all the components are functioning properly. Start by verifying the detection system. With many detection systems, function can be verified directly: chemiluminescent reagents can be quickly tested by adding enzyme conjugate to the prepared substrate in the darkroom and observing the production of light. In other systems, spots of diluted secondary antibody can be applied directly to membrane and carried through the detection step. If the secondary antibody shows up, the detection reagents are not at fault.

Backtracking further, the primary antibody can be spotted on membrane, the membrane blocked, incubated with the secondary antibody, and carried through the detection. This shows that the secondary antibody is able to detect the primary antibody. If this is not the problem, purified antigen or lysate can be serially diluted, dotted on the membrane, and carried through primary and secondary antibody incubations and detection. This shows the primary antibody is able to detect the target. If the problem still isn't apparent, then the transfer must be verified. The transfer of colored molecular weight markers does not always indicate efficient transfer of target proteins. It is best to verify transfer by use of a reversible stain like Ponceau S (Salinovich and Montelaro, 1986).

With the proliferation of high-sensitivity detection methods, high background is now probably the most common problem encountered in Western blotting. In trying to solve background problems, the first step is to stop and examine the offending blots carefully. Is the background occurring all over the blot (i.e., over the lanes and the areas between the lanes), or is it confined to the lanes themselves (i.e., extra bands, or in some cases, the entire lane showing up)?

Background over the entire blot suggests something general such as washing or blocking conditions. Check your procedures: Is your washing thorough and complete? Are you using sufficient volumes of wash solution? If you are already washing thoroughly, then it may be necessary to reassess your blocking conditions. Finally, greatly excessive antibody concentrations can cause generalized background: make sure you've optimized antibody concentrations.

Background confined to the lanes is more likely to be related to non-specific antibody binding. Again, be sure that you have optimized all your antibody concentrations. In order to pinpoint the problem, it may be a good idea to run a control blot with no primary antibody. If bands show up in the absence of primary antibody, the problem can be assigned to the secondary antibody; in most cases the concentration of secondary antibody is simply too high. Otherwise, your secondary antibody may have some specific affinity for something in your samples. If this is the case, the only choice is to switch to a different secondary antibody or even a different detection approach (e.g., Protein A or biotin/streptavidin).

With other problems the guiding principle is still the same: to try to glean as much information from the problem blot as possible, to isolate each step in the process, and change only one variable at a time. Holding each variable constant except for one makes each experiment decisive. This is the kind of situation in which detailed record-keeping is critical. When the performance of a system changes, carefully going back over records often will suggest the source of the trouble.

SETTING UP A NEW METHOD

When setting up a new method, it may appear that there is an impossible number of choices that need to be made all at once. Actually, it's not so difficult. Your decision to go with another method should be based on the properties of your protein of interest, the availability and nature of your samples, your needs for reprobing or quantitation, and the nature of your facilities. Read up on the relevant literature, and, at least in the beginning, base your protocol on a published method.

An important issue that needs to be addressed in setting up a new method is optimization of antibody concentrations. These concentrations will be different for every system. They can most easily be established through dot or slot blots: the target protein (either lysate or purified protein) is spotted on membrane and blocked. Detection is then carried out using varying dilutions of primary antibody. (To begin with, use the secondary antibody at the manufacturer's recommended dilution.) The maximum dilution of primary antibody that yields a usable signal should be your working dilution. The same process is repeated for the secondary antibody, using for the primary antibody the dilution you previously established. Again, the minimum concentration of secondary antibody that gives usable signal should be chosen. The use of minimum concentrations of primary and secondary antibodies helps ensure the greatest specificity with the minimum background (while at the same time conserving reagents).

For blocking and washing conditions, start by following a published method. If your model method was developed for the same protein you are looking at, then you can simply follow these conditions exactly. If you are looking at a new protein, 0.5% nonfat dry milk with 0.1% Tween-20 is probably the best blocking agent to start with. If you experience high background or other unexpected results, then you may want to evaluate other blockers, look at other washing conditions, consider loading less protein on your gels, or re-examine the optimization of antibody concentrations.

BIBLIOGRAPHY

Amersham Life Science. n.d. *A Guide to Membrane Blocking Conditions with ECL Western Blotting.* Tech Tip 136. Amersham Life Science Inc., Arlington Heights, IL.

Amersham Pharmacia Biotech. 1998. *ECL Western Blotting Analysis System.* Amersham Pharmacia Biotech, Piscataway, NJ.

Haugland, R. P., and You, W. W. 1998. Coupling of antibodies with biotin. *Meth. Mol. Biol.* 80:173–183.

Hoffman, W. L., and Jump, A. A. 1989. Inhibition of the streptavidin-biotin interaction by milk. *Anal. Biochem.* 181:318–320.

Linscott, W. 1999. *Linscott's Directory of Immunological and Biological Reagents*, 10th ed. Linscott, Mill Valley, CA.

Lissilour, S., and Godinot, C. 1990. Influence of SDS and methanol on protein electrotransfer to Immobilon P membranes in semidry blot systems. *Biotech.* 9:397–398, 400–401.

Lydan, M. A., and O'Day, D. H. 1991. Endogenous biotinylated proteins in *Dictyostelium discoideum. Biochem. Biophys. Res. Commun.* 174:990–994.

Salinovich, O., and Montelaro, R. C. 1986. Reversible staining and peptide mapping of proteins transferred to nitrocellulose after separation by sodium dodecylsulfate-polyacrylamide gel electrophoresis. *Anal. Biochem.* 156:341–347.

14

Nucleic Acid Hybridization

Sibylle Herzer and David F. Englert

Planning a Hybridization Experiment 401
The Importance of Patience 401
What Are Your Most Essential Needs? 401
Visualize Your Particular Hybridization Event 401
Is a More Sensitive Detection System Always Better? 403
What Can You Conclude from Commercial Sensitivity Data? 403
Labeling Issues 403
Which Labeling Strategy Is Most Appropriate for Your Situation? 403
What Criteria Could You Consider When Selecting a Label? 405
Radioactive and Nonradioactive Labeling Strategies Compared 409
What Are the Criteria for Considering Direct over Indirect Nonradioactive Labeling Strategies? 410
What Is the Storage Stability of Labeled Probes? 411
Should the Probe Previously Used within the Hybridization Solution of an Earlier Experiment Be Applied in a New Experiment? 412
How Should a Probe Be Denatured for Reuse? 412
Is It Essential to Determine the Incorporation Efficiency of Every Labeling Reaction? 412
Is It Necessary to Purify Every Probe? 413

Molecular Biology Problem Solver, Edited by Alan S. Gerstein.
ISBN 0-471-37972-7

Hybridization Membranes and Supports 413
What Are the Criteria for Selecting a Support for Your Hybridization Experiment? 413
Which Membrane Is Most Appropriate for Quantitative Experiments? 417
What Are the Indicators of a Functional Membrane? 417
Can Nylon and Nitrocellulose Membranes Be Sterilized? 417
Nucleic Acid Transfer 418
What Issues Affect the Transfer of Nucleic Acid from Agarose Gels? 418
Should Membranes Be Wet or Dry Prior to Use? 420
What Can You Do to Optimize the Performance of Colony and Plaque Transfers? 421
Crosslinking Nucleic Acids 422
What Are the Strengths and Limitations of Common Crosslinking Strategies? 422
What Are the Main Problems of Crosslinking? 423
What's the Shelf Life of a Membrane Whose Target DNA Has Been Crosslinked? 423
The Hybridization Reaction 424
How Do You Determine an Optimal Hybridization Temperature? 424
What Range of Probe Concentration Is Acceptable? 425
What Are Appropriate Pre-hybridization Times? 426
How Do You Determine Suitable Hybridization Times? ... 426
What Are the Functions of the Components of a Typical Hybridization Buffer? 427
What to Do before You Develop a New Hybridization Buffer Formulation? 430
What Is the Shelf Life of Hybridization Buffers and Components? 431
What Is the Best Strategy for Hybridization of Multiple Membranes? 432
Is Stripping Always Required Prior to Reprobing? 432
What Are the Main Points to Consider When Reprobing Blots? 433
How Do You Optimize Wash Steps? 434
How Do You Select the Proper Hybridization Equipment? 435
Detection by Autoradiography Film 436
How Does an Autoradiography Film Function? 436
What Are the Criteria for Selecting Autoradiogaphy Film? 438
Why Expose Film to a Blot at –70°C? 440

Helpful Hints When Working With Autoradiography Film ... 441
Detection by Storage Phosphor Imagers ... 441
How Do Phosphor Imagers Work? ... 441
Is a Storage Phosphor Imager Appropriate for Your Research Situation? ... 441
What Affects Quantitation? ... 443
What Should You Consider When Using Screens? ... 445
How Can Problems Be Prevented? ... 447
Troubleshooting ... 448
What Can Cause the Failure of a Hybridization Experiment? ... 448
Bibliography ... 453

PLANNING A HYBRIDIZATION EXPERIMENT

Hybridization experiments usually require a considerable investment in time and labor, with several days passing before you obtain results. An analysis of your needs and an appreciation for the nuances of your hybridization event will help you select the most efficient strategies and appropriate controls.

The Importance of Patience

Hybridization data are the culmination of many events, each with several effectors. Modification of any one effector (salt concentration, temperature, probe concentration) usually impacts several others. Because of this complex interplay of cause and effect, consider an approach where every step in a hybridization procedure is an experiment in need of optimization. Manufacturers of hybridization equipment and reagents can often provide strategies to optimize the performance of their products.

What Are Your Most Essential Needs?

Consider your needs before you delve into the many hybridization options. What criteria are most crucial for your research—speed, cost, sensitivity, reproducibility or robustness, and qualitative or quantitative data?

Visualize Your Particular Hybridization Event

Consider the possible structures of your labeled probes and compare them to your target(s). Be prepared to change your labeling and hybridization strategies based on your experiments.

What Do You Know About Your Target?

The sensitivity needs of your system are primarily determined by the abundance of your target, which can be approximated according to its origin. Plasmids, cosmids, phagemids as colony lifts or dot blots, and PCR products are usually intermediate to high-abundance targets. Genomic DNA is considered an intermediate to low-abundance target. Most prokaryotic genes are present as single copies, while genes from higher eukaryotes can be highly repetitive, of intermediate abundance, or single copy (Anderson, 1999). However, sensitivity requirements for single-copy genes should be considered sample dependent because some genes thought to be single copy can be found as multiples. Lewin (1993) provides an example of recently polyploid plants whose genomes are completely repetitive. The RNA situation is more straightforward; 80% of RNA transcripts are present at low abundance, raising the sensitivity requirements for most Northerns or nuclease protection assays (Anderson, 1999).

If you're uncertain about target abundance, test a series of different target concentrations (van Gijlswijk, Raap, and Tanke, 1992; De Luca et al., 1995). Manufacturers of detection systems often present performance data in the form of target dilution series. Known amounts of target are hybridized with a probe to show the lowest detection limit of a kit or a method. Mimic this experimental approach to determine your sensitivity requirements and the usefulness of a system. This strategy requires knowing the exact amount of target spotted onto the membrane.

What Do You Know about Your Probe or Probe Template?

The more sequence and structural information you know about your probe and target, the more likely your hybridization will deliver the desired result (Bloom et al., 1993). For example, the size and composition of the material from which you will generate your probe affects your choice of labeling strategy and hybridization conditions, as discussed in the question, *Which Labeling Strategy Is Most Appropriate for Your Situation?* GC content, secondary structure, and degree of homology to the target should be taken into account, but the details are beyond the scope of this chapter. (See Anderson, 1999; Shabarova, 1994; Darby, 1999; Niemeyer, Ceyhan, and Blohm, 1999; and *http://www2.cbm.uam.es/jlcastrillo/lab/protocols/hybridn.htm* for in-depth discussions.)

Is a More Sensitive Detection System Always Better?

Greater sensitivity can solve a problem or create one. The more sensitive the system, the less forgiving it is in terms of background. A probe that generates an extremely strong signal may require an extremely short exposure time on film, making it difficult to capture signal at all or in a controlled fashion.

Femtogram sensitivity is required to detect a single-copy gene and represents the lower detection limit for the most sensitive probes. Methods at or below femtogram sensitivity can detect 1 to 5 molecules, but this increases the difficulty in discerning true positive signals when screening low-copy targets (Klann et al., 1993; Rihn et al., 1995). Single-molecule detection is better left to techniques such as nuclear magnetic resonance or mass spectrometry.

The pursuit of hotter probes for greater sensitivity can be an unnecessary expense. Up to 56% of all available sites in a 486 nucleotide (nt) transcript could be labeled with biotinylated dUTP, but 8% was sufficient to achieve similar binding levels of Streptavidin than higher-density labeled probes (Fenn and Herman, 1990). Altering one or more steps of the hybridization process might correct some the above-mentioned problems. The key is to evaluate the true need, the benefits and the costs of increased sensitivity.

What Can You Conclude from Commercial Sensitivity Data?

Manufacturers can accurately describe the relative sensitivities of their individual labeling systems. Comparisons between labeling systems from different manufacturers are less reliable because each manufacturer utilizes optimal conditions for their system. Should you expect to reproduce commercial sensitivity claims? Relatively speaking, the answer is yes, provided that you optimize your strategy. However, with so many steps to a hybridization experiment (electrophoresis, blotting, labeling, and detection), quantitative comparisons between two different systems are imperfect. Side-by-side testing of different detection systems utilizing the respective positive controls or a simple probe/target system of defined quantities (e.g., a dilution series of a housekeeping gene) is a good approach to evaluation.

LABELING ISSUES

Which Labeling Strategy Is Most Appropriate for Your Situation?

Each labeling strategy provides features, benefits, and limitations, and numerous criteria could be considered for selecting the most appropriate probe for your research situation (Anderson, 1999; Nath and Johnson, 1998; Temsamani and Agrawal, 1996; Trayhurn, 1996; Mansfield et al., 1995; Tijssen, 2000). The questions raised in the ensuing discussions demonstrate why only the actual experiment, validated by positive and negative controls, determines the best choice.

The purpose of the following example is to discuss some of the complexities involved in selecting a labeling strategy. Suppose that you have the option of screening a target with a probe generated from the following templates: a 30 base oligo (30 mer), a double-stranded 800 bp DNA fragment, or a double-stranded 2 kb fragment.

30-mer

The 30-mer could be radioactively labeled at the 5′ end via T4 polynucleotide kinase (PNK) or at the 3′ end via Terminal deoxynucleotidyl transferase (TdT). PNK attaches a single molecule of radioactive phosphate whereas TdT reactions are usually designed to add 10 or less nucleotides. PNK does not produce the hottest probe, since only one radioactive label is attached, but the replacement of unlabeled phosphorous by ^{32}P will not alter probe structure or specificity. TdT can produce a probe containing more radioactive label, but this gain in signal strength could be offset by altered specificity caused by the addition of multiple nucleotides. A 30 mer containing multiple nonradioactive labels could also be manufactured on a DNA synthesizer, but the presence of too many modified bases may alter the probe's hybridization characteristics (Kolocheva et al., 1996).

800 bp fragment

The double-stranded 800 bp fragment could also be end labeled, but labeling efficiency will vary depending on the presence of blunt, recessed, or overhanging termini. Since the complementary strands of the 800 bp fragment can reanneal after labeling, a reduced amount of probe might be available to bind to the target. Unlabeled template will also compete with labeled probes for target binding reducing signal output further. However, probe synthesis from templates covalently attached to solid supports might overcome this drawback (Andreadis and Chrisey, 2000).

Random hexamer- or nanomer-primed and nick translation labeling of the 800 bp fragment will generate hotter probes than end labeling. However, they will be heterogeneous in size and specificity, since they originate from random location in the template. Probe size can range from about 20 nucleotides to the full-length template and longer (Moran et al., 1996; Islas, Fairley, and Morgan, 1998). However, the bulk of the probe in most random prime labeling reactions is between 200 and 500 nt.

If the entire 800 bp fragment is complementary to the intended target, a diverse probe population may not be detrimental. If only half the template contains sequence complementary to the target, then sensitivity could be reduced. Any attempt to compensate by increasing probe concentration could result in higher backgrounds. However, the major concern would be for the stringency of hybridization. Different wash conditions could be required to restore the stringency obtained with a probe sequence completely complementary to the target.

2 kb DNA Fragment

The incorporation of radioactive label into probes generated by random-primer labeling does not vary significantly between templates ranging from 300 bp to 2 kb, although the average size of probes generated from larger templates is greater (Ambion, Inc., unpublished data). Generating a probe from a larger template could be advantageous if it contains target sequence absent from a smaller template.

The availability of different radioactive and nonradioactive labels could further complicate the situation, but the message remains the same. Visualize the hybridization event before you go to the lab. Consider the possible structures of your labeled probes and compare them to your target(s). Be prepared to change your labeling and hybridization strategies based on your experiments.

What Criteria Could You Consider When Selecting a Label?

One perspective for selecting a label is to compare the strength, duration, and resolution of the signal. One could also consider the label's effect on incorporation into the probe, and the impact of the incorporated label on the hybridization of probe to target. The quantity of label incorporated into a probe can also affect the performance of some labels and the probe's ability to bind its target. Many experienced researchers will choose at least two techniques to empirically determine the best strategy to generate a new probe (if possible).

Signal Strength and Resolution

Signal strength of radioactive and nonradioactive labels is inversely proportional to signal resolution.

Radioactive

Isotope signal strength diminishes in the order: $^{32}P > ^{33}P > ^{35}S > ^{3}H$. When sensitivity is the primary concern, as when searching for a low-copy gene, ^{32}P is the preferred isotope. Tritium is too weak for most blotting applications, but a nucleic acid probe labeled with multiple tritiated nucleotides can produce a useful, highly resolved signal without fear of radiolytic degradation of the probe.

^{3}H and ^{35}S are used for applications such as in situ hybridization (ISH) where resolution is more essential than sensitivity. The resolution of ^{33}P is similar to ^{35}S, but Ausubel et al. (1993) cites an improved signal-to-noise ratio when ^{33}P is applied in ISH.

Nonradioactive

Signal strengths of nonradioactive labels are difficult to compare. It is more practical to assess sensitivity instead of signal strength. The resolution of nonradioactive signals is also more complicated to quantify because resolution is a function of signal strength at the time of detection, and most nonradioactive signals weaken significantly over time. Therefore the length of exposure to film must be considered within any resolution discussion. Background fluorescence or luminescence from the hybridization membrane has to be considered as well. Near-infrared dyes are superior due to low natural near-infrared occurrence (Middendorf, 1992). Some dyes emit in the far red ≥700 nm (Cy7, Alexa Fluor 549, allophycocyanin).

Older nonradioactive, colorimetric labeling methods suffered from resolution problems because the label diffused within the membrane. Newer substrates, especially some of the precipitating chemifluorescent substrates, alleviate this problem. Viscous components such as glycerol are often added to substrates to limit diffusion effects. Colorimetric substrates and some chemiluminescent substrates will impair resolution if the reaction proceeds beyond the recommended time or when the signal is too strong. Hence background can increase dramatically due to substrate diffusion.

Detection Speed

Mohandas Ghandi said that there is more to life than increasing its speed (John-Roger and McWilliams, 1994), and the same holds true for detection systems. Most nonradioactive systems deliver a signal within minutes or hours, but this speed is useless if the system can't detect a low-copy target. Searching for a single-copy gene with a ^{32}P labeled probe might require an exposure of several weeks, but at least the target is ultimately identified.

Signal Duration

Will you need to detect a signal from your blot several times over a period of hours or days? Are you pursuing a low-copy target that requires an exposure time of days or weeks? Would you prefer a short-lived signal to avoid stripping a blot prior to re-probing?

Some nonradioactive detection systems allow for the quick inactivation of the enzyme that generates the signal, eliminating the stripping step prior to re-probing (Peterhaensel, Obermaier, and Rueger, 1998). The effects and implications of stripping are discussed in greater detail later in the chapter.

The practical lifetime of common radiolabels is several days to weeks, and is dependent on the label, the ligand, and its environment, as discussed in Chapter 6, "Working Safely with Radioactive Materials." Some nonradioactive systems based on alkaline phosphatase can generate signals lasting 10 days without marked reduction (personal observation). Some chemifluorescent systems generate a fluorescent precipitate capable of producing a cumulative signal, much like isotopes.

The functional lifetime of fluorescent labels will vary with the chemical nature of the fluorescent tag and the methodology of the application. For example, signal duration of a fluorescent tag could be defined by the number of times the chromophore can be excited to produce a fluorescent emission. Some tags can only be excited/scanned once or a few times, while others are much more stable. Consult the manufacturer of the labeled product for this type of stability information. In systems where an enyzme catalyzes the production of a reagent required for fluorescence, the enzyme's half-life and sufficient presence of fresh substrate can limit the duration of the signal.

Will the Label Be Efficiently Incorporated into the Probe?

The effects of label size, location, and linkage method on the incorporation of nucleotides into DNA or RNA are enzyme-dependent and can be difficult to predict. Small side chains can inhibit nucleic acid synthesis (Racine, Zhu, and Mamet-Bratley, 1993), while larger groups such as biotin might have little or no effect (Duplaa et al., 1993; Richard et al., 1994). In general, nucleotides labeled with isotopes of atoms normally present in nucleotides (^{32}P, ^{33}P, ^{3}H, ^{14}C) will be incorporated by DNA and RNA polymerases more efficiently than nucleotides labeled with isotopes of nonnative atoms. Commercial polyermases are frequently engineered to overcome such incorporation bias.

Some applications will exploit impaired incorporation of labeled nucleotides (Alexandrova et al., 1991). Fluorescein attached to position 5 of cytosine in dCTP inhibits terminal transferase and causes addition of only one to two labeled dNTPs at the 3′ end of DNA. Fluorescein- or biotin-riboUTP have been similarly applied (Igloi and Schiefermayr, 1993).

If no specific data exist regarding incorporation efficiency of your labeled nucleotide–labeling enzyme combination, contact the manufacturer of both products. They will likely be able to provide you with a starting point from which you can optimize your labeling reactions.

Will the Label Interfere with the Probe's Ability to Bind to the Target?

Hybridization efficiency can be altered by a label's chemical structure, its location within the probe, the linker that connects the label to the ligand, and the quantity of label within the probe. Isotopes of elements present in nucleic acids in vivo might not directly alter the probe's structure, but as described below (and in Chapter 6), a label's radioactive emissions can fragment a probe. The importance of label location is demonstrated by comparing the hybridization efficiencies of probes labeled with Cy5™ originating at either C5 or the primary amine attached to C4 of dCTP. Probes labeled throughout their sequence with the C5-linked label hybridize efficiently, and are commonly applied in microarray applications (Lee et al., 2000). Probes containing the label attached to the amine group at C4 do not hybridize efficiently to their targets. The purpose of the C4 amine-label is the addition of a single molecule of Cy5 dCTP to the 3′ end of a sequencing primer (Ansorge et al., 1992). A molecular model to accurately predict the effects of labels on analog conformation, hydrogen bonding, stacking interactions, and hybrid helical geometry has been proposed (Yuriev, Scott, and Hanna, 1999).

The C5 position is also preferred for dUTP (Petrie, 1991; Oshevskii, Kumarev, and Grachev, 1989). C5 is such an attractive labeling site because it does not contribute to base-pairing by hydrogen bonding, and at least some linkers seem to allow positioning of the label attached in position 5 so that helix formation is not impaired. But bulky tags linked to pyrimidines in the C5 position still interfere to some degree with hybridization because of steric hindrance. Other sites on purines and pyrimidines have been used as tag or label attachment points. However, they have only been shown to work as primers, not for internal labeling strategies (Srivastava, Raza, and Misra, 1994).

Linker length and sequence have been shown to have a major impact as well. Very short linkers (4–10 atoms) inhibit incorporation and affect hybridization (Haralambidis, Chai, and Tregear, 1987). Above a certain linker length (>20 atoms), incorporation and hybridization are impaired, and a model of steric hindrance has been postulated to explain this effect (Zhu, Chao, and Waggoner, 1994).

Overlabeling

With few exceptions, radioactive and nonradioactive probes should not be internally labeled to 100% completion. Strong beta-emitters (^{32}P, ^{33}P) will degrade extensively labeled probes. A random primer-generated probe labeled with ^{32}P-dCTP to a specific activity of 10^9 cpm/μg will have an average size of about 300 to 500 nucleotides immediately after labeling. After storage at 4°C for 24 hours, the average size falls to 100 nucleotides or less (Amersham Pharmacia Biotech, unpublished observations). Maximally diluting the probe immediately after labeling can reduce this radiolytic degradation.

High densities of large nonradioactive tags can interfere with duplex formation and strand extension due to steric hindrance (Zhu and Waggoner, 1997; Lee et al., 1992; Day et al., 1990; Mineno et al., 1993). Although linker chains that connect label to nucleotide are designed to minimize interference (Petrie et al., 1991), steric hindrance cannot be completely circumvented.

High label densities can also cause quenching. Quenching effects for fluorescein densities greater than 1 in 10 have been described (Makrigiorgos, Chakrabarti, and Mahmood, 1998). Manufacturers of nonradioactive labeling kits optimize protocols to avoid interference from the label. Consult your manufacturer before you alter a recommended procedure.

RADIOACTIVE AND NONRADIOACTIVE LABELING STRATEGIES COMPARED

The decision to apply radioactive or nonradioactive labeling and detection systems can be based on issues of sensitivity, high-throughput, cost, safety, and ease of use, to name a few criteria discussed in this chapter. While it is feasible on paper to evaluate your research needs against these criteria, the decision must usually be determined at the lab bench. In lieu of the need to test individual systems, several studies compared the sensitivity of nonradioactive probes to the ^{32}P gold standard: Yang et al. (1999),

Plath, Peters, and Einspanier (1996), Nass and Dickson (1995), Moore and Margolin (1993), Puchhammer-Stoeckl, Heinz, and Kunz (1993), Engler-Blum et al. (1993), Bright et al. (1992), Kanematsu et al. (1991), Lion and Haas (1990), Jiang, Estes, and Metcalf (1987), Tenberge et al. (1998), Holtke et al. (1992), Pollard-Knight et al. (1990), Hill and Crampton (1994), Dubitsky, Brown, and Brandwein (1992), and Nakagami et al. (1991).

What Are the Criteria for Considering Direct over Indirect Nonradioactive Labeling Strategies?

Direct labeling strategies utilize probes that are directly conjugated to a dye or an enzyme, which generates the detection signal. Indirect labeling systems utilize probes that contain a hapten that will bind to a secondary agent generating the detection signal; the probe itself does not generate signal. Typical secondary agents are dye- or enzyme-linked antibodies, and enzyme-linked avidin complexes.

Sensitivity

Comparing the sensitivities of indirect and direct strategies is a difficult process. The fluorescent tags or dyes incorporated directly into probes usually have lower sensitivity. Detection limits will vary with tag or label incorporation efficiency, amplification level introduced by the secondary agent, and amplification level added by substrate or dye. In one instance, simply increasing the duration of the labeling reaction within a direct labeling strategy produced more sensitive probes than an indirect labeling strategy which in previous experiments had produced the more sensitive probe (Herzer, P., unpublished observations).

Comparisons are further complicated because direct and indirect labeling or detection strategies demand very different hybridization, washing, and detection procedures. Additionally the performances of these different strategies can vary with the size and structure of a probe or template from which the probe will be synthesized, further complicating any prediction of performance.

Manufacturers of labeling and detection systems can usually provide sensitivity comparisons of their different products that are qualitatively, if not always quantitatively reproducible.

Flexibility

A directly labeled probe can be detected after hybridization and washes; no further blocking or antibody steps are required. Direct, nonradioactive techniques limit choices for hybridization, wash buffer, and detection options. Optimization of your signal-to-noise ratio might be more difficult. Indirect nonradioactive detection systems are usually compatible with the common hybridization and wash buffers, but subsequent antibody incubation and detection steps can be difficult to optimize.

What Is the Storage Stability of Labeled Probes?

Radioactive

The effect of high and low emissions from radioactive labels, and methods to minimize their impact are discussed in Chapter 6. Ideally probes labeled with ^{32}P, ^{33}P, or ^{35}S should be prepared fresh for each experiment. If you choose to store a radiolabeled probe, the unincorporated label should be removed prior to storage. Damage from radioactive emissions can be minimized by dilution and the addition of free radical scavengers such as ethanol or reducing agents, but the probe must then be re-purified before reuse. As discussed in Chapter 6, it is crucial to fast-freeze radiolabeled probes to avoid complications from clustering.

Nonradioactive

The chemical nature of the tag will dictate specific storage conditions, but in general, nonradioactively labeled probes may be aliquoted and stored in the dark in –20 or –70°C non-frost-free freezers. Multiple freeze–thaws should be avoided. Stability varies depending on storage buffer formulation and the nature of the tag (e.g., fluorescent label and enzyme), and can vary from one month to one year.

Nucleic acids labeled by direct crosslinking of enzymes are supposed to be stable if stored in 50% glycerol at –20°C for several months, but this cannot be guaranteed. Since it is difficult to quantitate the remaining enzyme activity after storage, it is recommended that fresh probes be prepared to ensure that results over time are comparable. For probes labeled with chromophores or fluorophores, it is crucial to contact the manufacturer for the most contemporary storage information. In a system dependent upon a enzyme-labeled antibody, the storage conditions must maintain the integrity of the antibody, the enzyme and the probe.

If you plan to apply any probe (radioactive or not) over the long term, a positive control that can be used to evaluate the probe's effectivenss is highly recommended. Probe stability is also a function of the required sensitivity. If an old preparation of a labeled probe generates the desired signal, the probe was sufficiently stable.

Should the Probe Previously Used within the Hybridization Solution of an Earlier Experiment Be Applied in a New Experiment?

Does Sufficent Probe Remain?

Most blotting applications require and use probe in excess over target and depending on the amount of probe bound to the blot, and on decay and decomposition effects, sufficient probe might remain in the hybridization solution. A dot blot of a dilution series of the target DNA can determine if sufficient probe remains.

Label Potency

Consider the many issues regarding the lifetime and storage stability of labels and tags mentioned above. I (S.H.) have successfully reused ^{32}P random primer-labeled probes up to five days after the initial hybridization experiment. Storage of radiolabeled probes at –20°C in hybridization buffer for a few days or at 4°C overnight is usually not problematic. Reuse of radiolabeled probes is not recommended for high-sensitivity and/or quantitative applications. The storage issues discussed above for radioactive and nonradioactive labels should also be considered here.

It may be worthwhile to re-purify some probes from the buffer for optimal storage. Peptide nucleic acids (PNA) probes are an example where the probe expense may justify the expense of re-purification.

How Should a Probe be Denatured for Reuse?

Probes are stored in hybridization buffer prior to reuse. Such buffers may contain components that will aid the denaturation of probe (e.g., formamide) so that boiling is not required. Heating to temperatures above T_m is sufficient, since at T_m half of the nucleic acid is denatured. Boiling can also destroy buffer components such as blocking reagents, SDS, volume excluders, and label. Heating the hybridization buffer containing the probe to temperatures of 10 to 20°C above the hybridization temperature would be ideal, but 20 to 30°C below the boiling point has to suffice.

Is It Essential to Determine the Incorporation Efficiency of Every Labeling Reaction?

Labeling reactions are straightforward, and with the advent of commercial labeling kits, unlikely to fail. Hence we do not measure incorporation efficiency of every probe that we label. We only begin to question labeling efficiency when hybridization results are unsatisfactory, a point at which it might be too late to determine incorporation efficiency.

Before skipping any control steps, consider the implications. Minimally, measure incorporation efficiency when working with a new technique, a new probe, a new protocol, or a new kit. Radiolabeled probes need to be purified or at least Trichloroacetic acid (TCA) precipitated to determine labeling efficiency, as discussed in Chapter 7, "DNA Purification." Determining the efficiency of nonradioactive labeling reactions can be more time-consuming, often involving dot blots and/or scanning of probe spots. Follow manufacturer recommendations to determine labeling efficiency of nonradioactive probes.

Is It Necessary to Purify Every Probe?

Unincorporated nucleotides, enzyme, crosslinking reagents, buffer components, and the like, may cause high backgrounds or interfere with downstream experiments. Hybridization experiments where the volume of the probe labeling reaction is negligible in comparison to the hybridization buffer volume do not always require probe cleanup. If you prefer to minimize these risks, purify the probe away from the reaction components.

While there are some labeling procedures (i.e., probes generated by random primer labeling with ^{32}P-dCTP), where unpurified probe can produce little or no background (Amersham Pharmacia Biotech, unpublished observations), such ideal results can't be guaranteed for every probe. When background is problematic, researchers have the option to repurify the probe preparation. Admittedly, this approach wouldn't be of much use if the experiment producing the background problem required a five day exposure. (Purification options are discussed in Chapter 7, "DNA Purification.")

HYBRIDIZATION MEMBRANES AND SUPPORTS

What Are the Criteria for Selecting a Support for Your Hybridization Experiment?

Beyond the information listed below and your personal experience, the most reliable approach to determine if a membrane can be used in your application is to ask the manufacturer for application and or quality control data. Whether a new membrane formulation will provide you with superior results is a matter that can usually be decided only at the bench, and the results can vary for different sets of targets, probes, and detection strategies.

Physical Strength

Nitrocellulose remains popular for low to medium sensitivity (i.e., screening libraries) applications and for situations that require minimal handling. The greater mechanical strength of nylon makes it superior for situations that require repeated manipulation of your blot. Nylon filters may be probed 10 times or more (Krueger and Williams, 1995; Li, Parker, and Kowalik, 1987). Even though nitrocellulose may be used more than once, brittleness, loss of noncovalently bound target during stripping, and decreased stability in harsh stripping solutions make nitrocellulose a lesser choice for reusable blots. Glass supports and chips can be stripped, but stripping efficiency and aging of target on these supports may impair reuse of more than two to three cycles of stripping and reprobing. Supported nitrocellulose is sturdier and easier to handle than pure nitrocellulose, but remember that it needs to be used in the proper orientation.

Binding Capacity

Nylon and PVDF (polyvinylidene difluoride) membranes bind significantly more nucleic acid than nitrocelluose; hence they can generate stronger signals after hybridization. Nucleic acids can be covalently attached to nylon but not to nitrocellulose, as discussed below. Positively charged nylon offers the highest binding capacities. As is the case with detection systems of greater sensitivity, the greater binding capacity of positively charged membranes could increase the risk of background signal. However, optimization of hybridization conditions, such as probe concentration and hybridization buffer composition, will usually prevent background problems. If such optimization steps do not prevent background, a switch to another membrane type, such as to a neutral nylon membrane, might be required. If your signal is too low, try a positively charged nylon membrane. Positively charged nylon is often chosen for nonradioactive applications to ensure maximum signal strength. The quantity of positive charges (and potential for background) can vary by 10-fold between manufacturers. The lower binding capacity of nitrocellulose decreases the likelihood of background problems under conditions that generate a detectable signal.

Thickness

Most membranes are approximately 100 to 150 μm thick. Thickness influences the amount of buffer required per square centimeter. Thicker membranes soak up more buffer, wet more slowly, and dry application of thicker filters to the surface of a gel can be more difficult.

Pore Size

Pore size limits the size of the smallest fragment that can be bound and fixed onto a membrane, but bear in mind that pore size is an average value. In general, 0.45 μm micron pore sizes can bind oligonucleotides down to 50 bases in length, but the more common working limit is 100 to 150 nucleotides or base pairs. Membranes comprised of 0.22 μm micron pores are recommended for work with the smallest single- and double-stranded DNA fragments. Custom manufacturing of membranes with 0.1 μm pore size is also available. Table 14.1 compares membrane characteristics.

Specialized Application

Microarrays

Glass slides stand apart from membrane supports because glass allows for covalent attachment of oriented nucleic acid, is nonporous and offers low autofluorescence. On a nonporous support, buffer volumes can be kept low, which decreases cost and allows increased probe concentration. Unlike nylon and nitrocellulose membranes, background isn't problematic under these aggressive hybridization conditions. Probes are labeled with different dyes and allow detection of multiple targets in a single hybridization experiment; nylon arrays are often restricted to serial or parallel hybridization, although examples of simultaneous detection on nylon membranes can be found in the literature. (Some references for multiple probes on nylon are Kondo et al. (1998), Holtke et al. (1992), and Bertucci et al. (1999).) These features make glass slides ideal for nonradioactive detection in micro arrays.

Macroarrays

Background problems, high buffer volumes, and hence cost, limit the usefulness of nonradioactive labels for macroarrays on nylon filters. Macroarrays employ thin charged or uncharged nylon membranes to reduce buffer consumption but suffer from low sensitivity due to the high autofluorescence of nylon. Stronger signals derived from enzyme-substrate driven signal amplification compromise resolution and quantitation. Radioactive labels such as ^{33}P are preferred for macroarrays (Moichi et al., 1999; Yano et al., 2000; Eickhoff et al., 2000).

Table 14.1 Characteristics of Membranes Used in Nucleic Acid Blotting Applications

Membrane	Physical Characteristics	Strip or Reprobe	Benefits	Limitations	Recommended Use
Nitrocellulose	Hydrophilic membrane of low solvent resistance and physical strength	Not recommended	Low cost, low background, easy to block	More difficult to handle, flammable, low binding capacity not good for nonrad systems because it binds proteins, not for RNA	Radioactive colony/ plaque lifts, library screens, low sensitivity applications
Nitrocellulose supported		Possibly 2–6 times via gentle conditions	Low cost, low background, easy to block	As for unsupported nitrocellulose	As for unsupported nitrocellulose
PVDF	Hydrophobic membrane of fair solvent resistance/ physical strength with a thermal stability <135C	Yes, strip with SDS, water, formamide	Intermediate binding capacity and hence sensitivity without nitrocellulose drawbacks of brittleness	Fairly new and hence not so well documented method that might be more difficult to optimize	some nonradioactive applications, some mixed protein/ nucleic acid applications
Nylon neutral	Hydrophilic, hygroscopic membrane of good solvent resistance and physical strength	Up to 10 times if under conditions that don't hydrolyze the target		Irreversibly binds many stains, staining often lowers signal	Where nitrocelluose inappropriate and background is problematic with charged nylon
Nylon negatively charged	Hydrophilic, hygroscopic membrane of good solvent resistance and physical strength	As for uncharged nylon	Lowest background of all nylon membranes	Poor specificity of signal/retention	
Nylon positively charged	Hydrophilic, hygroscopic membrane of good solvent resistance and physical strength	As for uncharged nylon	Highest sensitivity; usually optimal for most nonradioactive systems, but some supercharged membranes can interfere with nonradioactive detection systems	Can give high backgrounds if not optimzed properly	

Which Membrane Is Most Appropriate for Quantitative Experiments?

The size of the nucleic acid being transferred, the physical characteristics of the membrane, and the composition of transfer buffer affect the transfer efficiency. There is no magic formula guaranteeing linear transfer of all nucleic acids at all times. Linearity of transfer needs to be tested empirically with dilution series of nucleic acid molecular weight markers.

What Are the Indicators of a Functional Membrane?

Membranes will record every fingerprint, drop of powder, knick, and crease. Always handle membranes with plastic forceps and powder-free gloves.

Membranes should be dry and uniform in appearance. They should be wrinkle- and scratch-free since mechanical damage may lead to background problems in these affected areas. Membranes should wet evenly and quickly. If membranes do appear blotchy or spotty, or seem to have different colors, it is best not to use them. Membranes are hygroscopic, light sensitive, and easily damaged, but as long as membranes are properly stored, may remain functional for years. Please note that manufacturers only guarantee potency for shorter time periods, usually six to twelve months. If the vitality of the membrane is in doubt, a quick dot blot or test of the binding capacity may help. Manufacturers can provide guidelines for assessing binding capacity. Including an untreated, target-free piece of membrane to evaluate background in a given hybridization buffer or wash system can help to troubleshoot background problems.

Can Nylon and Nitrocellulose Membranes Be Sterilized?

Researchers performing colony hyrbidizations often ask about membrane sterilization. While membranes might not be supplied guaranteed to be sterile, they are typically produced and packaged with extreme care, minimizing the likelihood of contamination.

Theoretically it is possible to autoclave membranes, but cycles should be very short (two minutes at 121°C in liquid cycle). Note that such short durations cannot guarantee sterility. Membranes should be removed as soon as the autoclave comes down to a safe temperature, and dried at room temperature. Multiple membranes should be separated by single sheets of Whatman paper. Note that filters may turn brown, become brittle, may shrink and warp and become difficult to align with plates, but this does not interfere with probe hybridization.

Treatment of membranes with 15% peroxide or 98% ethanol at room temperature after crosslinking can also sterilize filters. Peroxide may be more harmful to nucleic acid and filter chemistry over time.

NUCLEIC ACID TRANSFER

What Issues Affect the Transfer of Nucleic Acid from Agarose Gels?

This discussion will focus on the transfer of nucleic acids from agarose gels onto a membrane via passive transfer. Details on the transfer of DNA from polyacrylamide gels are presented in Westermeier (1997).

Active or Passive Techniques

Vacuum, electrophoretic, and downward gravity transfer methods are fast (less than 3 hours) and efficient (greater than 90% transfer). Transfer efficiency depends on thickness and percentage of the gel and nucleic acid concentration or size. Transfer time increases with percentage of agarose, gel thickness, and fragment size. Capillary blotting of RNA larger than 2.5 kb takes more than 12 hours, and downward transfer only 1 to 3 hours (Ming et al., 1994; Chomczynski, 1992; Chomczynski and Mackey, 1994). Speed, low cost, no crushing of gel, and efficient alkaline transfer of RNA are the main reasons why downward transfer is gaining popularity for RNA transfer (Inglebrecht, Mandelbaum, and Mirkov, 1998).

Transfer Buffer

Manufacturers of filter or blotting equipment provide transfer protocols that serve as a starting point for transfer buffer formulation. If nucleic acids are of unusual size or sequence, modified protocols might be required. RNA, small DNA fragments (<100 bp), and nitrocellulose membranes usually require greater salt concentrations. Keep in mind that RNA has a very low affinity for nitrocellulose even at high salt.

The effects of pH on transfer efficiency and subsequent detection of target are many and complex. Transfer buffer pH can directly affect the stabilities of the membrane and the nucleic acid target. Nitrocellulose and some nylon membranes are not stable at pH > 9, and nitrocellulose will not bind DNA at pH above 9 (Ausubel et al., 1993). Some nylon membranes are not stable at acidic pH (Wheeler, 2000). Transfer buffer pH can also affect signal output and background levels, especially when working with nylon membranes (Price, 1996; McCabe et al., 1997).

Transfer buffer pH can also affect the surface charge of the membrane. Nylon membranes are polyamides. The net charge of unmodified nylon is zero, but the polyamide backbone will become more positive when lowering the pH. Different side groups are introduced into the nylon precursors for the purpose of increasing the positive or negative charge of the membrane. These side chains may alter the membrane's response to the pH of the transfer buffer, which might ultimately affect the ability of a probe to bind to the target nucleic acid. When using an acidic or alkaline transfer buffer, you may want to verify the expected impact of pH on a particular membrane. For further effects of pH and salt concentration, see Khandjian (1985).

Alkaline transfer conditions will fragment and denature nucleic acids, and these effects have been exploited to crosslink DNA after transfer. Prolonged exposure of RNA to mildly alkaline conditions (pH > 9) will degrade RNA, but Inglebrecht, Mandelbaum, and Mirkov (1998) applied alkaline pH for short periods to enhance the transfer of large, problematic RNA. Some membrane manufacturers warn against alkaline transfer of RNA and DNA because of nonuniform results. If the gel is depurinated prior to alkaline or nonalkaline transfer, omission of the neutralization step prior to transfer can reduce signal. Without a neutralization step, depurination continues in the gel.

Depurination

Breakdown of nucleic acids via depurination increases transfer efficiency. Transfer of targets larger than 5 kb, agarose concentrations greater than 1%, and gels thicker than 0.5 cm improve upon depurination. Depurination beyond recommended times will result in reduced sensitivities on hybridization.

Stains

Gels and/or membranes can be stained in order to monitor transfer efficiency, but it is impossible to make an absolute statement regarding whether stains interfere with transfer and subsequent hybridization. Intercalating dyes, such as ethidium bromide or methylene blue, can influence transfer and hybridization efficiency (Thurston and Saffer, 1989; Ogretmen et al., 1993), yet others report no effect of ethidium bromide utilized in Southern hybridization experiments (Booz, 2000). In another instance, ethidium bromide interfered with transfer onto supercharged nylon membrane (Amersham Pharmacia Biotech, unpublished observation). DNA stains are usually intercalating cations; hence intercalation will be affected by salt concentration. Therefore salt concentration of the transfer buffer might also affect transfer and subsequent hybridization. Tuite and Kelly (1993) also show the interference of methylene blue staining upon subsequent hybridization.

Some newer dyes (SYBR® Gold and SYBR® Green, Molecular Probes Inc.) are promoted as noninterfering stains. Otherwise, in light of the inconsistencies described above, it is best to destain the gel prior to transfer, or to stain a marker lane only. Visualization of DNA on membranes by UV shadowing has been done, but concerns exist about insufficient sensitivity and overfixation of nucleic acids and (Thurston and Saffer, 1989; Herrera and Shaw, 1989).

Staining details are provided in Wilkinson, Doskow, and Lindsey (1991), Wade and O'Conner (1992), Correa-Rotter, Mariash, and Rosenberg, 1992) and at *http://www.mrcgene.com/met-blue.htm*, *http://www.cbs.umn.edu/~kclark/protocols/transfer.html*, *http://www.bioproducts.com/technical/visualizingdnainagarosegels.shtml.*

Physical Perturbations

Air bubbles between gel and membrane, between membrane and filters, and between gel and support will interfere with transfer. Crushed gel sections trap nucleic acids, as does a gel whose surface has dried out. Moving a membrane in contact with a gel after transfer has begun causes stamp or shadow images and/or fuzzy bands.

Should Membranes Be Wet or Dry Prior to Use?

It is best to follow the recommendations from the manufacturer of your particular blotting equipment or membrane; strategies from different suppliers are not always identical.

In general, capillary transfer can benefit from pre-equilibration of membrane and gel. Free floating of gel and membrane in excess (transfer) buffer pre-equilibrates them to the conditions necessary for good transfer, and can reduce transfer time. Another factor to consider is ease of membrane application; some researchers prefer applying a wet membrane to the gel, but this is a matter of personal preference.

If pre-wetting is preferred, nitrocellulose as well as nylon should be pre-wet in distilled water first. Both membranes will wet more quickly and evenly if no salt is present.

Most membranes need not be wet for dot blots. Dots may spread more if the membrane has been pre-wet. Dots and/or slot blot-applied samples will soak more evenly onto dry membranes. Uneven dot spreading due to unevenly wet membrane or damp membrane can lead to asterisk shapes instead of circles or squares.

What Can You Do to Optimize the Performance of Colony and Plaque Transfers?

Single colonies or plaques usually contain millions of target copies, so transfer can afford to be less efficient. Cell lysis and DNA denaturation are achieved in a sodium hydroxide/SDS step. Fixation can also be achieved in this same step when using positively charged membranes. The blotting process is finished by a neutralization step and a filter equilibration step into salt buffers such as SSC prior to fixation. Transfer may be followed with a proteinase K digestion to remove debris and reduce background (Kirii, 1987; Gicquelais et al., 1990). Proteinase K treatment will reduce background signal when using nonradioactive detection systems, especially those based on alkaline phosphatase. Bacterial debris can also be removed mechanically by gentle scrubbing with equilibration buffer-saturated tissue wipes.

Ideally colonies or plaques should be no larger than 1 mm in diameter; colonies smaller than 0.5 mm deliver a more focused signal (*http://www.millipore.com/analytical/pubdbase.nsf/docs/TN1500ENUS.html*). Filters should be "colony side up" during denaturing/neutralization steps. Two different methods have been described for filter treatment: the bath method, where filters are floated or submerged in the buffers, and the wick method, where 3 MM Whatman paper is saturated with buffers. The wick method yields clearer, more focused dots; the "bath" method is less likely to lead to only partial denaturation and loss of signal. Newer protocols skip the denaturing/neutralization steps in favor of a microwaving step (*http://www.ambion.com/techlib/tb/tb_169.html*) or an autoclaving/crosslinking protocol (*http://www.jax.org/~jcs/techniques/protocols/ColonyLifts.html*). These techniques, though difficult to optimize, save time. However, microwaving can warp membranes, making it difficult to align filters with the original agar plate.

CROSSLINKING NUCLEIC ACIDS

What Are the Strengths and Limitations of Common Crosslinking Strategies?

Four different methods for crosslinking nucleic acids to membrane are commonly applied, but the efficiency will vary with the target and the type of membrane.

UV Crosslinking

UV light photoactivates uracil (U) or thymine (T) of RNA and DNA, respectively, such that they react with amine groups on the nylon membrane. Therefore short nucleic acids (<100 bases) with high GC content may bind less efficiently. If the duration of UV exposure is too long, or the UV energy output too high, the hybridization potential of the target is reduced, and so is any subsequent detection signal. Depending on the UV crosslinker and membrane used, membranes can be wet or dry, but settings will depend on the percentage of moisture on the membrane. Hence wet and dry crosslinking times or energy settings are not interchangeable. Nitrocellulose is flammable and may combust during UV crosslinking.

Crosslinking on transilluminators tends to produce inconsistent results because the delivered energy (in microjoules or Watts × time) fluctuates with these instruments. When crosslinking on a UV transilluminator, a 254 nm emission is required, and the optimal time needs to be determined empirically. Because the light source in a UV transilluminator is not calibrated for a preset energy output, one cannot predict how to compensate for an aging UV bulb by increasing the time of crosslinking. Exposing the nucleic acid side (side of membrane in direct contact with gel surface) to a multiple-user transilluminator increases the chance of target degradation and contamination.

Baking

Baking membranes at 80°C drives all water from the nucleic acid and membrane until the hydrophobic nucleotide bases form a hydrophobic bond to the aromatic groups on the membrane. As little as 15 minutes at 80°C may be sufficient. Vacuum baking is used for nitrocellulose to reduce the risk of combustion. Excessive temperature (>100°C) or extended exposure to heat (two hours) will destroy a membrane's ability to absorb buffers efficiently, leading to background problems, loss of signal, and membrane damage.

Alkaline Transfer

Alkaline transfer onto positively charged nylon membranes produces covalent attachment of the nucleic acid, but the process is slow (Reed and Mann, 1985). Transfers of short duration (few minutes versus hours) will not produce covalent attachment. Short transfer time applications, such as slot blots, dot blots, or colony filter lifts should be followed by a fixation step to secure linkage to the membrane. Opinions diverge whether crosslinking after longer alkaline transfer times is necessary. Some researchers skip crosslinking to avoid loss of signal due to overfixation. Others crosslink because loss of nucleic acids due to incomplete fixation is feared.

Alkali Fixation after Salt Transfer

DNA may also be covalently immobilized onto positively charged nylon by laying this membrane onto 0.4M NaOH—soaked 3MM Whatman paper for 20–60 minutes. The exact time needs to be determined empirically.

What Are the Main Problems of Crosslinking?

Avoid rinsing membranes prior to to crosslinking, especially with water. Washing with large volumes of low salt solutions, such as 2× SSC, is also risky. Ideally fix nucleic acids first, then stain, wash, and so forth.

UV crosslinking and baking are nonspecific fixation techniques, so any biopolymers present on the filter have the potential to bind, increasing the risk of background and errant signals. Therefore filters should be kept free of dirt and debris. Brown and/or yellow stains observed after alkaline transfer did not interfere with signal or add to background (personal observation). Standard electrophoresis loading dyes do not interfere with transfer and/or fixation.

What's the Shelf Life of a Membrane Whose Target DNA Has Been Crosslinked?

Membranes can be stored between reprobings for a few days in plastic bags or Saran wrap in the refrigerator in 2× SSC. For storage lasting weeks or months, dried blots, kept in the dark, are preferable (note that blots need to be stripped of their probe(s) prior to drying). Dry, dark conditions will minimize microbial contamination and nucleic acid degradation. Dried membranes may be stored in the dark at room temperature in a desiccator at 4°C, or at –20°C in the presence of desiccant.

One reference cited decreased shelf lives for storage at room temperature (Giusti and Budowle, 1992). Blots maintained dry (desiccant for long-term storage), dark, and protected from mechanical damage may be stored safely for 6 to 12 months.

THE HYBRIDIZATION REACTION

The hybridization step is central to any nucleic acid detection technique. Choices of buffer, temperature, and time are never trivial because these effectors in combination with membrane, probe, label, and target form a complex network of cause and effect. Determining the best conditions for your experiment will always require a series of optimization experiments; there is no magic formula. The role of the effectors of hybridization, recommended starting levels, and strategies to optimize them will be the focus of this section. Readers interested in greater detail on the intricacies and interplay of events within hybridization reactions are directed to Anderson (1999), Gilmartin (1996), Thomou and Katsanos (1976), Ivanov et al. (1978), and Pearson, Davidson, and Britten (1977).

How Do You Determine an Optimal Hybridization Temperature?

Hybridization temperature depends on melting temperature (T_m) of the probe, buffer composition, and the nature of the target:hybrid complex. Formulas to calculate the T_m of oligos, RNA, DNA, RNA-DNA, and PNA-DNA hybrids have been described (Breslauer et al., 1986; Schwarz, Robinson, and Butler, 1999; Marathias et al., 2000). Software that calculates T_m is described by Dieffenbach and Dveksler (1995).

The effects of labels on melting temperatures should be taken into consideration. While some claim little effect of tags as large as horseradish peroxidase on hybrid stability/T_m (Pollard-Knight et al., 1990a), others observed T_m changes with smaller base modifications (Pearlman and Kollman, 1990). It will have to suffice that nonradioactive tags may alter the hybridization characteristics of probes and that empiric determination of T_m may be quicker than developing a formula to accurately predict hybridization behavior of tagged probes. Hybridization temperatures should also take into account the impact of hybridization temperature on label stability. Alkaline phosphatase is more stable at elevated temperatures than horseradish peroxidase. Thermostable versions of enzymes or addition of thermal stabilizer such as trehalose

(Carninci et al., 1998) may provide alternatives to hybridization at low temperatures.

When switching from a DNA to an RNA probe, hybridization temperatures can be increased due to the increased T_m of RNA-DNA heteroduplexes. Because of concerns about instability of RNA at elevated temperatures, an alternative approach with RNA probes is the use of a denaturing formamide or urea buffer that allows hybridization at lower temperature.

A good starting point for inorganic (nondenaturing) buffers are hybridization temperatures of 50 to 65°C for DNA applications and 55 to 70°C for RNA applications. Formamide buffers offer hybridization at temperatures as low as 30°C, but temperatures between 37 and 45°C are more common. Enzyme-linked probes should be used at the lowest possible temperature to guarantee enzyme stability.

After hybridization and detection has been performed at the initially selected hybridization temperature, adjustments may be required to improve upon the results. A hybridization temperature that is too low will manifest itself as a high nonspecific background. The degree by which the temperature of subsequent hybridizations should be adjusted will depend on other criteria discussed throughout this chapter (GC content of the probe and template, RNA vs. DNA probe, etc.), and thus hybridization temperature can't be exactly predicted. Most hybridization protocols employ temperatures of 37°C, 42°C, 50°C, 55°C, 60°C, 65°C, and 68°C.

Note that sometimes a clean, strong, specific signal that is totally free of nonspecific background cannot be obtained. Background reduction, especially through the use of increased hybridization temperatures, will result in the decrease of specific hybridization signal as well. There is often a trade-off between specific signal strength and background levels. You may need to define in each experiment what amount of background is acceptable to obtain the necessary level of specific hybridization signal. If the results are not acceptable, the experiment might have to be redesigned.

What Range of Probe Concentration Is Acceptable?

Probe concentration is application dependent. It will vary with buffer composition, anticipated amount of target, probe length and sequence, and the labeling technique used.

Background and signal correlate directly to probe concentration. If less probe than target is present, then the accuracy of band quantities is questionable.

In the absence of rate-accelerating "fast" hybridization buffers, probe concentration is typically 5 to 10 ng/ml of buffer. Another convention is to apply 2 to 5 million counts/ml of hybridization buffer, which may add up to more than 10 ng/ml if the probe was end-labeled, as compared to a random primer-generated probe. The use of rate accelerators or "fast" hybridization buffers requires a reduction in probe concentration to levels of 0.1 to 5 ng/ml of hybridization buffer.

Another approach to select probe concentration is based on the amount of target. A greater than 20× excess of probe over target is required in filter hybridization (Anderson, 1999). Solution hybridization may not require excess amounts for qualitative experiments. To determine if probe is actually present in excess over target, perform replicate dot or slot blots containing a dilution series of immobilized target and varying amounts of input probe (Anderson, 1999). If probe is present in excess, the signal should reflect the relative ratios of the different concentrations of target. If you do not observe a proportional relationship between target concentration and specific hybridization signal at any of the probe concentrations used, you may need to increase your probe concentration even higher. Probe concentration cannot be increased indefinitely; a high background signal will eventually appear.

What Are Appropriate Pre-hybridization Times?

Prehybridization time is also affected by the variables of hybridization time. For buffers without rate accelerators, prehybridization times of at least 1 to 4 hours are a good starting point. Some applications may afford to skip prehybridization altogether (Budowle and Baechtel, 1990). Buffers containing rate accelerators or volume excluders usually do not benefit from prehybridization times greater than 30 minutes.

How Do You Determine Suitable Hybridization Times?

Hybridization time depends on the kinetics of two reactions or events: a slow nucleation process and a fast "zippering" up. Nucleation is rate-limiting and requires proper temperature settings (Anderson, 1999). Once a duplex has formed (after "zippering"), it is very stable at temperatures below melting, given that the duplex is longer >50 bp. Hybridizing overnight works well for a wide range of target or probe scenarios. If this generates a dissatisfactory signal, consider the following.

There are several variables that affect hybridization time. Double-stranded probes (i.e., an end-labeled 300 bp fragment) require longer hybridization times than single-stranded probes (end-labeled oligonucleotide), because reassociation of double-stranded probes in solution competes with annealing events of probes to target. At 50% to 75% reassociation, free probe concentration has dwindled to amounts that make further incubation futile. Hybridization time for a double-stranded probe can therefore be deduced from its reassociation rate (Anderson, 1999). Glimartin (1996) discusses methods to predict hybridization times for single-stranded probes, as does Anderson (1999). Other variables of hybridization time include probe length and complexity, probe concentration, reaction volume, and buffer concentration.

Buffer formulations containing higher concentrations (≥10 ng/ml) of probe and/or rate accelerators or blots with high target concentrations may require as little as 1 hour for hybridization. Prolonged hybridization in systems of increased hybridization rate will lead to background problems. The shortest possible hybridization time can be tested for by dot blot analysis. Standard buffers usually require hybridization times between 6 and 24 hours. Plateauing of signal sets the upper limit for hybridization time. Again, optimization of hybridization time by a series of dot blot experiments, removed and washed at different times, is recommended. Plaque or colony lifts may benefit from extended hybridization times if large numbers of filters are simultaneously hybridized.

What Are the Functions of the Components of a Typical Hybridization Buffer?

Hybridization buffers could be classified as one of two types: denaturing buffers, which lower the melting temperatures (and thus hybridization temperatures) of nucleic acid hybrids (i.e., formamide buffers), and salt/detergent based buffers, which require higher hybridization temperatures, such as sodium phosphate buffer (as per Church and Gilbert, 1984).

Denaturants

Denaturing buffers are preferred if membrane, probe, or label are known to be less stable at elevated temperatures. Examples are the use of formamide with RNA probes and nitrocellulose filters, and urea buffers for use with HRP-linked nucleic acid probes. Imperfectly matched target:probe hybrids are hybridized in formamide buffers as well.

For denaturing, 30% to 80% formamide, 3 to 6 M urea, ethylene glycol, 2 to 4 M sodium perchlorate, and tertiary alkylamine chloride salts have been used. High-quality reagents, such as deionized formamide, sequencing grade or higher urea, and reagents that are DNAse- and/or RNAse-free are critical.

Formamide concentration can be used to manipulate stringency, but needs to be >20%. Hybrid formation is impaired at 20% formamide but not at 30 or 50% (Howley et al., 1979). 50% to 80% formamide may be added to hybridization buffers. 50% is routinely used for filter hybridization. 80% formamide formulations are mostly used for in situ hybridization (ISH) where temperature has the greatest influence on overall stability of the fixed tissue and probe, and in experiments where RNA:DNA hybrid formation is desired rather than DNA:DNA hybridization. In 80% formamide, the rate of DNA:DNA hybridization is much lower than RNA:DNA hybrid formation (Casey and Johnson 1977). Phosphate buffers are preferred over citrate buffers in formamide buffers because of superior buffering strength at physiological pH.

In short oligos 3M tetramethylammonium chloride (TMAC) will alter their T_m by making it solely dependent on oligonucleotide length and independent of GC content (Bains, 1994; Honore, Madsen, and Leffers, 1993). This property has been exploited to normalize sequence effects of highly degenerate oligos, as are used in library screens. Note that some specificity may be lost.

Salts

Binding Effects

Hybrid formation must overcome electrostatic repulsion forces between the negatively charged phosphate backbones of the probe and target. Salt cations, typically sodium or potassium, will counteract these repulsion effects. The appropriate salt concentration is an absolute requirement for nucleic acid hybrid formation.

Hybrid stability and sodium chloride concentration correlate in a linear relation in a range of up to 1.2M. Stability may be increased by adding salt up to a final concentration of 1.2M, or decreased by lowering the amount of sodium chloride. It is the actual concentration of free cations, or sodium, that influences stability (Nakano et al., 1999; Spink and Chaires, 1999). Final concentrations of 5 to 6× SSC or 5 to 6× SSPE (Sambrook, Fritsch, and Maniatis, 1989), equivalent to approximately 0.8 to 0.9M sodium chloride and 80 to 90mM citrate buffer or 50mM sodium phosphate buffer, are common starting points for hybridization buffers. At 0.4 to 1.0M sodium chloride, the hybridization rate of

DNA:DNA hybrids is increased twofold. Below 0.4M sodium chloride, hybridization rate drops dramatically (Wood et al., 1985). RNA:DNA and RNA:RNA hybrids require slightly lower salt concentrations of 0.18 to 1.0M to increase hybridization by twofold.

pH Effects

Incorrect pH may impair hybrid formation because the charge of the nucleic acid phosphate backbone is pH dependent. The pH is typically adjusted to 7.0 or from 7.2 to 7.4 for hybridization experiments. Increasing concentrations of buffer substances may also affect stringency. EDTA is sometimes added to 1 to 2mM to protect against nuclease degradation.

Detergent

Detergents prevent nonspecific binding caused by ionic or hydrophobic interaction with hydrophobic sites on the membrane and promote even wetting of membranes. 1% to 7% SDS, 0.05% to 0.1% Tween-20, 0.1% *N*-lauroylsarcosine, or Nonidet P-40 have been used in hybridization buffers. Higher concentrations of SDS (7%) seem to reduce background problems by acting as a blocking reagent (Church and Gilbert, 1984).

Blocking Reagents

Blocking reagents are added to prevent nonspecific binding of nucleic acids to sites on the membrane.

Proteinaceous and nucleic acid blocking reagents such as BSA, BLOTTO (nonfat dried milk), genomic DNA (calf thymus, herring, or salmon sperm), and poly A may be used. Denhardt's solution is often referred to as a blocking reagent, but it is really a mix of blocking reagents and volume excluder or rate accelerator. Screening tissue samples with nucleic acid probes labeled with enzyme-linked avidin might require additional blocking steps because of the presence of endogenous biotin within the sample. Vector Laboratories, Inc., manufacturers a solution for blocking endogenous biotin.

The best concentration of each of the blocking reagents for individual applications needs to be determined empirically. If nonspecific binding is observed, then increase the concentration of blocking agent or switch to a different blocking agent. Concentrations of BSA range from 0.5% to 5%; 1% is a common starting point. Other blocking agents include nonfat dry milk (BLOTTO) (1–5%), 0.1 to 1mg/ml sonicated, denatured genomic

DNA (calf thymus or salmon sperm), or 0.1 to 0.4 mg/ml yeast RNA.

Hybridization Rate Accelerators

Agents that decrease the time required for hybridization are large, hydrophilic polymers that act as volume excluders. That is, they limit the amount of "free" water molecules, effectively increasing the concentration of probe per ml of buffer without actually decreasing the buffer volume. Common accelerators are dextran sulfate, ficoll, and polyethylene glycol. There are no hard and fast rules, but test a 10% solution of these polymers as accelerants. Rate accelerators can increase the hybridization rate several-fold, but if background is problematic, reduce the concentration to 5%. The performance of dextran sulfate (and perhaps other polymers whose size distribution changes between lots) can vary from batch to batch, so the concentration of this and perhaps other accelerators might have to be adjusted after ordering new materials.

Higher concentrations (30–40%) of Ficoll 400, polyethylene glycol, and dextran sulfate are difficult to dissolve, and microwaving or autoclaving may help. Carbohydrate polymers such as Ficoll and dextran sulfate will be ruined by standard autoclave temperatures; 115°C should be the temperature maximum, and allow solutions to cool slowly. Pipetting of stock solutions of any of these viscous polymers can be difficult. Pouring solutions into tubes or metric cylinders followed by direct dilution with aqueous buffer components may be easier than pipetting. An alternative approach to increase hybridization rate is the use of high salt concentrations and/or lower hybridization temperatures. This simply allows faster annealing of homologous probe/target duplexes that are significantly less than 100% homologous.

What to Do before You Develop a New Hybridization Buffer Formulation?

Check for Incompatibilities

Not every combination of the above components will be chemically compatible. Membranes blocked with milk may form precipitates in the presence of hybridization buffers containing high concentrations of SDS, as found in Church and Gilbert (1984). Most sodium, potassium, and ammonium salts are soluble, but mixing soluble magnesium chloride from one buffer component with phosphate buffers produces insoluble magnesium phosphate.

A proteinaceous blocking reagent could be salted out by ammonium sulfate.

Stock solutions of protein blocking agents may contain azide as a preservative. Undiluted azide may inhibit the horseradish peroxidase used in many nonradioactive detection systems.

Change One Variable at a Time

Unless you change to a totally different buffer system, optimization is usually faster if you alter one variable incrementally and monitor for trends.

Hybridization is an experiment within an experiment. The calculation of theoretical values that closely resemble your research situation may require more work than empiric determination, especially when selecting hybridization temperature and time.

Record-Keeping

At the very least, include a positive control to monitor your overall experimental performance. As described elsewhere in this chapter, the better you control for the different steps (labeling, transfer, etc.) in a hybridization reaction, the better informed your conclusions will be.

Consider equipment-related fluctuation when modifying a strategy. Glass and plastic heat at different rates, and heat exchange in water is quicker than in air. So the duration of washes may need to be prolonged if you switch from sealed polyethylene sleeves incubating in a water bath to roller bottles heated in a hybridization oven.

What Is the Shelf Life of Hybridization Buffers and Components?

Most hybridization buffers are viscous at room temperature, and floccular SDS precipitates are often observed that should go into solution upon pre-warming to hybridization temperature. Colors vary from colorless to very white to yellow. The yellowish tint often comes from the nonfat dried milk blocking agent.

An analysis of different hybridization buffers stored at room temperature for a year showed that the most common problem was formation of precipitates that would not go into solution when heated. No difference in scent or color of the buffer could be observed (S. Herzer, unpublished observations).

Blocking reagents were much less stable. DNA, nonfat dried milk and BSA were stable for a few weeks at 4°C, and stable for three to six months when frozen. A foul smell appeared in stored solutions of protein blocking reagents, most likely due to microbial contamination.

What Is the Best Strategy for Hybridization of Multiple Membranes?

When simultaneously hybridizing several blots in a tub, box, or bag, the membranes can be separated by meshes, which are usually comprised of nylon. Additional buffer will be required to compensate for that soaked up by the mesh. The mesh should measure at least 0.5 cm larger than the blot. Meshes should be rinsed according to manufacturers instructions (with stripping solution if possible) before reuse because they may soak up probe from previous experiments. When working with radioactive labels, check meshes with a Geiger counter before reuse. Multiple filters may also be hybridized without separating meshes. Up to 40 20 × 20 cm could be hybridized in one experiment without meshes (S. Herzer, unpublished observation).

Filter transfer into hybridization roller bottles can be difficult. Dry membranes are not easy to place into a hybridization tube/roller bottle. Pre-wetting in hybridization buffer or 2× SSC may help. Membranes may be rolled around sterile pipettes and inserted with the pipette into the roller bottle. If several filters need to be inserted into the tube, consider inserting them one by one, because uniform and even wetting with prehybridization solution is important. If tweezers are to be used to handle filters, use blunt, nonridged plastic (metal is more prone to damage membrane) tweezers. Avoid scraping or wrinkling of the membrane. A second approach is to pre-wet the filters and stack them alternating with a mesh membrane, roll them up (like a crepe), and insert this collection into the roller bottle. A third approach is to insert filters into 2× SSC and then exchange to prewarmed prehybridization buffer. Rotate roller bottles slowly, allowing tightly wound filters to uncurl without trapping air between tube and filter, or between multiple filters.

Is Stripping Always Required Prior to Reprobing?

If a probe is stripped away, some target might be lost. If the probe is not stripped away prior to reprobing, will the presence of that first probe interfere with the hybridization by a second probe? There are too many variables to predict which strategy will generate your desired result. If faced with a situation where your prefer not to remove an earlier probe, consider the following options.

If different targets are to be probed, you can sometimes circumvent stripping of radioactively labeled probes by letting the signal decay. Make sure that a positive control for probe A does not light up with probe B if stripping has been skipped. Some non-radioactive systems may allow simple signal inactivation rather than stripping. Horseradish peroxidase activity can be inactivated by incubating the blot in 15% H_2O_2 for 30 minutes at room temperature (Amersham Pharmacia Biotech, Tech Tip 120). Other protocols circumvent stripping by employing different haptens or detection strategies for each target (Peterhaensel, Obermaier, and Rueger, 1998).

What Are the Main Points to Consider When Reprobing Blots?

Considering the amount of work involved in preparing a high quality blot, reuse of blots to gain additional information makes sense. As discussed previously, not all membranes are recommended for reuse. Nylon membranes are more easily stripped and reprobed. If you plan on reusing a blot many times, there are a few guidelines you could consider:

1. No stripping protocol is perfect; some target is always lost. Therefore start out by detecting the least abundant target first.
2. The number of times a blot can be restripped and reprobed cannot be predicted.
3. Never allow blots to dry out before stripping away the probe. Dried probes will not be removed by subsequent stripping procedures.
4. Store the stripped blot as discussed above in the question, *What's the Shelf Life of a Membrane Whose Target DNA Has Been Crosslinked?*
5. Select the most gentle approach when stripping for the first time in order to minimize target loss. Regarding the harshness of stripping procedures, formamide < boiling water < SDS < NaOH, where formamide is the least harsh. NaOH is usually not recommended for stripping Northern blots.
6. Excess of probe or target on blots can form complexes that are difficult to remove from a blot with common stripping protocols (S. Herzer unpublished observation). Avoid high concentrations of target and/or probe if possible when reuse of the blot is crucial.
7. UV crosslinking is preferred when blots are to be reprobed because they withstand harsher stripping conditions.

8. A comparison of stripping protocol efficiencies suggests that NaOH at 25°C led to a fourfold higher loss of genomic DNA compared to formamide at 65°C or a 0.1% SDS at 95°C (Noppinger et al., 1992). Formamide was found to be very ineffectual in stripping probes of blots (*http://www.millipore.com/analytical/pubdbase.nsf/docs/TN056.html*).

How Do You Optimize Wash Steps?

What Are You Trying to Wash Away?

Washes take advantage of the same salt effects described above for hybridization buffers. During removal of unbound or nonspecifically bound probe, sequential lowering of salt concentrations will wash away unwanted signal and background, but may also wash away specific signal if washing is too stringent. Since the required stringency of wash steps is often not known prior to the first experiment, always begin with low-stringency washes, and monitor wash efficiency whenever possible. You can always wash more, but you can never go back after washing with buffer whose stringency is too high.

When increasing the stringency of washes, ask yourself whether you are trying to remove nonspecific or specific background. It is easy to confuse the requirement of a more stringent wash with just more washing. An overall high background with a mismatched probe may not benefit from higher-temperature or lower-salt concentration in the wash steps because you are already at the limit of stringency. Instead, extended washes at the same stringency may be used to remove additional background signal. To summarize, increase the duration (time and/or number) of washing steps to remove more material of a particular stringency; increase temperature and/or decrease salt concentration if further homologous materials need to be removed.

The Wash Solutions

After removing the bulk of the hybridization buffer, a quick rinse of the membrane with wash buffer to remove residual hybridization buffer can drastically improve reproducibility and efficiency of subsequent wash steps. Efficient washing requires excess buffer. At least 1 to 2 ml/cm^2 of membrane or to 30% to 50% of total volume in roller bottles are required for each wash step. Washes may be repeated up to three times for periods of 5 to 30 minutes per wash.

Low-stringency washes start out at 2× SSC, 1% SDS and room temperature to 65°C; intermediate stringency can vary from

0.5× SSC to 1× SSC/0.5% SDS and room temperature to 70°C; high-stringency washes require 0.1% SDS/0.1× SSC at higher temperatures. Some of the newer wash buffers may include urea or other denaturants to increase the stringency (*http://www.wadsworth.org/rflp/Tutorials/DNAhybridization.html*); concentrations similar to those used in the hybridization buffer may be used. Detergent is added to ensure even wetting of filters.

Nonradioactive protocols often call for re-equilibration steps of blots in buffers that provide optimal enzyme activity or antibody binding. Contact the manufacturer of the detection system before you change these conditions.

Monitor Washing Efficiency

Where practical, it is recommended to measure the efficiency of the washing steps. Radioactive applications can be analyzed with handheld probes to check for localized rather than diffuse signal on a blot. Nonradioactive applications may benefit from a pre-experiment where a series of membrane samples containing dot blots is hybridized and washed where a sample is removed before each increase in wash stringency and signal-to-noise ratio is compared. It is crucial to include a negative control to ensure that detected signal is actually specific.

How Do You Select the Proper Hybridization Equipment?

Boxes (plastic or otherwise), plastic bags, and hybridization oven bottles are the common options. Buffer consumption in boxes is higher than in bags or bottles, but these larger volumes can help reduce background problems. Larger capacity also makes it feasible to simultaneously manipulate multiple filters, whereas bags accommodate one filter each.

Hybridization bottles can accommodate multiple membranes, but the membranes tend to stick together much more than in boxes, and the number of filters incubated in a bottle even when using separating meshes will be lower than in a box of the same volume. As described earlier under *What Is the Best Strategy for Hybridization of Multiple Membranes*, membranes are more easily inserted into hybridization bottles after rolling them around clean pipettes. Washing in boxes is more efficient than in bottles or bags, so an increase in number or duration of wash steps might be necessary with bottles or bags.

When working with radioactive probes, contamination of hybridization bottles and loss of probe is minimized by treating the glassware with a siliconizing agent. Bottle caps need to be tightly sealed, nonporous, and fit snugly into the tube. Note that most hybridization buffers and wash solutions are prone to foaming upon gas exchange between the environment and heated air/buffer when the cap on top of the tube is removed, so open roller bottles in a safe area over absorbent paper. Plan for the possibility of minor spills and contaminations when working with plastic bags/sleeves, which don't always seal completely.

DETECTION BY AUTORADIOGRAPHY FILM

How Does An Autoradiography Film Function?

Autoradiography film is composed of a polyester base covered with a photographic emulsion of silver halide crystals. The emulsion may lie on one or both sides of the plastic base, and is usually covered with a material to protect the emulsion against scratches and other physical perturbation.

Photons of light and radioactive emissions can reduce a portion of the ionic silver in a silver halide crystal to silver atoms, forming a catalytic core (the latent image) that, upon development, causes the precipitation of the entire crystal. These precipitated crystals are the grains that form the images seen on the film.

One photon of light produces one silver atom, but a single silver atom in a crystal is unstable and will revert to a silver ion. A minimum of two silver atoms in a crystal are required to prevent reversion to the ionic form. In a typical emulsion, several photons of visible light must interact with an individual silver halide crystal in rapid succession to produce a latent image. In contrast, the energy of a single beta particle or gamma ray can produce hundreds of crystals capable of development into an image (Laskey, 1980 and Amersham International, 1992, *Guide to Autoradiography*).

Indirect Autoradiography

Indirect autoradiography involves the exposure of sample to film at −70°C in the presence of an intensifying screen (Laskey, 1980; Bonner and Laskey, 1974; Laskey and Mills, 1977). An intensifying screen is a flat plate coated with a material such as calcium tungstate, which, when bombarded with radiation, will phosphoresce to produce photons of light. The plates are typically placed on the inside of one side or both of a film cassette. In this way, the film is sandwiched in between. Indirect autoradiography creates a composite signal consisting of radioactive and photon emissions. Exposure at –70°C is essential; a photon of light will generate only a single unstable silver atom (in a silver halide crystal) that will rapidly revert to a silver ion. –70°C stabilizes a single silver atom long enough to allow hits by additional photons of light, producing stable silver atoms and hence visible grains on the film.

Fluorographic chemicals are also utilized for indirect autoradiography. A fluorographic reagent is a solution (organic or aqueous) containing fluors, which will soak into a gel or accrete onto a membrane (Laskey and Mills, 1975; Chamberlain, 1979). When dried, the gel or membrane will have an even layer of fluors impregnated onto the surface. The fluors that are in proximity to the radioactivity fixed on the matrix will be activated by the radiation. These fluors give off light upon being activated, enhancing the signal coming from the radioactive sample. Fluorography requires film exposure at –70°C for the same reason as required by intensifying screens.

The additional sensitivity provided by intensifying screens is offset by a loss of resolution because the signals generated from the screen disperse laterally. In addition, the use of screens and fluorographic reagents also compromises the quantitative response of the film. Two or more silver atoms within a silver halide crystal are required to generate a visible grain on film, but a photon of light will generate only a single unstable silver atom that will rapidly revert to a silver ion. Because larger quantities of radioactivity are more likely than smaller quantities to produce sufficient photons to generate stable silver atoms, lesser amounts of radioactivity are under-represented when working with screens and flours.

When working with radioactive labels, this problem can be corrected by a combination of exposure at –70°C and sensitizing the film with a controlled pre-flash of light of the appropriate duration and wavelength (Laskey and Mills, 1975). Pre-flashing provides stable pairs of silver atoms to many crystals within the emulsion. The appropriate duration and intensity of the flash is crucial to restoring the linear response of the film (Amersham Review Booklet, 23).

Direct Autoradiography

Direct autoradiography refers to the exposure of sample to film at room temperature without use of intensifying screens or reagents.

What Are the Criteria for Selecting Autoradiogaphy Film?

Sensitivity and Resolution

There are two major aspects of film to bear in mind. There is sensitivity, or how much the investigator can see, and resolution, or how well defined the area of activity is. In most cases, higher sensitivity (less time for an image to come up on the film) rather than resolution is crucial. Resolution is more crucial to applications such as DNA sequencing, when probing for multiple bands indicative of mobile genetic clements and repetitive sequence, and when analyzing tissue sections, where location of activity is critical.

Sensitivity and resolution of films are based on the size and packing density of the silver halide crystals. A film is said to be more sensitive if its silver grains are larger (J. DeGregaro, Kodak Inc., Personal communication); Helmrot and Carlsson (1996) suggest that grain shape also affects sensitivity. Higher resolution is achieved when the grains are packed less densely in the emulsion. Some films eliminate the protective anti-scratch coating to improve sensitivity to labels that produce weak energy emissions.

Double and Single Coatings

Most double-coated films contain blue light-sensitive emulsion on both sides of the plastic base, allowing for added sensitivity with and without intensifying screens, albeit at the expense of resolution. High energy emitters such as ^{32}P and ^{125}I can be detected without screens, although the ^{125}I story is more complicated as described below. The emissions from medium emitters (^{14}C and ^{35}S) are essentially completely absorbed by the first emulsion layer, negating any benefit by the second emulsion. However, the use of a specialized intensifying screen (Kodak Transcreen LE) and double-coated film (Kodak Biomax MS) can increase the sensitivity and speed of detection of signal from ^{3}H, ^{14}C, ^{33}P and ^{35}S (J. DeGregaro, Kodak Inc., Personal communication).

Single-coated films allow for greater resolution. Radioactive and nonradioactive signals continue to spread (much like an expanding baloon) as they travel to the second emulsion of a double-coated film, resulting in a bleeding or fuzzy effect. Some emulsion formulations also allow for added speed and sensitivity.

Label

Weak Emitters

Very weak beta emitters such as tritium usually require special films and/or intensifying screens, as described above in the discussion about double-coated films. The tritium beta emission travels only a few microns through material. So, if the film has a coating over the emulsion, the beta particle will not come in contact with the silver grains. In cases of direct autoradiography, that is, without any fluorographic enhancement of signal, tritium samples are best recorded on film without a coating over the emulsion.

If you have the luxury of using fluorographic reagents (described above) and tritium, however, standard autoradiography film (single or double-coated) will work fine, since the film will be picking up the photons of light instead of the betas. This will generally tend to give much faster exposure times, about less than a week, although usually there will be a loss of resolution. This is not recommended for tissue section work, since definition would be compromised by the scattering photons. In this case a liquid nuclear emulsion can be applied.

Medium Emitters

When working with "medium" beta emitters, such as ^{35}S, ^{14}C, and ^{33}P, commonly available single- and double-coated autoradiographic film works well. However, there is no added sensitivity provided by the second emulsion layer without the use of specialized intensifying screens mentioned above. Fluorographic reagents will enhance the signals coming from these isotopes as well, but the impact is less dramatic than observed with tritium. Exposure times can vary greatly. They usually range 6 to 120 hours.

If you're considering the simultaneous use of a fluorographic reagent and an intensifying screen, perform a first experiment with the intensifying screen alone. The presence of a layer of fluorographic material can also attenuate a signal before it reaches the phosphor surface of the screen (Julie DeGregaro, Kodak Inc., Personal communication).

High-Energy Emitters

The most commonly used high-energy beta emitter is ^{32}P. Using standard autoradiographic film (single or double coated), it is not uncommon to have an image within a few minutes to a few hours. Because ^{32}P has such a high energy, the beta particles hitting the film can expose surrounding silver halide crystals and thus result in very poor resolution. At lower levels of counts in a given sample, ^{32}P does benefit from the use of intensifying screens.

^{125}I is a more complex isotope than those described above because it has gamma-ray emission, and a very low-energy X-ray emission. The low-energy X rays have an energy emission similar to tritium. Specialty films used by investigators working with tritium can also easily detect ^{125}I. The high energy gamma rays will pass through the film and are less likely to expose the silver halide crystals. Standard film might detect a portion of the ^{125}I, but most of the signal will not be detected. Specialty films (i.e., Kodak BioMax MS) exist that will detect gamma rays.

The gamma rays from the ^{125}I are best detected by a standard autoradiography film with intensifying screens. Gamma rays are penetrating radiation, and as such are less likely to collide with anything in their path. In combination with intensifying screens on both sides of the cassette, you'll get a good signal from ^{125}I. A single Kodak Transcreen (HE) can also be applied to detect ^{125}I. The use of intensifying screens usually results in some loss of resolution.

Nonradioactive Emissions

Chemiluminescent signals and intensifying screens have a lambda max of light output (Durrant et al., 1990; Pollard-Knight, 1990a). Most double-coated films and intensifying screens are appropriate for chemiluminescent applications. Films dedicated to direct autoradiography are not always responsive to blue and ultraviolet light. They should not be used in fluorography, with intensifying screens, or with most chemiluminescent-based detection systems.

Speed of Signal Detection

The composition of some emulsions are designed for rapid signal generation.

Why Expose Film to a Blot at –70°C?

As described above, a single silver atom in a silver halide crystal is unstable and will revert to a silver ion. At low temperatures this reversion is slowed, increasing the time available to capture a second photon to produce a stable pair of silver atoms. When using intensifying screens or fluorographic reagents to decrease exposure times, keeping the film with cassette at –70°C can enhance the signal several fold. One report indicates that exposure at –20°C might be equally useful (Henkes and Cleef, 1988).

Chemiluminescent detection systems are enzyme driven, and should never be exposed to film at –70°C. Instead, nonradioactive signals can receive a short term boost by heating or microwaving the detection reaction within the membrane (Kobos et al., 1995; Schubert et al., 1995). Since enzymes will not survive this thermoactiviation, long-term signal accumulation is lost. Heating steps that dry the membrane while the probe is attached also make it impossible to strip away that probe. For these reasons thermoactivation is considered a last resort.

Helpful Hints When Working with Autoradiography Film

Static electricity can produce background signals on film. A solution to this problems has been proposed by Register (1999). The use of fluorescent crayolas to mark the orientation of filters in the cassette has been described (Lee and Wevrik, 1997). A protocol for data recovery from underdeveloped autoradiographs has also been described (Owunwanne, 1984).

DETECTION BY STORAGE PHOSPHOR IMAGERS (David F. Englert)

Research has pushed the need for convenience and quantification to a point where autoradiography on film may no longer suffice.

How Do Phosphor Imagers Work?

Storage phosphor imaging is a method of autoradiography that works much like X-ray film. Energy from the ionizing radiation of radioisotopic labels is stored in inorganic crystals that are formed into a thin planar screen. The energy stored in the crystals can be released in the form of light when the crystals are irradiated with intense illumination. After contact exposure to the sample the screen is scanned in a storage phosphor imager with a focused laser beam, and the light emission (at a wavelength different from that of the laser) is recorded with a sensitive light detector. An image is constructed from the raster scan of the screen and is stored for viewing and analysis. The pixel values in the image are linearly proportional to the radioactivity in the sample, and spatial relationships between labeled materials can be determined within the spatial resolution of the system.

Is a Storage Phosphor Imager Appropriate for Your Research Situation?

Speed, Sensitivity, Resolution

Storage phosphor imaging is convenient for autoradiography with most radioisotopes used in biological research. It provides faster results than film autoradiography, and quantitative results in electronic form can be obtained much more readily with storage phosphor imaging than with film. Because of the relatively large dynamic range with storage phosphor imaging, one has much greater latitude with the exposure time, and usually a single exposure will provide acceptable results with storage phosphor imaging. With film, it may be necessary to perform more than one exposure to get the dynamic range of the activity in the sample to correspond to the film's more limited dynamic range.

Better resolution can usually be obtained with film, so when very good resolution is more important than quantitative results, film autoradiography (or autoradiography with emulsions) may be a better choice. For imaging tritium, special storage phosphor screens are necessary which are much less durable than other screens. Thus storage phosphor autoradiography of tritium can be expensive compared to film.

Dynamic Range

Dynamic range is the intensity range over which labels can be quantified in a storage phosphor image. This is equal to the net signal from the highest activity that can be measured (at the level of saturation) divided by the signal from the lowest activity that can be detected or measured. The noise level of the measurement determines the lowest signal that can be detected or measured. The noise level can be assessed with standard statistical tests for hypothesis testing, but generally, the lowest detectable signal is that which can be readily seen in an image with appropriate adjustment of image scaling and contrast levels.

The dynamic range of storage phosphor imaging is generally in the range of 10^4 to 10^5. The dynamic range of X-ray film is somewhat greater than two orders of magnitude or about 100 times less than storage phosphor imaging. This is important for two reasons: (1) a larger range of intensities can be quantified in a single image with storage phosphor imaging, and (2) a user has much greater latitude for the exposure time. The result is that one is much more likely to capture the desired information in a first exposure without saturating the image.

The dynamic range of computer monitors is only about 8 bit or 256 levels of gray, which is far less than the dynamic range that may exist in a storage phosphor image. The image data must be transformed in some way to match the dynamic range of the image data to the display device. The software provided with the storage phosphor imager usually allows one to adjust the way the image data are transformed.

The transformation may be linear, in which case all the detail of the intensity variations may not be visible because the increments of intensity of the computer display are larger than the increments of intensity in the image. The transformation between the image data and the computer display may be nonlinear, for example, exponential. Nonlinear transformation has an effect similar to a logarithmic scale on a graphical plot. Namely, intensity variations are evident over a large dynamic range, but the scale is compressed, providing a distorted view of the intensities in the image. It is also possible to clip the lowest or highest intensities in the image, for example, so that all intensities below a certain level are displayed as white, and the image background is eliminated from view. Alternatively, intensities above a certain level may be displayed as black, and high intensities effectively saturate the display. The software tools usually allow one to adjust the computer display interactively to optimize the display to emphasize the desired information in the image.

Although these manipulations of the image display may cause an apparent loss of image information, all the information is usually retained in the image file, so quantitative analysis of the image will provide accurate information, regardless of what is displayed on the computer monitor. Note that conventional photo-editing software may store modified versions of the image file in which there may be loss of information or distortions of the original information.

Quantitative Capabilities

With proper use of the analytical software, storage phosphor imaging provides accurate quantitative results. Although the response may appear nonlinear at very low activity because of inaccurate estimation of the background level or at very high activity because of saturation of the image, the response of storage phosphor imaging is linear over its entire dynamic range between these extremes. Other aspects of quantitating data by phosphor imaging are discussed below.

What Affects Quantitation?

Is the Reproducibility of Phosphor Imaging Instrumentation Sufficient for Microarray Applications Such as Expression Profiling?

Although there is some risk that local damage to the storage phosphor screen could affect results, storage phosphor imaging with a system that is in good condition will contribute insignificantly to the measurement error. Phosphor imaging is appropriate for microarray analysis.

Can One Accurately Compare the Results Obtained with Different Screens in the Same Experiment?

Different screens may have slightly different responses to the same level of activity, and the exposure times with different screens are difficult to control accurately. Therefore calibration is required for accurate comparison of results obtained on two or more screens. Since the response of storage phosphor imaging is linear, this is a simple matter. Calibration standards can be included during the exposure of all the screens, and the quantitative results within each image can be normalized to (divided by) the signal measured from the calibration standards. This normalization can be performed with a spreadsheet program or may be performed with the analytical software provided with the scanner. Of course, the normalization is only as accurate as the calibration standards. Several nominally identical standards can be used on each screen to determine the error associated with the standards.

Can Storage Phosphor Imaging Provide Results in Absolute Units such as Disintegrations per Minute or Moles of Analyte?

The units of the results reported by the analytical software are arbitrary and have no physical meaning except that they are proportional to the light intensity emitted from the screen during the scanning process. However, calibration standards can be included with samples in the exposure cassette to linearly transform the arbitrary units to units that have significance in a particular experiment. For example, aliquots of a solution containing a radioactive tracer could be dispensed within the same physical matrix as the sample and included with the sample. Other aliquots could be counted by liquid scintillation counting to determine the actual activity in disintegrations per minute. Then quantitative results obtained from the storage phosphor image can be multiplied by a factor to obtain results in disintegrations per minute. Either a spreadsheet program or the scanner software may be used to perform the calibration. For accurate calibration it is important that the calibration standards be within the same physical matrix as the sample, since detection efficiency depends on the sample matrix, especially for relatively low-energy radioisotopes.

Suppose That the Amount of Activity in Part of the Sample Exceeds the Range of the Instrument. What Effect Does This Have on Quantification and How Does One Know That This Has Occurred? Can Accurate Results Be Obtained If This Occurs?

High levels of activity in some part of the sample can result in signal levels greater than the instrument was designed to measure. This is referred to as "saturation." Pixel values in this part of the sample will usually be set to some maximum value, and if the activity in this part of the image is quantified, the results obtained will underestimate the true level of activity. Some instruments paint any pixels that saturate red to warn the user that saturation has occurred.

Saturation is a concern only if the user wishes to quantify the activity in the part of the image that saturated. Accurate results can be obtained by exposing the sample again for a shorter period of time. Another solution is to scan the storage phosphor screen again. When the screen is scanned by the laser, much, but not all, of the signal is erased, and a second scan will result in an image with intensities three to five times less than in the first scan. Parts of the sample that were saturated in the first image may not be saturated in the second image.*

What Should You Consider When Using Screens?

Does the Sensitivity of Storage Phosphor Imaging Increase Indefinitely with Increasing Screen Exposure Times?

No. Energy is stored in the storage phosphor screen throughout the exposure, but there is also a slow decay of the stored energy during the exposure. After a long exposure time, a relatively large amount of signal will be stored in the phosphor, and the decay of this stored signal becomes nearly as great as the accumulation of new signal. Hence the net increase of signal is small. The net increase in signal becomes marginal after a few days, but longer exposures are sometimes used.

**Editor's note: Some manufacturers strongly urge not to rescan the storage phosphor screen because subsequent scans will not produce quantitative data. A third alternative would be to rescan at different voltages where applicable.*

Is There Any Advantage to Exposing Storage Phosphor Screens at Low Temperatures?

There is a small improvement in signal intensity if the storage phosphor screen is kept at low temperature during exposure, probably because the slow decay of the stored signal is slower at lower temperature. This can be beneficial for very long exposures (more than one week), but it has little practical value for most routine work. Because of the marginal effect and the potential for damage due to condensation on the screen, low-temperature exposure should be considered only as a last resort.

Are the Storage Phosphor Screens Used for Tritium Reusable? What Precautions Can One Take to Get Multiple Uses from These Screens?

Because tritium screens are not coated to protect the storage phosphor crystals (any coating would "protect" the crystals from the weak beta radiation of tritium), the screens cannot be cleaned and are readily contaminated or damaged. Nevertheless, some investigators have been able to use the screens multiple times. To reuse tritium screens, samples must be very dry, must not stick to the screen, and must not contain loose material that could adhere. The screens should be stored in a dry place. To check for contamination between uses, the screens should be left in an exposure cassette for the same period of time that one would use to expose a sample and then scanned. Any contamination should be quantified to assess whether it is significant compared to the level of signal expected with a sample.

What Limits Resolution with Storage Phosphor Imaging? Do Some Screens Provide Better Resolution Than Others?

Resolution is limited largely by the isotropic spread of radiation within the storage phosphor screen. As is the case with autoradiography film, resolution is generally better with lower-energy radioisotopes, since their radiation is less penetrating. For example, resolution is better with ^{33}P than with ^{32}P (although the sensitivity with ^{32}P is better due to its higher energy and shorter half-life). Resolution is better with thinner layers of phosphor on the screen, and with thinner protective coatings. The resolution of screens varies between manufacturers, and between the screen types available from a single manufacturer. Resolution is also affected by the quality of the instrumentation, although the newer confocal scanners provide very good resolution and do not limit the resolution that can be achieved in autoradiography.

What Practices Should the Laboratory Use to Ensure That Storage Phosphor Screens Are Completely Erased before Exposure to a Sample?

Storage phosphor screens are erased by exposure to white light, and light boxes with bright fluorescent bulbs are usually used after scanning to completely erase the residual image. Since one cannot always be sure that the previous user has adequately erased the screen, it is a good practice to always erase a screen with white light just before beginning an exposure. This practice also minimizes any background signal on the screen due to prolonged storage in the presence of cosmic radiation or slight contamination of the screen surface.

How Can Problems Be Prevented?

Can These Machines Accidentally Generate Misleading Data?

Storage phosphor imagers could generate misleading data if the screen was contaminated or incompletely erased so that artifactual signals appear in the image. Storage phosphor imagers, like other imaging systems, can generate misleading or confusing results depending on how the image data are displayed on the computer monitor or in an exported or printed image. Important details might be overlooked or significant artifacts might be intentionally hidden by manipulation of the image display.

What Causes the Background with Storage Phosphor Imaging and How Can It Be Reduced?

Some of the background in storage phosphor images is due to instrument noise or very slight stimulated emission of light from the storage phosphor in the absence of stored energy. This component of the background is a property of the system and cannot be reduced. Another component of the background is due to the absorption of cosmic radiation during the exposure. Shielding the exposure cassette from cosmic radiation with lead bricks during the exposure can reduce this component of the background. This measure is worthwhile only for very long exposure times. For exposures up to a few days long, the background due to cosmic radiation is not very significant.

What Is "Flare" in Storage Phosphor Imaging? What Effect Does This Have on Results? How Can It Be Minimized?

Flare is an optical artifact due to the collection of light from adjacent regions of the screen during scanning. It can cause errors if regions of high activity are close to regions of low activity. For example, in images of high-density arrays used for expression profiling, the activity resulting from a highly expressed gene could increase the apparent activity in nearby spots. Flare is an instrument effect that is evident in older storage phosphor imagers but is largely eliminated by the use of confocal optics in newer instruments. With confocal optics, light is collected only from the region (pixel) of the image that is currently being excited by the laser.

Is It Crucial to Avoid Exposing the Storage Phosphor Screen to Bright Light after Exposure and before Imaging?

Ambient light will erase the latent image on a storage phosphor screen. After exposure to radioactive samples, exposure of the storage phosphor screen to ambient light (e.g., the bright fluorescent lighting in many laboratories) should be minimized. Transfer the screen to the scanner without delay. Turn off overhead fluorescent lighting, and work in dim light to retain the maximum signal on the screen.

TROUBLESHOOTING

What Can Cause the Failure of a Hybridization Experiment?

What is the difference in appearance of hybridization data between an experiment where the probe-labeling reaction failed due to inactive polymerase, and an experiment where the gel filtration column trapped the labeled probe? Will the data above look different in a Northern hybridization where the mRNA was stored in a Tris buffer whose pH increased beyond 8.0 when stored in the cold, or in a Northern where the transfer failed? The answer is no. Where hybridization produced a weak signal, was it due to overly stringent hybridization conditions, insufficient quantity of probe, a horseradish peroxidase-linked probe that lost activity during six weeks of storage?

The take-home lessons from the above discussions and the information presented in Table 14.2 are two:

- Problems at any one or combination of steps can generate inadequate hybridization data.
- Problems at different stages of a hybridization experiment can generate data that appear identical.

Scrupulous record-keeping, thorough controls, an open mind, and a stepwise approach to troubleshooting as discussed for

Western blots (Chapter 13) will help you identify the true cause of a disappointing hybridization result. A gallery of images of hybridization problems is provided in Figures 14.1–14.9, and inhibitors of enzymes used to label probes are listed at *http//:www.wiley.com/go/gerstein.*

Table 14.2 Potential Explanations for a Failed Hybridization Experiment

Type of Failure	Possible Causes
Probe Labeling	Template quality Template quantity Reaction components; enzyme, nucleotides, etc. Label integrity
Probe Purification	Inappropriate purification strategy Failed purification reaction
Target-related	Target quantity and quality Target transfer Crosslinking
Hybridization failure	Probe quantity Hybridization conditions; prehybridization, blocking, hybridization buffer, washing
Detection failure	Film Developer Imaging instrumentation

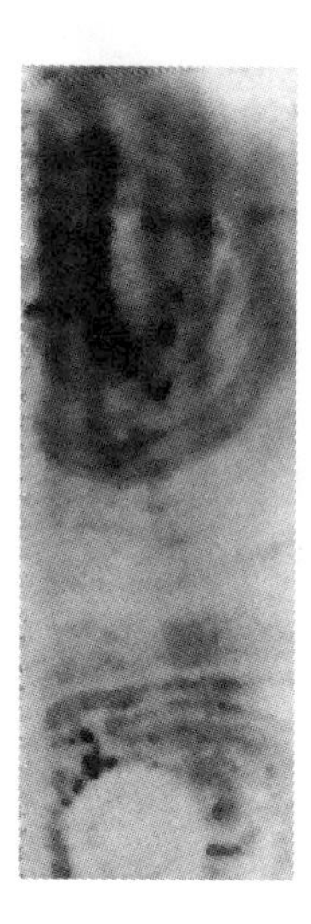

Figure 14.1 Human genomic Southern blot hybridized with the proto-oncongene N-ras DNA probe (1.5 kb), labeled using the ECL random prime system. Exposed to Hyperfilm™ ECL for 30 minutes. Poorly dissolved agarose during preparation of the gel has swirls of high background. Ensure that the agarose is completely dissolved before casting the gel, or invert the gel before blotting. Published by kind permission of Amersham Pharmacia Biotech, UK Limited.

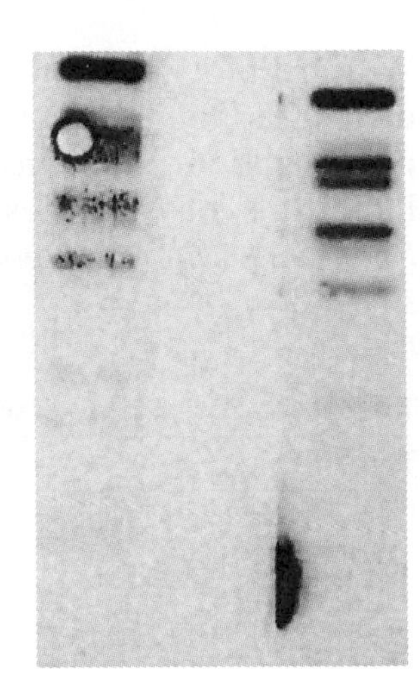

Figure 14.2 Lambda Hind III Southern blot hybridized with a lambda DNA probe, labeled using ECL direct. Exposed to Hyperfilm™ ECL for 60 minutes. Air bubbles trapped between the gel and the membrane have prevented transfer of the nucleic acid; the result is no visible signal. These may be removed by rolling a clean pipette or glass rod over the surface. Published by kind permission of Amersham Pharmacia Biotech, UK Limited.

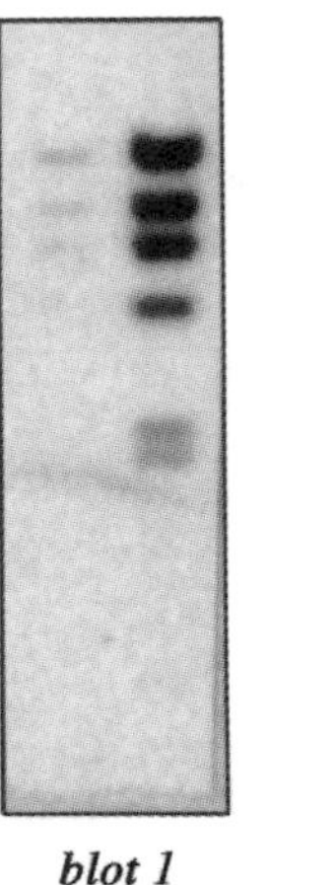

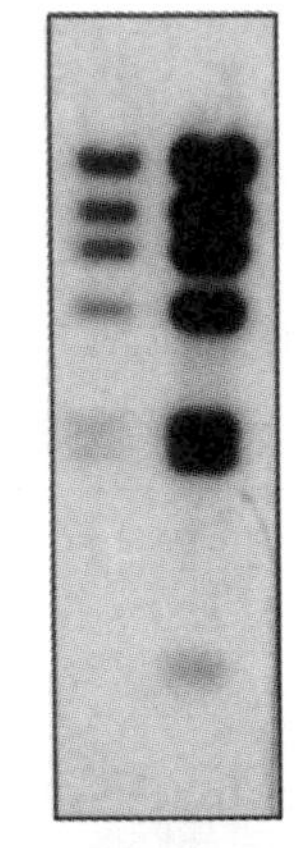

blot 1 *blot 2*

Figure 14.3 Lambda Hind III Southern blot (1 ng and 100 pg loadings) hybridized with a lambda DNA probe using ECL direct. Exposed to Hyperfilm™ ECL for 30 minutes. Blot 1 Hybond™—C pure; Blot 2 Hybond™—N+. Published by kind permission of Amersham Pharmacia Biotech, UK Limited.

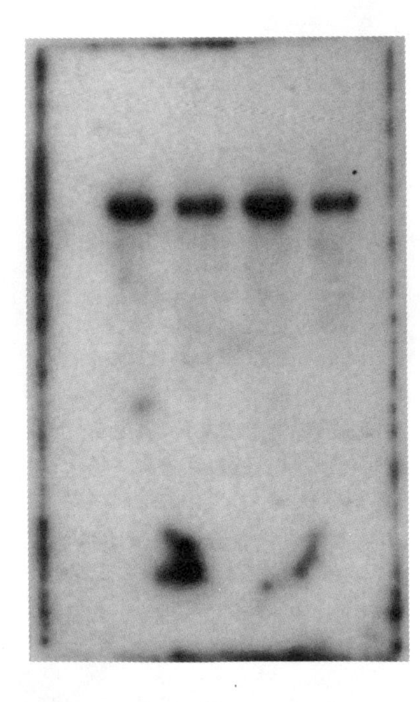

Figure 14.4 Human genomic Southern blot hybridized with the proto-oncogene *N*-ras DNA probe (1.5 kb), labeled using [alpha-^{32}P] dCTP and Megaprime™ labeling (random primer-based) system. Exposed to Hyperfilm™ MP for 6 hours. Membrane damage at the cut edges has caused the probe to bind; subsequent stringency washes are unable to remove the probe. Similar results are obtained with non-radioactive labeling and detection systems. Membranes should be prepared using a clean, sharp cutting edge. Published by kind permission of Amersham Pharmacia Biotech, UK Limited.

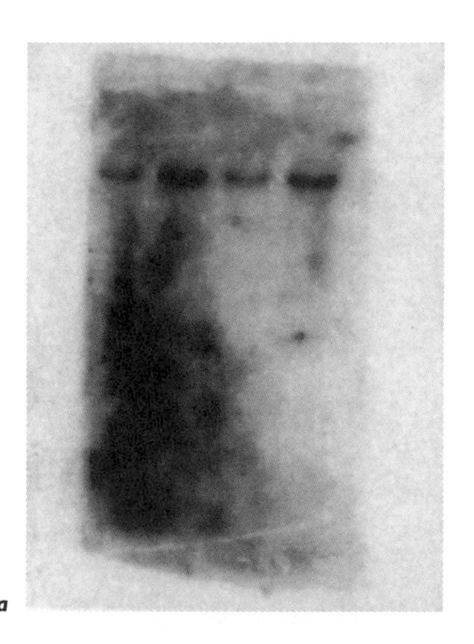

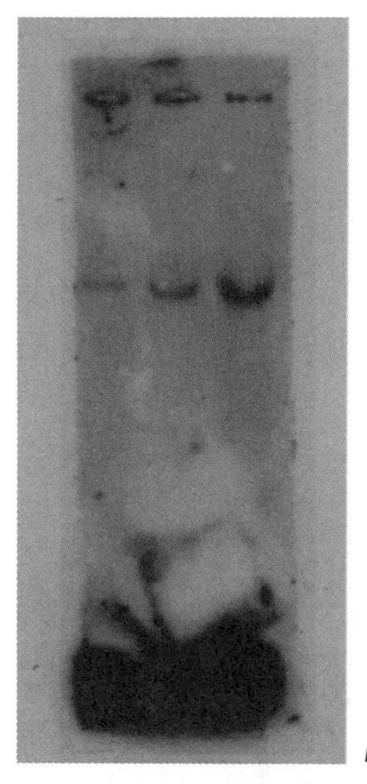

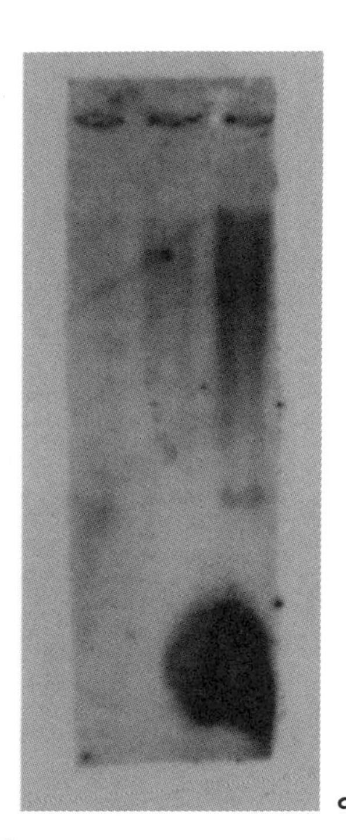

Figure 14.5*a* Human genomic Southern blot hybridized with the proto-oncogene *N*-ras DNA probe (1.5 kb) labeled using [alpha-^{32}P] dCTP and Megaprime™ labeling (random primer-based) system. Exposed to Hyperfilm™ MP for 6 hours. Labeled probe has been added directly onto the blot to cause this effect. Labeled probe should be added to the hybridization buffer away from the blot or mixed with 0.5 to 1.0 ml of hybridization buffer before addition. **Figure 14.5*b*, 5*c*** Human genomic Southern blot hybridized with *N*-ras insert labeled via ECL™ Direct labeling system. Exposed to Hyperfilm ECL for 1 hour. These probes were also directly added to the membrane, rather than first added to hybridization buffer. Published by kind permission of Amersham Pharmacia Biotech, UK Limited.

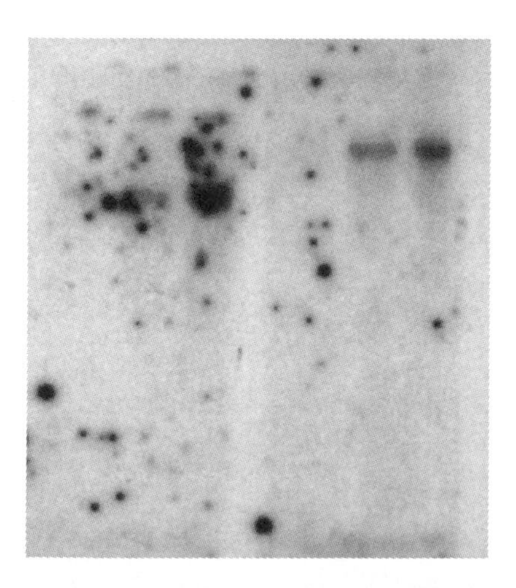

Figure 14.6 Human genomic Southern blot hybridized with the proto-oncogene *N*-ras DNA probe (1.5 kb) labeled using [alpha-^{32}P] dCTP and Megaprime™ labeling (random primer-based) system. Exposed to Hyperfilm™ MP for 6 hours. There are two probable causes of this "spotted" background: (1) Excess unincorporated labeled nucleotide in the probe solution. Always check the incorporation of the radioactive label before using the probe and purify as required. (2) Particulate matter present in the hybridization buffer. Ensure that all buffer components are fully dissolved before used. Published by kind permission of Amersham Pharmacia Biotech, UK Limited.

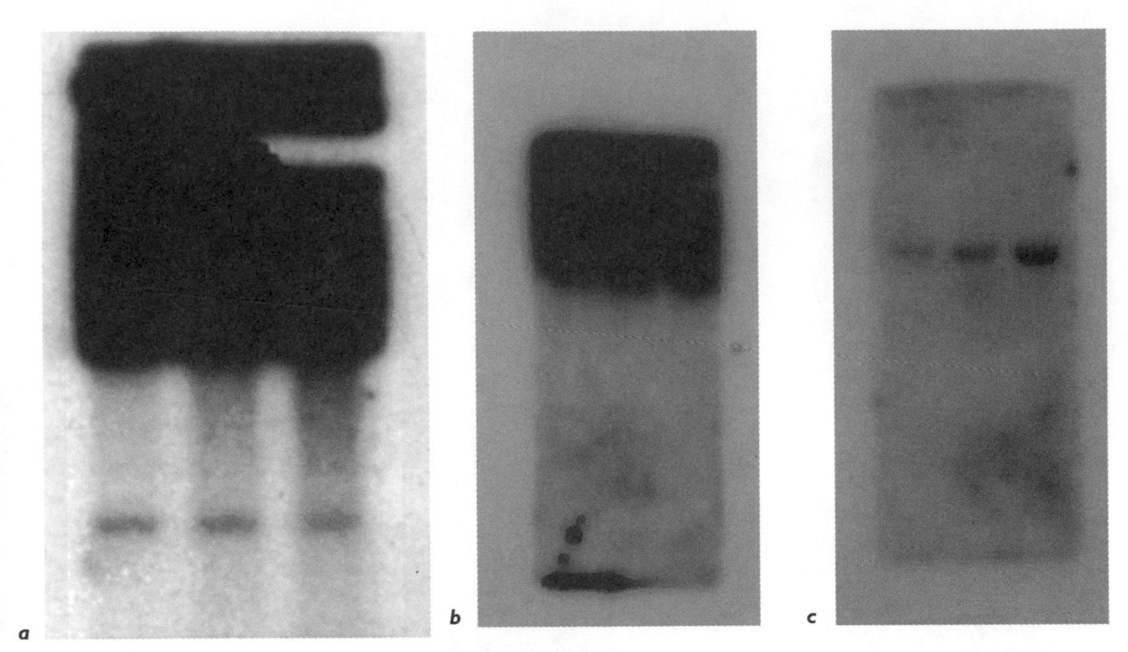

Figure 14.7*a* Human genomic DNA probe (0.8 kb), labeled using the ECL™ Direct system. Exposed to Hyperfilm™ ECL for 30 minutes. The heavy blot background nearest to the cathode has two possible causes: dirty electrophoresis equipment or electrophoresis buffer. Similar results are obtained with radioactive probes. Ensure that the electrophoresis tanks are rinsed in clean distilled water after use. Do not reuse electrophoresis buffers. **Figure 14.7*b*** Human genomic Southern blots on Hybond N$^+$ detected with ^{32}P labeled *N*-ras insert using [alpha-^{32}P] dCTP and Megaprime™ labeling (random primer-based) system. Exposed to Hyperfilm™ MP overnight. Electrophoresis was carried out in old TAE buffer. **Figure 14.7*c*** represents same samples as in Figure 14.7*b*, but after electrophoresis tank had been cleaned and filled with fresh TAE buffer. Published by kind permission of Amersham Pharmacia Biotech, UK Limited.

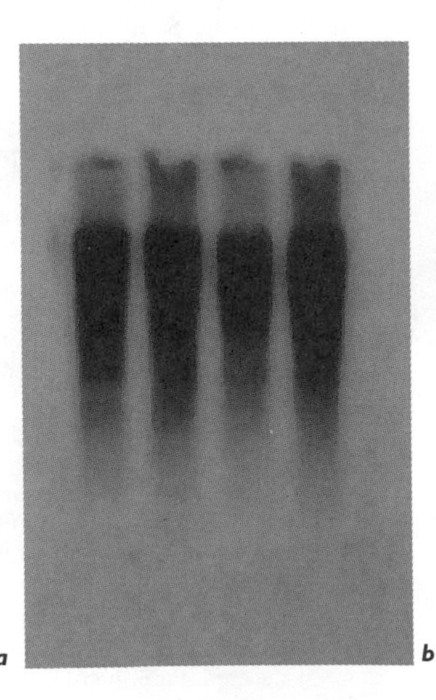

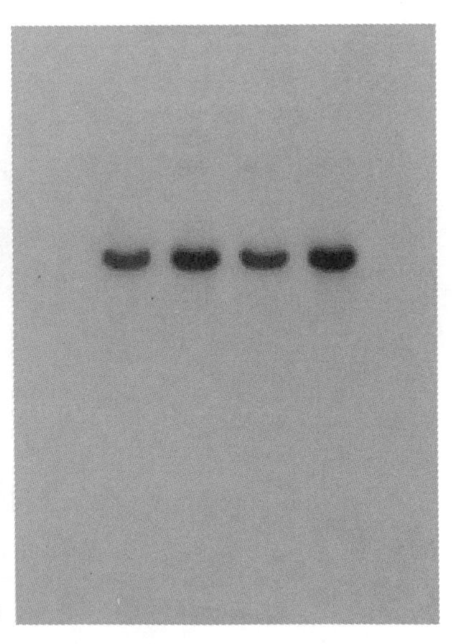

Figure 14.8 Human genomic Southern blots on Hybond N$^+$ detected with ^{32}P labeled *N*-ras insert using [alpha-^{32}P] dCTP and Megaprime™ labeling (random primer-based) system. **Figure 14.8*a*, 14.8*b*** Importance of controlling temperature during hybridization. (Figure 14.8*a*) The temperature of the water bath fell during an overnight hybridization, reducing the stringency and increasing the level of nonspecific hybridization. (Figure 14.8*b*) The temperature was properly controlled, and only specific homology is detected. Published by kind permission of Amersham Pharmacia Biotech, UK Limited.

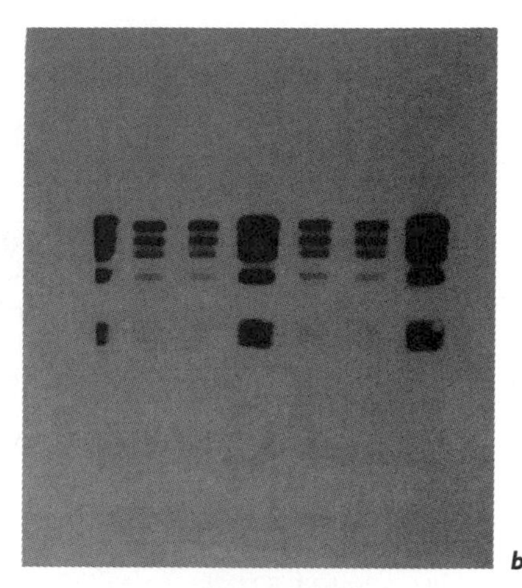

a b

Figure 14.9 Hind III fragments of lambda DNA were blotted onto Hybond™ ECL, and probed with lambda DNA labeled via the ECL™ Detection system. **Figure 14.9 (*a*)** Blocking agent excluded from hybridization buffer. **(*b*)** Blocking agent present in hybridization buffer. Published by kind permission of Amersham Pharmacia Biotech, UK Limited.

BIBLIOGRAPHY

Alexandrova, L. A., Lukin, M. A., Victorova, L. S., and Rozovskaya, T. A. 1991. Enzymatic incorporation of fluorescent labels into oligonucleotides. *Nucl. Acids Symp. Series* 24:277.

Amersham International. 1992. *Guide to Autoradiography*. Amersham, Buckingham shire U.K.

Amersham Pharmacia Biotech. 1992. Tech Tip 120: Sequential Labeling of Western Blots with Enhanced Chemiluminescence.

Amersham Review Booklet 23, Efficient Detection of Biomolecules by Autoradiography, Fluorography or Chemiluminescence. Amersham International plc. Buckinghamshire, England. 1993.

Anderson, M. L. M., 1999. *Nucleic Acid Hybridization*. Springer, New York.

Andreadis, J. D., and Chrisey, L. A. 2000. Use of immobilized PCR primers to generate covalently immobilized DNAs for in vitro transcription/translation reactions. *Nucl. Acids Res*. 28:5.

Ansorge, W., Zimmermann, J., Erfle, H., Hewitt, N., Rupp, T., Schwager, C., Sproat, B., Stegemann, J., and Voss, H. 1993. Sequencing reactions for ALF (EMBL) automated DNA sequencer. European Molecular Biology Laboratory. *Meth. Mol. Cell. Biol*. 23:317–356.

Ausubel, M., Brent, R., Kingston, R. E., Moore, D. D., Seidman, J. G., and Struhl, K. 1993. *Current Protocols in Molecular Biology*. Wiley, New York.

Bains, W. 1994. Selection of oligonucleotide probes and experimental conditions for multiplex hybridization experiments. *Genet. Anal. Tech. Appl*. 11:49–62.

Bertucci, F., Bernard, K., Loriod, B., Chang, Y. C., Granjeaud, S., Birnbaum, D., Nguyen, C., Peck, K., and Jordan, B. R. 1999. Sensitivity issues in DNA array-based expression measurements and performance of nylon microarrays for small samples. *Human Mol. Genet*. 8:1715–1722.

Bloom, L. B., Otto, M. R., Beechem, J. M., and Goodman, M. F. 1993. Influence of 5′-nearest neighbors on the insertion kinetics of the fluorescent nucleotide analog 2-aminopurine by Klenow fragment. *Biochem*. 32:11247–11258.

Bonner, W. M., Laskey, R. A. 1974. A film detection method for tritium-labelled proteins and nucleic acids in polyacrylamide gels. *Eur. J. Biochem.* 46:83–88.
Booz, M. L. 2000. Personal communication. Bio-Rad Inc.
Breslauer, K. J., Frank, R., Blocker, H., and Marky, L. A. 1986. Predicting DNA duplex stability from the base sequence. *Proc. Nat. Acad. Sci. U.S.A.* 83:3746–3750.
Bright, B. D., Kumar, R., Jailkhani, B., Srivastava, R., and Bhan, M. K. 1992. Nonradioactive polynucleotide gene probe assay for the detection of three virulence toxin genes in diarrhoeal stools. *Indian J. Med. Res.* 95:121–124.
Budowle, B., and Baechtel, F. S. 1990. Modifications to improve the effectiveness of restriction fragment length polymorphism typing. *Appl. Theor. Electrophor.* 1:181–187.
Carninci, P., Nishiyama, Y., Westover, A., Itoh, M., Nagaoka, S., Sasaki, N., Okazaki, Y., Muramatsu, M., and Hayashizaki, Y. 1998. Thermostabilization and thermoactivation of thermolabile enzymes by trehalose and its application for the synthesis of full length cDNA. *Proc. Nat. Acad. Sci. U.S.A.* 20:520–524.
Casey, J., Davidson, N. 1977. Rates of formation and thermal stabilities of RNA:DNA and DNA:DNA duplexes at high concentrations of formamide. *Nucl. Acids Res.* 4:1539–1552.
Chamberlain, J. P. 1979. Fluorographic detection of radioactivity in polyacrylamide gels with the water-soluble fluor, sodium salicylate. *Anal. Biochem.* 98:132–135.
Chomczynski, P. 1992. One-hour downward alkaline capillary transfer for blotting of DNA and RNA. *Anal. Biochem.* 201:134–139.
Chomczynski, P., and Mackey, K. 1994. One-hour downward capillary blotting of RNA at neutral pH. *Anal. Biochem.* 221:303–305.
Church, G. M., and Gilbert, W. 1984. Genomic sequencing. *Proc. Nat. Acad. Sci. U.S.A.* 81:1991–1995.
Correa-Rotter, R., Mariash, C. N., and Rosenberg, M. E. 1992. Loading and transfer control for Northern hybridization. *Biotech.* 12:154–158.
Darby, I A. 1999. *In situ Hybridization Protocols*. Humana Press, Clifton, NJ.
Day, P. J., Bevan, I. S., Gurney, S. J., Young, L. S., and Walker, M. R. 1990. Synthesis in vitro and application of biotinylated DNA probes for human papilloma virus type 16 by utilizing the polymerase chain reaction. *Biochem. J.* 267:119–123.
DeGregaro, J., Kodak Inc., personal communication, 2000.
De Luca, A., Tamburrini, E., Ortona, E., Mencarini, P., Margutti, P., Antinori, A., Visconti, E., and Siracusano, A. 1995. Variable efficiency of three primer pairs for the diagnosis of *Pneumocystis carinii* pneumonia by the polymerase chain reaction. *Mol. Cell Probes* 9:333–340.
Dieffenbach, C. W., and Dveksler, G. S., eds. 1995. *PCR Primer: A Laboratory Manual.* Cold Spring Harbor Laboratory Press, Cold Spring Harbor, NY.
Dubitsky, A., Brown, J., and Brandwein, H. 1992. Chemiluminescent detection of DNA on nylon membranes. *Biotech.* 13:392–400.
Duplaa, C., Couffinhal, T., Labat, L., Moreau, C., Lamaziere, J. M., and Bonnet, J. 1993. Quantitative analysis of polymerase chain reaction products using biotinylated dUTP incorporation. *Anal. Biochem.* 212:229–236.
Durrant, I., Benge, L. C., Sturrock, C., Devenish, A. T., Howe, R., Roe, S., Moore, M., Scozzafava, G., Proudfoot, L. M., and Richardson, T. C. 1990. The application of enhanced chemiluminescence to membrane-based nucleic acid detection. *Biotech.* 8:564–570.
Eickhoff, H., Schuchhardt, J., Ivanov, I., Meier-Ewert, S., O'Brien, J., Malik, A., Tandon, N., Wolski, E. W., Rohlfs, E., Nyarsik, L., Reinhardt, R., Nietfeld, W.,

and Lehrach, H. 2000. Tissue gene expression analysis using arrayed normalized cDNA libraries. *Genome Res*. 10:1230–1240.

Engler-Blum, G., Meier, M., Frank, J., and Muller, G. A. 1993. Reduction of background problems in nonradioactive Northern and Southern blot analyses enables higher sensitivity than ^{32}P-based hybridizations. *Anal. Biochem.* 210:235–244.

Fenn, B. J., and Herman, T. M. 1990. Direct quantitation of biotin-labeled nucleotide analogs in RNA transcripts. *Anal. Biochem*. 190:78–83.

Gicquelais, K. G., Baldini, M. M., Martinez, J., Magii, L., Martin, W. C., Prado, V., Kaper, J. B., and Levine, M. M. 1990. Practical and economical method for using biotinylated DNA probes with bacterial colony blots to identify diarrhea-causing *Escherichia coli*. *J. Clin. Microbiol*. 28:2485–2490.

Gilmartin, P. M. 1996. Nucleic Acid Hybridization. Essential Data. Wiley/Bios Scientific, New York.

Giusti, A. M., and Budowle, B. 1992. Effect of storage conditions on restriction fragment length polymorphism (RFLP) analysis of deoxyribonucleic acid (DNA) bound to positively charged nylon membranes. *J. Forensic Sci.* 37:597–603.

Haralambidis, J., Chai, M., and Tregear, G. W. 1987. Preparation of base-modified nucleosides suitable for non-radioactive label attachment and their incorporation into synthetic oligodeoxyribonucleotides. *Nucl. Acids Res.* 15:4857–4876.

Henke, J., Henke, L., and Cleef, S. 1988. Comparison of different X-ray films for ^{32}P-autoradiography using various intensifying screens at –20 degrees C and –70 degrees C. *J. Clin. Chem. Clin. Biochem*. 26:467–468.

Herrera, R. E., and Shaw, P. E. 1989. UV shadowing provides a simple means to quantify nucleic acid transferred to hybridization membranes. *Nucl. Acids Res*. 17:8892.

Herzer, P., Amersham Pharmacia Biotech, Piscataway, 2000, unpublished results.

Herzer, S., Amersham Pharmacia Biotech, Piscataway, NJ, 1999, unpublished results.

Herzer, S., Amersham Pharmacia Biotech, Piscataway, NJ, 2000–2001, unpublished results.

Hill, S. M., and Crampton, J. M. 1994. Synthetic DNA probes to identify members of the *Anopheles gambiae* complex and to distinguish the two major vectors of malaria within the complex, *An. gambiae s.s.* and *An. arabiensis*. *Am. J. Trop. Med. Hyg*. 50:312–321.

Holtke, H. J., Ettl, I., Finken, M., West, S., and Kunz, W. 1992a. Multiple nucleic acid labeling and rainbow detection. *Anal. Biochem*. 207:24–31.

Holtke, H. J., Sagner, G., Kessler, C., Schmitz, G. 1992b. Sensitive chemiluminescent detection of digoxigenin-labeled nucleic acids: A fast and simple protocol and its applications. *Biotech*. 12:104–113.

Honore, B., Madsen, P., and Leffers, H. 1993. The tetramethylammonium chloride method for screening of cDNA libraries using highly degenerate oligonucleotides obtained by backtranslation of amino-acid sequences. *J. Biochem. Biophys. Meth*. 127:39–48.

Howley, P. M., Israel, M. A., Law, M. F., and Martin, M. A. 1979. A rapid method for detecting and mapping homology between heterologous DNAs. Evaluation of polyoma virus genomes. *J. Biol. Chem*. 254:4876–4883.

Igloi, G. L., and Schiefermayr, E. 1993. Enzymatic addition of fluorescein- or biotin-riboUTP to oligonucleotides results in primers suitable for DNA sequencing and PCR. *Biotech*. 15:486–488, 490–492, 494–497.

Ingelbrecht, I. L., Mandelbaum, C. I., and Mirkov, T. E. 1998. Highly sensitive northern hybridization using a rapid protocol for downward alkaline blotting of RNA. *Biotech*. 25:420–423, 425–426.

Islas, L., Fairley, C. F., and Morgan, W. F. 1998. DNA synthesis on discontinuous templates by human DNA polymerases: Implications for non-homologous DNA recombination. *Nucl. Acids Res.* 26:3729–3738.

Ivanov, I., Antanov, P., Markova, N., and Markov, G. 1978. "Maturation" of DNA duplexes. *Mol. Biol. Rep.* 4:67–71.

John-Roger, and McWilliams, P. 1991. *Do It! Let's Get Off Our Butts.* Prelude Press, Los Angeles, CA.

Jiang, X., Estes, M. K., and Metcalf, T. G. 1987. Detection of hepatitis A virus by hybridization with single-stranded RNA probes *Appl. Environ. Microbiol.* 53:2487–2495.

Kanematsu, S., Hibi, T., Hashimoto, J., and Tsuchizaki, T. 1991. Comparison of nonradioactive cDNA probes for detection of potato spindle tuber viroid by dot-blot hybridization assay. *J. Virol. Meth.* 35:189–197.

Khandjian, E. W. 1985. Optimized hybridization of DNA blotted and fixed to nitrocellulose and nylon membranes. *Bio/Technolgy* 5:165–167.

Kirii, Y., Danbara, H., Komase, K., Arita, H., and Yoshikawa, M. 1987. Detection of enterotoxigenic *Escherichia coli* by colony hybridization with biotinylated entertoxin probes. *J. Clin. Microbiol.* 25:1962–1965.

Klann, R. C., Torres, B., Menke, J. B., Holbrook, C. T., Bercu, B. B., and Usala, S. J. 1993. Competitive polymerase chain reaction quantitation of c-erbA beta 1, c-erbA alpha 1, and c-erbA alpha 2 messenger ribonucleic acid levels in normal, heterozygous, and homozygous fibroblasts of kindred S with thyroid hormone resistance. *J. Clin. Endocrinol. Metab.* 77:969–975.

Kobos, R. K., Blue, B. A., Robertson, C. W., and Kielhorn, L. A. 1995. Enhancement of enzyme-activated 1,2-dioxetane chemiluminescence in membrane-based assays. *Anal. Biochem.* 224:128–133.

Kolocheva, T. I., Maksakova, G. A., Zakharova, O. D., and Nevinsky, G. A. 1996. The algorithm of estimation of the K_m values for primers in DNA synthesis catalyzed by human DNA polymerase alpha. *FEBS Lett.* 399:113–116.

Kondo, Y., Fujita, S., Kagiyama, N., and Yoshida, M. C. 1998. A simple, two-color fluorescence detection method for membrane blotting analysis using alkaline phosphatase and horseradish peroxidase *DNA Res.* 5:217–220.

Krueger, S. K., and Williams, D. E. 1995. Quantitation of digoxigenin-labeled DNA hybridized to DNA and RNA slot blots. *Anal. Biochem.* 229:162–169.

Laskey, R. A. 1980. The use of intensifying screens or organic scintillators for visualizing radioactive molecules resolved by gel electrophoresis. *Methods Enzymol.* 65:363–371.

Laskey, R. A., and Mills, A. D. 1975. Quantitative film detection of ^{3}H and ^{14}C in polyacrylamide gels by fluorography. *Eur. J. Biochem.* 56:335–341.

Laskey, R. A., and Mills, A. D. 1977. Enhanced autoradiographic detection of ^{32}P and ^{125}I using intensifying screens and hypersensitized film. *FEBS Lett.* 82:314–316.

Lee, L. G., Connell, C. R., Woo, S. L., Cheng, R. D., McArdle, B. F., Fuller, C. W., Halloran, N. D., and Wilson, R. K. 1992. DNA sequencing with dye-labeled terminators and T7 DNA polymerase: Effect of dyes and dNTPs on incorporation of dye-terminators and probability analysis of termination fragments. *Nucl. Acids Res.* 20:2471–2483.

Lee, M.-L. T., Kuo, F. C., Whitmore, G. A., and Sklar, J. 2000. Importance of replication in microarray gene expression studies: Statistical methods and evidence from repetitive cDNA hybridizations. *Proc. Natl. Acad. Sci. U.S.A.* 97: 9834–9839.

Lee, S., and Wevrick, R. 1997. "Glow in the Dark" crayons as inexpensive autoradiography markers. Technical Tips Online. May.

Lewin, B. 1993. *Genes V.* Oxford University Press, Oxford, U.K., pp. 668–669.

Li, J. K., Parker, B., and Kowalik, T. 1987. Rapid alkaline blot-transfer of viral dsRNAs. *Anal. Biochem.* 163:210–218.
Lion, T., and Haas, O. A. 1990. Nonradioactive labeling of probe with digoxigenin by polymerase chain reaction. *Anal. Biochem.* 188:335–337.
Makrigiorgos, G. M., Chakrabarti, S., and Mahmood, A. 1998. Fluorescent labelling of abasic sites: A novel methodology to detect closely-spaced damage sites in DNA. *Int. J. Radiat. Biol.* 74:99–109.
Mansfield, E. S., Worley, J. M., McKenzie, S. E., Surrey, S., Rappaport, E., and Fortina, P. 1995. Nucleic acid detection using non-radioactive labeling methods. *Mol. Cell Probes* 9:145–156.
Marathias, V. M., Jerkovic, B., Arthanari, H., and Bolton, P. H. 2000. Flexibility and curvature of duplex DNA containing mismatched sites as a function of temperature. *Biochem.* 39:153–160.
McCabe, M. S., Power, J. B., de Laat, A. M., and Davey, M. R. 1997. Detection of single-copy genes in DNA from transgenic plants by non-radioactive Southern blot analysis. *Mol. Biotechnol.* 7:79–84.
Middendorf, L. R., Bruce, J. C., Bruce, R. C., Eckles, R. D., Grone, D. L., Roemer, S. C., Sloniker, G. D., Steffens, D. L., Sutter, S. L., Brumbaugh, J. A., et al. 1992. Continuous, on-line DNA sequencing using a versatile infrared laser scanner/electrophoresis apparatus. *Electrophoresis* 12:487–494.
Mineno, J., Ishino, Y., Ohminami, T., and Kato, I. 1993. Fluorescent labeling of a DNA sequencing primer. *DNA Seq.* 4:135–141.
Ming, Y. Z., Di, X., Gomez-Sanchez, E. P., and Gomez-Sanchez, C. E. 1994. Improved downward capillary transfer for blotting of DNA and RNA. *Biotech.* 16:58–59.
Moichi, M., Yoshida, S., Morita, K., Kohara, Y., and Ueno, N. 1999. Identification of transforming growth factor beta regulated genes in *C. elegans* by differential hybridization of arrayed cDNAs. *Proce. Nat. Acad. Sci. U.S.A.* 96: 15020–15025.
Moore, N. J., and Margolin, A. B. 1993. Evaluation of radioactive and nonradioactive gene probes and cell culture for detection of poliovirus in water samples. *Appl. Environ. Microbiol.* 59:3145–3146.
Moran, S., Ren, R. X., Sheils, C. J., Rumney IV, S., and Kool, E. T. 1996. Nonhydrogen bonding "terminator" nucleosides increase the 3′-end homogeneity of enzymatic RNA and DNA synthesis. *Nucl. Acids Res.* 24:2044–2052.
Nakagami, S., Matsunaga, H., Oka, N., and Yamane, A. 1991. Preparation of enzyme-conjugated DNA probe and application to the universal probe system. *Anal. Biochem.* 198:75–79.
Nakano, S., Fujimoto, M., Hara, H., and Sugimoto, N. 1999. Nucleic acid duplex stability: Influence of base composition on cation effects. *Nucl. Acids Res.* 27:2957–2965.
Nass, S. J., and Dickson, R. B. 1995. Detection of cyclin messenger RNAs by nonradioactive ribonuclease protection assay: A comparison of four detection methods. *Biotech.* 19:772–776, 778.
Nath, J., and Johnson, K. L. 1998. Fluorescence in situ hybridization (FISH): DNA probe production and hybridization criteria. *Biotech. Histochem.* 73:6–22.
Niemeyer, C. M., Ceyhan, B., and Blohm, D. 1999. Functionalization of covalent DNA-streptavidin conjugates by means of biotinylated modulator components. *Bioconjug. Chem.* 10:708–719.
Noppinger, K., Duncan, G., Ferraro, D., Watson, S., and Ban, J. 1992. Evaluation of DNA probe removal from nylon membrane. *Biotech.* 13:572–575.
Ogretmen, B., Ratajczak, H., Kats, A., Stark, B. C., and Gendel, S. M. 1993. Effects of staining of RNA with ethdium bromide before electrophoresis on performance of northern blots. *Biotech.* 14:932–935.

Oshevskii, S. I., Kumarev, V. P., and Grachev, S. A. 1989. Inclusion of biotinylated analogs of dUTP and dCTP in DNA by DNA-polymerases: Cloning DNA fragments, containing biotinylated deoxyribouridine in *E. coli*. *Bioorg. Khim.* 15:1091–1099.

Owunwanne, A. 1984. Radiochemical technique for intensification of underexposed autoradiographs. *Anal. Biochem.* 138:74–77.

Pearlman, D. A., and Kollman, P. A. 1990. The calculated free energy effects of 5-methyl cytosine on the B to Z transition in DNA. *Biopolymers* 29:1193–1209.

Pearson, W. R., Davidson, E. H., and Britten, R. J. 1977. A program for least squares analysis of reassociation and hybridization data. *Nucl. Acids Res.* 4:1727–1737.

Peterhaensel, C., Obermaier, I., and Rueger, B. 1998. Non-radioactive Northern blot analysis of plant RNA and the application of different haptens for reprobing. *Anal. Biochem.* 264:279–283.

Petrie, C. R., Adams, A. D., Stamm, M., Van Ness, J., Watanabe, S. M., and Meyer Jr., R. B. 1991. A novel biotinylated adenylate analogue derived from pyrazolo[3,4-d]pyrimidine for labeling DNA probes. *Bioconj. Chem.* 2:441–446.

Plath, A., Peters, F., and Einspanier, R. 1996. Detection and quantitation of specific mRNAs by ribonuclease protection assay using denaturing horizontal polyacrylamide gel electrophoresis: A radioactive and nonradioactive approach. *Electrophoresis* 17:471–472.

Pollard-Knight, D., Read, C. A., Downes, M. J., Howard, L. A., Leadbetter, M. R., Pheby, S. A., McNaughton, E., Syms, A., and Brady, M. A. 1990a. Nonradioactive nucleic acid detection by enhanced chemiluminescence using probes directly labeled with horseradish peroxidase. *Anal. Biochem.* 185:84–89.

Pollard-Knight, D., Simmonds, A. C., Schaap, A. P., Akhavan, H., and Brady, M. A. 1990b. Nonradioactive DNA detection on Southern blots by enzymatically triggered chemiluminescence. *Anal. Biochem.* 185:353–358.

Price, D. C. 1996. Chemiluminescent substrates for detection of restriction fragment length polymorphism. *Sci. Justice* 36:275–282.

Puchhammer-Stoeckl, E., Heinz, F. X., and Kunz, C. 1993. Evaluation of 3 nonradioactive DNA detection systems for identification of herpes simplex DNA amplified from cerebrospinal fluid. *Virol. Meth.* 43:257–266.

Racine, J. F., Zhu, Y., and Mamet-Bratley, M. D. 1993. Mechanism of toxicity of 3-methyladenine for bacteriophage T7. *Mutat. Res.* 294:285–298.

Reed, K. C., and Mann, D. A. 1985. Rapid transfer of DNA from agarose gels to nylon membranes. *Nucl. Acids Res.* 13:7207–7221.

Register, K. B. 1999. Elimination of background due to discharge of static electricity from X-ray film. *Biotech.* 26:434–435.

Richard, F., Vogt, N., Muleris, M., Malfoy, B., and Dutrillaux, B. 1994. Increased FISH efficiency using APC probes generated by direct incorporation of labeled nucleotides by PCR. *Cytogenet. Cell Genet.* 65:306–311.

Rihn, B., Coulais, C., Bottin, M. C., and Martinet, N. 1995. Evaluation of nonradioactive labelling and detection of deoxyribonucleic acids. Part I: Chemiluminescent methods. *J. Biochem. Biophys. Meth.* 30:91–102.

Sambrook, J., Fritsch, E. F., and Maniatis, T. 1989. *Molecular Cloning: A Laboratory Manual*, 2nd ed. Cold Spring Harbor Laboratory, Cold Spring Harbor, NY.

Schubert, F., Knaf, A., Moller, U., and Cech, D. 1995. Non-radioactive detection of oligonucleotide probes by photochemical amplification of dioxetanes. *Nucl. Acids Res.* 23:4657–4663.

Schwarz, F. P., Robinson, S., and Butler, J. M. 1999. Thermodynamic comparison of PNA/DNA and DNA/DNA hybridization reactions at ambient temperature. *Nucl. Acids Res.* 27:4792–4800.

Shabarova, Z., and Bogdanov, A. 1994. *Advanced Organic Chemistry of Nucleic Acids*. Wiley, New York.
Spink, C. H., and Chaires, J. B. 1999. Effects of hydration, ion release, and excluded volume on the melting of triplex and duplex DNA. *Biochem.* 38:496–508.
Srivastava, S. C., Raza, S. K., and Misra, R. 1994. 1,N^6-etheno deoxy and ribo adenosine and 3,N^4-etheno deoxy and ribo cytidine phosphoramidites. Strongly fluorescent structures for selective introduction in defined sequence DNA and RNA molecules. *Nucl. Acids Res.* 22:1296–1304.
Temsamani, J., and Agrawal, S. 1996. Enzymatic labeling of nucleic acids. *Mol. Biotechnol.* 5:223–232.
Tenberge, K. B., Stellamanns, P., Plenz, G., and Robenek, H. 1998. Nonradioactive in situ hybridization for detection of hydrophobin mRNA in the phytopathogenic fungus Claviceps purpurea during infection of rye. *Eur. J. Cell Biol.* 75:265–272.
Thomou, H., and Katsanos, N. A. 1976. The theory of the deoxyribonucleic acid–ribonucleic acid hybridization reaction. *Biochem. J.* 153:241–247.
Thurston, S. J., and Saffer, J. D. 1989. Ultraviolet shadowing nucleic acids on nylon membranes. *Anal. Biochem.* 178:41–42.
Tijssen, P. 2000. *Hybridization with Nucleic Acid Probes*, vols 1 and 2. Laboratory Techniques in Biochemistry and Molecular Biology. Elsevier Science, New York.
Trayhurn, P. 1996. Northern blotting. *Proc. Nutr. Soc.* 55:583–589.
Tuite, E. M., and Kelly, J. M. 1993. Photochemical interactions of methylene blue and analogues with DNA and other biological substrates. *J. Photochem. Photobiol. B.* 21:103–112.
van Gijlswijk, R. P., Raap, A. K., and Tanke, H. J. 1992. Quantification of sensitive non-isotopic filter hybridizations using the peroxidase catalyzed luminol reaction. *Mol. Cell Probes* 6:223–230.
Wade, M. F., and O'Conner, J. L. 1992. Using a cationic carbocyanine dye to assess RNA loading in Northern gel analysis. *Biotech.* 12:794–796.
Wallace, R. B., Schold, M., Johnson, M. J., Dembek, P., and Itakura, K. 1981. Oligonucleotide directed mutagenesis of the human beta-globin gene: A general method for producing specific point mutations in cloned DNA. *Nucl. Acids Res.* 9:879–894.
Westermeier, R. 1997. *Electrophoresis in Practice: A Guide to Methods and Applications of DNA and Protein Separations*, 2nd ed. Wiley, New York.
Wheeler, C. 2000. Personal communication. Amersham Pharmacia Biotech Research And Development Department, Bucks, England, U.K.
Wilkinson, M., Doskow, J., and Lindsey, S. 1991. RNA blots: Staining procedures and optimization of conditions. *Nucl. Acids Res.* 19:679.
Wood, W. I., Gitschier, J., Lasky, L. A., and Lawn, R. M. 1985. Base composition-independent hybridization in tetramethyl ammonium chloride: A method for oligonucleotide screening of highly complex gene libraries. *Proc. Nat. Acad. Sci. U.S.A.* 82:1585–1588.
Yang, H., Wanner, I. B., Roper, S. D., and Chaudhari, N. 1999. An optimized method for in situ hybridization with signal amplification that allows the detection of rare mRNAs. *J. Histochem. Cytochem.* 47:431–446.
Yano, N., Endoh, M., Fadden, K., Yamashita, H., Kane, A., Sakai, H., and Rifai, A. 2000. Comprehensive gene expression profile of the adult human renal cortex: Analysis by cDNA array hybridization. *Kidney Int.* 57:1452–1459.
Yuriev, E., Scott, D., and Hanna, M. M. 1999. Effects of 5-[S-(2,4-dinitrophenyl)-thio]-2′-deoxyuridine analog incorporation on the structure and stability of DNA hybrids: Implications for the design of nucleic acid probes. *J. Mol. Recognit.* 12:337–345.

Zhu, Z., Chao, J., Yu, H., and Waggoner, A. S. 1994. Directly labeled DNA probes using fluorescent nucleotides with different length linkers. *Nucl. Acids Res.* 22:3418–3422.

Zhu, Z., and Waggoner, A. S. 1997. Molecular mechanism controlling the incorporation of fluorescent nucleotides into DNA by PCR. *Cytometry* 28:206–211.

15

E. coli Expression Systems

Peter A. Bell

Expression Vector Structure 462
What Makes a Plasmid an Expression Vector? 462
Is a Stronger Promoter Always Desirable? 463
Why Do Promoters Leak and What Can You Do about It? 464
What Factors Affect the Level of Translation? 464
What Can Affect the Stability of the Protein in the Cell? 464
Which Protein Expression System Suits Your Needs? 465
Track Record 465
What Do You Know about the Gene to Be Expressed? ... 465
What Do You Know about Your Protein? 468
Advertisements for Commercial Expression Vectors Are Very Promising. What Levels of Expression Should You Expect? 470
Which *E. coli* Strain Will Provide Maximal Expression for Your Clone? 471
Why Should You Select a Fusion System? 471
When Should You Avoid a Fusion System? 472
Is It Necessary to Cleave the Tag off the Fusion Protein? 474
Will Extra Amino Acid Residues Affect Your Protein of Interest *after* Digestion? 475
Working with Expression Systems 475
What Are the Options for Cloning a Gene for Expression? 475

Molecular Biology Problem Solver, Edited by Alan S. Gerstein.
ISBN 0-471-37972-7

Is Screening Necessary Prior to Expression? 476
What Aspects of Growth and Induction Are Critical to Success? 478
What Are the Options for Lysing Cells? 479
Troubleshooting 480
No Expression of the Protein 480
The Protein Is Expressed, but It Is Not the Expected Size Based on Electrophoretic Analysis 480
The Protein Is Insoluble, Now What? 481
Solubility Is Essential. What Are Your Options? 482
The Protein Is Made, but Very Little Is Full-Length; Most of It Is Cleaved to Smaller Fragments 483
Your Fusion Protein Won't Bind to Its Affinity Resin 484
Your Fusion Protein Won't Digest 485
Cleavage of the Fusion Protein with a Protease Produced Several Extra Bands 485
Extra Protein Bands Are Observed after Affinity Purification 485
Must the Protease Be Removed after Digestion of the Fusion Protein? 486
Bibliography 486

Over the past decade the variety of hosts and vector systems for recombinant protein expression has increased dramatically. Researchers now select from among mammalian, insect, yeast, and prokaryotic hosts, and the number of vectors available for use in these organisms continues to grow. With the increased availability of cDNAs and protein coding sequencing information, it is certain that these and other, yet to be developed systems will be important in the future. Despite the development of eukaryotic systems, *E. coli* remains the most widely used host for recombinant protein expression. *E. coli* is easy to transform, grows quickly in simple media, and requires inexpensive equipment for growth and storage. And in most cases, *E. coli* can be made to produce adequate amounts of protein suitable for the intended application. The purpose of this chapter is to guide the user in selecting the appropriate host and troubleshooting the process of recombinant protein expression.

EXPRESSION VECTOR STRUCTURE

What Makes a Plasmid an Expression Vector?

Vectors for expression in *E. coli* contain at a minimum, the following elements:

Table 15.1 Characteristics of Popular Prokaryotic Promoters

Promoter		Regulation/Inducer (Concentration)	Strength
LacUV5	Lactose operon	*lacI*/IPTG (0.1–1 mM)	Strong
Trp	Tryptophan operon	*trpR* 3-beta-indoleacrylic acid	Strong
Tac	Hybrid of –35 Trp and –10 lac promoter	*lacI*/IPTG (0.1–1 mM)	Strong
P_L	Phage lambda	Lambda *cI* repressor/heat	Strong
Phage T5	T5 phage	*lacI* (2 operators)/ IPTG (0.1–1 mM)	Strong
Arabinose	Arabinose operon	*AraBAD*/arabinose (1 μm–10 mM)	Variable
T7	T7 phage RNA polymerase	*lacI*/IPTG (0.1–1 mM)	Very strong

- A transcriptional promoter.
- A ribosome binding site.
- A translation initiation site.
- A selective marker (e.g., antibiotic resistance).
- An origin of replication.

In general, things that affect these can affect the level of protein expression. At a minimum, transcription promoters in *E. coli* consist of two DNA hexamers located –35 and –10 relative to the transcriptional start site. Together these elements mediate binding of the about 500 kDa multimeric complex of RNA polymerase.

Suppliers of vectors for expression have selected highly active, and inducible promoter sequences, and there is usually little need to be concerned until a problem is encountered in expression. A list of the commonly used promoters and their regulation is shown in Table 15.1.

Is a Stronger Promoter Always Desirable?

A strong promoter may not be best for all situations. Overproduction of RNA may saturate translation machinery, and maximizing RNA synthesis may not be desirable or necessary. A weaker promoter may actually give higher steady-state levels of soluble, intact protein than one that is rapidly induced.

Why Do Promoters Leak and What Can You Do about It?

Most promoters will have some background activity. Promoters regulated by the lactose operator/repressor will drive a small amount of transcription in the absence of added inducer (e.g., IPTG). To minimize this leakage, 10% glucose can be added to the medium to repress the lactose induction pathway, the growth temperature can be reduced to 15 to 30°C, and a minimal medium that contains no trace amounts of lactose can be used. Promoter leakage is only a problem when the expressed protein is highly toxic to the cells.

The tightly regulated T7 promoter has very low background due to the low levels of T7 RNA polymerase made in the absence of inducer (in specifically engineered host cells such as BL21 (DE3)/pLysS). It has been estimated that the fold induction of transcription in the T7 driven pET vector system is greater than 1000, while the magnitude of induction obtained with lac repressor regulated promoters is generally about 50-fold.

What Factors Affect the Level of Translation?

Translation can be affected by nucleotides adjacent to the ATG initiator codon, the amino acid residue immediately following the initiator, and secondary structures in the vicinity of the start site. Most commercially available vectors for expression use optimal ATG and Shine-Dalgarno sequences. Secondary structures in the mRNA contributed by the gene of interest can prevent ribosome binding (Tessier et al., 1984; Looman et al., 1986; Lee et al., 1987). In addition, the downstream box AAUCACAAAGUG found after the initiator codon in many bacterial genes can also enhance translation initiation. Conversion of the amino terminal sequence of the gene of interest to one that comes close to this consensus may improve the rate of translation of the mRNA (Etchegaray and Inouye, 1999).

What Can Affect the Stability of the Protein in the Cell?

One of the first steps in protein degradation in *E. coli* is the catalyzed removal of the amino terminal methionine residue. This reaction, catalyzed by methionyl aminopeptidase, occurs more slowly when the amino acid in the +2 position has a larger side chain (Hirel et al., 1989; Lathrop et al., 1992). When the methionine residue is intact, the protein will be stable to all but endopeptidase cleavage. Tobias et al. (1991) have determined the relationship between a protein's amino terminal amino acid and its stability in bacteria, that is, the "N-end rule." They reported protein half-lives of only 2 minutes when the following amino acids were present at the amino terminus: Arg, Lys, Phe, Leu, Trp, and Tyr. In contrast, all other amino acids conferred half-lives of >10 hours when present at the amino terminus of the protein examined. This suggests that one should examine the sequence to be expressed for the residue in the +2 position. If the residue is among those that destabilize the protein, it may be worth the effort to change this residue to one that confers stability.

WHICH PROTEIN EXPRESSION SYSTEM SUITS YOUR NEEDS?

Track Record

What systems are currently used in the laboratory or by others in the field? If the protein coding sequence of interest is well characterized, and the protein or its close relatives have been expressed successfully by others in the field, it is wise to try the same expression system. Go with what has worked in the past. If nothing else, results obtained using the familiar system will serve as a starting point. As an example, most of the recombinant expression of mammalian src homology SH2 protein interaction domains has been done using the pGEX vector series, and similar examples of preferred systems are found in other fields of research. If little is known about the protein to be expressed, it is best to take stock of what information there is before entering the lab. Before beginning any experimentation, it is wise to answer the following question:

What Do You Know about the Gene to Be Expressed?

Source

In general, simple globular proteins from prokaryotic and eukaryotic sources are good candidates for expression in *E. coli.* Monomeric proteins with few cysteines or prosthetic groups (e.g., heme and metals) and of average size (<60 kDa) will likely give good production. Secreted eukaryotic proteins and membrane-bound proteins, especially those with several transmembrane domains, are likely to be problematic in *E. coli.* Solubility of recombinant proteins in *E. coli* can also be estimated by a mathematical analysis of the amino acid sequences (Wilkinson and Harrison, 1991).

Presence of a Start Codon

Some expression vectors provide the start codon for translation initiation, while others rely on the start codon of the gene you're trying to express. Note that in *E. coli*, 5 to 12 base pairs or less separate the ribosome binding site and the start codon. So you would incorporate this requirement into your cloning strategy when the start codon is provided by the gene you plan to express.

GC Content

Coding sequences with high GC (>70%) content may reduce the level of expression of a protein in *E. coli*. Check the sequence using a DNA analysis program.

Codon Usage

Codon usage may also affect the level of protein expression. If the gene of interest contains codons not commonly used in *E. coli*, low expression may result due to the depletion of tRNAs for the rarer codons. When one or more rare codons is encountered, translational pausing may result, slowing the rate of protein synthesis and exposing the mRNA to degradation. This potential problem is of particular concern when the sequence encodes a protein >60 kDa, when rare codons are found at high frequency, or when multiple rare codons are found over a short distance of the coding sequence. For example, rare codons for arginine found in tandem can create a recognition sequence for ribosome binding (e.g., _AGGAGG) that closely approximates a Shine-Dalgarno sequence UAAGGAGG. This may bind ribosomes nonproductively and block translation from the bona fide ribosome binding site (RBS) at the initiator codon further upstream. Nonetheless, the appearance of a rare codon does not necessarily lead to poor expression. It is best to try expression of the native gene, and then make changes if these seem warranted later. Strategies include mutating the gene of interest to use optimal codons for the host organism, and co-transforming the host with rare tRNA genes. In one example, introduction into the *E. coli* host of a rare arginine (AGG) tRNA resulted in a several-fold increase in the expression of a protein that uses the AGG codon (Hua et al., 1994). In another case, substitution of the rare arginine codon AGG with the *E. coli*-preferred CGU improved expression (Robinson et al., 1984). Other work has shown that rare codons account for decreased expression of the gene of interest in *E. coli* (Zhang, Zubay, and Goldman, 1991; Sorensen, Kurland, and Pederson, 1989). Rare codons may have an even more dramatic effect on translation when they occur close to the initiator codon (Chen and Inouye, 1990). While codon usage is not the only or most important factor, be aware that it may influence translation efficiency.

Secondary Structure

Secondary structures that occur near the start codon may block translation initiation (Gold et al., 1981; Buell et al., 1985), or serve as translation pause sites resulting in premature termination and truncated protein. These can be found using DNA or RNA analysis software. Structures with clear stem structures greater than eight bases long may be disrupted by site-specific mutation or by making all or a portion of the coding sequence synthetically.

Depending on the size of the gene, and the importance of obtaining high-expression levels, it may be worth synthesizing the gene. This has been generally done by synthesizing overlapping oligonucleotides that when annealed can be extended using PCR and ligated to form the full-length coding sequence. There are several examples where this approach has been used to optimize codon usage for *E. coli* (Koshiba et al., 1999; Beck von Bodman et al., 1986). In addition, if one takes on the work and expense of synthesizing a gene, secondary structures in the predicted RNA that might stall translation can be removed, and sites for restriction endonucleases can be introduced.

Size of a Gene or Protein

As a rule, very large (>100 kDa) and very small (<5 kDa) proteins are more difficult to express in *E. coli*. Small polypeptides with little secondary structure tend to be rapidly degraded in *E. coli*. Degradation can be minimized by expressing such short oligopeptides as concatemers with proteolytic or chemical cleavage sites in between the monomeric units (Hostomsky, Smrt, and Paces, 1985). Short peptides are also successfully expressed as fusion proteins. Fusion with GST, MalB or other larger, well-folded partners will tend to stabilize a short peptide, making expression possible and purification relatively simple. One publication has shown MBP to be superior to other large fusion proteins at stabilizing short polypeptides (Kapust and Waugh, 1999). At the other extreme, proteins that are above 60 kDa are best made using smaller affinity tags, such as FLAG, his_6, or on their own, without any fusion. While there is no clear upper limit, the larger the protein, the lower the yield is likely to be.

What Do You Know about Your Protein?

Cysteines

There are many things that *E. coli* does not do well, or at all. If the protein of interest is naturally multimeric, or requires post-translational modifications for activity, *E. coli* as an expression host may be a poor choice. Disulfide bonds, formed between two cysteines in an expressed protein, are made inefficiently in the reducing environment of the *E. coli* cytoplasm (Bessette et al., 1999; Derman et al., 1993). If the protein is produced, and can be purified from *E. coli*, in vitro oxidation of the cysteines may be tried (Dodd et al., 1995). Alternatively, the gene of interest can be cloned in a vector that includes a signal sequence (e.g., OmpA, geneIII, and phoA) that will direct the recombinant protein to the relatively oxidizing environment of the periplasm of *E. coli*, where disulfide formation is more efficient. Strains of *E. coli* that are deficient in thioredoxin reductase (trxB) permit proper disulfide formation in the cytoplasm (Derman et al., 1993; Yasukawa et al., 1995). Subsequent work has produced strains that lack both trxB and glutathione oxidoreductase and give better rates of disulfide formation than those seen in native *E. coli* periplasm (Bessette et al., 1999).

Membrane Bound

If the protein to be expressed is naturally associated with membrane and/or has at least one transmembrane domain, addition of a secretion signal to the amino terminus may help to maximize expression of functional protein. Signal sequences, about 20 residues long are derived from proteins that naturally are secreted into the periplasmic space, such as pelB, OmpA, OmpT, MalE, alkaline phosphatase (phoA), or geneIII of filamentous phage (Izard and Kendall, 1994). Protein with an amino terminal signal will be directed to the inner membrane of *E. coli*, and the carboxy terminal portion of the protein will be translocated into the periplasmic space. Depending on the hydrophobicity of the protein of interest, it may not translocate entirely into the periplasm but remain associated with the inner membrane. Secretion may help stabilize proteins from proteolytic attack (Pines and Inouye, 1999), or at least can reduce aggregation of hydrophobic proteins in the cytoplasm, and minimize inclusion body formation. Because of the reducing environment of the periplasmic space, proteins that contain one or more disulfide bonds are best secreted.

The presence of an N-terminal signal sequence appears to

be necessary but not sufficient to direct a target protein to the periplasm. Translocation across the outer membrane and into the growth medium is inefficient. In most cases target proteins found in the growth medium are the result of damage to the cell envelope and do not represent true secretion (Stader and Silhavy, 1990). Translocation across the inner cell membrane of *E. coli* is incompletely understood (reviewed by Wickner, Driessen, and Hartl, 1991), and the efficiency of export will depend on the individual target protein. Currently the export cannot be predicted based on protein sequence, although some generalizations have been made about the sequence immediately following the signal peptide (Boyd and Beckwith, 1990; Yamane and Mizushima, 1988). Therefore it is possible to find target proteins in the cytoplasm (with uncleaved signal sequence) or in the periplasm in partially processed form, in place of or in addition to the expected periplasmic processed species. In some cases the proportion of protein that is exported can be increased by lowering the temperature 15 to 30°C during induction.

Post-translational Modification

E. coli does not glycosylate or phosphorylate proteins or recognize proteolytic processing signals from eukaryotes, so take this into account when designing the cloning strategy. If proteolytic processing is needed, it is best to express only the coding sequences for the fully processed protein. If the protein of interest requires glycosylation for activity, and full activity is important in the final use, consider a eukaryotic host, such as Pichia, insect cells, or mammalian cells.

Is the Protein Potentially Toxic?

Consider whether the protein of interest is likely to have a toxic effect on the host cell. Where the function of the protein is known, this can be guessed at with some accuracy. For example, nonspecific proteases, nucleases, or pore-forming membrane proteins might all be expected to have some toxic effect on *E. coli*. Expression of toxic proteins may be very low, and there will be strong selective pressure on cells to eliminate the gene of interest by point mutation to change the translation frame, insertion of a stop codon, or change in an amino acid residue critical to the protein's function. Larger deletion of parts of the plasmid may also be seen. If there is a suggestion that the gene product will be toxic, use an expression vector with a tightly regulated promoter (e.g., T7, pET vectors). Minimize propagation of the cells to avoid opportunities for mutation and recombination.

Must Your Protein Be Functional?

Each requirement placed on a recombinant protein will affect the choice of expression system. If a protein is to be used only to prepare antibody, it need not be soluble or active, and the production of inclusion bodies (aggregates of improperly folded protein) in *E. coli* may be all that is needed. Alternatively, if a protein's biological activity will be assayed, or if it is to be used in structural studies (NMR, crystallography, etc.), a properly folded and soluble form will be required.

Will Structural Changes (Additional or Fewer Amino Acids) Affect Your Application?

Depending on the way that a gene is inserted in an expression vector, additional sequences may be added to the clone, and these may lead to extra amino acid residues at the N- or C-termini of the final expressed protein. In many cases these will have no deleterious effect, but if structural studies or precise comparisons to a native protein are to be done, it is wise to eliminate amino acids added by cloning steps. PCR amplification is the most commonly used method to generate inserts for expression, and proper design of PCR primers can eliminate most or all additional residues in the protein.

Is the Sequence of Your Protein Recognized by Specific Proteases?

If you plan to express your gene in a fusion vector that provides an internal protease cleavage site for removal of the affinity tag (discussed below), check that your native protein is not recognized by the protease. Most proteases are highly specific, but thrombin has a variety of secondary cleavage sites (Chang, 1985).

Advertisements for Commercial Expression Vectors Are Very Promising. What Levels of Expression Should You Expect?

There are several systems available for protein expression in mammalian, insect, yeast, and *E. coli*. While it is impossible to predict the yields of protein from these systems for any given protein, some rough guidelines can be given. For any vector it is possible that no expression will be seen! Reported yields in stably transfected mammalian cells are in the range of 1 to 100 μg/10^6

cells. Insect cell systems will yield between 5 and 200 mg/L of culture (Schmidt et al., 1998), Pichia can produce up to 250 mg/L (Eldin et al., 1997), and reported yields in *E. coli* range from 50 μg to over 100 mg/L. Usually yields of from 1 to 10 mg/L can be expected from *E. coli*. Higher yields, up to a gram or more per liter, can be had using fermentation vessels where oxygen and pH levels can be controlled throughout the cell growth. The above-mentioned values are guidelines; they are entirely dependent on the protein to be expressed. It is always best to test one or more systems in parallel to select the best solution.

Nonbiological synthesis of protein is now possible as an alternative to production in a host organism (Kochendoerfer and Kent, 1999). Oligopeptides are synthesized and then assembled by chemical ligation to give full-length protein. The method has the potential to synthesize gram quantities of >30 kDa proteins, and such preparations would of course be free of host contaminants that might interfere with function or use in diagnostic or therapeutic applications. Unfortunately, chemical synthesis of proteins is not widely available.

Which *E. coli* Strain Will Provide Maximal Expression for Your Clone?

The choice of an expression host depends on the promoter system to be used. Promoters that depend on *E. coli* RNA polymerase can be expressed in most common cloning strains, while T7 promoter vectors must be used in *E. coli* that co-express T7 RNA polymerase (e.g., strains that contain the DE3 lysogen) (Dubendorff and Studier, 1991). Strains that are protease deficient (Bishai, Rappuoli, and Murphy, 1987) or overexpress chaparones have been shown to be useful for some proteins (Georgiou and Valax, 1996; Gilbert, 1994). At a minimum, a recombination deficient strain is advisable. Vendors of the commercially available *E. coli* expression vectors generally will recommend a host for use in expression. As with many questions related to protein expression, the results will depend on the nature of the protein of interest. A given gene may give high yields of intact protein in most strains, while the next would show no product except in a protease deficient host.

Why Should You Select a Fusion System?

Increased Yields

There are several reasons that one would choose to use a fusion system. Translational initiation from the amino terminal fusion partner may be more efficient than the start contributed by the protein of interest, so larger amounts of protein can be obtained as a fusion. In addition smaller proteins (<20kDa), or subfragments of larger ones often benefit from association with a stable fusion partner, due in part to improved folding or protection from proteolysis. Fusion with GST, MBP, and thioredoxin may be useful for this purpose.

Simplified Purification and Detection

Most of the commonly available fusion partners double as affinity tags, and these make isolation of the protein of interest relatively simple. Protein can often be purified to >90% in a single step. In contrast to conventional chromatographic techniques, little or no information about the sequence, pI, or other physical characteristics of the protein is needed in order to perform the purification. Novice chromatographers or those who have not developed methods for purification of the native protein are advised to begin with an affinity system.

Detection of fusion proteins is a simple matter, since antibodies and colorimetric substrates are available for several of the more common fusion partners. Thus, if there is no established method to detect the protein, detection of the fusion partner can be the most convenient way to assay for the presence of the protein in cells and throughout purification and assay of the protein of interest.

When Should You Avoid a Fusion System?

Since affinity tags make purification relatively simple, and tags can be removed by proteolyic cleavage, use of a tag usually makes sense. If, on the other hand, a nonfusion vector has been used in earlier work, and one wishes to compare results with older data, use the nonfusion system. If there is an established method for purification and a biochemical assay or antibody available to detect the protein of interest, an affinity partner or tag for detection may simply be unnecessary. Ask again what use the protein will be put to. If the end application is likely to be sensitive to the presence of the tag (e.g., NMR, crystallography, therapeutics), and other conditions above are met, there is reason to avoid the tag.

If a fusion affinity tag is desired, several are available. Table 15.2 summarizes some of the characteristics of the most widely used fusion partners.

Table 15.2 Commercially Available Fusion Systems

Tag	Tag Size	Purification	Detection	Cleavage
Calmodulin/CBP	(CBP, 4 kDa)	Calmodulin-agarose EGTA for elution	Biotinylated calmodulin and streptavidin alkaline phosphatase	Thrombin, enterokinase
Chitin binding domain (CBD)	*Bacillus circulans* chitin binding domain (CBD, 52 amino acid residues)	chitin beads	Anti-CBD antibody	Used with intein. On-column cleavage is induced at 4°C by DTT or 2-mercaptoethanol.
E-tag	1.4 kDa	Anti-E sepharose	Anti-E antibodies	NA
FLAG®	1 kDa	Anti-Flag resin	Anti-FLAG antibodies	Enterokinase
Glutathione *S*-transferase GST	26.5 kDa homodimer GST forms a 58 kDa homodimer with two GSH binding sites. The affinity of the enzyme for GSH is approximately 0.1 μM.	Glutathione sepharose/ Glutathione Agarose	Anti-GST antibodies, CDNB substrate	Thrombin Factor Xa PreScission™ protease
HA (hemagglutinin)	~1 kDa YPYDVPDYA	NA	Anti-HA antibodies	
His_6	1 kDa	NTA-agarose, Iminodiacetic acid-sepharose	Anti-His_6 antibodies	Enterokinase, if desired
Maltose binding protein	42.5 kDa K_d of MBP for maltose is 3.5 μM; for maltotriose, 0.16 μM (Miller et al., 1983)	Amylose beads	Anti-MBP	Factor Xa
Myc tag	10 amino acids from human c-Myc EQKLISEEDL	Anti-Myc antibody resin	Anti-Myc antibodies (9E10)	NA
Nus-tag	E. coli NusA protein, 495 amino acids	NA	None	Thrombin
Pinpoint™	12.5 kDa peptide biotinylated *in vivo* (Samols et al., 1988)	Monomeric avidin resin (SoftLink™ soft release avidin resin)	Avidin/strep tavidin conjugates	Factor Xa
S-tag	15 amino acid peptide (S-tag) with strong affinity ($K_d = 10^{-9}$ M) for a 104 amino acid fragment of	S-protein agarose beads	S-protein FITC conjugate	Thrombin, enterokinase

Table 15.2 (*Continued*)

Tag	Tag Size	Purification	Detection	Cleavage
	pancreatic ribonuclease A.			
Strep-tag	A 10 amino acid sequence that binds streptavidin	Streptavidin bead	Streptavidin conjugates	
Z-domain	Two Z domains add a 14 kDa peptide	IgG-sepharose		Factor Xa

Susceptibility To Cleavage Enzymes

As discussed below, some fusion systems allow for the removal of the affinity tag by specific proteolytic or chemical cleavage. Before beginning any experiment, examine the sequence of the protein to be cloned and expressed. The protein of interest may have a binding site for one of the proteases listed in Table 15.3, and if so, this site should be avoided, or a different expression system might be required. Most proteases used for cleavage of fusion protein are quite specific, with theoretical frequencies of 10^{-6}. However, it is best to check as a matter of course.

Is It Necessary to Cleave the Tag off the Fusion Protein?

For many proteins, cleavage is not needed. If the goal of the work is to raise an antibody, the whole fusion protein can be used successfully as antigen—provided that antibodies to the tag do not interfere in the application. If, on the other hand, the protein is to be used in structural studies, or where the function of recombinant protein will be compared with native protein, it may be necessary to remove the fusion tag.

Systems have been developed that use chemical (Nilsson et al., 1985) or specific proteolytic cleavage to separate the protein of interest from the fusion tag. The proteases have the advantage that cleavage is done at near neutral pH and at 4 to 37°C. In addition to proteolytic cleavage, the use of self-splicing inteins has been developed and commercialized by New England Biolabs. In this latter case fusion proteins with chitin-binding domain are bound to high molecular weight chitin chromatography media and incubated in the presence of a reducing agent, generally overnight. Protein splicing takes place, leaving the protein of interest in the flow through, while chitin and the spliced peptide remain bound.

Table 15.3 Characteristics of Popular Fusion Protein Cleavage Enzymes

Protease	Cleavage Site	Comment
Thrombin	?VPR^GS secondary cleavage sites exist; (Chang, 1985)	Widely used, works at 1:1000–1:2000 mass ratio relative to target protein. Purified from bovine sources and may include other proteins.
Factor Xa protease	IEGR^	Leaves defined N-terminus. Works at 1:500–1:1000 mass ratio relative to target protein. Recognition site with proline immediately following Arg residue will not be cleaved.
Enterokinase	DDDDK^	Leaves defined N-terminus. Recombinant.
rTEV	ENLYFQ^G	Recombinant endopeptidase from the Tobacco Etch Virus.
Intein-mediated self-cleavage		No added protease required. Leaves defined N-terminus
PreScission protease	LEVLFQ^GP	Rhinoviral 3C protease expressed as GST fusion protein. Optimal activity at 4°C.

Recognition sites for enzymes commonly used to cleave fusion proteins, and their advantages/disadvanatges are listed in Table 15.3

Will Extra Amino Acid Residues Affect Your Protein of Interest *after* Digestion?

Depending on the protease, and the way in which the protein of interest was cloned in the expression vector, there may be one or more nonnative residues left at the amino terminal of the protein of interest following cleavage. Whether or not this poses a problem depends entirely on the protein and the use to which it will be put. Even the most demanding applications may not be negatively affected by the presence of extra amino terminal residues. Wherever possible, it is best to design a cloning strategy that at least minimizes the number of these residues, and if relatively inoccuous residues (e.g., glycine, serine) can be introduced, all the better.

WORKING WITH EXPRESSION SYSTEMS

What Are the Options for Cloning a Gene for Expression?

In some cases the protein of interest is already cloned in another vector, for example, in a clone isolated from a cDNA expression library. If the frame of the insertion is known, and compatible restriction sites are found in the expression vector(s) selected, the insert can be cloned directly. In some cases excision from a lambda vector can generate a plasmid vector ready for expression of the insert, without any manipulation at all.

More commonly PCR is used to amplify the target sequence using oligonucleotide primers that have 15 to 20 bases of homology with the 5′ and 3′ ends of the target. These primers will have in addition tails that encode restriction enzyme sites compatible with the expression vector. The PCR products can be digested with the appropriate restriction enzymes, purified, and ligated into an appropriately prepared vector.

The efficiency of cloning can be improved if two different restriction enzyme sites are available. This will allow for directional cloning of inserts into the vector, and all of the clones screened should have the insert in the desired orientation. Please refer to Chapter 9, "Restriction Endonucleases" for a discussion on double digestion strategies. If PCR is used to generate the insert, then primers must be designed appropriately. It is important to leave 4 to 6 random bases at the 5′ end of each PCR primer. These provide a spacer at the ends of the PCR product and allow the restriction enzymes to digest the DNA more efficiently. While *in vitro* ligation is still the most widely used method, ligation independent cloning (LIC) (Li and Evans, 1997) has the advantage that no DNA ligase is required (though an exonuclease activity is), and efficiencies are comparable to those obtained with conventional ligation with T4 DNA ligase.

Is Screening Necessary Prior to Expression?

There are no guarantees that the gene to be expressed will be present in the cell after transformation. As discussed above, most expression vectors are prone to produce small amounts of the protein even in the absence of inducing agent, which can prove toxic to the host. Alternatively, host cells can cause deletions and rearrangements in the expression vector. Either way, it is usually a very good idea to confirm the presence of the inserted gene prior to expression experiments.

Unless a library of clones is to be prepared, the efficiency of ligation and transformation is rarely an issue. Screening of a dozen clones for the presence of an insert should be sufficient to identify one or more positive candidate clones.

The first step is generally to prepare several plasmid DNA minipreps and digest the DNA with the same enzyme(s) used in cloning to generate the insert. Products should be analyzed by agarose gel electrophoresis to determine if DNA of the predicted size was inserted in the vector. As an alternative, PCR can be done using as template a small scraping from a colony on the plate. Amplification of the plasmid DNA contained in the cells using the same primers used in cloning, or primers that anneal to flanking vector sequences, should show a band of the predicted size. This latter method does not confirm the presence of the restriction sites used in cloning, but has the advantage of being rapid.

Once the presence of an insert of the correct size is confirmed, the DNA sequence at the cloning junctions should be determined. It is not uncommon for a primer sequence to be synthesized with an error—whether by faulty design or at the hands of the oligo supplier. DNA sequencing to confirm the cloning junctions should be done in parallel with a small-scale expression experiment, in which a 1 to 2 ml culture is grown and induced according to a standard protocol. It is important to include a culture that is transformed with the parent expression vector as a negative control in this screening experiment. Following centrifugation, the cell pellet should be suspended in SDS-PAGE loading buffer, and a small amount loaded on an SDS-acrylamide gel. The viscosity of the whole cell lysate (caused by the release of genomic DNA) may make gel loading difficult. However, addition of extra 1× loading buffer, DNaseI (10 μg/ml), extended heating of the sample, or sonication should alleviate the problem.

After electrophoresis, the gel should be stained (e.g., Coomassie Blue) to visualize the proteins in the whole cell lysate. If expression is good, an induced band will clearly be seen at the predicted molecular weight, and this will be absent in the no-insert control culture. If no band is visible and the restriction digestion/DNA sequencing data indicate that all is well, don't despair. Perform an immunoblot of an SDS-acrylamide gel. Screen for the presence of the protein of interest or use an antibody directed against the affinity or epitope tag if one has been used. Use of both N- and C-terminal specific antibodies is ideal in troubleshooting. Be sure to include both positive and negative controls in the immunoblot. Alternatives to immunblotting include ELISA or specific biochemical assays for the protein of interest.

If an antibody is not available for Western blotting, and you have a procedure to purify your protein, attempt the purification. This can visualize a protein that is present in quantities insufficient to stand out on a PAGE gel of a total cell extract.

Once expression of a protein of the predicted molecular weight is found, minimize propagation of the cells. Serial growths under conditions that permit expression may lead to plasmid loss or rearrangements.

Once analysis is complete, glycerol stocks of positive clones should be prepared. This can be done by streaking culture residue from the DNA miniprep on a plate to get a fresh colony, by reusing the colony that was originally picked, or by re-transforming *E. Coli* with isolated miniprep DNA. In either case a fresh colony should be used to prepare a 2 to 4 ml log-phase culture for the purpose of making a glycerol stock. Be sure to keep protein expression repressed during this step by reduced temperature, use of minimal medium, or adapting it to the vector in use.

What Aspects of Growth and Induction Are Critical to Success?

Aeration, Temperature

The best expression results are had when cultures are grown with sufficient aeration and positive selection for the plasmid. For small-scale experiments, use 2 ml of medium (e.g., LB, SOC or 2XTY) in a 15 ml culture tube. Vigorous shaking (>250 rpm) should be used to maintain aeration. Appropriate antibiotics, such as ampicillin should be added to recommended levels. At larger scales, Ehrlenmeyer flasks should be used. Flasks with baffles improve aeration and $\frac{1}{8}$ to $\frac{1}{2}$ of the total volume of the flask should be occupied by medium. Good results may be obtained using 250 ml to 1 L in a 2 L baffled flask.

Scaling Up

When scaling up growth, monitor the light scattering at 590 or 600 nm. Note that a culture with OD_{600} of one corresponds to about 5×10^8 cells/ml, though this number will vary depending on the strain of *E. coli* used. Two rules of thumb are particularly important: minimize the time in each stage of growth, and monitor both cell density and protein expression at each stage.

From a colony or glycerol stock, begin a small overnight culture (e.g., 2–5 ml) in a selective medium under conditions that repress expression. Don't allow the culture to overgrow. This starter culture is then used to inoculate a larger volume of medium at a volume ratio of about 1:100 (pre-warming the media is a good idea). Monitor the growth by absorbance at 600 nm, and keep the cell density low (OD_{600} below 1). Once the growth has been scaled to give sufficient starter for the final growth vessel, make an inoculum of about 1%. Monitor the OD every 30 minutes or so, and remove aliquots for analysis by SDS-PAGE, immunoblotting, or functional assay. After an initial lag following the inoculation, the density of the cells should double every 20 to 40 minutes. A graph of the OD coupled with an immunoblot is very useful in selecting optimal conditions for the growth. Once the culture reaches a late log phase (usually about OD_{600} of 0.8–1.2), induction is done by the addition of the appropriate inducing agent. Continue to monitor growth and take aliquots. It is not unusual that cells expressing a foreign protein will either stop growing or show a 10% to 20% decrease in density following induction. While it is common to grow for 1 to 3 hours postinduction prior to harvest, this induction period can vary depending on temperature and other conditions. So it is best determined empirically.

What Are the Options for Lysing Cells?

E. coli are easily broken by several methods including decavitation, shearing, and the action of freeze–thaw cycles. The choice of method depends on the scale of growth, and the type of equipment available (reviewed in Johnson, 1998). For most lab-scale experiments, sonication, or freeze–thaw will be the most practical choices. Ultrasonic distruptors are available from many vendors, but all operate on the conversion of electrical energy through piezoelectric transducers into ultrasonic waves of 18 to 50 kHz. The vibration is transferred to the sample by a titanium tip, and the energy released causes decavitation and shearing of the cells. Several models are available that are microprocessor controlled, programmable, and allow very reproducible cell lysis. It is important to keep the sample on ice and avoid frothing. This latter problem is caused by a probe that is not immersed sufficiently in the sample, or by excessive power. If bubbles begin forming and accumulating on the surface, stop immediately, reposition the probe, and reduce output. Once a sample has been turned to foam, sonication will be ineffective, and there is little to do but start again. Even if frothing is not seen, treatment beyond that needed to cause cell lysis can result in physical damage to the protein of interest. The addition of protease inhibitors to the cell suspension immediately prior to cell lysis is an important precaution, and several commercial cocktails are available for this purpose.

Freeze–thaw, particularly in conjunction with lysozyme and DNase treatment, is one of the mildest procedures to break *E. coli*. Cells are simply resuspended in buffer (PBS, Tris-pH 8.0) containing 10 μg/ml hen egg lysozyme, and the sample is cycled between a dry ice–alcohol bath and a container of tepid water. Generally, 5 to 10 cycles is sufficient to break nearly all of the cells.

As the cells lyse and DNA is liberated, it may be necessary to add DNase to 10 μg/ml to reduce the viscosity of the preparation. Commercial or homemade detergent preparations including *N*-octylamine are also very effective at lysing cells and simple to use.

Whatever method is used, lysis should be monitored. Microscopic examination is the best option. Retain some of the starting suspension, and compare to the lysate. Phase contrast optics will permit direct visualization, though staining can be used as well. Lysis will be evidenced by a slight darkening of the suspension, or clearing, and under the microscope, cells will be broken with membrane fragments or small vesicles present.

Other physical lysis methods include the use of a French Press, Manton-Gaulin, and other devices that place cells under rapid changes of pressure or shear force. These are very effective and reproducible, but generally, they are best used when the original culture volume is >1 L, since most of these cell disruptors have minimum volume requirements.

TROUBLESHOOTING

No Expression of the Protein

If one has checked for small-scale expression as discussed above, there should have been a detected band on a stained gel or immunoblot. If neither are seen, sequencing of the cloning junctions and entire insert should be undertaken to confirm that no frame shifts, stop codons, or rearrangements have occurred. Purification can be tried in parallel to see if even very low levels are made. A slight band on SDS-PAGE of the expected protein will make clear that the cloning went as planned, but the biology of expression is at fault. Varying temperature, time of induction, and the type of plasmid or fusion system can all be tried. In the end some proteins may not express well in *E. coli*, and they should be tried in other organisms.

The Protein Is Expressed, but It Is Not the Expected Size Based on Electrophoretic Analysis

On SDS-PAGE the net charge on the protein of interest will affect mobility. Highly charged proteins will tend to bind less SDS and will have retarded mobility. Proteins rich in proline may also exhibit dramatically slower mobility in SDS-PAGE. If the protein has a calculated pI in the range of 5 to 9, and is not strongly biased in amino acid composition, then a protein that shows multiple bands or a strong species far from the predicted molecular weight is likely due to something other than an artifact of SDS-PAGE. Probing immunoblots with the appropriate antibodies to N- and/or C-terminal tags of the protein is particularly useful at this stage. Try to identify the halves or pieces of the protein on stained gels and immunoblots to locate likely points in the coding sequence where proteolytic cleavage and/or translation termination may occur. Cleavage at the junction between the protein of interest and the fusion partner (if any) that is used is often seen. Addition of protease inhibitors should be routine in all work, and protease-deficient strains should be tried in parallel or as a next step. If these measures fail, try re-cloning in another vector with a different fusion tag or tags, and promoter.

The Protein Is Insoluble. Now What?

Many heterologous proteins expressed in *E. coli* will be found as insoluble aggregates that are failed folding intermediates (Schein, 1989). Such inclusion bodies are seen as opaque areas in micrographs of *E. coli* that express the protein of interest. Depending on the purpose of expression, the production of inclusion bodies may be a welcome occurrence. If for example, the recombinant protein is to be used solely for the production of antibodies, inclusion bodies may be isolated to high purity by differential centrifugation and used directly as an antigen. If the protein is relatively small, the inclusion bodies may be isolated as above, and refolded with good efficiency. Other (particularly large) proteins will not refold well, and if production of functional protein is required, then an alternative must be found. Before proceeding, it is best to answer the following questions.

Are You Sure Your Protein Is Insoluble?

A first consideration is whether the protein is truly insoluble, or the cells were simply not lysed. Here is where microscopic examination will be of great use. Examine a cell lysate under phase contrast microscopy or after staining. Are intact cells visible? After it sediments, is the pellet large and similar in appearance to the original cell pellet? Is the post-lysis supernatant clear? Any of the above may indicate that cells are not completely disrupted. The protein of interest may be soluble but trapped in intact cells.

If cells are lysed as measured by microscopy, analyze whole cell lysate, clarified lysate, and post-lysis pellet by SDS-PAGE, followed by staining or immunoblotting. If cells are lysed as measured by microscopy, and the protein of interest is found in the post-lysis pellet, it is likely that it is being made in an insoluble form. While most use a relatively low-speed centrifugation step at around 10,000 × *g*, it is best to do a 100,000 × *g* spin to sediment all aggregates before drawing any conclusion about insolubility. Another indication is microscopic examination of cells under high power (>400×). If inclusion bodies are being made, and expression levels are high, optically dense areas in the *E. coli* cells will be seen. These inclusion bodies may occupy more than half of the cell.

Must Your Protein Be Soluble?

The accumulation of proteins in inclusion bodies is not necessarily undesirable. Insolubility has three important advantages:

1. Inclusion bodies can represent the highest yielding fraction of target protein.
2. Inclusion bodies are easy to isolate as an efficient first step in a purification scheme. Nuclease-treated, washed inclusion bodies are usually 75% to 95% pure target protein.
3. Target proteins in inclusion bodies are generally protected from proteolytic breakdown.

Isolated inclusion bodies can be solubilized by a variety of methods in preparation for further purification and refolding. If the application is to prepare antibodies, inclusion bodies can be used directly for injection after suspension in PBS and emulsification with a suitable adjuvant (e.g., Fischer et al., 1992). If the target protein contains a his_6-tag, purification can be performed under denaturing conditions. The purified protein can be eluted from the resin under denaturing conditions and then refolded.

Solubility Is Essential. What Are Your Options?

Prevent Formation of Insoluble Bodies

A number of approaches have been used to obtain greater solubility, including induction of protein expression at 15 to 30°C (Burton et al., 1991), use of lower concentrations of IPTG (e.g., 0.01–0.1 mM) for longer induction periods, and/or using a minimal defined culture medium (Blackwell and Horgan, 1991).

Solubilize and Refold

Solubilization and refolding methods usually involve the use of chaotropic agents, co-solvents or detergents (Marston and

Hartley, 1990; Frankel, Sohn, and Leinwand, 1991; Zardeneta and Horowitz, 1994). A strategy that has been successful for some proteins is to express as a his_6-tagged fusion, bind under denaturing conditions, and refold while protein is still bound to the resin by running a gradient from 6M to 0M guanidine-HCl in the presence of reduced (GSH) and oxidized (GSSH) glutathione. Once folding has occurred, elution is done with imidazole as usual. Some researchers enhance refolding of enzymes by the addition of substrate or a substrate analogue during gradual removal of denaturant by dialysis (Zhi et al., 1992; Taylor et al., 1992).

The Protein Is Made, but Very Little Is Full-Length; Most of It Is Cleaved to Smaller Fragments

It is important to distinguish among proteolytic breakdown, translation termination, and cryptic translation start sites within the gene of interest. Proteolytic breakdown is most likely to occur at exposed domains of the protein. Examine the pattern of breakdown products by SDS-PAGE, estimate their sizes, and compare the result with the predicted amino acid sequence. Keep an eye out for bends or surface-exposed regions, and any sequences that conform to those for known proteases. While protease inhibitors such as PMSF should be present in the sample prior to cell lysis, expand the group of protease inhibitors and test their effect. Also consider the pattern of expression seen when growth is monitored before and after induction. If there is a switch between intact and fragmented protein after induction, it is likely that proteolysis is the culprit.

Translation Termination

There is little clear-cut evidence for inappropriate translation termination, but in at least one case a stretch of 20 serine residues was suggested to cause premature termination in *E. coli* (Bula and Wilcox, 1996). If a truncated protein is definitely seen, DNA sequencing in the expected termination region should be done to confirm that no cryptic stop codons exist.

Cryptic translation initiation may be seen as well (Preibisch et al., 1988). Cryptic translation initiation can occur within an RNA coding sequence when a sequence resembling the ribosome binding site (AAGGAGG) occurs with the appropriate spacing (typically 5 to 20 nucleotides upstream of an AUG (Met) codon. These smaller products can be problematic when attempting to purify full-length proteins. If some expression of full-length protein is seen, a useful strategy may be to try dual tag affinity purification, in which the gene of interest is expressed in a vector that encodes two affinity tags, one each at the C- and N-termini. Sequential purification using both affinity tags can give reasonable yields of full-length protein whatever the original cause (Kim and Raines, 1994; Pryor and Leiting, 1997).

Your Fusion Protein Won't Bind to Its Affinity Resin

A lysate is produced, and contacted with the affinity medium. The protein of interest is present in the cell and clarified lysate, as shown by SDS-PAGE, but after purification of the lysate over the medium, all of the protein is found in the flow-through. The presence of a large amount of protein in the eluate after an attempt to bind to the affinity medium does not prove an inability to bind. If there is a very large excess of protein, it may appear that none is binding, when in fact the column has simply been overloaded. Try to wash and elute the protein from the affinity medium before drawing a conclusion. One simple test is to remove 10 to 50 μl of the purification medium after binding and washing, and then boil the sample in an equal volume of 1 × SDS-PAGE loading dye. Gel analysis may show binding of the protein to the resin. Consideration of the amount loaded on the column and the expected capacity of the purification medium will sort out the various causes. If in fact expression is clearly seen in the lysate applied to the purification medium, there are other explanations:

1. The affinity medium was not equilibrated properly, or the protein folded to mask the residues responsible for binding to the affinity medium. Purification in the presence of detergents (e.g., 0–1% Tween-20), or mild chaotropes (e.g., 1–3 M guanidine-HCl or urea) may unmask these residues and enable binding.
2. Your fusion protein won't elute from its affinity resin. Protein may apparently bind to the resin, as measured by the presence of an SDS-PAGE gel band after boiling a sample of the washed resin. Little or no protein of interest may be eluted, however, when the loaded resin is contacted with eluting agent. In this latter case the protein may interact nonspecifically with the base matrix, or the protein precipitated during contact with the resin and is trapped. Addition of detergent, of varying ionic strength and pH, may improve the situation.

Your Fusion Protein Won't Digest

If expression is otherwise good, and the protein is not digested to *any* extent, one should confirm by DNA sequencing that the protease site is intact. Checking the activity of the protease in parallel experiments using a known and well-behaved protein will give some confidence that the protease itself is not to blame. If the site is present, the protease has activity, and buffer conditions are close to those specified for the protease, it may be that the fusion protein folds so that the protease site is inaccessible. Additives that alter the structure slightly, including salts and detergents may unmask the site; see Ellinger et al. (1991). Alternatively, recloning to create a flexible linker flanking the protease site has been shown to increase the efficiency of digestion with Thrombin (Guan and Dixon, 1991) and presumably other proteases.

Cleavage of the Fusion Protein with a Protease Produced Several Extra Bands

Cryptic Sites

The specificity of any protease is inferred from its natural substrates, and there is reason to believe that cryptic sites are also cleaved. (Nagai, Perutz, and Poyart, 1985; Eaton, Rodriguez, and Vehar, 1986; Quinlan, Moir, and Stewart, 1989; Wearne, 1990).

Excess Protease

If multiple bands are seen by SDS-PAGE, a titration of the amount, time and temperature of digestion should be done. Often reducing time or temperature will minimize cleavage at secondary sites, while retaining digestion at the desired site.

Extra Protein Bands Are Observed after Affinity Purification

E. coli host chaperone protein GroEL, with an apparent molecular weight of about 57 to 60 kDa on SDS-PAGE, is often found to co-purify with a protein of interest (Keresztessy et al., 1996) This may be caused by misfolding or by a recombinant protein that is trapped at an intermediate folding stage. High salt concentration (1–2 M), non-ionic detergents, and ligand or co-factors (e.g., ATP or GTP) may be effective in removing chaperones from the protein of interest. Often chaperones and other contaminating proteins are seen following affinity purification; they are best removed by conventional chromatography such as ion exchange.

Their co-purification can be minimized by inducing the culture at a lower density (e.g., OD_{600} = 0.3 vs. 1.0) or by reducing temperature.

Must the Protease Be Removed after Digestion of the Fusion Protein?

The removal of the protease is not necessary for many applications. Generally, protease is added at a ratio of 1:500 or lower relative to the protein of interest, so protease may not interfere with downstream applications. Biochemical assays and antibody production may not require removal, while structural studies, or assays where other proteins are added to the protein of interest in a biochemical assay indicate that a further purification be performed.

The commonly used serine proteases, thrombin and factor Xa, can be removed from a reaction mixture by contacting the digested protein/protease with an immobilized inhibitor such as benzamidine-sepharose (Sundaram and Brandsma, 1996). This purification is not complete due to the equilibrium binding of the inhibitor to the protease, but the majority of the protease can be removed in this way. Better yet, a different purification method like ion-exchange or hydrophobic interaction chromatography can be used to separate the protein of interest from both the protease and other cleavage products including the affinity tag.

Some commercially available proteases (Table 15.3) include affinity tags that can be used effectively to remove the protease from the sample. Biotinylated thrombin can be removed with high efficiency due to the extreme affinity of biotin for avidin or streptavidin-agarose beads. Other proteases containing affinity tags include PreScission protease; a fusion of GST with human rhinoviral 3C protease.

BIBLIOGRAPHY

Beck von Bodman, S., Schuler, M. A., Jollie, D. R., and Sligar, S. G. 1986. Synthesis, bacterial expression, and mutagenesis of the gene coding for mammalian cytochrome b5. *Proc. Nat. Acad. Sci. U.S.A.* 83:9443–9447.

Bessette, P. H., Aslund, F., Beckwith, J., and Georgiou, G. 1999. Efficient folding of proteins with multiple disulfide bonds in the *Escherichia coli* cytoplasm. *Proc. Nat. Acad. Sci. U.S.A* 96:13703–13708.

Bishai, W. R., Rappuoli, R., and Murphy, J. R. 1987. High-level expression of a proteolytically sensitive diphtheria toxin fragment in *Escherichia coli*. *J. Bact.* 169:5140–5151.

Blackwell, J. R., and Horgan, R. 1991. A novel strategy for production of a highly expressed recombinant protein in an active form. *FEBS Lett.* 295:10–12.

Boyd, D., and Beckwith, J. 1990. The role of charged amino acids in the localization of secreted and membrane proteins. *Cell* 62:1031–1033.

Buell, G., Schulz, M. F., Selzer, G., Chollet, A., Movva, N. R., Semon, D., Escanez, S., and Kawashima, E. 1985. Optimizing the expression in *E. coli* of a synthetic gene encoding somatomedin-C (ICF-I). *Nucl. Acids Res.* 13:1923–1938.

Bula, C., and Wilcox, K. W. 1996. Negative effect of sequential serine codons on expression of foreign genes in *Escherichia coli*. *Prot. Expr. Purif.* 7:92–103.

Burton, N., Cavallini, B., Kanno, M., Moncollin, V., and Egly, J. M. 1991. Expression in *Escherichia coli*: Purification and properties of the yeast general transcription factor TFIID. *Prot. Expr. Purif.* 2:432–441.

Chang, J.-Y. 1985. Thrombin specificity: Requirement for apolar amino acids adjacent to the cleavage site of polypeptide substrate. *Eur. J. Biochem.* 151:217–224.

Chen, G. F., and Inouye, M. 1990. Suppression of the negative effect of minor arginine codons on gene expression; preferential usage of minor codons within the first 25 codons of the *Escherichia coli* genes. *Nucl. Acids Res.* 18:1465–1473.

Derman, A. I., Prinz, W. A., Belin, D., and Beckwith, J. 1993. Mutations that allow disulfide bond formation in the cytoplasm of *Escherichia coli*. *Science* 262:1744–1747.

Dodd, I., Mossakowska, D. E., Camilleri, P., Haran, M., Hensley, P., Lawlor, E. J., McBay, D. L., Pindar, W., and Smith, R. A. 1995. Overexpression in *Escherichia coli*: Folding, purification, and characterization of the first three short consensus repeat modules of human complement receptor type 1. *Prot. Expr. Purif.* 6:727–736.

Dubendorff, J. W., and Studier, F. W. 1991. Controlling basal expression in an inducible T7 expression system by blocking the target T7 promoter with lac repressor. *J. Mol. Biol.* 219:45–59.

Eaton, D., Rodriguez, H., and Vehar, G. A. 1986. Proteolytic processing of human factor VIII: Correlation of specific cleavages by thrombin, factor Xa, and activated protein C with activation and inactivation of factor VIII coagulant activity. *Biochem.* 25:505–512.

Eldin, P., Pauza, M. E., Hieda, Y., Lin, G., Murtaugh, M. P., Pentel, P. R., and Pennell, C. A. 1997. High-level secretion of two antibody single chain Fv fragments by *Pichia pastoris*. *J Immunolog. Meth.* 201:67–75.

Ellinger, S., Mach, M., Korn, K., and Jahn, G. 1991. Cleavage and purification of prokaryotically expressed HIV gag and env fusion proteins for detection of HIV antibodies in the ELISA. *Virol.* 180:811–813.

Etchegaray, J. P., and Inouye, M. 1999. Translational enhancement by an element downstream of the initiation codon in *Escherichia coli*. *J. Biol. Chem.* 274:10079–10085.

Fischer, L., Gerard, M., Chalut, C., Lutz, Y., Humbert, S., Kanno, M., Chambon, P., and Egly, J. M. 1992. Cloning of the 62-kilodalton component of basic transcription factor BTF2. *Science* 257:1392–1395.

Frankel, S., Sohn, R., and Leinwand, L. 1991. The use of sarkosyl in generating soluble protein after bacterial expression. *Proc. Nat. Acad. Sci. U.S.A.* 88:1192–1196.

Georgiou, G., and Valax, P. 1996. Expression of correctly folded proteins in *Escherichia coli*. *Curr. Opin. Biotechnol.* 7:190–197.

Gilbert, H. F. 1994. Protein chaperones and protein folding. *Curr. Opin. Biotechnol.* 5:534–539.

Gold, L., Pribnow, D., Schneider, T., Shinedling, S., Singer, B. S., and Stormo, G. 1981. Translational initiation in prokaryotes. *Ann. Rev. Microbiol.* 35:365–403.

Guan, K. L., and Dixon, J. E., 1991. Eukaryotic proteins expressed in *Escherichia coli*: An improved thrombin cleavage and purification procedure of fusion proteins with glutathione *S*-transferase. *Anal. Biochem.* 192:262–267.

Hirel, P. H., Schmitter, M. J., Dessen, P., Fayat, G., and Blanquet, S. 1989. Extent of N-terminal methionine excision from *Escherichia coli* proteins is governed by the side-chain length of the penultimate amino acid. *Proc. Nat. Acad. Sci. U.S.A.* 86:8247–8251.
Hostomsky, Z., Smrt, J., and Paces, V. 1985. Cloning and expression in *Escherichia coli* of the synthetic proenkephalin analogue gene. *Gene* 39:269–274.
Hua, Z., Wang, H, Chen, D., Chen, Y., and Zhu, D. 1994. Enhancement of expression of human granulocyte-macrophage colony stimulating factor by argU gene product in *Escherichia coli*. *Biochem. Mol. Biol. Int.* 32:537–543.
Izard, J. W., and Kendall, D. A. 1994. Signal peptides: Exquisitely designed transport promoters. *Mol. Microbiol.* 13:765–773.
Johnson, B. 1998. Breaking up isn't hard to do: A cacophony of sonicators, cell bombs, and grinders. *Scientist* 12:23.
Kapust, R. B., and Waugh, D. S. 1999. *Escherichia coli* maltose-binding protein is uncommonly effective at promoting the solubility of polypeptides to which it is fused. *Prot. Sci.* 8:1668–1674.
Kelley, R. F., and Winkler, M. E. 1990. Folding of eukaryotic proteins produced in *Escherichia coli*. *Genet. Eng.* (*NY*) 12:1–19.
Keresztessy, Z., Hughes, J., Kiss, L., and Hughes, M. A. 1996. Co-purification from *Escherichia coli* of a plant beta-glucosidase-glutathione *S*-transferase fusion protein and the bacterial chaperonin GroEL. *Biochem. J.* 314:41–47.
Kim, J. S., and Raines, R. T. 1994. Peptide tags for a dual affinity fusion system. *Anal. Biochem.* 219:165–166.
Kochendoerfer, G. G., and Kent, S. B. 1999. Chemical protein synthesis. *Curr. Opin. Chem. Biol.* 3:665–671.
Koshiba, T., Hayashi, T., Miwako, I., Kumagai, I., Ikura, T., Kawano, K., Nitta, K., and Kuwajima, K. 1999. Expression of a synthetic gene encoding canine milk lysozyme in *Escherichia coli* and characterization of the expressed protein. *Prot. Eng.* 12:429–435.
Lathrop, B. K., Burack, W. R., Biltonen, R. L., and Rule, G. S. 1992. Expression of a group II phospholipase A2 from the venom of *Agkistrodon piscivorus piscivorus* in *Escherichia coli*: Recovery and renaturation from bacterial inclusion bodies. *Prot. Expr. Purif.* 3:512–517.
Lee, N., Zhang, S. Q., Cozzitorto, J., Yang, J. S., and Testa, D. 1987. Modification of mRNA secondary structure and alteration of the expression interferon alpha 1 in *Escherichia coli*. *Gene* 58:77–86.
Li, C., and Evans, R. M. 1997. Ligation independent cloning irrespective of restriction site compatibility *Nucl. Acids Res.* 25:4165–4166.
Looman, A. C., Bodlaender, J., de Gruyter, M., Vogelaar, A., and van Knippenberg, P. H. 1986. Secondary structure as primary determinant of the efficiency of ribosomal binding sites in *Escherichia coli*. *Nucl. Acids Res.* 14:5481–5496.
Marston, F. A., and Hartley, D. L. 1990. Solubilization of protein aggregates. *Meth. Enzymol.* 182:264–282.
Miller III, D. M., Olson, J. S., Pflugrath, J. W., and Quiocho, F. A. 1983. Rates of ligand binding to periplasmic proteins involved in bacterial transport and chemotaxis. *J. Biol. Chem.* 258:13665–13672.
Nagai, K., and Thogersen, H. C. 1984. Generation of beta-globin by sequence-specific proteolysis of a hybrid protein produced in *Escherichia coli*. *Nature* 309:810–812.
Nagai, K., Perutz, M. F., and Poyart, C. 1985. Oxygen binding properties of human mutant hemoglobins synthesized in *Escherichia coli*. *Proc. Nat. Acad. Sci. U.S.A.* 82:7252–7255.
Nilsson, B., Holmgren, E., Josephson, S., Gatenbeck, S., Philipson, L., and Uhlen,

M. 1985. Efficient secretion and purification of human insulin-like growth factor I with a gene fusion vector in *Staphylococci. Nucl. Acids Res.* 13: 1151–1162.
Pines, O., and Inouye, M. 1999. Expression and secretion of proteins in *E. coli. Mol. Biotechnol.* 12:25–34.
Preibisch, G., Ishihara, H., Tripier D., and Leineweber, M. 1988. Unexpected translation initiation within the coding region of eukaryotic genes expressed in *Escherichia coli. Gene* 72:179–186.
Pryor, K. D., and Leiting, B. 1997. High-level expression of soluble protein in *Escherichia coli* using a His_6-tag and maltose-binding-protein double-affinity fusion system. *Prot. Expr. Purif.* 10:309–319.
Quinlan, R. A., Moir, R. D., and Stewart, M. 1989. Expression in *Escherichia coli* of fragments of glial fibrillary acidic protein: Characterization, assembly properties and paracrystal formation. *J. Cell Sci.* 93:71–83.
Robinson, M., Lilley, R., Little, S., Emtage, J. S., Yarranton, G., Stephens, P., Millican, H., Eaton, M., Humphries, G. 1984. Codon usage can affect efficiency of translation of genes in Escherichia coli. *Nucl. Acids Res.* 12:6663–6671.
Sambrook, J., Fritsch, E. F., and Maniatis, T. 1989. *Molecular Cloning: A Laboratory Manual*, 2nd ed. Cold Spring Harbor Laboratory, Cold Spring Harbor, NY.
Samols, D., Thornton, C. G., Murtif, V. L., Kumar, G. K., Haase, F. C., and Wood, H. G. 1988. Evolutionary conservation among biotin enzymes. *J. Biol. Chem.* 263:6461–6464.
Schein C. H. 1989. Production of soluble recombinant proteins in bacteria. *Biotechnol.* 7:1141–1149.
Schmidt, M., Tuominen, N., Johansson, T., Weiss, S. A., Keinänen, K., and Oker-Blom, C. 1998. Baculovirus-mediated large-scale expression and purification of a polyhistidine-tagged rubella virus capsid protein. *Prot. Expr. Purif.* 12:323–330.
Sorensen, M. A., Kurland, C. G., and Pedersen, S. 1989. Codon usage determines translation rate in *Escherichia coli. J. Mol. Biol.* 207:365–377.
Stader, J. A., and Silhavy, T. J. 1990. Engineering *Escherichia coli* to secrete heterologous gene products. *Meth. Enzymol.* 185:166–187.
Sundaram, P., and Brandsma, J. L. 1996. Rapid, efficient, large-scale purification of unfused, nondenatured E7 protein of cottontail rabbit papillomavirus. *J. Virol. Meth.* 57:61–70.
Taylor, M. A., Pratt, K. A., Revell, D. F., Baker, K. C., Sumner, I. G., and Goodenough, P. W. 1992. Active papain renatured and processed from insoluble recombinant propapain expressed in *Escherichia coli. Prot. Eng.* 5:455–459.
Tessier, L. H., Sondermeyer, P., Faure, T., Dreyer, D., Benavente, A., Villeval, D., Courtney, M., and Lecocq, J. P. 1984. The influence of mRNA primary and secondary structure on human IFN-gamma gene expression in *E. coli. Nucl. Acids Res.* 12:7663–7675.
Tobias, J. W., Shrader, T. E., Rocap, G., and Varshavsky, A. 1991. The N-end rule in bacteria. *Science* 254:1374–1377.
Wearne, S. J. 1990. Factor X_a cleavage of fusion proteins: Elimination of non-specific cleavage by reversible acylation. *FEBS Lett.* 263:23–26.
Wickner, W., Driessen, A. J., and Hartl, F. U. 1991. The enzymology of protein translocation across the *Escherichia coli* plasma membrane. *Ann. Rev. Biochem.* 60:101–124.
Wilkinson, D. L., and Harrison, R. G. 1991. Predicting the solubility of recombinant proteins in *Escherichia coli. Biotechnol.* 9:443–448.
Yamane, K., and Mizushima, S. 1988. Introduction of basic amino acid residues after the signal peptide inhibits protein translocation across the cytoplasmic

membrane of *Escherichia coli*: Relation to the orientation of membrane proteins. *J. Biol. Chem.* 263:19690–19696.
Yasukawa, T., Kanei-Ishii, C., Maekawa, T., Fujimoto, J., Yamamoto, T., and Ishii, S. 1995. Increase in solubility of foreign proteins in *E. coli* by coproduction of the bacterial thioredoxin. *J. Biol. Chem.* 270:25328–25331.
Zardeneta, G., and Horowitz, P. M. 1994. Detergent, liposome, and micelle-assisted protein refolding. *Anal. Biochem.* 223:1–6.
Zhang, S. P., Zubay, G., and Goldman, E. 1991. Low-usage codons in *Escherichia coli*, yeast, fruit fly and primates. *Gene* 105:61–72.
Zhi, W., Landry, S. J., Gierasch, L. M., and Srere, P. A. 1992. Renaturation of citrate synthase: Influence of denaturant and folding assistants. *Prot. Sci.* 1:522–529.

16

Eukaryotic Expression

John J. Trill, Robert Kirkpatrick, Allan R. Shatzman, and Alice Marcy

Section A: A Practical Guide to Eukaryotic Expression 492
Planning the Eukaryotic Expression Project 493
What Is the Intended Use of the Protein and What Quantity Is Required? 493
What Do You Know about the Gene and the Gene Product? 496
Can You Obtain the cDNA? 497
Expression Vector Design and Subcloning 498
Selecting an Appropriate Expression Host................ 501
Selecting an Appropriate Expression Vector 506
Implementing the Eukaryotic Expression Experiment 511
Media Requirements, Gene Transfer, and Selection 511
Scale-up and Harvest 514
Gene Expression Analysis 515
Troubleshooting 517
Confirm Sequence and Vector Design 517
Investigate Alternate Hosts 519
A Case Study of an Expressed Protein from cDNA to Harvest 519
Summary 521
Section B: Working with Baculovirus 521
Planning the Baculovirus Experiment 521

Molecular Biology Problem Solver, Edited by Alan S. Gerstein.
ISBN 0-471-37972-7

Is an Insect Cell System Suitable for the Expression of Your Protein? 521
Should You Express Your Protein in an Insect Cell Line or Recombinant Baculovirus? 522
Procedures for Preparing Recombinant Baculovirus 524
Criteria for Selecting a Transfer Vector 524
Which Insect Cell Host Is Most Appropriate for Your Situation? 525
Implementing the Baculovirus Experiment 527
What's the Best Approach to Scale-Up? 527
What Special Considerations Are There for Expressing Secreted Proteins? 527
What Special Considerations Are There for Expressing Glycosylated Proteins? 528
What Are the Options for Expressing More Than One Protein? 529
How Can You Obtain Maximal Protein Yields? 529
What Is the Best Way to Process Cells for Purification?... 530
Troubleshooting 530
Suboptimal Growth Conditions 530
Viral Production Problems 531
Mutation 531
Solubility Problems 532
Summary 532
Bibliography 533

SECTION A: A PRACTICAL GUIDE TO EUKARYOTIC EXPRESSION

Recombinant gene expression in eukaryotic systems is often the only viable route to the large-scale production of authentic, post-translationally modified proteins. It is becoming increasingly easy to find a suitable system to overexpress virtually any gene product, provided that it is properly engineered into an appropriate expression vector. Commercially available systems provide a wide range of possibilities for expression in mammalian, insect, and lower eukaryotic hosts, each claiming the highest possible expression levels with the least amount of effort. Indeed, many of these systems do offer vast improvements in their ease of use and rapid end points over technologies available as recently as 5 to 10 years ago. In addition methods of transferring DNA into cells have advanced in parallel enabling transfection efficiencies approaching 100%. However, one still needs to carefully consider the most appropriate vector and host system that is compatible with a particular expression need. This will largely depend on the type of protein being expressed (e.g., secreted, membrane-bound, or intracellular) and its intended use. No one system can or should be expected to meet all expression needs.

In this section we will attempt to outline the critical steps involved in the planning and implementation of a successful eukaryotic expression project. Planning the project will begin by answering pertinent questions such as what is known about the protein being expressed, what is its function, what is the intended use of the product, will the protein be tagged, how much protein is needed, and how soon will it be needed. Based on these considerations, an appropriate host or vector system can be chosen that will best meet the anticipated needs.

Considerations during the implementation phase of the project will include choosing the best method of gene transfer and stable selection compared to transient expression and selection methods for stable lines, and clonal compared to polyclonal selection. Finally, we will discuss anticipated outcomes from various methods, commonly encountered problems, and possible solutions to these problems.

PLANNING THE EUKARYOTIC EXPRESSION PROJECT

What Is the Intended Use of the Protein and What Quantity Is Required?

Protein quantity is an important consideration, since substantial time and effort are required to achieve gram quantities while production of 10 to 100 milligrams is often easily obtained from a few liters of cell culture. Therefore we tend to group the expressed proteins into the following three categories: target, reagent, and therapeutic protein. This is helpful both in choosing an appropriate expression system and in determining how much is enough to meet immediate needs (Table 16.1).

Targets

Protein targets represent the majority of expressed proteins used in classical pharmaceutical drug discovery, which involves the configuration of a high-throughput screen (HTS) of a chemical or natural product library in order to find selective antagonists or agonists of the protein's biological activity. Protein targets include enzymes (e.g., kinases or proteases), receptors (e.g., 7

Table 16.1 Categories of Expressed Proteins

Class of Protein	Examples	Expression Amount	Appropriate System
Target	Enzymes and receptors	For screening: 10 mg For structural studies: 100 mg	Stable insect Baculovirus Mammalian Yeast
Reagent	Modifying enzymes Enzyme Substrates	<10 mg	Stable insect Baculovirus Mammalian Yeast
Therapeutic	Therapeutic Monoclonal antibody (mAb) Cytokine Hormone	g/L	Mammalian (CHO, myelomas)

transmembrane, nuclear hormone, integrin), and their ligands and membrane transporters (e.g., ion channels). In basic terms, sufficient quantities of a protein target need to be supplied in order to run the HTS. The actual amounts depend on the size of a given library to be screened and the number of hits that are obtained, which will then need to be further characterized. As a rule of thumb, for purified proteins such as enzymes and receptor ligands, amounts around 10 mg are usually needed to support the screen. For nonpurified proteins such as receptors, one needs to think in terms of cell number and the growth properties of the cell line. For most cell lines, screens are configured by plating between 100,000 to 300,000 cells per milliliter. By way of example, a typical screen of one million compounds in multiwell formats (e.g., 96, 384, or 1536 well) could use between 0.5 to 1.5×10^9 cells. The smaller the volume of the screen, the fewer cells will be required.

Because protein targets require a finite amount of protein, one has the flexibility of choosing from virtually any expression system. Consequently the selection of the system for producing a target protein really depends on considerations other than quantity. The most important goal is to achieve a product with the highest possible biological activity. This will enable a screen to be configured with the least amount of protein and will give the best chance of establishing a screen with the highest possible signal to background ratio. Other considerations include the type of protein being expressed (e.g., intracellular, secreted, and membrane-associated proteins). As discussed below, stable cell systems tend to be more amenable to secreted and membrane-associated proteins, while intracellular proteins are often produced very efficiently from lytic systems such as baculovirus. Whatever system is used, it should be scaled appropriately to meet the needs of HTS.

A subset of target proteins are those that are used for structural studies. In order to grow crystals that are of sufficient quality to yield high-resolution structures, it is particularly important to begin with properly folded, processed, active protein. Proteins used for structural studies are often supplied at very high concentrations (>5 mg/ml) and must be free of heterogeneity. Glycosylation is often problematic because its addition and trimming tends to be heterogenous (Hsieh and Robbins, 1984; Kornfeld and Kornfeld, 1985). As a result it is often necessary to enzymatically remove some or all of the carbohydrate before crystals can be formed. As a starting point, one often needs approximately 10 mg of absolutely pure protein so that crystallization conditions can be tested and optimized, with the total protein requirement often exceeding 100 mg.

In order to avoid the issue of glycosylation in structural studies altogether, one can express the protein in a glycosylation-deficient host (Stanley, 1989). Alternatively one can remove glycosylation sites by site-directed mutagenesis prior to expression. However, these are very empirical methods that do not often work well for a variety of reasons, including the need in some cases to maintain glycosylation for proper solubility. Thus, for direct expression of a nonglycosylated protein, a first-pass expression approach would likely involve a bacterial system in which high level expression of nonglycosylated protein is more readily attained.

Reagents

A second category of expressed proteins is reagents. These are proteins that are not directly required to configure a screen but are needed to either evaluate compounds in secondary assays or to help produce a target protein itself. Examples of reagent proteins include full-length substrates that are replaced by synthetic peptides for screening. Enzyme substrates themselves are often cleaved to produce biologically active species whose activities can be assessed in vitro. Reagent proteins can also include processing enzymes that are required for the in vitro activation of a purified protein (e.g., cleavage of a zymogen or phosphorylation by an upstream activating kinase). Also included in this category are gene orthologues from species other than the one being used in the screen, whose expression will be used to support animal studies and to determine the cross-species selectivity or activity of selected compounds.

Reagent proteins are usually required in much lower amounts than target proteins. Some can even be purchased commercially in sufficient quantities to meet the required need. Others, because of price or the required quantity, may necessitate recombinant expression. But, since only small quantities are usually required (<10 mg), it is possible to choose an expression system with features that will favor efficient and rapid expression. Furthermore the expression scale can be minimized. The bottom line is that reagent proteins should be the least resource intensive to produce. One should avoid trying to overproduce reagent proteins or scaling them to quantities that will never be used.

Therapeutics

In contrast to reagent proteins, therapeutic protein agents are the most demanding in terms of resource. Therapeutic proteins have intrinsic biological properties like medical drugs. The ultimate objective for expression of a therapeutic protein is the production of clinical-grade protein approaching or exceeding gram per liter quantities. For most expression systems this is not readily achievable. Other than bacterial and yeast expression, the most robust system for producing these levels is the Chinese hamster ovary (CHO) system. Due to the lack of proper post-translational modifications (e.g., glycosylation) in bacteria and yeast, CHO cell expression is often the only choice to achieve sufficient expression. Examples of therapeutic proteins, produced in CHO cells, include humanized monoclonal antibodies (Trill, Shatzman, and Ganguly, 1995), tPA (tissue plasminogen activator; Spellman et al., 1989), and cytokines (Sarmiento et al., 1994). In many cases months are spent selecting and amplifying lines with appropriate growth properties and expression levels to meet production criteria.

What Do You Know about the Gene and the Gene Product?

Information about the gene product or for that matter, its homologues or orthologues, enables one to make an educated guess as to what is the best eukaryotic expression system to use. Is there anything published in the literature about the gene, or is it completely uncharacterized? Do we know in what tissue the gene is expressed, based on either Northern blot analysis or by quantitative or semiquantitative RT-PCR measures? Other factors to determine are whether the protein to be expressed is secreted, cytosolic, or membrane-bound. If it is a receptor, is it a homodimer, heterodimer, multimeric, single, or multispanning transmembrane receptor or anchored to the surface (e.g., through a glycosyl phosphatidylinositol phosphate (GPI linkage).

Fortunately we usually have the luxury of working with genes that are at least partially characterized by their biological properties. But what about the genes of unknown origin or function? In this new age of genomics, many of the genes we obtain are "like" genes, belonging to large families of related genes that share only a minimal percentage of homology with a known gene. Despite these similarities there is often no way to know whether the same expression and purification methods used for one orthologue or homologue will be effective for another. Thus one is immediately faced with the challenging prospect of having to consider multiple expression strategies in order to get the protein expressed and purified to sufficient levels in an active form, in addition to not knowing what activity to look for.

Can You Obtain the cDNA?

Before embarking on an expression project you will need to locate a cDNA copy of the gene of interest. It is also possible in theory to express genomic DNA containing introns, provided that the expression host will recognize the proper splice junctions. In practice, however, this is not often the most efficient route to expression because it is not usually known how the introns will affect expression levels or whether the desired splice variant will be expressed. Furthermore most mammalian genes are interrupted by multiple intron sequences that can span many kilobases in length. This can make subcloning of genomic DNA considerably more difficult than for the corresponding cDNA.

The three most common ways to obtain a known gene of interest include purchase from a distributor of clones from the Integrated Molecular Analysis of Genomes and their Expression (IMAGE) consortium (*http://image.llnl.gov/*), requests from a published source such as an academic lab, or RT-PCR cloning from RNA derived from a cell or tissue source. IMAGE clones can be found by performing a BLAST search of an electronic database such as GenBank, which can be accessed at the National Library of Medicine PubMed browser (*http://www.ncbi.nlm.nih.gov/PubMed/*). From there you can quickly determine if a sequence is present, if it is full length, publications related to this gene, and possible sources of the gene (tissue sources, personal contacts, etc). Most expressed sequence tags (EST's) matching the gene of interest are available as IMAGE clones. The trick is to find one that is full length. It is easy to determine if an EST is likely to contain a full-length sequence if it is derived from a directional oligo dT primed library and sequenced from the 5′ end by searching for an ATG and an upstream stop codon. Once you identify a full-length EST, you should then be able to obtain the corresponding IMAGE clone from Incyte Genomics, LifeSeq Public Incyte clones (*http://www.incyte.com/reagents/index.shtml*), Research Genetics (*http://www.resgen.com*), or the American Type Culture Collection (*ATCC, http://www.atcc.org*). If the gene is published, you can also try contacting the author who cloned it in order to obtain a cDNA clone. Most labs, including both academic and pharmaceutical/biotech companies, will honor a request for a cDNA clone if it is published. Alternatively, you may consider deriving the gene de novo by RT-PCR using the sequence obtained above.

Depending on the size, abundance, and tissue distribution of the mRNA, a PCR approach could be straightforward or complex. One may isolate RNA from tissue, generate cDNA from the RNA using reverse transcriptase, design PCR primers to perform PCR, and fish out the gene of interest. Alternatively, one may simply purchase a cDNA library from which to PCR amplify the gene. Several vendors carry a wide array of high-quality cDNA libraries derived from human and animal tissues. For example, cDNA libraries for virtually every major human or murine tissue/organ can be obtained from Invitrogen (*http://www.invitrogen.com./catalog_project/index.html*) or Clontech (*http://www.clontech.com/products/catalog/Libraries/index.html*). These companies obtain their samples from sources under Federal Guidelines.*

Expression Vector Design and Subcloning

Perhaps the most critical step in the process of expressing a gene is the vector design and subcloning. As much an art as a science, it nevertheless requires complete precision. In many cases you will need to amplify the gene by PCR from RNA. If the gene is in a library, you may also need to trim the 5′ and 3′ UTR (untranslated region) and to add restriction sites and/or a signal sequence if one is not already present. You may also want to add

**Editor's note: In addition to the planning recommended by the authors, it is wise to ask commercial suppliers of expression systems about the existence of patents relating to the components of an expression vector (i.e., promoters) or the use of proteins produced by a patented expression vector/system.*

epitope tags for detection and purification (e.g., His_6 tag). When PCR is involved, the gene will eventually need to be entirely re-sequenced in order to rule out PCR-induced mutations that can occur at a low frequency. If mutations are found, they will need to be repaired, thereby adding to the time required to generate the final expression construct. The best practice is to start with a high-fidelity polymerase with a proofreading (3′–5′ exonuclease activity) function to avoid PCR errors.

Sequence Information

If you are lucky enough to obtain a DNA from a known source, a new litany of questions will need to be answered. Is a sequence and restriction map available? Do you know what vector the gene has been cloned into? Has the gene been sequenced in its entirety? How much do you trust the source from which you have received the gene? It is usually best to have the gene re-sequenced so that you know the junctions and restriction sites and can assure yourself that you are indeed working with the correct gene. What do you do if there are differences between your sequence and the published sequence? You will need to decide if the difference is due to a mutation, an artifact from the PCR reaction, a gene polymorphism, or an error in the published sequence. A search of an EST database coupled with a comparison with genes of other species can help distinguish whether the error is in the database or due to a polymorphism. Alternatively, sequencing multiple, independently derived clones can also help answer these questions.

Control Regions

We now have a gene with a confirmed sequence. But which control regions are present? Does the gene contain a Kozak sequence, 5′-GCCA/GCCAUGG-3′, required to promote efficient translational initiation of the open reading frame (ORF) in a vertebrate host (Kozak, 1987) or an equivalent sequence 5′-CAAAACAUG-3′ for expression in an insect host (Cavener, 1987)? If this sequence is missing, it is essential to add it to your expression vector. It is also advisable to trim the gene to remove any unnecessary sequences upstream of the ATG. The 5′ noncoding regions may contain sequences (e.g., upstream ATG's or secondary structures) that may inhibit translation from the actual start. A noncoding sequence at the 3′ end may destabilize the message.

Epitope Tags and Cleavage Sites

Another sequence you might need to add to your gene is an epitope tag or a fusion partner with or without a protease cleavage site. This will aid in the identification of your protein product (via Western blot, ELISA, or immunofluorescence) and assist in protein purification. Among the various epitope tags available are FLAG® (DYKDDDDK) (Hopp et al., 1988), influenza hemaglutinin or HA (YPYDVPDYA) (Niman et al., 1983), His_6 (HHHHHH) (Lilius et al., 1991), and c-myc (EQKLISEEDL) (Evan et al., 1985). The more popular protease cleavage sites, used to remove the tag from the protein, are thrombin (VPR'GS) (Chang, 1985), factor Xa (IEGR'; Nagai and Thogersen, 1984), PreScission protease (LEVLFQ'GR; Cordingley et al., 1990), and enterokinase (DDDDK'; Matsushima et al., 1994) One may also use larger fusion partners such as the Fc region of human IgG1 or GST. It is crucial to choose a protease that is not predicted to cleave within the protein itself, but this does not preclude spurious cleavages.

The benefits and drawbacks of utilizing epitope tags are discussed in greater detail below in the section, "Gene Expression Analysis."

Subcloning

Your gene is now ready to be cloned into an expression vector of your choice, provided that you have already decided what system to use. This will traditionally involve the use of restriction enzymes to precisely excise the gene on a DNA fragment, which is subsequently ligated into a donor expression vector at the same or compatible sites. If appropriate unique restriction sites are not located in flanking regions they can be added by PCR (incorporating the sequence onto the end of the amplification primer), or by site-directed mutagenesis.

Recent technological advances also offer the possibility of subcloning without restriction enzymes. These new age cloning systems are based on recombinase-mediated gene transfer. Invitrogen offers ECHO™ and Gateway™ cloning technologies, while Clontech markets the Creator™ gene cloning and expression system. Recombinases essentially perform restriction and ligation in a single step, thereby eliminating the time-consuming process of purifying restriction fragments for subcloning and ligating them. These new systems are particularly advantageous when transferring the same gene into multiple expression vectors for expression in different host systems.

Selecting an Appropriate Expression Host

Expressed Protein Issues

The properties of the protein and its intended usage will also have a direct impact on which expression system to choose. Since many eukaryotic proteins undergo post-translational modifications (phosphorylation, signal-sequence cleavage, proteolytic processing, and glycosylation), which can affect function, circulating half-life, antigenicity, and the like, these issues must be addressed when choosing an expression host. These steps have a direct influence on the quality of protein produced. For instance, it has been demonstrated that there is a clear difference in the glycosylation patterns between various mammalian and insect systems. Insect cells lack the pathways necessary to produce glycoproteins containing complex *N*-linked glycans with terminal sialic acids (Ailor and Betenbaugh, 1999; Kornfeld and Kornfeld, 1985), and the absence of sialic acid residues can strongly influence the in vivo pharmacokinetic properties of many glycoproteins (Grossmann et al., 1997). Using tPA as a model system, it has also been shown that glycosylation patterns differ within different mammalian cell types (Parekh et al., 1989).

The expression strategies for both targets and reagents are the same. We desire a purified protein, cell membranes for a binding assay, or attached cell lines for a cell-based assay. The determining factor for selecting a host system depends on the quantity of the protein needed, what signaling components are necessary in the host line, and the degree to which endogenously expressed host proteins generate background responses (e.g., for receptors). For example, insect cell lines often provide a null background for mammalian signaling components, which enable lower basal level activation and high signal to background in cell-based assays.

If the protein is a target and will be used in a cell-based assay, one needs to make a high expressing cell line. In most cases the higher the expression is, the better is the result. But this is not always the case for cytoplasmic or membrane anchored proteins where the expressed protein can be toxic. In these cases it might be better to achieve lower expression or to use some type of regulated promoter vector system as discussed in the following section.

If the desired protein is to be a therapeutic and used to supply clinical trials, the choices are very well documented. There are numerous examples of commercial therapeutic proteins being produced in *E. coli* and yeast. However, if the protein contains numerous disulfide linkages, or requires extensive post-translational modifications (i.e., folding of antibody heavy and light chains), one needs to consider expression in a mammalian cell line. The gene needs to be cloned into a plasmid system allowing for some type of amplification so that the protein can be expressed at very high levels. In addition one needs to be cognizant of GMP, GLP, and FDA guidelines for the entire expression, selection or amplification process.

The inability to obtain homogeneously pure protein for crystallization is a frequently encountered problem due to the heterogeneous carbohydrate content of many eukaryotic proteins (Grueninger-Leitch et al., 1996). In the past *E. coli* expression systems were exclusively used to produce material for crystallization in order to avoid having glycosylation at all. Recently there have been an increasing number of examples where crystals were generated using baculovirus-expressed protein (Cannan et al., 1999; Sonderman et al., 1999). Another approach has been to use the glycosylation-deficient mutant CHO cell line, Lec3.2.8.1, (Stanley, 1989; Butters et al., 1999; Casasnovas, Larvie, and Stehle, 1999; Kern et al., 1999). In these cases the incomplete or underglycosylation has allowed the formation of high-resolution, diffractable crystals.

Transient Expression Systems

Transient systems are used for the rapid production of small quantities of heterologous gene products and are often suitable to make "reagent" category proteins. The cell lines of choice include the following;

- COS cells (COS-1, ATCC CRL 1650; COS-7 ATCC CRL 1651; see Gluzman, 1981). These are derived from the African green monkey cell line, CV-1, which was infected with an origin-defective SV40 genome. Upon transfection with a plasmid containing a functional SV40 origin of replication, the combination of SV40 replication origin (donor) and SV40 large T-antigen (host cell) results in high copy extrachromosomal replication of the transfected plasmid (Mellon et al., 1981).
- Human embryonic kidney (HEK) 293 cells (ATCC CRL 1573). An immortalized cell line derived from human embryonic kidney cells transformed with human adenovirus type 5 DNA. This cell line contains the adenovirus E1A gene, which transactivates CMV promoter-based plasmids, and this results in increased expression levels. This cell line is widely used to express 7 trans membrane G-protein-coupled receptors (GPCRs) (Ames et al., 1999; Chambers et al., 2000).

In our own experiments involving transient expression systems, we have consistently found that COS cells yield approximately 50% higher expression than HEK 293 cells. (Trill, 2000, unpublished). To take monoclonal antibodies (mAbs) as an example, transient systems such as COS can allow one to examine multiple constructs in two to three days at expression levels ranging from 100 ng/ml to 2 μg/ml. Stable cell lines can yield over 200-fold more protein, but it is often a time-consuming process to achieve those levels, often taking six months to a year to accomplish (Trill, Shatzman, and Ganguly, 1995).

Viral Lytic Systems

Viral lytic systems offer the advantage of rapid expression combined with high-level production. The most popular of the viral lytic systems utilizes baculovirus.

The baculovirus expression system is based on the manipulation of the circular *Autographa californica* virus genome to produce a gene of interest under the control of the highly efficient viral polyhedrin promoter. Engineered viruses are used to infect cell lines derived from pupal ovarian tissue of the fall army worm, *Spodoptera frugiperda* (Vaughn et al., 1977). This lytic system is most useful for the high-level expression of enzymes and other soluble intracellular proteins. Secreted proteins can also be obtained from this system but are more difficult to scale to large volumes due to the rapid onset of the lytic cycle. Cell lines include Sf9, Sf21, and *T. ni* (available as High Five™) cells are from *Trichoplusia ni* egg cell homogenates. Refer to Section B for more detail on baculovirus expression.

Adenovirus expression has also increased in popularity of late. This may be due in part to its use for in vivo gene delivery in animal systems and limited use in experimental gene therapy (Robbins, Tahara, and Ghivizzani, 1999; Ennist, 1999; Grubb et al., 1994). The advantages of this system include a broad host specificity and the ability to use the same expression vector to infect different host cells for contemporaneous animal studies (von Seggern and Nemerow, 1999). Commercial vectors are available for generating recombinant viruses such as the AdEasy™ system sold by Stratagene. This system simplifies the process of generating recombinant viruses since it relies on homologous recombination in *E. coli* rather than in eukaryotic cells (He et al., 1998). The main limitations of this system include moderate to low expression levels and the need to maintain a dedicated tissue culture space in order to avoid crosscontamination with other host cells. Other animal viruses of interest, including Sindbis, Semliki Forest virus, and the adeno-associated virus (AAV), share many of the same advantages as adenovirus, including broad host specificity (Schlesinger, 1993; Olkkonen et al., 1994; Bueler, 1999). None of these virus expression systems are discussed in detail in this chapter because they do not currently represent mainstream methods for large-scale protein production as is evident from the limitations discussed.

Stable Expression Systems

Stable expression systems are preferred when one desires a continuous source and high levels of expressed heterologous protein. The actual levels of expression largely depend on which host cells are used, what type of plasmids are used, and where the genes are integrated into the host genome (i.e., whether they are influenced by chromosomal position effects).

What are the cell line choices? If it is a mammalian system, the most common choices are as discussed next.

Mouse

Mouse cells such as L-cells (ATCC CCL 1), Ltk$^-$ cells (ATCC CCL 1.3), NIH 3T3 (ATCC CRL 1658), and the myeloma cell lines, Sp2/0 (ATCC CRL 1581), NSO (Bebbington et al., 1992) and P3X63.Ag8.653 (ATCC CRL 1580). These myeloma cell lines have the advantages of suspension growth in serum-free medium and their derivation from secretory cells makes them well-suited hosts for high-level protein production. Because of the presence of the endogenous dihydrofolate reductase (DHFR) gene, none of these cells can be amplified through the use of methotrexate (Schimke, 1988). However, as shown by Bebbington et al. (1992), NSO cells can be amplified using the glutamine synthetase system.

Rat

Rat cell lines, RBL (ATCC CRL 1378), derived from a basophillic leukemia, have been used to express 7TM G-protein-coupled receptors (Fitzgerald et al., 2000; Santini et al., 2000), while the myeloma cell line YB2/0 (ATCC CRL 1662), has been used in the high-level production of monoclonal antibodies (Shitara et al., 1994).

Human

Human cell lines that are frequently used include HEK 293, HeLa (ATCC CCL 2), HL-60 (ATCC CCL 240), and HT-1080 (ATCC CCL 121).

Hamster

Chinese hamster ovary (CHO) cells, such as CHO-K1 (ATCC CCL 61), and two different DHFR$^-$ cell lines DG44 (Urlaub et al., 1983) or DUK-B11 (Urlaub and Chasin, 1983) in which the gene of interest can be amplified via the selection/amplification marker DHFR (Kaufman, 1990). CHO cells have been used to express a large variety of proteins ranging from growth factors (Madisen et al., 1990; Ferrara et al., 1993), receptors (Deen et al., 1988; Newman-Tancredi, Wootton, and Strange, 1992), 7TM G-protein-coupled receptors (Ishii et al., 1997; Juarranz et al., 1999), to monoclonal antibodies (Trill, Shatzman, and Ganguly, 1995).

Also of significance are engineered derivatives of these lines. One example is a CHO cell line containing the adenovirus E1A gene. Cockett, Bebbington, and Yarronton (1991) first established a CHO cell line stably expressing the adenovirus E1A gene, which trans-activates the CMV promoter. Transfection of a human procollagenase gene into this CHO cell line produced a 13-fold increase in stable expression compared with that of CHO-K1. This is significant because an E1A host cell line can be used to rapidly produce sufficient material for early purification and testing without the need for amplification. Stably expressing clones produced from this host can be obtained in as little as two weeks and yield 10 to 20 mg/L of expressed protein.

Baby Hamster Kidney (BHK) Cells (ATCC CCL 10)

BHK cells have also been used to express a variety of genes (Wirth et al., 1988).

Drosophila

Drosophila S2 is a continuous cell line derived from primary cultures of late stage, 20 to 24 hours old, *D. melanogaster* (Oregan-R) embryos (Schneider, 1972). The cell line is particularly useful for the stable transfection of multiple tandem gene arrays without amplification. High copy number genes can be expressed in a tightly regulated fashion under the control of the copper-inducible *Drosophila* metallothionein promoter (Johansen et al., 1989). This cell line is particularly useful for the inducible expression of secreted proteins. S2 cells also grow well in serum-free, conditioned medium, simplifying the purification of expressed proteins.

Yeast Expression Systems (Pichia pastoris and Pichia methanolica)

The main advantages of yeast systems over higher eukaryotic tissue culture systems such as CHO include their rapid growth rate to high cell densities and a well-defined, inexpensive media. Main disadvantages include significant glycosylation differences of secreted proteins comprised of high mannose, hyperglycosylation consisting of much longer carboydrate chains than those found in higher eukaryotes, and the absence of secretory components for processing certain higher eukaryotic proteins (reviewed in Cregg, 1999). Because of these limitations, yeast systems will not be discussed in full detail in this chapter. More information on *Pichia* expression can be found in the following references: Higgins and Cregg (1998), Cregg, Vedvick, and Raschke (1993), and Sreekrishna et al. (1997).

We all have our preferences for what are the best cell lines to use. Therefore, when setting up an expression laboratory, one should consider obtaining a variety of host cell lines. Listed are a few examples of cell lines that have been routinely used and reasons for their selection: CHO-DG-44 and *Drosophila* S2 (available from Invitrogen), based on consistency in growth, high-level expression, and ability to be easily adapted to serum-free growth in suspension; COS for transient expression; HEK 293, a versatile human cell line which can be used for both transient (but not as good as COS) and stable expression; and Sf9 a host cell for baculovirus infection, a system best suited for internalized proteins rather than secreted proteins. A majority of these cell lines can be grown in serum-free suspension culture, a property that facilitates ease of use and product purification as well as reducing cost.

Selecting an Appropriate Expression Vector

Once an appropriate host system has been chosen, it's time to find a suitable expression vector. For each of the host systems described above, there are a wide variety of vectors to choose from.

A typical expression vector requires the following regulatory elements necessary for expression of your gene: a promoter, translational initiator codon, stop codon, a polyadenylation signal, a selectable marker, and several prokaryotic elements such as a bacterial antibiotic selection marker and an origin of replication for plasmid maintenance. (The presence of prokaryotic elements is for shuttling between mammalian and prokaryotic hosts.) There are numerous choices for each regulatory element, but unfortunately there is no blueprint on which combinations will yield the highest expressing plasmid.

Table 16.2 Promoter Strength Table

Promoter	Source	Strength	Reference
EF-1α	Human elongation factor 1α	40–160	Mizushima and Nagata (1990)
CMV	Human cytomegalovirus immediate-early gene	4	Boshart et al. (1985)
RSV	Rous sarcoma virus LTR	2	Gorman et al. (1982)
SV40 late	Simian virus 40 Late gene	1.1	Wenger, Moreau, and Nielsen (1994)
SV40 early	Simian virus 40 Early gene	1	
Adeno major late	Adenovirus major late promoter	0.4	Mansour, Grodzicker, and Tjian (1986)
Beta-globin	Mouse beta-globin promoter	0.2	Hamer, Kaehler, and Leder (1980)
Beta-actin	Human beta-actin promoter	ND	Ng et al. (1985)

Note: SV40 early promoter strength set as 1 for comparative purposes, and the numbers indicate how much stronger these promoters are.

Promoters

Promoters are DNA sequences that recruit cellular factors and RNA polymerase to activate transcription of a particular gene. They must contain a transcriptional start site, a CAAT box, and TATA box. Examples of various mammalian promoters are given in Table 16.2.

The promoter strength is based on a compilation of comparative experiments where various promoters were compared in transient experiments using the R1610 cell line (Thirion, Banville, and Noel, 1976). The strength of EF-1α and CMV was derived from a comparison to the RSV LTR involving stable expression of various monoclonal antibodies and tPA (Trill, 1998 unpublished). The EF-1α promoter (available from Invitrogen) is by far the strongest promoter and a good choice if you want quick high-level expression.

Polyadenylation Regions

Polyadenylation occurs at a consensus sequence, AAUAAA, and results in increased mRNA stability. Cleavage after the U by poly A polymerase adds a string of adenylate residues (Wahle and Keller, 1992). As with the promoters, there are a number of sources of polyadenylation regions. Several examples are shown in Table 16.3.

Table 16.3 Polyadenylation Regions

Poly A Region	Source	Efficiency
BGH	Bovine growth hormone	3
SV40 late	Simian virus 40	2
TK	Herpes simplex virus thymidine kinase	1.5
SV40 early	Simian virus 40	1
Hep B	Hepatitis B surface antigen	1

Note: SV40 early poly(A) region strength set as 1 for comparative purposes, and the numbers indicate how much more efficient these polyadenylation regions are. The data above and polyadenylation regions are referenced in Pfarr et al. (1985, 1986).

Drug Selection Markers

Choice number three: What drug selection markers should one use? These genes provide resistance to a particular selective drug, and only cells in which the plasmid has been integrated will survive selection. Some effective choices are Blasticidin (Izumi et al., 1991), Histidinol, (Hartman and Mulligan, 1988), Hygromycin B (Gritz and Davies, 1983), Geneticin® (G418) (Colbere-Garapin et al., 1981), Puromycin (de la Luna et al., 1988), mycophenolic acid (Mulligan and Berg, 1981), and Zeocin™ (Mulsant et al., 1988). Whatever marker you decide to use, remember, you will need to determine the effective concentration of drug for each cell line you use. Second, if you are on a tight budget, there is a huge disparity in cost of these drugs. Also there are environmental concerns regarding waste disposal of the conditioned growth medium containing some of these drugs.

Amplification

Finally, if expression is unacceptably low, one solution is to amplify your gene copy number. Two such amplification systems are the use of dihydrofolate reductase (DHFR) as a drug selection marker in the presence of methotrexate, a competitive inhibitor of DHFR (Kaufman, 1990) and inhibition of the enzyme glutamine synthetase (GS) by methionine sulfoxide (MSX) (Bebbington et al., 1992).

Amplification through the DHFR gene is by far the more popular of the two systems. DHFR catalyzes the conversion of folate to tetrahydrofolate, which is necessary in the synthesis of glycine, thymidine monophosphate, and the biosynthesis of purines. If the transfected plasmid contains a DHFR gene, use of the CHO DG-44 and DUK-B11 cell lines allows one to initially select cells in medium devoid of nucleotides and then to amplify gene copy number by selection with increasing concentrations of methotrexate (Geisse et al., 1996). In the majority of the cases, amplification of the gene copy number results in increased expression.

The glutamine synthetase system can be used as a dominant selectable marker in cell lines that contain GS activity, in glutamine-free growth medium. GS catalyzes the formation of glutamine from glutamate and ammonia. CHO-K1 and NSO are the more widely used cell lines for this method of selection, but myeloma cells offer a distinct advantage over CHO cells because of their low levels of endogenous GS activity. Myeloma cells transfected with a plasmid containing a gene of interest and the GS gene are often selected with low levels of MSX (up to 100 μM), while CHO cells are amplified using higher levels of MSX (up to 1 mM) (Bebbington et al., 1992; Cockett, Bebbington, and Yarronton, 1990).

Regulating Expression

What happens if overexpression of a gene results in a protein which is toxic to the host cell? There are a number of inducible promoters and regulated expression systems available that allow one the ability to control when and how much of the toxic protein is produced. Examples of such promoter-based systems include the Mouse mammary tumor virus (MMTV) promoter which is induced using dexamethasone (James et al., 2000), the *Drosophila* metallothionein promoter which is induced by addition of metal (e.g., cupric sulfate; Johansen et al., 1989), or the mifepristone-dependent plasmid-based gene switch system (Wang et al., 1994). The addition of inducers allows flexible control of expression in these systems. However, inducers such as heavy metals can also interfere with purification efforts, especially if your protein contains an epitope tag. For example, the use of the standard IMAC (immobilized metal affinity chromatography) method for the direct capture of His-tagged proteins from *Drosophila* culture medium is inefficient due to the presence of free copper, which interferes with binding. However, we recently found that when copper-supplemented medium containing an expressed His-tagged protein is loaded directly onto chelating sepharose, the protein binds efficiently to the resin via copper (Lehr et al., 2000). Furthermore this interaction is of greater affinity than that of free copper alone, which can be washed away under low-salt conditions.

Other methods for achieving regulated expression include the Ecdysone-inducible system, based on the heterodimeric ecdysone receptor of *Drosophila* (Christopherson et al., 1992), and the tetracycline-regulated expression system, based on two regulatory elements derived from the tetracycline-resistance operon of the *E. coli* Tn10 transposon (Gossen and Bujard, 1992).

Single- or Double-Vector Systems?

What type of vector system will we use to house all of these regulatory elements? We can use a two-vector system in which the gene is contained on one plasmid while the selection marker is on the second. *Drosophila S2* cells are an example of a host where a two-vector system is preferable. In this case, varying the proportions of the two plasmids enables one to modulate the number of gene copies inserted onto the chromosome from just a few to more than a thousand (Johansen et al., 1989). Higher gene copies tend to correlate with higher expression levels. Thus the two-vector system can add to the flexibility of the expression outcome. Double-vector systems are also used for mAb expression where the heavy chain and the DHFR gene are located on one vector, and the light chain and the selectable marker are located on the second vector (Trill, Shatzman, and Ganguly, 1995). Alternatively, one could also use a single plasmid where the drug selection cassette and the amplification gene are located on the same plasmid (Aiyar et al., 1994). Again, using mAbs as an example, we can use a single plasmid that contains both the heavy and light chain cDNAs along with the selection and amplifiable drug markers (Trill, Shatzman, and Ganguly, 1995).

Which vector system should you use? This really depends on how much effort you want to expend in your plasmid cloning and transfection and how quickly you need your protein. With two plasmids, it means two separate clonings and two plasmids to sequence. You will also need to co-transfect both plasmids in a ratio that will favor optimal expression. This ratio may need to be empirically determined. A single plasmid, containing two different genes of interest necessitates a unique cloning strategy due to the decrease in unique restriction sites for the cloning process. It also means designing gene-specific bi-directional sequencing primers because of the duplication of regulatory elements.

Summary

There are a large assortment of commercially available mammalian and insect expression vectors to choose from. The majority of the mammalian vectors have common regulatory elements. Most use the CMV promoter to drive expression, contain a polylinker region to clone in your gene of interest, and use a drug selection marker, most often Neomycin. One of the most popular is pCDNA3.1 sold by Invitrogen. Variations in these vectors include different choices of epitope TAGS for detection using an antibody or through the intrinsic fluorescence of the green fluorescence protein (GFP) and its derivatives. There are also bicistronic vectors that use a single expression cassette containing both the gene of interest and selection marker, separated by an internal ribosome entry site (IRES) from the encephalomyocarditis virus, to promote translation from a bicistronic transcript. In addition there are vectors containing signal sequences designed to aid secretion. Finally, to circumvent the need to develop multiple vectors for each system you use, you can obtain a single expression vector enabling protein expression in bacterial, insect, and mammalian cells from a single plasmid, such as the pTriEx expression vector marketed by Novagen.

It is advisable that one take the time to find a vector that is optimized for a particular host or, if one is not available, to construct a new vector and optimize it for each system that you intend to use. Take the time to create your own polylinker region with convenient, unique restriction sites so that you can easily exchange regulatory elements. CMV is perhaps one of the most versatile promoters available. You will also need to incorporate a resistance marker under the control of its own promoter and including a polyadenylation site. The choice of a selection marker will depend on considerations such as the cost of the drug, the efficiency of its action in a particular host, and environmental concerns for disposal.

IMPLEMENTING THE EUKARYOTIC EXPRESSION EXPERIMENT

Media Requirements, Gene Transfer, and Selection

Stable cell line generation, especially for a therapeutic protein, is a long, labor-intensive process that takes anywhere from six to nine months to complete. Therefore it is essential that one pay close attention to the methods employed to maintain, transfect, and select the cell lines.

Serum

When possible, try to adapt your cells for growth in a chemically defined, serum-free growth medium. Serum contains numerous undefined components, is costly to use, may contain adventitious agents, and varies from lot to lot. Serum-free medium, available from a number of suppliers, offers several advantages. It allows cell culture to be performed with a defined set of conditions leading to a more consistent performance, possible increases in growth, increased productivity, and easier purification and downstream processing.

If you must use a serum-containing medium, be sure to have the serum lot tested for mycoplasma and other adventitious agents, such as BVDV (bovine viral diarrhea virus). If possible, order gamma-irradiated serum and ask for a certificate of analysis. With the recent concern over bovine spongiform encephalitis (BSE) disease in cattle from the United Kingdom, it is also wise to request serum from regions where BSE is not present (e.g., United States and New Zealand). This extra precaution further adds to the high cost of serum.

Antibiotics

Many researchers supplement their growth media with antibiotics such as penicillin, streptomycin, and antifungals such as amphotericin B. While this is effective in preventing either bacterial or fungal contamination, it does nothing to prevent contamination from mycoplasma or viruses. Furthermore antibiotics can mask poor cell culture sterile technique, lead to drug-resistant bacteria, and increase the risk of mycoplasma contamination. In short, there is no substitute for proper sterile technique, which should eliminate the need to add antibiotics in the first place.

On the subject of sterility, it is also prudent to have your cell lines tested monthly for mycoplasma. Trypsin, for use in removing attached cell lines, should be free of mycoplasma, PPV (porcine parvovirus), and PRRS (porcine respiratory and reproductive syndrome virus). If your medium will be used to support growth of production cell lines expressing therapeutic agents, it is also advisable to consult the FDA guidelines for the use of medium containing animal products.

If one of your cultures should become contaminated with mycoplasma, the best cure is to dispose of the cell lines in question. If this is not an alternative, there are a number of reports indicating that mycoplasma has been eradicated through the use of MRA (mycoplasma removal agent (ICN), a quinolone derivative) (Uphoff, Gignac, and Drexler, 1992; Gignac et al., 1992), either ciprofloxacin (Gignac et al., 1991; Schmitt et al., 1988) or enrofloxacin (Fleckenstein, Uphoff, and Drexler, 1994) both of which are fluoroquinolone antibiotics and BM-cyclin (Roche

Molecular Biochemicals, a combination of tiamulin and minocycline) (Uphoff, Gignac, and Drexler, 1992). However, this is a time-consuming, cost-intensive process that may result in irreversible damage to your cell cultures.

Transfection

The most contemporary methods for transfection of foreign genes into cells employ either cationic lipid reagents or electroporation (Potter, Weir, and Leder, 1984). The former relies on different liposome formulations of cationic or polycationic lipids (as per the manufacturer) that complex with DNA facilitating its uptake into cells. The procedure is simple, very rapid, and can be used for a large variety of cell types. It is the method of choice for transient transfections, especially into COS cells, and is by far the most preferred method for transfecting attached cell lines.

Electroporation, which relies on an electric pulse to reversibly permeabilize the cell's plasma membrane, creates transient pores on the surface of the cell that allow plasmid DNA to enter. This technique is also very rapid, and the protocols are straightforward and can be used in a variety of cell types. Electroporation can be used on suspension cell lines and attached cells, which have been detached from the plate. Electroporation is most efficient when the DNA is linearized prior to transfection (Trill, unpublished). It also offers the unique advantage that a majority of the DNA is integrated in single copies at single sites without any rearrangements (Boggs et al., 1986; Toneguzzo et al., 1988). This is significant when assessing stability and chromosomal location of the gene within the cells and the expressed protein.

Clonal or Polyclonal Selection?

There are advantages and disadvantages to selecting cells as bulk populations over their selection as clones through limit dilution, colony formation, or fluorescence-activated cell sorting (FACS). On the one hand, polyclonal lines can be derived much more quickly than clonal lines, and a reasonable expression level can be achieved in many cases. On the other, there are also many inherent problems with this method. For example, expression levels tend to be diluted by a population of nonproducers within the selected population. These cells contain the transfected plasmid and an intact, fully functioning drug selection gene, but have somehow lost expression from the gene of interest. Within such populations, the risk is great that nonproducers will eventually overgrow the producers, further diluting expression levels.

This problem is compounded by the tendency of overexpressing cells to grow slower than low or nonexpressors.

In general, it is preferable to select clones rather than polyclonal populations in order to achieve the highest reproducible expression. However, the isolation of clonal lines is considerably more time-consuming and labor-intensive. In addition you will need to evaluate expression from tens to hundreds of clonal cell lines rather than a polyclonal population from a single flask.

Whatever selection method you should choose, you will need to do some type of experimentation to assess such cell line characteristics as growth, viability, and protein expression.

Scale-up and Harvest

The final task prior to purification of the recombinant protein is to convert your cell culture into a "factory" for the production of the desired recombinant protein. Again, the type of system that you employ depends largely on the intended use of the protein and how much will be needed. Other deciding factors include cost and complexity of use. Benchtop fermentation systems can be purchased from a number of companies, and each system has its own distinct pro's and con's.

The following systems are restricted to volumes of one liter or less of culture due to limitations in O_2 transfer. These include the following:

- Attachment cell culture using T-flasks, roller bottles, and other carriers such as Cytodex™ (Amersham Pharmacia Biotech), CultiSpher® (HyClone), and Fibra-Cel® disks (New Brunswick Scientific).
- Spinner flasks for use with a stir plate apparatus. One can use suspension cell lines or attached cells grown on carrier surfaces.
- Shake flasks in systems that range from individual platforms placed into incubators to self-contained chambered shakers allowing independent control of temperature and CO_2 gassing. Shake flask systems are mainly used for growth of suspension cell lines.

Medium volumes, more complex than above, include the following:

- CellCube® (marketed by Corning) is a closed loop perfusion system for the culture of attachment-dependent cell lines.
- Wave Bioreactor (*www.wavebiotech.com*) consisting of a fixed rocker base and a disposable plastic Cellbag. This system can

be used in volumes from 100 ml to 10 L for both suspension cells and cells on carriers.

Ideal for larger volumes (ranging from 1 to 10,000 L), although more complex and costly, are the following:

- Stirred tank bioreactors come in all shapes and sizes. They have modular designs, can be upgraded and are versatile, allowing one to control dissolved O_2, airflow, temperature, impeller type and speed, pH, nutrient addition, and vessel size. One can also perform two-compartment fermentation through the use of dialysis membranes separating cells from the medium. You can vary the mode of culture using either a fed-batch or perfusion process to maximize protein expression. These bioreactors are best suited for growing suspension cell cultures. However, a fibrous-bed of polyester disks may be employed as a matrix for high-density growth of cells immobilized on the disks for use in the stirred tank bioreactors.
- Hollow fiber bioreactors are composed of a matrix of hollow fibers that separate the bulk of the culture medium from the cell mass by means of hollow-fiber walls, allowing production of high-density cultures of viable cells in the extracapillary space. Cells are nourished by nutrients circulating in the ICS (intracapillary space) medium that readily diffusc across the hollow-fiber membrane. This system is ideal for production of secreted proteins, specifically monoclonal antibodies, and can be used for both suspension- and attachment-dependent cells.

Gene Expression Analysis

Following gene transfer, the time has come to determine how successful your expression efforts have been. This is done by analysis of either cells or cell lysates in the case of intracellular or membrane proteins, or conditioned medium in the case of secreted proteins. It is presumed at this point that you have specific detection reagents for the expressed protein, that the protein is tagged for detection, or that there is a specific functional assay in place for detecting the protein's biological activity. If the expressed protein is fairly well characterized, there are likely to be commercial antibodies for Western blot analysis and/or enzyme-linked immunosorbant assay (ELISA) detection.

The Pro's and Con's of Tags

If the expressed protein is not well characterized or completely novel, then it is useful to have an epitope tag (e.g., FLAG, HA,

His_6, c-myc as described above) fused to the expressed protein. This will enable detection of protein expression in the absence of specific reagents and will aid in purification. Various tag detection reagents are commercially available through various vendors. In the case of receptors, tagging can be particularly useful when trying to determine if the receptor is expressed onto the cell surface. For example, HA (hemagglutinin) tagging has been used to detect cell surface staining of 7TM receptors (Koller et al., 1997). In our experience we have relied extensively upon the use of immunoglobulin Fc fusions as a reporter to monitor expression. Fc fusions are easy to detect both by ELISA and Western blotting using commercially available reagents. Recently we have employed the Origen technology (IGEN, *www.igen.com*) based electrochemiluminescence detection method (Yang et al., 1994) which we have adapted for the direct detection of Fc expression from individual colonies (Trill, 2001). Following expression of Fc fusions, one can often utilize Fc fusion proteins directly in screens. Alternatively, a protease cleavage site can be engineered for removal of the Fc following purification.

The expression of novel or uncharacterized proteins requires special consideration for detection. On the one hand, there is likely to be very little known about what regions of the protein are important for function. Thus one would ideally like not to have additional residues such as tags, which could potentially interfere with folding (e.g., activity) or expression. However, since there are usually no specific detection reagents or functional assays available, it is often necessary to add a tag anyway in order to detect and purify the protein. Alternatively, one could consider the production of antibodies raised to antigenically pronounced regions. Certain vendors will do both the peptide synthesis and immunization. However, this will take several weeks to months, and there is no guarantee that high titer or neutralizing antibodies will be obtained. Since turnaround time is usually a critical parameter for expression projects, most researchers will take the chance of adding an epitope tag for initial expression. At the same time, if cost and resource are not prohibitive, it is also safest to express both tagged and untagged versions and to prepare peptide antiserum in the process.

Most commercial expression vectors contain modular regions for the optional incorporation of tags. This is a convenient way to fuse tags to an expressed protein. However, the options for tag fusions in commercial vectors are frequently limited to C-terminal tags, which are more prone to clipping through the action of carboxy peptidases in the cell. Furthermore the fusions in most commercial vectors also include a significant number of extraneous residues derived from the polylinker region that could affect protein folding and activity. Thus, while the one-size-fits-all commercial vectors are generally suitable for initial expression, you will probably want to design more precise fusion constructs with either N-terminal or C-terminal tags.

Tags can provide other benefits for expression. For example, some researchers have found that the use of large soluble tags such as GST can enhance the solubility of certain proteins, which favors the production of active protein (Davies, Jowett, and Jones, 1993; Weiss et al., 1995; Ciccaglione et al., 1998). We have also observed that the additional C-terminal immunoglobulin Fc fusions sometimes result in the enhanced production of certain proteins secreted from CHO cells. Thus the addition of tags for detection and purification remains an empirical process, as does the choice of a system in which to express the protein.

Functional Assays

In many cases the most efficient way to screen for expression is not through direct detection of the protein itself but through some kind of functional assay for the expressed protein's biological activity (e.g., apoptotic, chemotactic, proliferative, or enzymatic). Crude cellular extracts or conditioned medium containing secreted proteins can sometimes be directly screened for biological activity. Functional assays are particularly useful when screening for the expression of a receptor whose ligand is known. In this case, clones can be directly screened for cellular responses to added ligand. Calcium-mobilization and cAMP assays are two of the most commonly used methods of detecting signal transduction through G-coupled-protein receptors.

TROUBLESHOOTING

Finally, after weeks or even months of selection, you have isolated clonal cell lines that should be expressing large quantities of protein. However, Western, ELISA, or functional assays are performed, and they show that little or even no protein is being expressed. What can and should you do now? There are many possible explanations for why you fail to detect a protein.

Confirm Sequence and Vector Design

The first thing to do if you haven't already done so is to double-check the original design of the expression vector and the confirmed sequences. In some cases an overlooked point mutation or mistake in the original design is the problem. Ideally such problems are best uncovered before weeks of work have been devoted to selecting lines. It cannot be overstated that one should make every effort to check and double-check sequences and vector designs in order to ensure that this never happens. It is also a good idea to first confirm expression by performing transient assays (e.g., in COS cells).

Once you have ruled out problems with the expression construct, there are a few obvious places where problems may be occurring. First, one could perform Northern blot analysis or RT-PCR to determine if any message is produced. The second possibility is that the protein tag is being proteolytically removed. Many C-terminal tags are prone to clipping as mentioned above. If clipping is the problem and there is no other way of detection, it will be difficult to prove that your protein is being expressed and even more difficult to purify it. However, you might be lucky and find that the expression levels are high enough to enable detection through direct staining of SDS-PAGE gels either with Coomassie Blue or silver stain. If direct detection is ambiguous, then you will either have to wait for specific peptide antibodies to detect the untagged protein or have to modify the expression to limit proteolytic digestion (e.g., removal of arginine-serine rich sequences that may be the target of proteolysis). Baculovirus, being a lytic system, is particularly prone to protease problems. In some cases researchers have even resorted to adding protease inhibitors directly to the infection in order to inhibit proteolysis as it is occurring (Pyle et al., 1995). This is not highly recommended, however, since the protease inhibitors also tend to inhibit cellular and viral functions. On the other hand, the addition of protease inhibitors upon harvest and lysis is imperative in order to prevent such proteolysis during purification.

Secreted proteins present their own particular set of issues related to processing and trafficking the protein out of the cell. If a protein is not naturally secreted to high levels, one may find that the native signal peptide sequence does not guide efficient secretion into the ER (endoplasmic reticulum). In these cases one may consider replacing the native signal sequence for a known efficient signal peptide sequence. For the *Drosophila* S2 system, we have utilized a signal sequence derived from chaperone protein HSC3 (*Drosophila* BIP) (Rubin et al., 1993). This sequence has been adapted into commercial vectors sold by Invitrogen.

Investigate Alternate Hosts

The choice of an expression host is often a critical parameter for efficient expression, but it is not usually possible to predict which system will work for a particular protein. Certain hosts may contain the necessary processing machinery while others do not. Thus it is often worthwhile to switch to a new system if expression is not initially detected. One can learn a great deal by performing transient expression assays in different hosts to narrow the field of compatible host systems. This is best done first, before all the time and effort is expended in the selection of stable cell lines.

Finally, one of the most difficult problems that you can face is expression of an inactive protein. This is particularly troublesome when expression levels are good and the protein appears fully soluble. In many cases the protein requires additional processing that is not supplied by the host cell. Alternatively, the host cell may lack a particular cofactor or signaling component that is necessary to establish activity. For example, G-coupled-protein receptors signal through specific G-proteins, interacting directly through one of several different G alpha subunits. The absence of a specific G-protein subunit could impair receptor function when expressed in certain hosts. Fortunately this specific problem can be ameliorated by co-transfection with one of several promiscuous G-protein subunits that will couple functionally with a broader range of receptors (Offermanns and Simon, 1995). However, not all cofactors are quite so well characterized to enable their supplementation. In most cases, if the cofactor is not endogenous to that host, then expression of active protein will not be directly possible. Again, exploring a number of different cellular hosts will often be the best approach to achieving the desired product.

A Case Study of an Expressed Protein from cDNA to Harvest

It is easy to explain how one goes about expressing a particular gene of interest, but how does this relate to real laboratory situations? The following example of a gene, which we will call ABCD, may help illustrate this.

Information concerning the gene has been published, and its sequence is also contained in the GenBank database. The gene contains 349 amino acids, including the signal peptide. Northern blot analysis indicates that the gene is highly expressed in the vascular endothelium. It is a secreted, cysteine-rich, glycosylated protein that has both chemoattractant and mitogenic activity. We wish to use this protein as a reagent in screening assays, as a comparison to a homologue we previously expressed. We do not have the gene for this protein, but analysis of in-house cDNA libraries indicates that the gene is present in one of our clones.

The homologue, designated ABCD-Like, was originally expressed in a baculovirus expression system using a N-terminal His_6 epitope tag, separated from the gene of interest with a factor Xa cleavage site. This strategy was based on published reports of similar proteins. Expression levels were very low, which led to purification problems. We were then forced to consider alternative strategies. ABCD-Like was recloned into a mammalian expression vector as a C-terminal Fc fusion protein and expressed in the CHO-DG-44 cell line, which had been adapted for suspension growth in a serum-free medium. We were able to express ABCD-Like at very high levels.

Let's revisit our original three questions, and determine what steps we need to take. We know what the gene is, we know where to find it, and we know a number of facts about ABCD, including its intended use. Finally, we have an idea of what expression system to use based on previous work with the homologue ABCD-Like.

Using the sequence we located in GenBank, a PCR primer is designed to trim the 5′ end of the gene and add a unique restriction site. To the 3′ end of the gene, sequences encoding a Factor Xa cleavage site and a unique restriction site are likewise introduced. The generated PCR fragment could be cloned into our pCDN/Fc vector (Aiyar et al., 1994) as an Fc fusion protein. Upon positive sequencing results, the resulting plasmid, pCDN-ABCD/Fc is linearized and electroporated into our CHO cell line and selected for resistance to maintenance medium without nucleosides, since our plasmid contains the mouse dhfr gene. The colonies that arise are assayed using a Fc sandwich assay with an Origen analyzer, and the high expressors are expanded. A single clone is eventually scaled up into flottles (a cross between a flask and a bottle). A flottle is often referred to as a modified Fernbach Flask, and is available from Corning. The clone is grown for 13 days to produce enough medium for purification and testing. N-terminal sequence analysis of the purified protein revealed the correct mature protein sequence, indicating that processing had occurred. Western blot analysis revealed the presence of two smaller bands. N-terminal analysis of these bands indicated that the protein was cleaved several amino acids before the N-terminus of the Fc region.

The entire process, from inception to purification, took less than three months to complete. There still was the problem of determining how to eliminate the extraneous cleavage products. Analysis of the amino acid sequence revealed a "possible" arginine-rich protease cleavage site. Site directed mutagenesis was performed to eliminate the suspect amino acids. Subsequent re-expression in CHO, using the aforementioned techniques, demonstrated that the correct uncleaved protein was obtained.

SUMMARY

The expression of recombinant proteins in eukaryotic systems represents an important technological advance in the study of the biological function of proteins. This technology enables the isolation of authentic, post-translationally modified proteins in large quantities without having to purify them from a native source. In pharmaceutical research and development, recombinant proteins are used to supply high-throughput drug screens, functional studies, structural biology, and therapeutic agents. In this chapter we have discussed the process by which one goes about finding a gene for expression of a protein, choosing an appropriate expression host, choosing an appropriate vector for that host, cloning the gene into the vector, transfection of the recombinant vector into the host, isolating cells that are expressing the protein, and scaling protein expression for purification. We have also discussed several possible pitfalls commonly encountered and suggestions on how best to fix these problems. The practical considerations on these topics discussed in this chapter are intended to help guide one through the vast array of possible expression systems that one has to choose from including many commercial systems that bring recombinant protein expression technology to virtually anyone who wants to use it.

SECTION B: WORKING WITH BACULOVIRUS

PLANNING THE BACULOVIRUS EXPERIMENT

Is an Insect Cell System Suitable for the Expression of Your Protein?

The first choice for recombinant overexpression of a plain vanilla cytoplasmic protein is nearly always *E. coli*. For many of the remaining proteins that are membrane bound, covalently modified, secreted, or components of multiprotein complexes, expression in eukaryotic cells is the system of choice. Expression in cells from higher organisms can also be a solution for proteins that, when expressed in bacteria, are insoluble or are expressed as truncated products due to proteolysis, premature translational termination, or the presence of rare codons (Pikaart and Felsenfeld, 1996). In some instances it may be useful to express soluble protein from a eukaryotic source as a "gold standard" to compare with refolded protein from a bacterial source.

The most commonly used eukaryotic cells for recombinant protein expression are derived from mammalian or insect tissues and utilize either viral- or plasmid-based vehicles to transduce your gene of interest. This section will address using baculovirus infection of insect cells as a way to provide modest levels (1–10 mg/L) of proteins in a reasonably quick time frame (7–10 days).

Although recombinant baculoviruses are most often used to infect cultured insect cells and caterpillars, a more recent development has been their use as transfer vectors for mammalian cells (Condreay, 1999; Kost and Condreay, 1999). Several types of mammalian cells are capable of baculovirus uptake and transient expression of recombinant genes, but are incapable of producing progeny virus. This technique has proved particularly valuable for introducing genes into cells that are notoriously difficult to transfect using more traditional methods. It is likely that recombinant baculoviruses incorporating a more specific uptake mechanism by an established receptor-ligand pair will make this approach more common in the future.

Should You Express Your Protein in an Insect Cell Line or Recombinant Baculovirus?

Insect cell expression is relegated to the creation of a cell line or to a lytic infection with recombinant baculovirus infected cells. General descriptions for the creation of stable insect cell lines are given by McCarroll and King (1997), Ivey-Hoyle (1991), and Benting et al. (2000). Invitrogen and Novagen sell reagents to produce such lines and provide detailed manuals available on their Web sites. The most important differences in the two approaches lies in the level of attention needed to maintain the various cell lines, in the elapsed time before it is possible to evaluate expression of a given construct, and in the relative ease of expressing multiprotein complexes (Table 16.4). There are several instances where expression in a baculovirus system makes sense as a first choice. Since insect cells can be infected with multiple different baculoviruses, each expressing an individual protein, this system requires no additional time to analyze the expression of

Table 16.4 Comparison of Protein Expression Systems

	Insect Cell Line	Baculovirus	*E. coli*
Nature of expression	Inducible or constitutive cell line	Lytic viral infection	Inducible
Modifications • Glycosylation • Myristylation • Palmitoylation • Sulfation • Isoprenylation • GPI linker addition	System of choice, since cells are not dying at the time of highest expression	Modifications may not occur efficiently, since cells are dying at the time of highest expression	Not present
Codon preference	Bias against certain codons in *Drosophila*	Very little codon bias	Bias against certain codons
Expression of >1 protein	Time-consuming	Straightforward	Can be done
Ease of cell culture	Easy	Cells need careful attention; plaque assay must be mastered	Easy
How soon after making my expression plasmid will I have 1 L of cells to examine?	3–4 weeks after transfection	7–10 days after transfection	1 day after transformation
Storage	Frozen cells	Virus at 4°C or Frozen stocks	Frozen cells or plasmid DNA

multiple-protein complexes. A comparable insect cell line may require months for the sequential isolation of clonal cell lines that express more than one protein. Baculoviral genes show little evidence of codon bias (Ayres et al., 1994; Levin and Whittome, 2000), and expression in this system may be preferable with genes that contain numerous rare codons for *Drosophila*. If a gene has suspected cellular toxicity, a cell line may be unattainable, making baculovirus a more suitable expression choice. Perhaps the most common reason for using the baculovirus approach is the rapidity with which recombinant protein can be obtained. For the expression of a soluble cytoplasmic protein, it is possible to obtain protein from as much as a few liters of baculovirus infected cells within 7 to 10 days from the initial transfection. The analysis of a comparable amount of cells from a cell line would take 3 to 4 weeks from the initial transfection.

Expression from an insect cell line is preferable for proteins that are secreted or require a modification such as glycosylation or acylation. Most protein expression from baculovirus late or very late promoters occurs just prior to cell lysis, and as a result the cellular machinery for protein export or modification may be compromised.

Procedures for Preparing Recombinant Baculovirus

This chapter will not discuss the common protocols available for baculovirus expression. References that contain good protocols for cell culture and handling of virus are King and Possee (1992), O'Reilly, Miller, and Luckow (1992), and Murphy et al. (1997). Additionally manuals for cell culture and baculovirus expression can be obtained from the Web sites at Invitrogen, Novagen, Clontech, and Life Technologies. Miller (1997) has details of baculovirus biology.

Criteria for Selecting a Transfer Vector

Epitope Tags

Baculoviruses are most easily formed by homologous recombination between viral DNA containing a lethal deletion and a transfer vector plasmid containing the gene of interest flanked by viral sequences. There are dozens of baculovirus transfer vectors commercially available, and manufacturers are coming up with new ones all the time. Good sources of vectors are Novagen, Pharmingen, Clontech and Invitrogen; check their Web sites for new ones that are not described in the catalogs. Commercial vectors often include sequences for "tags" that are useful for monitoring protein expression by immunoblot analysis. If the protein needs to be purified, the inclusion of an epitope tag that can be bound to an affinity resin (e.g., anti-Flag antibody resin for the Flag® epitope or a metal chelate resin for His_6 tagged proteins) will minimize the processing steps needed to obtain homogeneous recombinant protein.

Choice of Promoter

Most proteins are expressed from transfer vectors containing the very strong p10 or polyhedrin promoters that are most active very late (20–72 hours postinfection). Since expression from these promoters occurs at a time when such modifications as glycosylation are compromised because of the cytopathic effects of the viral infection, modified proteins are best expressed using the moderately strong basic protein or 39K promoters that are active at slightly earlier times (12–24 hours postinfection) (Hill-Perkins and Possee, 1990; Murphy et al., 1990; Jarvis and Summers, 1989; Sridhar et al., 1993; Pajot-Augy et al., 1999).

Cloning Strategy

An alternative approach to obtaining a recombinant baculovirus is available from Life Technologies (Bac-To-Bac™). Instead of recombination occurring in the insect cell, the recombinant viral DNA is recovered from *E. coli* and subsequently transfected into insect cells (Luckow et al., 1993). A disadvantage of this system is that it requires the manufacturer's limited set of transfer vectors. In addition there are less commonly used procedures for making the viral recombinant DNA in vitro (Ernst, Grabherr, and Katinger, 1994; Peakman, Harris, and Gewart, 1992) or in yeast (Patel, Nasmyth, and Jones, 1992). The ability to make recombinants in vitro is essential for creating baculovirus expression libraries, and the in vitro procedure may be required for the expression of proteins that are toxic to insect cells.

Control Elements

Although insect cells have the ability to splice RNA, often just the open reading frame with a minimal amount of untranslated flanking regions is inserted into the transfer vector. It is probably better to utilize baculoviral polyadenylation sequences (often present on the transfer vector) rather than substituting one such as the SV40 terminator (van Oers et al., 1999). Upstream sequences do have an influence on the rates of RNA transcription and/or protein translation, but no pattern has yet emerged (Luckow and Summers, 1988). There is limited evidence for a consensus base context around the initiating ATG (AAAATGA: Ranjan and Hasnain, 1994; Ayers et al., 1994), although experiments with transfected cells suggest a preference for A or T immediately downstream of the initiation codon (Chang, Kuzion, and Blissard, 1999). This apparent lack of a highly preferred initiation sequence ("Kozak" sequence) makes it possible to transplant inserts from bacterial expression vectors directly into a baculovirus transfer vector. As an added benefit, the presence of bacterial sequences upstream of open reading frames may enhance baculovirus expression of the gene of interest (Peakman et al., 1992).

Which Insect Cell Host Is Most Appropriate for Your Situation?

Three cell lines are commonly used for baculovirus expression; Table 16.5 illustrates differences among them. Sf21 cells are ovarian cells derived from *Spodoptera frugiperda* (fall army worm) and Sf9 are a subclone of Sf21. *T. ni* (available as High Five™) cells are from *Trichoplusia ni* egg cell homogenates. For

Table 16.5 Commonly Used Cell Lines for Baculovirus Expression

	Sf9	Sf21	*T. ni* (High Five™)
Initial transfection	✓	✓	
Plaque assay	✓		
Expression of secreted proteins			✓
Expression of cytoplasmic proteins	✓	✓	✓
Adaptation to suspension culture	Easy	Easy	Challenging due to clumping
Media	Serum-containing and serum- or protein-free preparations	Same as Sf9	Same as Sf9 and some made specifically for these cells

initial transfections, Sf9 or Sf21 cells are best because they produce large amounts of virus. Plaque assays are best done with Sf9 or Sf21 cells for the same reason. Sf9 cells are preferred for plaque assays since the plaques on these cells have sharply defined edges with clearer centers compared to plaques on Sf21 monolayers. For expression, *T. ni* often produces more protein than the other two lines, but due to its adherent and clumping habit, it is more difficult to adapt to suspension culture (Saarinen et al., 1999). It is generally best to have two insect cell lines growing—either Sf9 or Sf21—and cells from *T. ni*.

It can't be stressed enough that success with the baculovirus system depends on healthy cells and that careful attention to providing optimal growth conditions will avoid many common expression problems. Insect cells are grown at 27 to 28°C in a non-CO_2 incubator. They can be grown at room temperature on the benchtop, but because of possible unanticipated temperature fluctuations, an incubator is preferred. The best temperature control requires an incubator equipped with cooling capability.

Unfortunately, these cells have a narrow range of densities at which they will grow—between around 1×10^6 and 4×10^6/ml (Sf9 and Sf21 in serum-containing media) or slightly lower densities for *T. ni* cells. Slightly higher densities can be obtained in serum-free media. Cells will cease growing if diluted too much, and they will begin to die if allowed to remain at the higher densities for more than a day or two. With a doubling time of around 24 hours, this means they must be split every two to three days. The cells are generally passaged continuously until there is a noticeable increase in the doubling time, or a decreased sensitivity to viral infection. This seems to occur when cells are grown in serum-free media after about 30 to 50 passages. At that time a fresh culture is started from frozen cells.

IMPLEMENTING THE BACULOVIRUS EXPERIMENT

What's the Best Approach to Scale-Up?

Cells are initially grown in 75 or 150 cm^2 flasks and transferred to suspension culture in spinner flasks or shake flasks for scale-up. An advantage of suspension culture is that cells are subjected to less handling, so they will attain higher densities than in stationary flasks. The cell volume should not exceed 50% of the flask volume as the oxygen demand increases greatly after infection. There are several types of media that can be used, including serum-free preparations and formulations specifically made for *T. ni* cells. Cells grown in the presence of serum may require a weaning period before being adapted for growth in serum-free media. For the expression of cytoplasmic proteins, all types of media will give adequate expression levels, but for secreted proteins a low-protein or serum-free preparation may be preferred. An important consideration for secreted proteins is that serum-free media often contains Pluronic (a detergent). If a downstream purification step requires a media concentration step, pluronic micelles will be concentrated as well, and this may affect subsequent chromatography efforts.

If large quantities of protein are required, it is worth comparing protein expression with a selection of both serum-supplemented and serum-free media preparations as part of an optimization effort. Unfortunately, no one media preparation seems to be optimal for all proteins. Many manufacturers of serum-free media occasionally have not been able to meet consumer demand at one time or another, so it is worth identifying an alternate commercial source for an acceptable serum-free preparation.

Virus stocks should be prepared in serum containing media or serum-supplemented serum-free media. The presence of pluronic in the growth medium may result in decreased virus production (Palomares, González, and Ramirez, 2000).

What Special Considerations Are There for Expressing Secreted Proteins?

In general, the levels of secreted proteins from baculovirus infected cells are low (less than 10 mg/L), but there are examples of proteins that are secreted at levels greater than 100 mg/L (Mroczkowski et al., 1994; George et al., 1997). Secreted proteins require a signal sequence for export to the media; commercial vectors (available from Stratagene, Pharmingen, and Novagen) that provide a signal sequence or the native signal sequence can be used. A bacterial signal peptide will also direct secretion of eukaroytic proteins in insect cells (Allet et al., 1997). It may be worth trying several different signal sequences, for no one sequence seems to work best for all proteins (Tessier et al., 1991; Mroczkowski et al., 1994; Golden et al., 1998). Of the commonly used cell lines, *T. ni* cells often produce higher levels of secreted proteins (Hink et al., 1991; Wickham and Nemerow, 1993; Mroczkowski et al., 1994).

Although the baculovirus system can quickly provide recombinant protein, it may not be the optimal approach to obtaining the highest levels of secreted protein possible. It is worth taking the time in parallel with baculovirus efforts to produce an insect cell line that overexpresses the gene of interest (Jarvis et al., 1990; Farrell et al., 1998). That way a backup expression system is in place in case the levels of protein from the baculovirus infection are intolerably low.

What Special Considerations Are There for Expressing Glycosylated Proteins?

Insect cells perform N-linked glycosylation at sites that are similarly targeted in mammalian cells, but in insect cells the modifications are of the high mannose type with inefficient trimming of the core sugar residues or just the trimannosyl core structure (reviewed in Altmann et al., 1999). There are several approaches available to obtain more complex glycosylation patterns typical of mammalian cell expression. Infection of cells from *Estigmene acrea* (available from Novagen) may produce a more mammalian-type of glycosylation pattern (Wagner et al., 1996a; Ogonah, 1996). Co-expression of a mammalian glycosyltransferase may result in a more complex glycosylation pattern (Wagner, 1996b; Jarvis and Finn, 1996; Jarvis, Kawar, and Hollister, 1998). Similarly, use of a Sf9 host cell that has been engineered to constitutively express a glycosyltransferase can be used for the same effect (Hollister, Shaper, and Jarvis, 1997). The addition of mannosamine to infected insect cells can increase the level of terminal N-acetylglucosamine structures in recombinant proteins (Donaldson et al., 1999).

What Are the Options for Expressing More Than One Protein?

A significant advantage to the baculovirus expression system is the ease of expressing multiple proteins. The ability to co-express proteins allows for the expression of heterodimers (Stern and Wiley, 1992; Graber et al., 1992) and even larger multiprotein complexes such as virus particles (Loudon and Roy, 1991). In one notable case, co-expression of seven herpesvirus proteins from seven different baculoviruses allowed replication of a plasmid containing a herpesvirus origin of replication (Stow, 1992). Cells can be simultaneously infected with multiple baculoviruses expresing different proteins, or recombinant baculoviruses can be made that have up to four separate promoters each regulating a different gene (Weyer, Knight, and Possee, 1990; Belyaev, Hails, and Roy, 1995). Vectors that express two or more proteins are available commercially (Pharmingen, Clontech, and Novagen). In contrast to mammalian cells, baculovirus infected insect cells do not make efficient use of an internal ribosomal entry site (IRES) sequence for the expression of multiple proteins (Finkelstein et al., 1999).

Co-expression also enables one to express modifiers of the target protein. Examples of this are co-expression of biotin ligase to obtain biotinylation (Duffy, Tsao, and Waugh, 1998), prohormone convertase to obtain proteolytically processed TGFβ1 (Laprise, Grondin, and Dubois, 1998), and signal peptidase to enhance processing efficiency for a secreted protein (Ailor and Betenbaugh, 1999).

How Can You Obtain Maximal Protein Yields?

Optimizing the host cell selection, cell density at infection, multiplicity of infection, type of media, and the time of harvest will allow maximal recovery of the protein of interest (Licari and Bailey, 1992; Power et al., 1994). All five conditions are interdependent, and it is possible that protein yields may be equal from a relatively low multiplicity of infection (moi) of dilute cells harvested after five to six days compared to high moi infection of a dense culture harvested after two days. If cells are being grown on a larger scale (e.g., in suspension cultures in 1 L spinner flasks), expression optimization should be done under such conditions. Although it may be convenient to examine infection conditions in small culture dishes such as a 24 well cluster dish, optimal parameters for cells growing in a stationary flask are likely to be very different from cells growing in suspension. A reasonable strategy to start an optimization procedure is to infect 200 ml of 1.5×10^6 cells/ml growing in three 500 ml spinner flasks with moi's of 0.1, 1, and 10, and then to remove 10 ml aliquots of cells every 24 hours for 5 days. For intracellular proteins, the cells should be lysed as they would for downstream purification, and both the soluble and insoluble fractions examined for the presence of the protein of interest.

What Is the Best Way to Process Cells for Purification?

For cytoplasmic proteins, cells are recovered by pelleting and washed with a buffer to remove media components. Infected cell pellets can be further processed or stored frozen until needed. Insect cells can be lysed by hypotonic lysis after incubation in a buffer lacking salt; disruption is completed by using a dounce homogenizer. Cells can also be lysed with a buffer containing a detergent such as Triton, CHAPS, or NP-40. Sonication should not be used as lysis conditions are difficult to control and reproduce from one preparation to another. It is important to keep the preparation on ice and perform cell lysis in the presence of a cocktail of protease inhibitors to avoid proteolysis. The lysate should be cleared by centrifugation at $100,000 \times g$ to remove large aggregates and insoluble material. Cleared lysates are then ready for chromatographic purification.

For nuclear proteins, nuclei are obtained following hypotonic lysis or detergent lysis and salt extracted to remove nuclear-associated proteins. Secreted proteins are generally recovered from cell-free clarified supernatants by direct adsorption to a chromatographic resin.

TROUBLESHOOTING

Western blot or a biochemical analysis of transfected cells should indicate expression of the gene of interest three to seven days after the transfection. It is rare that a protein is not expressed at all in baculovirus infected cells, and an observed lack of protein expression may be due to a variety of situations.

Suboptimal Growth Conditions

Many problems with baculovirus expression can be traced to suboptimal cell growth conditions. Healthy cells should show high viability (>98%) and have a doubling time of around 24 hours. If either of these conditions is not met, efforts should be directed toward getting a more robust cell stock. Start with frozen cells from the American Type Culture Collection (ATCC) or a commercial source, and use heat-inactivated serum that has been certified for insect cell culture (available from Life Technologies) in media without antibiotics. Grow cells initially in stationary flasks as it is easier to monitor their progress. For passaging in flasks, do not scrape or harshly pipette liquid over the cell monolayer. Instead, sharply rap the side of the flask to dislodge as many cells as possible. Remove the cells and media, and distribute these to new flasks containing additional fresh media; add back fresh media to the remaining cells that have adhered to the original flask for further growth. Once cells are growing in flasks in serum-containing media, the cells from several flasks can be pooled for growth in suspension and/or adaptation to serum-free media.

Viral Production Problems

A lack of protein expression may be due to inefficient production of virus in the initial transfection step. The virus may benefit from an amplification step by removing about 100 μl of the media from transfected cells and adding it to freshly plated uninfected cells in a T25 Flask. Cells from this infection should show evidence of viral cytopathic effect and demonstrate protein expression after three to four days. Transfections are generally performed with either a liposome mediated- or a calcium phosphate procedure provided as a "kit" with commercial viral DNA. It is important to follow the manufacturer's instructions carefully. Plasmid DNA should be very pure—preparations made with an anion exchange matrix or cesium chloride banding work well. The DNA should be sterilized by ethanol precipitation and resuspended in a sterile buffer. Viral DNA is very large and susceptible to shearing; use a sterile cut-off blunt pipette tip for transfers and never vortex it. Insect cells must be healthy (>98% viable) and actively growing in log phase when used for a transfection. If possible, transfections should be done in cells growing in serum-containing media to enhance the production of virus. Transfection conditions can be optimized with wild-type baculoviral DNA that produce distinctive polyhedrin in infected cells. Similarly optimization can be done with viral DNA from a baculovirus recombinant that encodes an easily assayed protein (e.g., beta-galactosidase). The presence of insert DNA incorporated into progeny virus can be determined by PCR or Southern blot analysis.

Mutation

A lack of expression may be due to an unwanted mutation or the presence of unintended upstream ATG sequences. The DNA encoding the open reading frame for the gene of interest in the transfer vector plasmid should be verified by sequencing to rule out this possibility. There is one report of translational initiation occurring at a non-ATG codon, AUU, in a baculovirus expressed protein (Beames et al., 1991). Occasionally larger transfer vectors (>8 kb) suffer deletions and will be unable to give rise to recombinant virus containing an intact gene of interest. Transfer plasmids should be digested with a few restriction enzymes to be sure this has not happened. The use of smaller transfer vectors (<6 kb) often eliminates such genetic instabilities.

Solubility Problems

In general, many recombinant proteins that are insoluble in *E. coli* become soluble when produced in insect cells. There are a few proteins that are completely insoluble in insect cells and will need to be denatured and refolded. More often, a protein will be partially soluble, and for these situations, infection at a low multiplicity (<1 virus/cell) and harvest at an early time (<36 hours) is usually beneficial. Co-expression with protein disulfide isomerase or a chaperonin molecule may enhance the percentage of nonaggregated secreted proteins (Hsu, Eiden, and Betenbaugh, 1994; Hsu et al., 1996; Ailor and Betenbaugh, 1998). Proteins that are susceptible to degradation may benefit from the addition of a signal sequence and export into the media (Mroczkowski et al., 1994). Use of viral DNA from a baculovirus lacking a viral protease (available from Novagen) may also help in the expression of proteins that are degraded in insect cells.

SUMMARY

Baculoviruses represent a versatile, relatively quick, minimal technology approach to recombinant gene expression, especially for proteins that are insoluble in *E. coli* or are covalently modified. All of these features make baculovirus expression an excellent complement to a bacterial expression system, especially for the production of proteins at levels <10 mg. If a cloned gene is on hand, the process of obtaining a recombinant baculovirus and analyzing the expression from approximately 1 L of infected cells can be completed in less than 2 weeks. Recombinant viruses can incorporate large amounts of DNA, making the expression of multiple genes from one virus possible. Additionally, insect cells can be infected with multiple viruses, allowing the expression of entire signaling pathways or protein/modifier combinations. A drawback to the use of this system is the inability to produce glycoproteins with complex N-linked glycans typical of mammalian cells. There are various approaches to increase this capability in insect cells, but none are truly optimal. Looking to the future, baculoviruses may have a utility as gene-delivery vehicles for protein expression not only in insect cells but also in a wide variety of mammalian cells.

BIBLIOGRAPHY

Ailor, E., and Betenbaugh, M. J. 1998. Overexpression of a cytosolic chaperone to improve solubility and secretion of a recombinant IgG protein in insect cells. *Biotechnol. Bioeng.* 58:196–203.

Ailor, E., and Betenbaugh, M. J. 1999. Modifying secretion and post-translational processing in insect cells. *Curr. Opin. Biotechnol.* 10:142–145.

Ailor, E., Pathmanathan, J., Jongbloed, J. D., and Betenbaugh, M. J. 1999. A bacterial signal peptidase enhances processing of a recombinant single chain antibody fragment in insect cells. *Biochem. Biophys. Res. Commun.* 255:444–450.

Aiyar, N., Baker, E., Wu, H. L., Nambi, P., Edwards, R. M., Trill, J. J., Ellis, C., and Bergsma, D. J. 1994. Human AT1 receptor is a single copy gene: characterization in a stable cell line. *Mol. Cell Biochem.* 131:75–86.

Allet, B., Bernard, A. R., Hochmann, A., Rohrbach, E., Graber, P., Magnenat, E., Mazzei, G. J., and Bernasconi, L. 1997. A bacterial signal peptide directs efficient secretion of eukaryotic proteins in the baculovirus expression system. *Prot. Expr. Purif.* 9:61–68.

Altmann, F., Staudacher, E., Wilson, I. B., and März, L. 1999. Insect cells as hosts for the expression of recombinant glycoproteins. *Glycoconj. J.* 16:109–123.

Ames, R. S., Sarau, H. S., Chambers, J. K., Willette, R. N., Aiyar, N. V., Romanic, A. M., Louden, C. S., Foley, J. J., Sauermelch, C. F., Coatney, R. W., Ao, Z., Disa, J., Holmes, S. D., Stadel, J. M., Martin, J. D., Liu, W. S., Glover, G. I., Wilson, S., McNulty, D. E., Ellis, C. E., Elshourbagy, N. A., Shabon, U., Trill, J. J., Hay, D. W. P., Ohlstein, E. H., Bergsma, D. J., and Douglas, S. A. 1999. Human urotensin-II: A potent vasoconstrictor and agonist for the orphan receptor GPR 14. *Nature* 401:282–286.

Ayres, M. D., Howard, S. C., Kuzio, J., Lopez-Ferber, M., Possee, R. D. 1994. The complete DNA sequence of *Autographa californica* nuclear polyhedrosis virus. *Virol.* 202:586–605.

Beames, B., Braunagel, S., Summers, M. D., and Lanford, R. E. 1991. Polyhedrin initiator codon altered to AUU yields unexpected fusion protein from a baculovirus vector. *Biotechniq.* 11:378–383.

Bebbington, C. R., Renner, G., Thomson, S., King, D., Abrams, D., and Yarronton, G. T. 1992. High level expression of a recombinant antibody from myeloma cells using a glutamine synthetase gene as an amplifiable selectable marker. *Bio/Technol.* 10:169–175.

Belyaev, A. S., Hails, R. S., and Roy, P. 1995. High level expression of five foreign genes by a single recombinant baculovirus. *Gene* 156:229–233.

Benting, J., Lecat, S., Zacchetti, D., and Simons, K. 2000. Protein expression in *Drosophila* Schneider cells. *Anal. Biochem.* 278:59–68.

Boggs, S. S., Gregg, R. G., Borenstein, N., and Smithies, O. 1986. Efficient transformation and frequent single-site, single-copy insertion of DNA can be obtained in mouse erythroleukemia cells transformed by electroporation. *Exp. Hematol.* 14:988–994.

Boshart, M., Weber, F., Jahn, G., Dorsch-Hasler, K., Fleckenstein, B., and Schaffner, W. 1985. A very strong enhancer is located upstream of an immediate early gene of human cytomegalovirus. *Cell* 41:521–530.

Bueler, H. 1999. Adeno-associated viral vectors for gene transfer and gene therapy. *Biol. Chem.* 380:613–622.

Butters, T. D., Sparks, L. M., Harlos, K., Ikemizu, S., Stuart, D. I., Jones, E. Y., and Davis, S. J. 1999. Effects of *N*-butyldeoxynojirimycin and the Lec3.2.8.1 mutant phenotype on *N*-glycan processing in Chinese hamster ovary cells: Application to glycoprotein crystallization. *Prot. Sci.* 8:1696–1701.

Cannan, S., Roussel, A., Verger, R., and Cambillau, C. 1999. Gastric lipase: Crystal structure and activity. *Biochim. Biophys. Acta* 1441:197–204.

Casasnovas, J. M., Larvie, M., and Stehle, T. 1999. Crystal structure of two CD46 domains reveals an extended measles virus-binding surface. *EMBO J.* 18: 2911–2922.

Cavener, D. R. 1987. Comparison of the consensus sequence flanking translational start sites in *Drosophila* and vertebrates. *Nucl. Acids Res.* 15:1353–1361.

Chambers, J. K., Macdonald, L. E., Sarau, H. M., Ames, R. S., Freeman, K., Foley, J. J., Zhu, Y., McLaughlin, M. M., Murdock, P., McMillan, L., Trill, J., Swift, A., Aiyar, N., Taylor, P., Vawter, L., Naheed, S., Szekeres, P., Hervieu, G., Scott, C., Watson, J. M., Moore, D., Emson, P., Faull, R. L. M., Waldvogel, H. J., Murphy, A. J., Duzic, E., Klein, C., Bergsma, D., Wilson, S., and Livi, G. P. 2000. A G protein-coupled receptor for uridine 5′-diphosphoglucose (UDP-glucose). *J. Biol. Chem.* 275:10767–10771.

Chang, J. Y. 1985. Thrombin specificity: Requirement for apolar amino acids adjacent to the thrombin cleavage site of polypeptide substrate. *Eur. J. Biochem.* 151:217–224.

Chang, M.-J., Kuzio, J., and Blissard, G. W. 1999. Modulation of translational efficiency by contextual nucleotides flanking a baculovirus initiator AUG codon. *Virol.* 259:369–383.

Christopherson, K. S., Mark, M. R., Bajaj, V., and Godowski, P. J. 1992. Ecdysteroid-dependent regulation of genes in mammalian cells by a *Drosophila* ecdysone receptor and chimeric transactivators. *Proc. Nat. Acad. Sci. U.S.A.* 89:6314–6318.

Ciccaglione, A. R., Marcantonio, C., Equestre, M., Jones, I. M., and Rapicetta, M. 1998. Secretion and purification of HCV E1 protein forms as glutathione-*S*-transferase fusion in the baculovirus insect cell system. *Virus Res.* 55:157–165.

Cockett, M. I., Bebbington, C. R., and Yarranton, G. T. 1991. The use of engineered E1A genes to transactivate the hCMV-MIE promoter in permanent CHO cell lines. *Nucl. Acids Res.* 19:319–325.

Colbere-Garapin, F., Horodniceanu, F., Kourilsky, P., and Garapin, A. C. 1981. A new dominant hybrid selective marker for higher eukaryotic cells. *J. Mol. Biol.* 150:1–14.

Condreay, J. P., Witherspoon, S. M., Clay, W. C., and Kost, T. A. 1999. Transient and stable gene expression in mammalian cells transduced with a recombinant baculovirus vector. *Proc. Nat. Acad. Sci. U.S.A.* 96:127–132.

Cordingley, M. G., Callahan, P. L., Sardana, V. V., Garsky, V. M., and Colonno, R. J. 1990. Substrate requirements of human rhinovirus 3C protease for peptide cleavage in vitro. *J. Biol. Chem.* 265:9062–9065.

Cregg, J. M. 1999. Expression in the Methylotrophic yeast *Pichia pastoris*. In Fernandez, J. M., and Hoeffler, J. P. eds., *Gene Expression Systems: Using Nature for the Art of Expression*. Academic Press, San Diego, CA, pp. 290–325.

Cregg, J. M., Vedvick, T. S., and Raschke, W. C. 1993. Recent advances in the expression of genes in *Pichia pastoris*. *Biotechnol.* 11:905–910.

Davies, A. H., Jowett, J. B., and Jones, I. M. 1993. Recombinant baculovirus vectors expressing glutathione-*S*-transferase fusion proteins. *Bio/Technol.* 11:933–936.

Deen, K. C., McDougal, J. S., Inacker, R., Folena-Wasserman, G., Arthos, J., Rosenberg, J., Maddon, P. J., Axel, R., and Sweet, R. W. 1988. A soluble form of CD4 (T4) protein inhibits AIDS virus infection. *Nature* 331:82–84.

de la Luna, S., Soria, I., Pulido, D., Ortin, J., and Jimenez, A. 1988. Efficient transformation of mammalian cells with constructs containing a puromycin-resistance marker. *Gene* 62:121–126.

Donaldson, M., Wood, H. A., Kulakosky, P. C., and Shuler, M. L. 1999. Use of mannosamine for inducing the addition of outer arm N-acetylglucosamine onto N-linked oligosaccharides of recombinant proteins in insect cells. *Biotechnol. Prog.* 15:168–173.

Duffy, S., Tsao, K.-L., and Waugh, D. S. 1998. Site-specific, enzymatic biotinylation of recombinant proteins in Spodoptera frugiperda cells using biotin acceptor peptides. *Anal. Biochem.* 262:122–128.

Ennist, D. L. 1999. Gene therapy for lung disease. *Trends Pharmacol. Sci.* 20:260–266.

Ernst, W. J., Grabherr, R. M., and Katinger, H. W. 1994. Direct cloning into the *Autographa californica* nuclear polyhedrosis virus for generation of recombinant baculoviruses. *Nucl. Acids Res.* 22:2855–2856.

Evan, G. I., Lewis, G. K., Ramsay, G., and Bishop, J. M. 1985. Isolation of monoclonal antibodies specific for human c-myc proto-oncogene product. *Mol. Cell. Biol.* 5:3610–3616.

Farrell, P. J., Lu, M., Prevost, J., Brown, C., Behie, L., and Iatrou, K. 1998. High-level expression of secreted glycoproteins in transformed lepidopteran insect cells using a novel expression vector. *Biotechnol. Bioeng.* 60:656–663.

Ferrara, N., Winer, J., Burton, T., Rowland, A., Siegel, M., Phillips, H. S., Terrell, T., Keller, G. A., and Levinson, A. D. 1993. Expression of vascular endothelial growth factor does not promote transformation but confers a growth advantage in vivo to Chinese hamster ovary cells. *J. Clin. Invest.* 91:160–170.

Finkelstein, Y., Faktor, O., Elroy-Stein, O., and Levi, B. Z. 1999. The use of bicistronic transfer vectors for the baculovirus expression system. *J. Biotechnol.* 75:33–44.

Fitzgerald, L. R., Dytko, G. M., Sarau, H. M., Mannan, I. J., Ellis, C., Lane, P. A., Tan, K. B., Murdock, P. R., Wilson, S., Bergsma, D. J., Ames, R. S., Foley, J. J., Campbell, D. A., McMillan, L., Evans, N., Elshourbagy, N. A., Minehart, H., and Tsui, P. 2000. Identification of an EDG7 variant, HOFNH30, a G-protein-coupled receptor for lysophosphatidic acid. *Biochem. Biophys. Res. Commun.* 273:805–810.

Fleckenstein, E., Uphoff, C. C., and Drexler, H. G. 1994. Effective treatment of mycoplasma contamination in cell lines with enrofloxacin (Baytril). *Leukemia* 8:1424–1434.

Geisse, S., Gram, H., Kleuser, B., and Kocher, H. P. 1996. Eukaryotic expression systems: A comparison. *Prot. Expr. Purif.* 8:271–282.

George, H. J., Marchand, P., Murphy, K., Wiswall, B. H., Dowling, R., Giannaras, J., Hollis, G., Trzaskos, J. M., and Copeland, R. A. 1997. Recombinant human 92-kDa type IV collagenase/gelatinase from baculovirus-infected insect cells: Expression, purification and characterization. *Prot. Expr. Purif.* 10:154–161.

Gignac, S. M., Brauer, S., Hane, B., Quentmeier, H., and Drexler, H. G. 1991. Elimination of mycoplasma from infected leukemia cell lines. *Leukemia* 5:162–165.

Gignac, S. M., Uphoff, C. C., MacLeod, R. A., Steube, K., Voges, M., and Drexler, H. G. 1992. Treatment of mycoplasma-contaminated continuous cell lines with mycoplasma removal agent (MRA). *Leuk. Res.* 16:815–822.

Golden, A., Austen, D. A., van Schravendijk, M. R., Sullivan, B. J., Kawasaki, E. S., and Osburne, M. S. 1998. Effect of promoters and signal sequences on the production of secreted HIV-1 gp120 protein in the baculovirus system. *Prot. Expr. Purif.* 14:8–12.

Gluzman, Y. 1981. SV40-transformed simian cells support the replication of early SV40 mutants. *Cell* 23:175–182.

Gorman, C. M., Merlino, G. T., Willingham, M. C., Pastan, I., and Howard, B. H. 1982. The Rous sarcoma virus long terminal repeat is a strong promoter when introduced into a variety of eukaryotic cells by DNA-mediated transfection. *Proc. Nat. Acad. Sci. U.S.A.* 79:6777–6781.

Gossen, M., and Bujard, H. 1992. Tight control of gene expression in mammalian cells by tetracycline-responsive promoters. *Proc. Nat. Acad. Sci. U.S.A.* 89:5547–5551.

Graber, S. G., Figler, R. A., Kalman-Maltese, V. K., Robishaw, J. D., and Garrison, J. C. 1992. Expression of functional G protein beta gamma dimers of defined subunit composition using a baculovirus expression system. *J. Biol. Chem.* 267:13123–13126.

Gritz, L., and Davies, J. 1983. Plasmid-encoded hygromycin B resistance: The sequence of hygromycin B phosphotransferase gene and its expression in *Escherichia coli* and *Saccharomyces cerevisiae*. *Gene* 25:179–188.

Grossmann, M., Wong, R., Teh, N. G., Tropea, J. E., East-Palmer, J., Weintrayb, B. D., and Szkudlinski, M. W. 1997. Expression of biologically active human thyrotropin (hTSH) in a baculovirus system: Effect of insect cell glycosylation on hTSH activity in vitro and in vivo. *Endocrinol.* 138:92–100.

Grubb, B. R., Pickles, R. J., Ye, H., Yankaskas, J. R., Vick, R. N., Engelhardt, J. F., Wilson, J. M., Johnson, L. G., and Boucher, R. C. 1994. Inefficient gene transfer by adenovirus vector to cystic fibrosis airway epithelia of mice and humans. *Nature* 371:802–806.

Grueninger-Leitch, F., D'Arcy, A., D'Arcy, B., and Chene, C. 1996. Deglycosylation of proteins for crystallization using recombinant fusion protein glycosidases. *Prot. Sci.* 5:2617–2622.

Hamer, D. H., Kaehler, M., and Leder, P. 1980. A mouse globin gene promoter is functional in SV40. *Cell* 21:697–708.

Hartman, S. C., and Mulligan, R. C. 1988. Two dominant-acting selectable markers for gene transfer studies in mammalian cells. *Proc. Nat. Acad. Sci. U.S.A.* 85:8047–8051.

He, T. C., Zhou, S., da Costa, L. T., Yu, J., Kinzler, K. W., and Vogelstein, B. 1998. A simplified system for generating recombinant adenoviruses. *Proc. Nat. Acad. Sci. U.S.A.* 95:2509–2514.

Higgins, D. R., and Cregg, J. M. 1998. *Pichia Protocols: Methods in Molecular Biology*, vol 103. Humana Press, Totowa NJ.

Hill-Perkins, M. S., and Possee, R. D. 1990. A baculovirus expression vector derived from the basic protein promoter of *Autographa californica* nuclear polyhedrosis virus. *J. Gen. Virol.* 71:971–976.

Hink, W. F., Thomsen, D. R., Davidson, D. J., Meyer, A. L., and Castellino, F. J. 1991. Expression of three recombinant proteins using baculovirus vectors in 23 insect cell lines. *Biotechnol. Prog.* 7:9–14.

Hollister, J. R., Shaper, J. H., and Jarvis, D. L. 1997. Stable expression of mammalian beta 1,4-galactosyltransferase extends the *N*-glycoslylation pathway in insect cells. *Glycobiol.* 8:473–490.

Hopp, T. P., Prickett, K. S., Price, V. L., Libby, R. T., March, C. J., Cerretti, D. P., Urdal, D. L., and Conlon, P. J. 1988. A short polypeptide marker sequence useful for recombinant protein identification and purification. *Bio/Technol.* 6:1204–1210.

Hsieh, P., and Robbins, P. 1984. Regulation of asparagine-linked oligosaccharide processing. *J. Biol. Chem.* 259:2375–2382.

Hsu, T. A., Eiden, J. J., and Betenbaugh, M. J. 1994. Engineering the assembly pathway of the baculovirus-insect cell expression system. *Ann. NY Acad. Sci.* 721:208–217.

Hsu, T. A., Watson, S., Eiden, J. J., and Betenbaugh, M. J. 1996. Rescue of immunoglobulins from insolubility is facilitated by PDI in the baculovirus expression system. *Prot. Expr. Purif.* 7:281–288.

Ishii, I., Izumi, T., Tsukamoto, H., Umeyama, H., Ui, M., and Shimizu, T. 1997. Alanine exchanges of polar amino acids in the transmembrane domains of a platelet-activating factor receptor generate both constitutively active and inactive mutants. *J. Biol. Chem.* 272:7846–7854.

Ivey-Hoyle, M. 1991. Recombinant gene expression in cultured *Drosophila melanogaster* cells. *Curr. Opin. Biotechnol.* 2:704–707.

Izumi, M., Miyazawa, H., Kamakura, T., Yamaguchi, I., Endo, T., and Hanaoka, F. 1991. Blasticidin *S*-resistance gene (bsr): A novel selectable marker for mammalian cells. *Exp. Cell. Res.* 197:229–233.

James, R. I., Elton, J. P., Todd, P., and Kompala, D. S. 2000. Engineering CHO cells to overexpress a secreted reporter protein upon induction from mouse mammary tumor virus promoter. *Biotechnol. Bioeng.* 67:134–140.

Jarvis, D. L., and Summers, M. D. 1989. Glycosylation and secretion of human tissue plasminogen activator in recombinant baculovirus-infected insect cells. *Mol. Cell. Biol.* 9:214–223.

Jarvis, D. L., Gleming, J. G. W., Kovacs, G. R., Summers, M. D., and Guarino, L. A. 1990. Use of early baculovirus promoters for continuous expression and efficient processing of foreign gene products in stably transformed lepidopteran cells. *Bio/Technol.* 8:950–955.

Jarvis, D. L., and Finn, E. E. 1996. Modifying the insect cell *N*-glycosylation pathway with immediate early baculovirus expression vectors. *Nature Biotechnol.* 14:1288–1292.

Jarvis, D. L., Kawar, Z. S., and Hollister, J. R. 1998. Engineering *N*-glycosylation pathways in the baculovirus–insect cell system. *Curr. Opin. Biotechnol.* 9:528–533.

Johansen, H., van der Straten, A., Sweet, R., Otto, E., Maroni, G., and Rosenberg, R. 1989. Regulated expression at high copy number allows production of a growth-inhibitory oncogene product in *Drosophila* Schneider cells. *Genes Dev.* 3:882–889.

Juarranz, M. G., Van Rampelbergh, J., Gourlet, P., De Neef, P., Cnudde, J., Robberecht, P., and Waelbroeck, M. 1999. Different vasoactive intestinal polypeptide receptor domains are involved in the selective recognition of two VPAC(2)-selective ligands. *Mol. Pharmacol.* 6:1280–1287.

Kaufman, R. J. 1990. Selection and coamplification of heterologous genes in mammalian cells. *Meth. Enzymol.* 185:537–566.

Kern, P., Hussey, R. E., Spoerl, R., Reinherz, E. L., and Chang, H. C. 1999. Expression, purification, and functional analysis of murine ectodomain fragments of CD8alphaalpha and CD8alphabeta dimers. *J. Biol. Chem.* 274:27237–27243.

King, L. A., and Possee, R. D. 1992. *The Baculovirus Expression System.* Chapman and Hall, London.

Kitts, P. A., and Possee, R. D. 1993. A method for producing recombinant baculovirus vectors at high frequency. *BioTechniq.* 14:810–817.

Koller, K. J., Whitehorn, E. A., Tate, E., Ries, T., Aguilar, B., Chernov-Rogan, T., Davies, A. M., Dobbs, A., Yen, M., and Barrett, R. W. 1997. A generic method for the production of cell lines expressing high levels of 7-transmembrane receptors. *Anal. Biochem.* 250:51–60.

Kornfeld, R., and Kornfeld, S. 1985. Assembly of asparagine-linked oligosaccharides. *Ann. Rev. Biochem.* 54:631–664.

Kost, T. A., and Condreay, J. P. 1999. Recombinant baculoviruses as expression vectors for insect and mammalian cells. *Curr. Opin. Biotechnol.* 10:428–433.

Kozak, M. 1987. An analysis of 5′-noncoding sequences from 699 vertebrate messenger RNA's. *Nucl. Acids Res.* 15:8125–8148.

Laprise, M. H., Grondin, F., and Dubois, C. M. 1998. Enhanced TGFbeta1 maturation in high five cells coinfected with recombinant baculovirus encoding the convertase furin/pace: Improved technology for the production of recombinant proproteins in insect cells. *Biotechnol. Bioeng.* 58:85–91.

Lehr, R. V., Elefante, L. C., Kikly, K. K., O'Brien, S. P., and Kirkpatrick, R. B. 2000. A modified metal-ion affinity chromatography procedure for the purification of histidine-tagged recombinant proteins expressed in *Drosophila* S2 cells. *Prot. Expr. Purif.* 19:362–368.

Leusch, M. S., Lee, S. C., and Olins, P. O. 1995. A novel host-vector system for direct selection of recombinant baculoviruses (bacmids) in *Escherichia coli*. *Gene* 160:191–194.

Levin, D. B., and Whittome, B. 2000. Codon usage in nucleopolyhedroviruses. *J. Gen. Virol.* 81:2313–2325.

Licari, P., and Bailey, J. E. 1992. Modeling the population dynamics of baculovirus-infected insect cells: Optimizing infection strategies for enhanced recombinant protein yields. *Biotechnol. Bioeng.* 39:432–441.

Lilius, G., Persson, M., Bulow, L., and Mosbach, K. 1991. Metal affinity precipitation of proteins carrying genetically attached polyhistidine affinity tails. *Eur. J. Biochem.* 198:499–504.

Loudon, P. T., and Roy, P. 1991. Assembly of five bluetongue virus proteins expressed by recombinant baculoviruses: Inclusion of the largest protein VP1 in the core and virus-like particles. *Virol.* 180:798–802.

Luckow, V. L., and Summers, M. D. 1988. Signals important for high-level expression of foreign genes in *Autographa californica* nuclear polyhedrosis virus expression vectors. *Virol.* 167:56–71.

Luckow, V. A., Lee, S. C., Barry, G. F., and Olins, P. O. 1993. Efficient generation of infectious recombinant baculoviruses by site-specific transposon-mediated insertion of foreign genes into a baculovirus genome propagated in *Escherichia coli*. *J. Virol.* 67:4566–4579.

Madisen, L., Lioubin, M. N., Marquardt, H., and Purchio, A. F. 1990. High-level expression of TGF-beta 2 and the TGF-beta 2(414) precursor in Chinese hamster ovary cells. *Growth Factors* 3:129–138.

Mansour, S. L., Grodzicker, T., and Tijan, R. 1986. Downstream sequences affect transcription from the adenovirus major late promoter. *Mol. Cell. Biol.* 6:2684–2694.

Matsushima, M., Ichinose, M., Yahagi, N., Kakei, N., Tsukada, S., Miki, K., Kurokawa, K., Tashiro, K., Shiokawa, K., Shinomiya, K., Umeyama, H., Inoues, H., Takahashi, T., and Takahashi, K. 1994. Structural characterization of porcine enteropeptidase. *J. Biol. Chem.* 269:19976–19982.

McCarroll, L., and King, L. A. 1997. Stable insect cell cultures for recombinant protein production. *Curr. Opin. Biotechnol.* 8:590–594.

Mellon, P., Parker, P., Gluzman, Y., and Maniatis, T. 1981. Identification of DNA sequences required for transcription of the human alpha 1-globin gene in a new SV40 host-vector system. *Cell* 27:279–288.
Merrington, C. L., King, L. A., and Possee, R. D. 1999. Baculovirus expression systems. In Higgins, S. J., and Hames, B. D., eds., *Protein Expression—A Practical Approach*. Oxford University Press, Oxford, U.K.
Miller, L. K. 1997. *The Baculoviruses*. Plenum Press, New York.
Mizushima, S., and Nagata, S. 1990. pEF-BOS, a powerful mammalian expression vector. *Nucl. Acids Res.* 18:5322.
Mroczkowski, B. S., Huvar, A., Lernhardt, W., Misono, K., Nielson, K., and Scott, B. 1994. Secretion of thermostable DNA polymerase using a novel baculovirus vector. *J. Biol. Chem.* 269:13522–13528.
Mulligan, R. C., and Berg, P. 1981. Selection for animal cells that express the *Escherichia coli* gene coding for xanthine-guanine phosphoribosyltransferase. *Proc. Nat. Acad. Sci. U.S.A.* 78:2072–2076.
Mulsant, P., Gatignol, A., Dalens, M., and Tiraby, G. 1988. Phleomycin resistance as a dominant selectable marker in CHO cells. *Somat. Cell Mol. Genet.* 14:243–252.
Murphy, C. I., Lennick, M., Lehar, S. M., Beltz, G. A., and Young, E. 1990. Temporal expression of HIV-1 envelope proteins in baculovirus-infected insect cells: Implications for glycoslyation and CD binding. *Genet. Anal. Tech. Appl.* 7:160–171.
Murphy, C. I., Piwnica-Worms, H., Grunwald, S., and Romanow, W. G. 1997. Expression of proteins in insect cells using baculovirus vectors. In Ausubel, F., Brent, R., Kingston, R., Moore, D., Seidman, J. G., Smith, J., and Struhl, K., eds., *Current Protocols in Molecular Biology*. Wiley, New York.
Nagai, K., and Thogersen, H. C. 1984. Generation of beta-globin by sequence-specific proteolysis of a hybrid protein produced in *Escherichia coli*. *Nature* 309:810–812.
Newman-Tancredi, A., Wootton, R., and Strange, P. G. 1992. High-level stable expression of recombinant 5-HT1A 5-hydroxytryptamine receptors in Chinese hamster ovary cells. *Biochem. J.* 285:933–938.
Niman, H. L., Houghton, R. A., Walker, L. E., Reisfeld, R. A., Wilson, I. A., Hogle, J. M., and Lerner, R. A. 1983. Generation of protein-reactive antibodies by short peptides is an event of high frequency: Implications for the structural basis of immune recognition. *Proc. Nat. Acad. Sci. U.S.A.* 80:4949–4953.
Ng, S. Y., Gunning, P., Eddy, R., Ponte, P., Leaviti, H., and Kedes, L. 1985. Evolution of the functional human beta-actin gene and its multi-pseudogene family: Conservation of noncoding regions of chromosomal dispersion of pseudogenes. *Mol. Cell. Biol.* 5:2720–2730.
O'Reilly, D. R., Miller, L. K., and Luckow, V. A. 1992. *Baculovirus Expression Vectors—A Laboratory Manual*. Freeman, New York.
Offermanns, S., and Simon, M. I. 1995. G alpha 15 and G alpha 16 couple a wide variety of receptors to phospholipase C. *J. Biol. Chem.* 270:15175–15180.
Ogonah, O. W., Freedman, R. B., Jenkins, N., Patel, K., and Rooney, B. C. 1996. Isolation and characterisation of an insect cell line able to perform complex N-linked glycosylation on recombinant proteins. *Nature Biotechnol.* 14:197–202.
Olkkonen, V. M., Dupree, P., Simons, K., Liljestrom, P., and Garoff, H. 1994. Expression of exogenous proteins in mammalian cells with the Semliki forest virus vector. *Methods Cell Biol.* 43: Pt A., 43–53.
Pajot-Augy, E., Bozon, V., Remy, J. J., Couture, L., and Salesse, R. 1999. Critical relationship between glycosylation of recombinant lutropin receptor

ectodomain and its secretion from baculovirus-infected cells. *Eur. J. Biochem.* 260:635–648.

Palomares, L. A., Gouzalez, M., and Ramirez, O. T. 2000. Evidence of Pluronic F-68 direct interaction with insect cells: Impact on shear protection, recombinant protein and baculovirus production, *Enz. Microb. Technol.* 26:324–331.

Parekh, R. B., Dwek, R. A., Rudd, P. M., Thomas, J. R., Rademacher, T. W., Warren, T., Wun, T. C., Hebert, B., Reitz, B., Palmier, M., Ramabhadran, T., and Tiemeier, D. C. 1989. *N*-glycosylation and in vitro enzymatic activity of human recombinant tissue plasminogen activator expressed in Chinese hamster ovary cells and a murine cell line. *Biochem.* 28:7644–7662.

Patel, G., Nasmyth, K., and Jones, N. 1992. A new method for the isolation of recombinant baculovirus. *Nucl. Acids Res.* 20:97–104.

Peakman, T. C., Charles, I. G., Sydenham, M. A., Gewart, D. R., Page, M. J., and Markoff, A. J. 1992. Enhanced expression of recombinant proteins in insect cells using a baculovirus vector containing a bacterial leader sequence. *Nucl. Acids Res.* 20:6111–6112.

Peakman, T. C., Harris, R. A., and Gewart, D. R. 1992. Highly efficient generation of recombinant baculoviruses by enzymatically mediated site-specific in vitro recombination. *Nucl. Acids Res.* 20:495–500.

Pfarr, D. S., Sathe, G., and Reff, M. E. 1985. A highly modular cloning vector for the analysis of eukaryotic genes and gene regulatory elements. *DNA* 4:461–467.

Pfarr, D. S., Rieser, L. A., Woychik, R. P., Rottman, F. M., Rosenberg, M., and Reff, M. E. 1986. Differential effects of polyadenylation regions on gene expression in mammalian cells. *DNA* 5:115–122.

Pikaart, M. J., and Felsenfeld, G. 1996. Expression and codon usage optimization of the erythroid-specific transcription factor cGATA-1 in baculoviral and bacterial systems. *Prot. Expr. Purif.* 8:469–475.

Potter, H., Weir, L., and Leder, P. 1984. Enhancer-dependent expression of human kappa immunoglobulin genes introduced into mouse pre-B lymphocytes by electroporation. *Proc. Nat. Acad. Sci. U.S.A.* 81:7161–7165.

Power, J. F., Reid, S., Radford, K. M., Greenfield, P. F., and Nielsen, L. K. 1994. Modeling and optimization of the baculovirus expression vector system in batch suspension culture. *Biotechnol. Bioeng.* 44:710–719.

Pyle, L. E., Barton, P., Fujiwara, Y., Mitchell, A., and Fiedge, N. 1995. Secretion of biologically active human proapolipoprotein A-1 in a baculovirus-insect cell system: Protection from degradation by protease inhibitors. *J. Lipid Res.* 86:2355–2361.

Ranjan, A., and Hasuain, S. E. 1994. Influence of codon usage and translation initiation codon context in the AcNPV-based expression system: Computer analysis using homologous genes. *Virus Genes* 9:149–153.

Robbins, P. D., Tahara, H., and Ghivizzani, S. C. 1998. Viral vectors for gene therapy. *Trends Biotechnol.* 16:35–40.

Rubin, D. M., Mehta, A. D., Zhu, J., Shoham, S., Chen, X., Wells, Q. R., and Palter, K. B. 1993. Genomic structure and sequence analysis of *Drosophila melanogaster* HSC70 genes. *Gene* 128:155–163.

Saarinen, M. A., Troutner, K. A., Gladden, S. G., Mitchell-Logean, C. M., and Murhammer, D. W. 1999. Recombinant protein synthesis in *Trichoplusia ni* BTI-Tn-5B1–4 insect cell aggregates. *Biotechnol. Bioeng.* 63:612–617.

Santini, F., Penn, R. B., Gagnon, A. W., Benovic, J. L., and Keen, J. H. 2000. Selective recruitment of arrestin-3 to clathrin coated pits upon stimulation of G protein-coupled receptors. *J. Cell. Sci.* 113:2463–2470.

Sarmiento, U. M., Riley, J. H., Knaack, P. A., Lipman, J. M., Becker, J. M., Gately, M. K., Chizzonite, R., and Anderson T. D. 1994. Biologic effects of recombinant human interleukin-12 in squirrel monkeys (*Sciureus saimiri*). *Lab. Invest.* 71:862–873.

Schimke, R. T. 1988. Gene amplification in cultured cells. *J. Biol. Chem.* 263:5989–5992.

Schlesinger, S. 1993. Alphaviruses-vectors for the expression of heterologous genes. *Trends Biotechnol.* 11:18–22.

Schmitt, K., Daubener, W., Bitter-Suermann, D., and Hadding, U. 1988. A safe and efficient method for elimination of cell culture mycoplasmas using ciprofloxacin. *J. Immunol. Meth.* 109:17–25.

Schneider, I. 1972. Cell lines derived from late embryonic stages of *Drosophila melanogaster*. *J. Embryol. Exp. Morph.* 27:353–365.

Shitara, K., Nakamura, K., Tokutake-Tanaka, Y., Fukushima, M., and Hanai, N. 1994. A new vector for the high level expression for chimeric antibodies in myeloma cells. *J. Immunol. Meth.* 167:271–278.

Sondermann, P., Jacob, U., Kutscher, C., and Frey, J. 1999. Characterization and crystallization of soluble human Fc gamma receptor II (CD32) isoforms produced in insect cells. *Biochem.* 38:8469–8477.

Spellman, M. W., Basa, L. J., Leonard, C. K., Chakel, J. A., O'Connor, J. V., Wilson, S., and van Halbeek, H. 1989. Carbohydrate structures of human tissue plasminogen activator expressed in Chinese hamster ovary cells. *J. Biol. Chem.* 264:14100–14111.

Sridhar, P., Panda, A. K., Pal, R., Talwar, G. P., and Hasnain, S. E. 1993. Temporal nature of the promoter and not the relative strength determines the expression of an extensively processed protein in a baculovirus system. *FEBS Lett.* 315:282–286.

Sreekrishna, K., Brankamp, R. G., Kropp, K. E., Blankenship, D. T., Tsay, J. T., Smith, P. L., Wierschke, J. D., Subramaniam, A., and Birkenberger, L. A. 1997. Strategies for optimal synthesis and secretion of heterologous proteins in the methylotrophic yeast *Pichia pastoris*. *Gene* 190:55–62.

Stanley, P. 1989. Chinese hamster ovary cell mutants with multiple glycosylation defects for production of glycoproteins with minimal carbohydrate heterogeneity. *Mol. Cell. Biol.* 9:377–383.

Stern, L. J., and Wiley, D. C. 1992. The human class II MHC protein HLA-DR1 assembles as empty alpha beta heterodimers in the absence of antigenic peptide. *Cell* 68:465–477.

Stow, N. D. 1992. Herpes simplex virus type 1 origin-dependent DNA replication in insect cells using recombinant baculoviruses. *J. Gen. Virol.* 73:313–321.

Tessier, D. C., Thomas, D. Y., Khouri, H. E., Laliberte, F., and Vernet, T. 1991. Enhanced secretion from insect cells of a foreign protein fused to the honeybee melittin signal peptide. *Gene* 98:177–183.

Thirion, J. P., Banville, D., and Noel, H. 1976. Galactokinase mutants of Chinese hamster somatic cells resistant to 2-deoxygalactose. *Genetics* 83:137–147.

Toneguzzo, F., Keating, A., Glynn, S., and McDonald, K. 1988. Electric field-mediated gene transfer: Characterization of DNA transfer and patterns of integration in lymphoid cells. *Nucl. Acids Res.* 16:5515–5532.

Trill, J. J. 2001. Quantification of expression levels from transient and stably transfected cells using an IGEN M-Series analyzer. Manuscript in preparation.

Trill, J. J. 2000. GlaxoSmithkline. Unpublished data.

Trill, J. J. 1998. GlaxoSmithkline. Unpublished data.

Trill, J. J., Shatzman, A. R., and Ganguly, S. 1995. Production of monoclonal antibodies in COS and CHO cells. *Curr. Opin. Biotechnol.* 6:553–560.

Uphoff, C. C., Gignac, S. M., and Drexler, H. G. 1992. Mycoplasma contamination in human leukemia cell lines. II. Elimination with various antibiotics. *J. Immunol. Meth.* 149:55–62.

Urlaub, G., and Chasin, L. A. 1983. Isolation of Chinese hamster ovary cell mutants deficient in dihydrofolate reductase. *Proc. Nat. Acad. Sci. U.S.A.* 77:4216–4220.

Urlaub, G., Kas, E., Carothers, A. M., and Chasin, L. A. 1983. Deletion of the diploid dihydrofolate reductase locus in cultured mammalian cells. *Cell* 33:405–412.

van Oers, M. M., Vlak, J. M., Voorma, H. O., and Thomas, A. A. 1999. Role of the 3′ untranslated region of baculovirus p10 mRNA in high-level expression of foreign genes. *J. Gen. Virol.* 80:2253–2262.

Vaughn, J. L., Goodwin, R. H., Tompkins, G. J., and McCawley, P. 1977. The establishment of two cell lines from the insect *Spodoptera frugiperda* (Noctuidae). *In vitro* 13:213–217.

von Seggern, D. J., and Nemerow, G. R. 1999. Adenoviral–vectors for protein expression. In Fernandez, J. M., and Hoeffler, J. P., eds., *Gene Expression Systems: Using Nature for the Art of Expression*. Academic Press, San Diego, CA, pp. 290–325.

Wagner, R., Geyer, H., Geyer, R., and Klenk, H.-D. 1996a. *N*-acetyl-beta glucosaminidase accounts for differences in glycosylation of influenza virus hemagglutin expressed in insect cells from a baculovirus vector. *J. Virol.* 70:4103–4106.

Wagner, R., Liedtke, S., Kretzschmar, E., Geyer, H., Geyer, R., and Klenk, H. D. 1996b. Elongation of the *N*-glycans of fowl plaque virus hemagglutinin expressed in *Spodoptera frugiperda* (Sf9) cells by coexpression of human beta 1,2-*N*-acetyl glucosaminyltransferase, I. *Glycobiol.* 6:165–175.

Wahle, E., and Keller, W. 1992. The biochemistry of 3′-end cleavage and polyadenylation of messenger RNA precursors. *Ann. Rev. Biochem.* 61:419–440.

Wang, Y., O'Malley, Jr, B. W., Tsai, S. Y., and O'Malley, B. W. 1994. A regulatory system for use in gene transfer. *Proc. Nat. Acad. Sci. U.S.A.* 91:8180–8184.

Weiss, S., Famulok, M., Edenhofer, F., Wang, Y. H., Jones, I. M., Groschup, M., and Winnacker, E. L. 1995. Overexpression of active Syrian golden hamster prion protein PrPc as a glutathione *S*-transferase fusion in heterologous systems. *J. Virol.* 69:4776–4783.

Wenger, R. H., Moreau, H., and Nielsen, P. J. 1994. A comparison of different promoter, enhancer, and cell type combinations in transient transfections. *Anal. Biochem.* 221:416–418.

Weyer, U., Knight, S., and Possee, R. D. 1990. Analysis of very late gene expression by *Autographa californica* nuclear polyhedrosis virus and the further development of multiple expression vectors. *J. Gen. Virol.* 71:1525–1534.

Wickham, T. J., and Nemerow, G. R. 1993. Optimization of growth methods and recombinant protein production in BTI-Tn-5B1–4 insect cells using the baculovirus expression system. *Biotechnol. Progr.* 9:25–30.

Wirth, M., Bode, J., Zettlmeissl, G., and Hauser, H. 1988. Isolation of overproducing recombinant mammalian cell lines by a fast and simple selection procedure. *Gene* 73:419–426.

Yang, H., Leland, J. K., Yost, D., and Massey, R. J. 1994. Electrochemiluminescence: A new diagnostic and research tool. *Bio/Technol.* 12:193–194.

Index

A_{210} measurements, 108
A_{230} measurements, 103
A_{260} : A_{280} ratio
in monitoring DNA purification, 190–191
in monitoring RNA purification, 202
for polymerase chain reactions, 312
in spectrophotometry, 106
A_{260} measurements, 103, 104, 105–106, 108, 276–277, 280
variations among, 278
A_{280} measurements, 103, 105–106, 108
A_{320} measurements, 104
A_{600} measurements, 102
A_{820} measurements, 280
Absorbance (A)
calculating, 275–276
concentration and, 104–105, 108–109
nucleotide, 272–273, 275
optical density versus, 275–277
path length and, 286
Absorbance accuracy, of spectrophotometers, 99, 102–103, 103–104
Absorbance range, in spectrophotometry, 105–106
Absorbance ratio, in spectrophotometry, 103, 104, 105–106
Absorbance unit (AU), 276–277
Absorbency index, 277
Absorption coefficient, 277
Absorption maxima, of nucleotides (table), 272
Absorptivity, 277
Accidental self-inoculation, in biosafety, 123
Accuracy. *See also* Calibration
of pH meters, 87–89, 90
of pipettes, 75
of spectrophotometers, 96–97, 98–100, 101–106
Achilles' heel cleavage (AC), 252
in genome digests, 252–255
Acid dissociation constant (K_a), 33
Acids. *See also* Strong acids; Weak acids
buffer absorption of, 38
buffering of, 32–33
as buffers, 33
Acid-type buffers, amine-type buffers versus, 32
Acknowledging orders, 19
Acrylamide
as neurotoxin, 335
polymerization of, 338, 339, 341– 343
pore size with, 339–341
safe disposal of, 335–336
safety issues with, 334–336
shelf life of, 336
storage of, 336
Acrylamide gel solutions, degassing of, 371
Acrylamide poisoning, 335
Acrylamido buffers, with gradient gels, 346–347
Acrylic plastic, as shielding, 163
Active nucleic acid transfer, 418
Activity tables, for restriction enzymes, 237–238
Additives
in polymerase chain reactions, 306, 307
for protein solubility, 355
Adenine methylases, in genomic digests, 250–252
Adeno-associated virus (AAD), in eukaryotic expression, 504
Adenoviruses, in eukaryotic expression, 503–504
Adjustable-volume pipettes, 67–68

Advantages of immobilized pH gradient gels, 347
Advertising claims, 13–14
Aeration, in gene expression, 478
Affinity purification, problems after, 485
Affinity resins, problems with, 484
Affinity tags, with fusion systems, 472–474, 474–475
Affinity techniques
  in plasmid purification, 183–184
  secondary reagents and, 385
Agar media, handling, 137–138
Agarose, 355
  DNA isolation from, 188, 190
  for pulsed field electrophoresis, 246, 247
  RNA isolation from, 203
Agarose electrophoresis, 355–357
Agarose gels, nucleic acid transfer from, 418–420
Agarose microbeads, 251–252
Agarose plugs, 251
Agarose preparations, table of, 356
Aging of glassware, 138–139
Air buoyancy, as affecting balance accuracy, 52
Air currents, as affecting balance accuracy, 53
Air displacement pipette, 67
Alcohols
  in complex digests, 240–241
  as disinfectants, 131
Alkali fixation, crosslinking via, 423
Alkaline lysis, in plasmid purification, 180–182
Alkaline phosphatase (AP)
  for detecting proteins, 376, 377
  problems with, 394
Alkaline transfer
  conditions for, 419
  crosslinking via, 423
Aluminum foil, as autoclave wrapping, 134
American Type Culture Collection, 117
Amine-type buffers, acid-type buffers versus, 32
Amino acids
  absorbance data and concentration and, 108–109
  in proteins, 470, 475
Ammonium compounds, as disinfectants, 132
Ammonium ions, in DNA purification, 170
Ampholytes, with gradient gels, 346–347
Amplification
  in eukaryotic expression, 508–509
  in Western blotting, 387–388
Analysis date, for radioisotopes, 148
Analytical balances, 51
Angle, of centrifuge rotors, 58–61
Anhydrous buffer salts, 35
Animals
  biosafety with, 128–130
  disposal of parts of infected, 129–130
Anion exchange (AIX)
  in DNA extraction, 178–179
  in plasmid purification, 181
Annular pH electrode, 80
Antibiotics, in eukaryotic expression, 512–513
Antibodies
  gene expression and, 477
  stripping and reprobing and, 388–389
  for Western blotting, 378, 381–382, 383–384
Anticoagulants, in DNA purification, 170–171
Aoyagi, Kazuo, 291
Applications, for products, 13
Aspergillus, safe handling of, 127
Aspiration, with pipettes, 68, 74, 77
Assays
  of eukaryotic expression, 515–517
  protein quantitation, 109–110
Attitude, in minimizing radiation exposure, 162
Authorized users, of radioisotopes, 144
Autobuffer recognition, in pH meters, 82–83
Autocalibration, of pH meters, 82–83
Autoclave bags, in waste decontamination, 140
Autoclaves. *See also* Sterilization
  agar media in, 137–138, 139
  carbohydrates in, 139
  condensation with, 135
  do-it-yourself, 137–140

glassware in, 138–139
indicator tape with, 135
microbial contamination and, 124
plastic materials in, 135
safety and, 120, 139
sterilizing membranes in, 417
time requirements for, 135
in waste decontamination, 139–140
wrapping for, 134
Autoclave settings, 133–134
Autographa californica genome, 503
Automatic temperature compensation (ATC), in pH meters, 84–85
Autoradiography film, in nucleic acid hybridization, 436–441
Avidin
in hybridization buffer, 429
problems with, 395
in Western blotting, 381–382, 386–387
Axenic cultures, for protozoa, 128

Baby hamster kidney (BHK) cell lines, eukaryotic expression with, 505
BAC (*bis*-acrylylcystamine), as crosslinker, 338
Background noise, with storage phosphor imagers, 447
Background stain, as problem, 362, 395–396
Backup cultures, for experiments, 124
Bacteria
in acrylamide polymerization, 344
disruption of, 217–218
minimizing degradation of RNA from, 215, 217–218
restriction enzymes from, 227
safe handling of, 126–127
total RNA isolation from, 208–209
Bacterial strains, maintaining, 125–126
Bacteriophages, safe handling of, 127
Baculovirus, 521–532
implementing experiment with, 527–530
insect cell line versus, 523
planning experiment with, 521–527
selecting insect cell system and, 525–527
troubleshooting experiment with, 530–532
Bad data, in research, 5
Baking membranes, 422
Balances, 51–55
calibrating, 55
factors affecting accuracy of, 51–55
in pipette testing, 72–73
selecting, 54–55
service calls for, 55
types of, 51
Banana plugs, in electrophoresis, 337
Bands
after affinity purification, 485–486
fuzzy, 357
in native PAGE, 348
with pre-stained proteins, 364–365
problems with, 396
reproducible, 353
from skin keratin, 368
BandShift Kit Instruction Manual (Hennighausen & Lubon), 35
Bandwidth resolution, of spectrophotometers, 97–98
Barometric pressure, in testing pipettes, 76
Baseline flatness, of spectrophotometers, 99–100
Bases. *See also* Strong bases
buffer absorption of, 38
buffering of, 32–33
Batch binding, in RNA purification, 211
Batch numbers, for products, 25
BCA assay, 109
Bead milling, cell disruption via, 217–218
Beckman-Coulter rotor, for centrifuges, 63, 65
Beer-Lambert law, 97, 100, 104–105, 108, 275, 277–278, 279
Bell, Peter A., 461
Below balance weighing, 53
Benzene, radioisotopes in, 147
Bequerel (Bq), 145–146*n*, 155
Beta emitters
autoradiography film and, 438–439
shielding for, 163
Big companies, 12–13
Binding capacity, of hybridization membranes, 414
Biochemical compatibility, of buffers, 35
Biocompare Web site, 14
Biohazards, 114–117

defined, 114
disposal of, 139–140
safety levels of, 115–117
Bioreactors, in eukaryotic expression, 514–515
Biosafety
in animal handling, 128–130
decontamination in, 130–132
emergencies in, 122–124
during experiments, 119–122
in handling human tissues, 130
for media preparation staff, 133
in microbe handling, 126–128
microbial contamination in, 124–125
protective clothing for, 118–119
Biosafety hoods, 116–117
Biosafety levels (BSLs), 115–117
Biosci Web site, 13, 14
Biotin-bearing proteins
amplification and, 387–388
problems with, 396
in Western blotting, 381–382
Biowaste, decontamination of, 139–140
Biowire Web site, 14
Bis-acrylamide
as crosslinker, 338, 339
safety issues with, 334–336
Biuret assay, 109
BLAST (Basic Local Alignment Search Tool) searches, 314, 318, 328
Blocking
problems with, 395
in Western blotting, 380–382
Blocking agents, 381–382
as hybridization buffers, 429–430
Blood, accidental self-inoculation with, 123
Blots, from Western blotting, 382–383
Blotting applications, membranes used in (table), 416. *See also* Southern blotting; Western blotting
"Blotto"
as blocking agent, 381
as hybridization buffer, 429–430
Blue-white screening assay, for restriction endonucleases, 235
Boiling, in plasmid purification, 180–182
Boil-over, during autoclaving, 137
Bonventre, Joseph A., 225
Booz, Martha L., 331
Bovine serum albumin (BSA)
as blocking agent, 381
as hybridization buffer, 429–430
as reducing agent, 239
Bradford assay, 109
Brakes, for centrifuges, 62–63
"Braking radiation," 152
Bremsstrahlung, 152
Brownlow, Eartell J., 113
Bruner, Brian, 197
Brush motors, in centrifuges, 64–66
BSL-1 agents, 115–116
eye protection against, 118
BSL-2 agents, 115, 116
eye protection against, 118
BSL-3 agents, 117
eye protection against, 118–119
BSL-4 agents, 115, 117
Budget managers, in project planning, 3
Buffer capacity, 34
Buffer concentration, 34
Buffer failure, 37–38
Buffers, 32–38. *See also* Lysis buffers
developing new hybridization, 430–431
in DNA purification, 170, 190
in drop dialysis, 258
effective ranges of, 33
in electrophoresis apparatus leaks, 368
filtration and, 37
with gradient gels, 346–347
hybridization equipment and, 435–436
hybridization temperature and, 425
hybridization times and, 427
microbial contamination of, 38
for native PAGE, 349–350
NIST, 83–84
for nucleic acid transfer, 418–419
operation of, 32–33
for peptide electrophoresis, 350
for pH meter calibration, 83–84, 89
in plasmid purification, 180–181
in polymerase chain reactions, 305–306, 317, 319
for polynucleotides, 283
probe concentration and, 426

in protein sample electrophoresis, 354
for pulsed field electrophoresis, 246, 247
in quantitating nucleotide solutions, 274
selecting, 33–35
for sequential double digests, 243–244
in simple digests, 238–239
for simultaneous double digests, 242
staining and, 362
from stock solutions, 36–37
storage lifetimes of, 37–38
substitutions among, 32
troubleshooting PCR, 320
types of hybridization, 427–430
unreliable, 35–37
uses of, 32
for Western blotting washing, 382
Buffer salts, 33
buffer reliability and, 35
for pH standards, 84
Burns, 120, 123–124

Calibration. *See also* Accuracy
of balances and scales, 55
of pH meters, 81, 82–83, 83–84, 87–89, 90, 92
of pipettes, 68, 70–77
of spectrophotometers, 96–97, 98–100
of storage phosphor imagers, 444
Callbacks, from suppliers, 27–28
Calomel reference systems, for pH meters, 77–78
Candida albicans, as biohazard, 115, 128
Carbohydrates, autoclaving of, 139
Carbon radioisotopes
autoradiography film and, 438, 439
as radioactive waste, 158
Casein, as blocking agent, 381
Catalyst concentration, in acrylamide polymerization, 343, 344
Catalyst potency, in acrylamide polymerization, 342–343
Cathodic drift, with gradient gels, 347
cDNA, in eukaryotic expression, 497–498
cDNA clones, in RNA purification, 200
cDNA libraries, 199, 210, 497–498
in RNA purification, 201
CellCube, in eukaryotic expression, 514
Cell cultures
for eukaryotic expression, 514
for pulsed field electrophoresis, 245–246
Cells. *See also* Host cells
isolating DNA from, 172–184
total RNA isolation from, 203–206
total RNA yield from, 201–202
Cellulose, in RNA purification, 210–212
Centers for Disease Control and Prevention (CDC), 115
Centrifugation, types of, 55–56
Centrifugation time, calculating, 59–61
Centrifuges, 55–67
brakes for, 62–63
hazardous materials in, 61
maintenance of, 62
refrigerated, 66
rotors for, 57, 58–61, 62–63, 64
selecting, 57
service calls for, 64–66
spills in, 66
troubleshooting, 64–67
tubes for, 61
walking, 66
Certificate of Analysis
for restriction endonucleases, 228, 232–233
with simple digests, 238
Cesium chloride (CsCl)
in plasmid purification, 182–183
in RNA purification, 204
Cesium gradients, in plasmid purification, 183
cGMP regulation, 22
Chaotropes, 175
in DNA precipitation, 174–175, 175–176
in RNA purification, 203–204
Charges, as affecting balance accuracy, 52
Chemical compatibility, of buffers, 34–35
Chemical contaminants, in DNA purification, 170–171
Chemical hazards, in microbiology labs, 120

Chemicals, data reliability and, 6–7
Chemical safety, during electrophoresis, 334–336
Chemiluminescent detection method
  amplification versus, 388
  with Western blotting, 375, 376, 379
Chemiluminescent labeling
  autoradiography film and, 440–441
  in hybridization experiments, 406
Chinese hamster ovary (CHO) cell lines, eukaryotic expression with, 505
Chloramphenicol, in plasmid purification, 182
Chlorinated water, organic compounds in, 45
Chloroform
  in complex digests, 240–241
  in DNA purification, 170
  in exonuclease contamination, 261
Chromatography, in plasmid purification, 180–182
Clarity, improving protein gel, 353
Class II restriction enzymes, 226
  characteristics of, 232
Class IIS restriction enzymes, 226
Clean sample preparation, for polymerase chain reactions, 312
Cleavage
  Achilles' heel, 252–255
  of DNA by restriction enzymes, 226–227, 229–230, 240
  fusion systems and, 474–475
  in genomic digests, 247–255
  problems with, 483–484, 485
  in star activity, 229–230
Cleavage enzymes
  fusion systems and, 474
  popular (table), 475
Cleavage sites, in genes, 500
Clonal selection, in eukaryotic expression, 513–514
Cloning
  with baculovirus, 525
  expression hosts for, 471
  of genes for expression, 475–476, 476–478
  methylation problems with, 231
  troubleshooting, 260–262
Closed footwear, for biosafety, 118
Cloth, as autoclave wrapping, 134
Clothing, decontamination of, 132
Coccidoides immitis, as biohazard, 128
Codons, in gene expression, 466–467
Cofactors, with simple digests, 238
Colony transfers, 421
Colored dyes, radioisotopes in, 147
Colorimetric assays, for oligonucleotides, 280
Colorimetric detection method, with Western blotting, 375, 376
Colorimetric labeling, in hybridization experiments, 406
Column chromatography, in RNA purification, 210–211
Combination electrodes, in pH meters, 86–87
Companies, 12
  big and small, 12–13
  communicating needs to, 18–19
  manufacturing by, 13, 14
  researching of, 13
Comparisons, side-by-side, 26
Compatibility, of buffers, 34–35
Complaints, 28–29
Complete digestion, determining, 244
Complex digests, 239–244
  double digests as, 242–244
  modifying reaction conditions in, 241–242
  PCR products in, 239–240
  substrates for, 239–241
Composite gels, protein resolution with, 347–348
Computers (PCS), with spectrophotometers, 95
Computer software
  for BLAST searches, 328
  for selecting primers, 327
Concentrated stock solutions, of polynucleotides, 287
Concentration, of polynucleotide solutions, 285–286
Condensation
  as affecting balance accuracy, 51
  in autoclaves, 135
Conditioning electrodes, 87–88
Constant current power, for electrophoresis, 351
Constant voltage power, for electrophoresis, 351–352
Contacting suppliers, 26–29
Containers, in centrifuges, 61

Contaminants
in complex digests, 240–241
in DNA purification, 168, 169, 170–171, 172–173
in experiments, 124–125
of nucleotides, 269
in polymerase chain reactions, 306
in spectrophotometry, 103
staining and, 362
Contaminated reagents, 39
Contamination
in acrylamide polymerization, 344
of culture media, 126
of hybridization equipment, 435–436
minimizing in polymerase chain reactions, 306–308
in radioactive shipments, 152–153
of restriction enzymes, 260–262
during RNA purification, 205, 206
Contamination monitoring, in radioactive work areas, 161
Continuous buffer systems, for native PAGE, 349–350
Control elements, with baculovirus, 525
Control regions, of genes, 499
Controls, in experiments, 21
Convection currents, as affecting balance accuracy, 54–55
Conversations with suppliers
initiating, 26
logging, 26–27
Coomassie stain, 359, 361, 362
Core histones, in complex digests, 241
COS cells, in eukaryotic expression, 502–503
Cosmotropes, 175
in DNA precipitation, 174–175
Costs
of DNA purification, 169
of restriction enzymes, 227–228
Count-rate meters, in monitoring radiation exposure, 159–160
Counts per second, dose rates versus, 160–161
Covalent affinity chromatography, in plasmid purification, 183–184
Coyer, Howard, 267
CRC Handbook of Chemistry, 84
Criticism, of research, 8–9
Cross-adsorption, secondary reagents and, 385
Crosslinkers, for PAGE, 338
Crosslinking nucleic acids, 422–424
membrane shelf life after, 423–424
methods of, 422–423
problems with, 423
Crush and soak procedure, DNA purification via, 187, 188
Cryptic translation, 483, 485
Culture media
for eukaryotic expression, 511–513
maintaining, 126
preparing, 132–140
Curie (Ci), 145–146*n*, 155
Customers
complaints by, 28–29
researchers as, 14–18
Custom products, 18–19
Cuvettes, 286
cleaning, 101
for quantitating dilute RNA, 218, 219
for spectrophotometers, 100, 101
Cycle efficiency, with polymerase chain reactions, 297–298
Cycling parameters
in polymerase chain reactions, 309, 310
troubleshooting PCR, 320
Cysteines, in protein expression, 468

Dadd, Andrew T., 49, 94
Data
bad, 5
compelling, 5
maximizing reliability of, 6–7
in project planning, 3
in research, 5
Databases, for protein identification, 367
Data gathering, in project planning, 3
DATD (N,N′-diallyltartardiamide), as crosslinker, 338
Davies, Michael G., 49, 94
Decontamination
in biosafety, 130–132
of RNase, 212–213
of waste, 139–140
Degassing, of acrylamide solutions, 371
Deionized water, 43
pH of, 44
Deionizing cartridges, 43

Delays, in project planning, 3
Delivery, frequency of, 18
Denaturants
  in DNA extraction, 173
  as hybridization buffers, 427–428
Denaturing, of probes, 412
Denaturing methods, in plasmid purification, 180–182
Density gradient, in centrifugation, 56
Deoxynucleotides
  nomenclature of, 269
  quantitating solutions of, 273–275
Department of Transportation (DOT), radioactive shipment regulations by, 153
DEPC-treated water, 214
Depurination, in nucleic acid transfer, 419
Deration curves, for centrifuges, 63, 64
Detection and analysis strategy
  choosing detection methods in, 375–380
  with electrophoresis, 357–363
  with polymerase chain reactions, 315, 316
Detection methods
  with fusion systems, 472
  in hybridization experiments, 436–441, 441–448
  table of, 375
Detergents
  in DNA extraction, 173
  as hybridization buffers, 427, 429
  for protein sample electrophoresis, 354–355
  in RNA purification, 203–204
Deuterium lamps
  maintaining, 107
  in spectrophotometry, 103
  working without, 107–108
Dharmaraj, Subramanian, 197
DHEBA (N,N′-dihydroxyethylene-bis-acrylamide), as crosslinker, 338
DHFR (dihydrofolate reductase), in eukaryotic expression amplification, 508–509
Dialysis, troubleshooting with, 257–259
Diatomaceous earths, in DNA extraction, 177
Dideoxynucleotides
  nomenclature of, 269
  quantitating solutions of, 273–275
Diethylpyrocarbonate (DEPC) treatment, RNase-free solutions via, 213–214
Digestion, problems with protein, 485–486
Diluted restriction enzymes, stability of, 236
Dilution, reducing restriction enzyme costs via, 228
DIN buffers, for pH meter calibration, 83, 84
Diode array spectrophotometers, 95, 96
Direct autoradiography, in nucleic acid hybridization, 437
Directions, reading, 19
Direct labeling strategies, 410–411
Discontinuous buffer systems, for native PAGE, 349–350
Disinfectants, 116–117
  types of, 131–132
Disintegrations, in radioactivity, 155
Dispensing, with pipettes, 68, 74–75, 77
Disposable cuvettes, for spectrophotometers, 100
Dissolved oxygen, eliminating in acrylamide polymerization, 342
Distance, in minimizing radiation exposure, 162
Distilled water, 42–43
  pH of, 44
Dithiothreitol (DTT), stripping via, 389
DNA. *See also* Nucleotides; Oligonucleotides; Polynucleotides; RNA
  complex digests of, 239–243
  in genomic digests, 244–255
  as hybridization target, 402
  simple digests of, 236–238
  in troubleshooting restriction enzymes, 255–259
DNA adenine methylases, in genomic digests, 250–252
DNA binding proteins, with Achilles' heel cleavage, 252–253
DNA concentration, in complex digests, 241–242

DNA ligase, faulty, 261–262
DNA methylation, in genomic digests, 248–255
DNA polymerases. *See also* Taq DNA polymerase
evaluating for PCR, 296–303
properties of (table), 300–301
DNA precipitation, 173–174
DNA purification, 168–191
filter cartridges for, 185–186
fundamental steps in, 172–174
maximizing quality of, 171–172
methods of, 174–180
monitoring quality of, 190–191
plasmid, 180–184
required for experiments, 168
strategies for, 168–172
via gels, 187–190
via silica resins, 186
via spun column chromatography through gel filtration resins, 184–185
DNA samples. *See also* Supercoiled DNA
in DNA purification, 171–172
extinction coefficients for, 108
maximizing shelf life of, 172
molecular weights of, 168
production of, 168
solving problems involving, 23–25
storage of, 172
washing glassware for, 137
DNA sequences, restriction enzymes for specific, 226–227
DNases
DNA contamination with, 169
in DNA extraction, 173
Documentation
in developing new hybridization buffers, 431
of media room requests, 136
in ordering custom products, 18, 19
Dose monitoring, in radioactive work areas, 161
Dose-rate meters, in monitoring radiation exposure, 159, 160–161
Dose rates, counts per second versus, 160–161
Dosimeters, monitoring radiation exposure with, 161
Dot/slot blots, in RNA purification, 200
Double-beam spectrophotometers, 95–96
Double-coated autoradiography film, 438
Double digests, 242–244
sequential, 243–244
simultaneous, 242–243
Double-junction combination electrode, 79
Double-stranded markers, in electrophoresis, 356, 364
Double-stranded nucleic acid polymers
nomenclature of, 281–282
solutions of, 287–288
Double-stranded probes, hybridization times and, 426–427
Double-vector systems, for eukaryotic expression, 510
Doubling efficiency, with polymerase chain reactions, 297–298
Dounce homogenizer, cell disruption via, 216
*Dpn*I restriction enzyme, in genomic digests, 250–252
Drafts, as affecting balance accuracy, 53
Drop dialysis, troubleshooting with, 257–259
Drosophila, 523
eukaryotic expression with, 505
Drug selection markers, for eukaryotic expression, 508
Drying
concentrating radioactive solutions via, 164–165
in DNA purification, 170
Dry items. *See* Nonliquids
Dry membranes
in hybridization, 432
in nucleic acid transfer, 420–421
Dyes
for nucleic acid transfer, 420
quantitating dilute RNA via, 219
radioisotopes in, 147
Dynamic range, 442
of storage phosphor imagers, 442–443

Ebola virus, as biohazard, 117
Ecdysone, in eukaryotic expression, 509–510

Efficiency, with polymerase chain reactions, 297–298
18MΩ water, 43–44
handling, 44
800-base pair DNA fragment, labeling in hybridization experiments, 404–405
Electrical hazards, in biosafety, 124
Electrical potential, in pH meters, 78–79
Electrical power, for electrophoresis, 350–353
Electrical safety, during electrophoresis, 336–337
Electrodes
in pH meters, 77–79, 85–87, 87–89, 92
preparing and conditioning, 87–88
testing, 92
Electroelution, DNA isolation via, 187, 189
Electroendosmosis (EEO), 355–356
Electronic rotor-stator homogenizers, cell disruption via, 216
Electrophoresis, 334–368. *See also* Gel electrophoresis; Pulsed field electrophoresis
agarose, 355–357
buffer system problems in, 349–350
chemical safety for, 334–336
degassing solutions for, 371
detection staining in, 358–363
electrical power for, 350–353
electrical safety for, 336–337
elution strategies in, 357, 358–359
gel clarity in, 353
gel standardization in, 363–368
native PAGE, 348–349
RNase-free, 213
sample preparation for, 353–355
selecting buffer systems for, 350–353
selecting PAGE gels for, 337–345, 345–348
troubleshooting, 368
Electrophoresis gels. *See also* Gel electrophoresis
DNA isolation from, 187–190
elution of nucleic acids and proteins from, 357, 358–359
pH gradients for, 366–368
selecting for PAGE, 337–345, 345–348
standardizing, 363–368
Electrophoresis work areas, safety in, 336–337
Electrostatic forces, as affecting balance accuracy, 52
ELISA (enzyme-linked immunoabsorbent assay), in eukaryotic expression, 515, 516
Elution, of nucleic acids and proteins from gels, 357, 358–359. *See also* Electroelution
Emergencies, biosafety, 122–124
Emergency lab shower, for biosafety, 119
Emissions, from radioisotopes, 145
Englert, David F., 399, 441
Entrepreneurs, scientists as, 12
Environmental contaminants, 124–125
Enzymatic reactions, for detecting proteins, 376
Enzymes
in DNA extraction, 173
in DNA purification, 170
in polymerase chain reactions (table), 300–301
Epitope tags
with baculovirus, 524
in genes, 500
Eppendorf standard operating procedure, for pipettes, 72–77
Equilibrium density centrifugation, 56
Equipment, 5–6
data reliability and, 6
in hybridization, 435–436
Erasure, of storage phosphor imagers, 447
Escherichia coli, 251, 521
accidental self-inoculation with, 123
as biohazard, 115
expression systems for, 462–486
expression vectors for, 462–463, 470–471, 478, 479
genealogy of strains of, 117
gene expression in, 465–467
lysis of, 479–480
methylation and, 231
preparing for pulsed field electrophoresis, 245–246
protein degradation in, 464
proteins in, 468–470

recognition sites in, 248
strains of, 125
troubleshooting protein expression in, 480–486
uses of, 462
Ethanol
DEPC treatment and, 213
as disinfectant, 131
in DNA extraction, 178, 189
radioisotopes in, 147
Ethidium bromide (EtBr)
in plasmid purification, 181
in RNA purification, 202
stained nucleic acids and, 363
troubleshooting in RNA purification, 222
Eukaryotic expression, 492–533. *See also Escherichia coli*; Gene expression; Prokaryotic promoters; Protein expression systems
baculovirus and, 511–532
case study of, 519–521
described, 492–493
implementing project with, 511–517
insect cell system for, 521–524
maximizing protein yield from, 529–530
planning project with, 493–511
purification after, 530
regulating, 509–510
troubleshooting, 517–521
European Community (EC), disease control centers of, 115
European Pharmacopoeia, spectrophotometers and, 98–99
Evaporation, during pipette testing, 73
Exonuclease assay, for restriction endonucleases, 234–235
Exonuclease contamination, of restriction enzymes, 260–261
Exonucleases, in complex digests, 240
Experience, in project planning, 3, 4
Experiments
contaminants in, 124–125
controls in, 21
handling hazardous materials during, 119–122
microbial mix-ups in, 124, 125
nucleic acid hybridization, 401–403
with radioisotopes, 153–155
Expertise, in data gathering, 5–6
Expiration dates, 22
Explosion-proof refrigerators, storing reagents in, 41
Exponential decay equation, modeling radioactive decay with, 148–149
Exposure times, with storage phosphor imagers, 445
Expression hosts
in eukaryotic expression troubleshooting, 519
selecting, 501–506
Expression systems. *See* Gene expression; Protein expression systems
Expression vectors
commercial, 470–471, 498–499
in eukaryotic expression, 498–500
selecting, 506–511
structure of, 462–465
Extinction coefficient (E), 108, 276. *See also* Molar extinction coefficient (ε)
calculating, 277
for oligonucleotides, 279–280
Extremely pure nucleotides, 269
Eye protection, for biosafety, 118–119
Eyewash station, for biosafety, 119

$F(ab')_2$ fragments, as secondary antibodies, 385–386
Face masks, for biosafety, 119
Facilities, in project planning, 2–3
Faint bands, troubleshooting in PCRs, 319
Federal Register 21 CFR parts 210, 211, and 820, 22
Femtogram sensitivity, in hybridization experiments, 403
Fergusson plots, 364
Fetal calf serum, as blocking agent, 381
Fibrous tissue, total RNA isolation from, 207
Fidelity, with polymerase chain reactions, 296, 297, 300, 302
Field measurements, with pH meters, 90
Fill holes, in pH meters, 80
Fill solutions, for pH meters, 79–80, 92

Film. *See* Autoradiography film
Filter cartridges, for DNA purification, 185–186, 189
Filters. *See also* Quartz filters
in colony and plaque transfers, 421
in hybridization, 432
in RNA purification, 206
Filtration. *See also* Gel filtration
of laboratory water, 45–46
in making buffers, 37
Fire, 120, 123–124
in electrophoresis work areas, 336–337
First-aid kit, for biosafety, 119
Fixed angle rotor, 59
Fixed sample preparation, for polymerase chain reactions, 311
Fixed-volume pipettes, 67–68
Flammable liquids, in biosafety, 123
Flammable storage refrigerators, storing reagents in, 41
Flare, with storage phosphor imagers, 447–448
Flasks, refrigerating reagents in, 41
Flexibility, in direct versus indirect labeling, 410–411
Fluorescent detection method, with Western blotting, 375, 376–377, 379
Fluorescent labeling, in hybridization experiments, 406
Fluorographic chemicals, autoradiography film and, 437, 439
Fluorometry, quantitating dilute RNA via, 219
Footwear, for biosafety, 118
Formamide, as hybridization buffer, 427–428
Franciskovich, Phillip P., 1
Free radicals, in secondary decomposition of radioisotopes, 156
Freeze and squeeze procedure, DNA purification via, 188, 190
Freeze-drying
in DNA purification, 171
in storing purified DNA, 172
Freezers, storing reagents in, 41
Freezing
in DNA purification, 171
in gene expression, 479–480
minimizing RNA degradation via, 214–215
of restriction endonucleases, 232
of secondary reagents, 384–385
in storing radioisotopes, 156–157
French Press lysis method, 480
Frequency of delivery, 18
Frost, in centrifuges, 62
Fungal contamination, in experiments, 124
Fungi
disruption of, 217
minimizing degradation of RNA from, 217
safe handling of, 126–127, 127–128
Fusion proteins, problems with, 484, 485–486
Fusion systems
amino acids and, 475
avoiding, 472–474
cleavage enzymes and, 474
commercially available (table), 473–474
selecting, 471–472
Fuzzy bands, in electrophoresis, 357

Gamma emitters
autoradiography film and, 439–440
shielding for, 163
GC content, of genes, 466
Geiger-Müller counter, in monitoring radiation exposure, 159
Gelatin, as blocking agent, 381
Gel casting, acrylamide and, 334
Gel electrophoresis. *See also* Electrophoresis; Electrophoresis gels; Pulsed field electrophoresis
in gene expression, 477
in Southern blotting, 244–245
Gel filtration
in DNA purification, 184–190
in plasmid purification, 181
Gel overlays, in electrophoresis, 341
Gel purification, 187–190
Genealogies, of host cells, 117
Gene array hybridization, in RNA purification, 200
Gene expression, 465–467
cloning for, 475–476
codon usage in, 466–467
gene GC content and, 466

gene size in, 467
protein size in, 467
reliable controls for, 309
screening for, 476–478
secondary structures and, 467
sources for, 465
start codons in, 466
Genes
cleavage sites in, 500
cloning for expression, 475–476
control regions of, 499
epitope tags in, 500
in eukaryotic expression, 496–497
screening for expression, 476–478
sequence information for, 499
subcloning of, 500
Gene size, in gene expression, 467
Genomic contamination, during RNA purification, 205, 206
Genomic digests, 244–255
creating rare or unique restriction sites and, 247–255
for pulsed field electrophoresis, 245–247
for Southern blotting, 244–245
Genomic DNA, as hybridization buffer, 429
Germicidal UV lights, 116
Gerstein, Alan S., 267
g forces, in centrifuges, 58–59, 61, 63
Glass fiber filters, in RNA purification, 206
Glass milk, 177
in DNA extraction, 176–178, 188
Glass slides, as membrane supports, 415
Glassware, aging of, 138–139
Global problems, in project planning, 3
Gloves
for biosafety, 119
in preparing RNase-free solutions, 213
Glutamine synthetase system, in eukaryotic expression amplification, 509
Glycosylated proteins, eukaryotic expression of, 528
Good fortune, taking advantage of, 4
Good Laboratory Practice (GLP), spectrophotometers and, 98
Grades, of reagents, 39
Gradient gels, 345–346
Gradowski, Anita, 267
Graduated cylinders, refrigerating reagents in, 41
Gratitude, 28
Gravimetric testing, of pipettes, 72–77
Gravity, as affecting balance accuracy, 53–54
Gravity-flow-based columns, in DNA purification, 179
Ground glass homogenizers, cell disruption via, 216
Guanidium acid-phenol procedure, in RNA purification, 204–206
Guanidium-cesium chloride procedure, in RNA purification, 204
Guanidium isothiocyanate, in RNA purification, 204
Guanidium salts
in DNA precipitation, 175–176, 177, 189
in RNA purification, 204–206
Guidelines and standards, for pipette calibration, 70–71

Haidaris, Constantine G., 113
Halogens, as disinfectants, 131
Hamster, eukaryotic expression with, 505
Handling, of Western blotting antibodies, 384. *See also* Sample handling
Hard tissue, total RNA isolation from, 208
Hazardous materials. *See also* Biohazards
in centrifuges, 61
in microbiology labs, 119–122
Hazardous reagents, storage of, 42
Heat stability, polymerase chain reactions and, 302–303
Heavy-duty protective gloves, for biosafety, 119
Heavy metals, as disinfectants, 131
Henderson-Hasselbalch equation, 33, 35
Heparin, in DNA purification, 170–171
Heparinase, in DNA purification, 171
Herzer, Sibylle, 167, 399
Heterogeneity, of synthetic polynucleotides, 283

High purity samples, data reliability and, 7
High speed centrifuges, 57
High throughput, with polymerase chain reactions, 299, 311
High-throughput screen (HTS), 493, 495
Histoplasma capsulatum, as biohazard, 128
Hollow fiber bioreactors, in eukaryotic expression, 515
Homogenization
in DNA purification, 171
troubleshooting RNA, 221
Homogenizers, cell disruption via, 216
Hoods
for acrylamide, 334
for biohazards, 116–117
for volatile nuclides, 163
Horseradish peroxidase (HRP), for detecting proteins, 376
Horse serum, as blocking agent, 381
Host cells, genealogies of, 117
Hot spots, in radioactive work areas, 161
Household refrigerators, storing reagents in, 41
Human cell lines, eukaryotic expression with, 504
Human embryonic kidney (HEK) cells, in eukaryotic expression, 502–503
Human tissues, precautions in handling, 130
Humidity
as affecting balance accuracy, 55
in pipette testing, 73
Hybridization, 424–436. *See also* Gene array hybridization; Northern hybridization
failure of, 448–449
multiple membranes in, 432
new buffers for, 430–431
of nucleic acids, 401–453
optimal temperature for, 424–425
prehybridization times in, 426
probe concentration for, 425–426
reprobing in, 432–433, 433–434
selecting equipment for, 435–436
shelf life of buffers for, 431–432
stripping in, 432–433
timing of, 426–427
troubleshooting, 448–453
washing in, 434–435
Hybridization bottles, 435–436
Hybridization buffers, 435–436
developing new, 430–431
types of, 427–430
Hybridization efficiency, labeling and, 408–409
Hybridization experiments
labeling in, 403–409, 409–413
planning, 401–403
sensitivity of, 403
signal duration in, 407
Hybridization membranes, 413–418
handling of, 417
multiple, 432
for quantitative experiments, 417
selecting, 413–416
sterilization of, 417–418
Hybridization rate accelerators, 430
Hydrogen ions, pH and, 80
Hydroxyapatite (HA), in DNA extraction, 179–180
Hygroscopic buffer salts, 35
Hygroscopic samples, as affecting balance accuracy, 55

Identical products, manufacture of, 14. *See also* Reproducibility
IEF gels, 346–347
pH gradients for, 366–368
Immunoglobulin (Ig)
reactivity of, 386
secondary antibodies and, 384, 385
Incomplete procedural information, for buffer pH adjustment, 37
Incorporation efficiency, of labels, 412–413
Indicator tapes, in autoclaving, 135
Indirect autoradiography, in nucleic acid hybridization, 436–437
Indirect labeling strategies, 410–411
Infections
disposal of animal parts with, 129–130
in experimental animals, 128–129
Inhibitor-free sample preparation, for polymerase chain reactions, 312
Inhibitors
of complex digests, 240–241
testing for, 257
troubleshooting in PCRs, 320

Ink, for marking lab materials, 136
Inoculating loops
proper handling of, 120–121
Inoculation, of experimental animals, 128–129
Insect cell system
baculovirus versus, 523
for eukaryotic expression, 521–524
selecting, 525–527
transfer vectors for, 524–525
Institutions, radioisotopes for, 144
Intact sample preparation, for polymerase chain reactions, 311
International Air Transport Association (IATA), radioactive shipment regulations by, 153
Iodine radioisotopes, 161
autoradiography film and, 438, 439–440
shielding for, 163
Ionic detergents, for native PAGE, 355
Ionic strength differences, in buffering, 36
IPG (immobilized pH gradient) gels, 346–347
pH gradients for, 366–368
Isoelectric focusing (IEF)
detergents for, 354–355
PAGE versus, 346–348
Isopotential point, with pH meters, 81, 82
Isopropanol
as disinfectant, 131
in DNA extraction, 189
Isopycnic centrifugation, 56
rotors for, 57
Isoschizomers, restriction enzymes as, 227

Junctions
cleaning, 91
in pH meters, 78–79, 80, 91

Kennedy, Michele A., 49, 67
Keratin, electrophoresis band from skin, 368
k-factor, of fixed angle rotors, 59–61
Kirkpatrick, Robert, 491
Kracklauer, Martin, 197
Kruger, Greg, 11

Lab coats, for biosafety, 118
Labeling, in hybridization experiments, 403–409, 409–413
Labeling strategies, 409–413
Label location, of radioisotopes, 146
Labels
for autoradiography film, 438–440
hybridization efficiency and, 408–409
hybridization temperature and, 424
incorporated into probes, 407–408
incorporation efficiency of, 412–413
potency of, 412
signal duration from, 407
Laboratory Acquired Infections (Collins & Kennedy), 115
Laboratory Information Management System (LIMS), 95
Laboratory measurements, with pH meters, 90
Lab shower, for biosafety, 119
Laemmli buffer system, for native PAGE, 349–350
Latex gloves, for biosafety, 119
Lead, as shielding, 163
Leaks
in electrophoresis apparatus, 368
in enzyme shipments, 259
in pipettes, 69, 71–72
from prokaryotic promoters, 464
in water systems, 46–47
Leishmania, as biohazard, 128
Leishmania donovani, as biohazard, 114
Leverage, through sales representatives, 17–18
Liability, reimbursement for, 28
Lifetime, of radioisotopes, 156–157
Ligation/recut assay, for restriction endonucleases, 235
Ligations, failed, 260–262
Light
autoradiography film and, 436
storage phosphor imagers and bright, 447–448
Light sources, for spectrophotometers, 95–96
Linear acrylamide, in RNA purification, 220
Linear DNA fragments, in complex digests, 240
Linker length, hybridization efficiency and, 409

Lipid-rich tissue, total RNA isolation from, 207–208
Liquids
  autoclaving of, 133, 134
  in electrophoresis safety, 337
Liquid scintillation counter (LSC), 155
Logging conversations, 26–27
Long PCR, 303
Long term monitoring, in radioactive work areas, 161
Low ionic strength samples, measuring pH of, 90
Low melting point (LMP) agarose, DNA isolation from, 190
Lowry assay, 109
Low speed centrifuges, 57
Low-stringency washes, in hybridization, 434–435
Luminescent labeling, in hybridization experiments, 406
Lumps, in agar media, 137–138
Lyophilization, concentrating radioactive solutions via, 164–165
Lyophilized nucleotides
  purity of, 269–270
  stability of, 270–271
Lysis
  in DNA extraction, 173, 175–176
  in gene expression, 479–480
  in plasmid purification, 180–182
  in RNA purification, 215
  of yeast cells, 209
Lysis buffers
  in DNA extraction, 173
  in DNA purification, 169
Lysozyme, cell disruption via, 217, 218

Macroarrays, of hybridization membranes, 415
Magnesium chloride, optimizing for PCR, 303, 304, 306
Magnetic forces, as affecting balance accuracy, 53
Magnetic samples, as affecting balance accuracy, 53
Mammalian cells
  disruption of, 216–217
  minimizing degradation of RNA from, 214–215, 216–217
Manton-Gaulin lysis method, 480
Manuals
  in pipette testing, 74
  in troubleshooting, 64
Manufacturers
  of custom products, 18, 19
  expiration dates from, 22
  purity of reagents from, 39
Manufacturing, 13, 14
  of polynucleotides, 282–283
Marcy, Alice, 491
Martin, Lori A., 197
Material requirements
  controlling, 7–8
  data reliability and, 6–7
Mean volume, of pipettes, 75
Mechanical stress
  in DNA extraction, 173
  in DNA purification, 169
Media preparation facilities, 132
Media preparation staff, 132–133
Media room, labware sterilization in, 136
Medium emitters, autoradiography film and, 439
Megabase DNA fragments, from genomic digests, 248–255
Membranes. *See also* Dry membranes; Hybridization membranes; Nitrocellulose membranes; Nylon membranes; PVDF (polyvinyl difluoride) membranes; Transfer membranes; Wet membranes
  baking, 422
  functional, 417
  multiple, 432
  in protein expression, 468–469
  stripping and, 389
  with Western blotting, 379–380
2-Mercaptoethanol (2-ME)
  radioisotopes in, 147
  stripping via, 389
Mercury, as disinfectant, 131
Methods and Reagents bulletin board, 13
Methylases, in genomic digests, 248–250, 250–251, 252–255
Methylation, in genomic digests, 248–255
Methylation sensitivity, of restriction enzymes, 231

Microarrays
of hybridization membranes, 415
storage phosphor imagers and, 443
Microbalances, 51
Microbes. *See also* Biosafety
accidental self-inoculation with, 123
decontamination of, 130–132
safe handling of, 126–128
Microbial contamination
of buffers, 38
of laboratory water, 45–46
in microbiology labs, 120–121
preventing, 124–125
Microbial culture spills, in biosafety, 123
Microbial mix-ups, in experiments, 124, 125
Microbial strains, maintaining, 125–126
Microbial suspensions, proper handling of, 122
Microelectrodes, in pH meters, 86
Microorganisms, as biohazards, 114–117
Migration, of nucleic acids in electrophoresis, 356
Millirem (mrem), 159
Minetti, Cica, 267
Misincorporation of nucleotides, troubleshooting, 321
Moisture
as affecting balance accuracy, 51
in refrigerated centrifuges, 66
Molar extinction coefficient (ε), calculating, 276–277
Molarity, of radioisotopes, 154
Molecular weight markers, for electrophoresis gels, 363–364
Molecular weights (MWs)
of DNA samples, 168
of polynucleotides, 285
pre-stained standards and protein, 364–365
of proteins, 363–368, 374
SDS-PAGE and, 345, 346
2-D gels and, 365–366
Western blotting and, 365
Moles
of radioisotopes, 154–155
storage phosphor imagers and, 444
Money
in doing research, 9. *See also* Costs
Monochromatic light, in spectrophotometry, 105
Monoclonal antibodies, for Western blotting, 384
"Moonsuits," in biosafety, 117
Motivation
for doing science, 8, 9
of sales representatives, 16
Mouse, eukaryotic expression with, 504
Mouth pipetting, 122
mRNA, purification of, 198–201, 212. *See also* Poly(A)-selected RNA
Multiple proteins, eukaryotic expression of, 529
Multiple-step procedures, with restriction enzymes, 260–262
Multiple-step reactions, in genomic digests, 248–255
Municipal water, organic compounds in, 45–46
Mutations, in baculovirus experiment, 531–532
Mycobacterium tuberculosis, as biohazard, 117, 128, 130
Mycoplasma contamination, in eukaryotic expression, 512–513

N,N′-Methylenebisacrylamide crosslinker, 339
National Center for Infectious Diseases (NCID), 115
National Institute of Standards and Technology (NIST)
calibration buffers from, 83–84
spectrophotometry standards from, 99
Native PAGE, 337, 348–349
detergents for, 354–355
electrical power for, 350–353
SDS-PAGE versus, 345–348
standardized gels for, 363
Near-vertical rotors, for centrifuges, 63
Negative controls, in polymerase chain reactions, 308–309
Neisseria gonorrhoeae, as biohazard, 116
Neoschizomers, restriction enzymes as, 227
Nernst equation, pH meters and, 77, 81–82

Neurotoxin, acrylamide as, 335. *See also* Toxicity
Neutral density filters, for spectrophotometers, 99
Nicking assay, for restriction endonucleases, 234
Nitrocellulose membranes
in hybridization experiments, 414, 416
sterilization of, 417–418
transfer buffers for, 418–419
UV crosslinking with, 422
with Western blotting, 379–380
Nitrogen gas, concentrating radioactive solutions via, 164–165
Noise, in spectrophotometers, 99–100
Nomenclature
for nucleotides, 268–269
of polynucleotides, 281–282
Nonfat dry milk
as blocking agent, 381
as hybridization buffer, 429
Nonliquids, autoclaving of, 133–134, 134
Non-NIST-traceable buffers, 83–84
Non-phenol based methods, of RNA purification, 206
Nonradioactive labeling
for autoradiography film, 440
in hybridization experiments, 405, 406, 409–413
stability of, 411
Nonrefillable electrodes, in pH meters, 87
Normal serum, as blocking agent, 381
Northern blot analysis, in eukaryotic expression troubleshooting, 518
Northern hybridization, in RNA purification, 198, 199–200, 203
Nose cones, of pipettes, 69, 70
*Not*I endonuclease, in genomic digests, 248–250
N-terminal signal sequences, in protein expression, 468–469
NTPs, nomenclature of, 269
Nuclear Regulatory Commission (NRC)
licenses to handle radioisotopes from, 143–144
radioactive shipment regulations by, 152
Nuclease protection, in RNA purification, 200, 203
Nuclease-rich tissue, total RNA isolation from, 208
Nucleases, DNA contamination with, 169, 172
Nucleation, hybridization times and, 426
Nucleic acid purification, 173–174. *See also* DNA purification
interference with, 169–171
methods of (table), 188–189
via centrifugation, 57
Nucleic acid-rich tissue, total RNA isolation from, 208
Nucleic acids. *See also* DNA samples; Plasmids; RNA
constant current separation of, 352
crosslinking, 422–424
elution from gels, 357, 358–359
ethidium bromide and, 363
hybridization of, 401–453
precipitation of, 173–174
spectrophotometry of, 96, 104, 105–106
stability of radiolabeled, 157–158
stains for (table), 360
Nucleic acid transfer, 418–421
from agarose gels, 418–420
dry versus wet membranes for, 420–421
optimizing, 421
Nucleotide concentration, in polymerase chain reactions, 303
Nucleotide quality, optimzing for PCR, 305
Nucleotides, 268–278. *See also* Oligonucleotides; Polynucleotides
absorption maxima of (table), 272
affecting PCRs, 303–305
monitoring degradation of, 272–273
nomenclature of, 268–269
purity of, 269–270, 272–273
quantitating volumes of, 273–275
stability of, 270–271, 275, 276
storing, 270–271
thermocycling of, 275, 276
troubleshooting in PCRs, 321
Nylon filters, as membrane supports, 415
Nylon membranes
in hybridization experiments, 414, 416

sterilization of, 417–418
transfer buffers for, 418–419

Obermoeller, Dawn, 197
Occupational Health and Safety Office (OSHA), decontamination rules from, 139–140
"Old-timers," learning from, 4
Oligo(dT)-cellulose
regeneration of, 211–212
in RNA purification, 210–212
Oligonucleotides, 278, 279–281
in Achilles' heel cleavage, 253
batch variation among, 284
extinction coefficients of, 108
labeling in hybridization experiments, 404
lyophilizing, 281
polynucleotides versus, 281–282
purity of, 279
quantitating, 279–280
in Southern blotting, 244
stability of, 280
storing, 280
One-step mRNA (poly(A) RNA) purification, 209
Optical density (*OD*), absorbance versus, 275–277
Orders, acknowledging, 19
Organic compounds, in water, 45–46
Organic solvents, in DNA precipitation, 175–176
O-rings, with pipettes, 69, 70
Overlabeling, in hybridization experiments, 409
Overlaying gels, in electrophoresis, 341
Overnight assay, for restriction endonucleases, 234

Paper, as autoclave wrapping, 134
Pareto principle, 15–16
Passive nucleic acid transfer, 418
Path length, absorbance and, 286
Pathogenic microbes, as biohazards, 114–117, 126–128
Patience
in doing research, 9
in hybridization experiments, 401
PCR cyclers, 309–310
PCR products, in complex digests, 239–240
PCRs (polymerase chain reactions), 292–322
BLAST searches for, 314, 328
buffers in, 305–306
computer software for, 327, 328
described, 292–293
evaluating DNA polymerases for, 296–303
long, 303
nucleotides affecting, 303–305
planning experiments using, 293–296, 296–315
plasmid DNA for, 327
positive and negative effectors of (table), 296, 297–300
primers affecting, 303–305
primers for, 327
sample preparation for, 311–312
troubleshooting, 315–322
Web sites for, 328–329
PCR strategies
developing, 293–296, 296–315
in RNA purification, 200
troubleshooting, 221–222
weak links in, 295–296
PDA (piperazine diacrylamide), as crosslinker, 338
Peer-reviewed journals, 5
Pelleting, 56, 57, 65
improving, 66–67
Pellets, problems with RNA, 220–221
Peptide electrophoresis, buffer systems for, 350
Percent *C* (%*C*), in electrophoresis, 341
Percent gels, 345–346
Percent *T* (%*T*), in electrophoresis, 341
Personal computers (PCS), with spectrophotometers, 95
Pfannkoch, Edward A., 31
pH
adjusting buffer, 37
of agar media, 137–138
buffer control of, 32–33, 34–35
buffer storage and, 38
of deionized water, 43
of 18MΩ water, 44
of hybridization buffers, 429
initial, 44
pK and, 33
for sequential double digests, 243

of stock solutions, 37
of transfer buffers, 419
pH adjustment
in buffering, 36
quantitating nucleotide solutions via, 274
pH calibration curve, 81
Phenol
in complex digests, 240–241
as disinfectant, 131
in DNA precipitation, 176
in DNA purification, 170
in drop dialysis, 258
in exonuclease contamination, 261
in RNA purification, 204–206
toxicity of, 120
Phenol derivatives, as disinfectants, 131
pH gradients, for electrophoresis gels, 366–368
pH meters, 77–94
accuracy of, 87–89, 90
autobuffer recognition with, 82–83
buffers for, 83–84, 89
calibrating, 81, 82–83, 83–84, 87–89, 90, 92
cleaning, 91
combination electrodes in, 86–87
electrodes in, 77–79, 85–87, 87–89
field measurements with, 90
fill holes for, 80
fill solutions for, 79–80
junctions in, 78–79, 80
maintenance of, 91–92
measurement with, 81–82, 90
microelectrodes in, 86
Nernst equation and, 77, 81–82
nonrefillable electrodes in, 87
operation of, 80–81
"ready" indicator with, 85
reference electrodes in, 77–78
refillable electrodes in, 87, 93
resolution of, 85
response time of, 93
sensing electrodes in, 77
service calls for, 94
temperature compensation for, 84–85
testing, 92
troubleshooting, 92–94
Phosphate buffers, in DNA purification, 170
Phosphate groups, in polynucleotides, 287
Phosphorus radioisotopes, 157–158, 161
autoradiography film and, 438, 439
as radioactive waste, 158
shielding for, 163
signal strength of, 406
Photometric accuracy, of spectrophotometers, 96–97
Photometric reproducibility, of spectrophotometers, 99
Photons, autoradiography film and, 436
Physical hazards, in microbiology labs, 120
pI
pH gradients and, 366–368
2-D gels for determining, 365–366
Pichia methanolica, eukaryotic expression with, 505–506
Pichia pastoris, eukaryotic expression with, 505–506
Pipettes, 67–77
calibrating, 68, 70–77
cleaning, 69, 70
leaks in, 69, 71–72
maintenance of, 68–69, 70–71
monitoring performance of, 71–77
parts of, 71
proper use of, 122
selecting, 67–68
techniques for using, 68
troubleshooting, 68–69, 77, 78
types of, 67
Pistons, with pipettes, 68, 70, 71, 77
pK, pH and, 33
Planning, 2–9
good fortune and, 4
Plant measurements, with pH meters, 90
Plant tissue
disruption of, 217
minimizing degradation of RNA from, 217
Plaque transfers, 421
Plasmids
centrifugation of, 63–64
in complex digests, 240
as expression vectors, 462–463
in polymerase chain reactions, 308, 309

preparation for PCRs, 327
purification of, 180–184
Plastic, as shielding, 163
Plastic cuvettes, for spectrophotometers, 100
Plastic materials, in autoclaves, 135
Plates, streaking of, 122
Poly(A)-selected RNA, purification of, 198–201, 209–210. *See also* mRNA
Polyacrylamide Gel Electrophoresis (PAGE)
buffers for, 349–350
Catalyst concentration, 343
Catalyst potency, 342
DNA isolation via, 187
electrical power for, 350–353
isoelectric focusing versus, 346–348
native versus SDS, 345–348
protein resolution with, 348
selecting gels for, 337–345, 345–348
successful native, 348–349
Polyadenylation regions, in eukaryotic expression, 507–508
Polyclonal antibodies, for Western blotting, 383
Polyclonal selection, in eukaryotic expression, 513–514
Polyethylene glycol (PEG), in plasmid purification, 181
Polymerization
of acrylamide, 338, 339
reproducible, 341–343
Polynucleotides, 281–288
length variation among, 284
manufacture of, 282–283
molecular weights of, 285
nomenclature of, 281–282
oligonucleotides versus, 281–282
phosphate groups in, 287
quantitating, 284–285, 285–286, 287
solutions of known concentration of, 285–286
storing, 287–288
structural uncertainty among, 283
Polysaccharide-rich tissue, total RNA isolation from, 207–208
Polystyrene cuvettes, for spectrophotometers, 100
Poorly labeled bottles, reagents in, 42
Pore size
in electrophoresis, 339–341
of hybridization membranes, 415
Positive controls, in polymerase chain reactions, 308–309
Positive displacement pipette, 67
Post PCR detection strategy, 316
Power. *See* Electrical power
Power supply parameters, for PAGE (table), 351
Prasauckas, Kristin A., 49, 55
Precision, of pipettes, 76
Prehybridization times, in hybridization, 426
Preparation, of reagents, 40
Preparing electrodes, 87–88
Primary antibody
problems with, 395
for Western blotting, 383–384
Primary decomposition, of radioisotopes, 156
Primer concentration, in polymerase chain reactions, 303–305
Primer matrix studies, 304–305
Primers
computer software for selecting, 313, 327
optimum lengths of, 313
in polymerase chain reactions, 303–305, 312–315
troubleshooting PCR, 320
Primer testing strategy, with polymerase chain reactions, 315
Priority check lists, for PCR projects, 293–295
Probe concentration, for hybridization, 425–426
Probes
denaturing of, 412
in hybridization experiments, 402, 407–408
purification of, 413
reuse of, 412
storage of, 411
Probe templates, in hybridization experiments, 402
Problem reduction, 2
Problem resolution, failed, 28
Problems. *See also* Global problems; Technical problems
with acrylamide polymerization, 342–343

with baculovirus experiment, 530–532
with big companies, 12–13
with buffers, 34–35
with centrifuge rotors, 62, 63
with centrifuges, 64–67
with centrifuge tubes, 61
with cloning, 231
with crosslinking, 423
data gathering for, 21–22
defining, 20
describing to suppliers, 27
with discontinuous buffer systems, 349–350
with electrophoresis protein sample preparation, 353–355
with *Escherichia coli* protein expression, 480–486
with eukaryotic expression, 517–521
explanations for, 20–21
in hybridization experiments, 448–453
from incomplete procedural information, 37
with methylation, 231
with pH meters, 91–92, 92–94
with pipettes, 68–77, 78
with polymerase chain reactions, 315–322
preventing, 19–20
with radioactive shipments, 151–153
with restriction enzymes, 255–262
with RNA purification, 220–222
six steps to solve, 20–23, 23–25
with small companies, 12–13
solving, 19, 20–23, 23–25
with storage phosphor imagers, 447–448
suppliers in solving, 25–26, 26–29
from water leaks, 46–47
with Western blotting, 389–396
Procedural information, for buffer pH adjustment, 37
Procedures
for buffer pH adjustment, 37
modifying, 19–20, 22–23
Products
changes in, 14
manufacture of identical, 14
modifications to, 27
ordering custom, 18–19
performance of, 13–14
reliability of, 14
testing applications of, 13
troubleshooting absence of PCR, 315–317, 319–320
troubleshooting wrong PCR, 318–319
Programming, cycling parameters and, 310
Projects
completing, 4, 296–315
defining, 2, 293–296
successful, 4
Prokaryotic promoters, 462–465
characteristics of (table), 463
leakage from, 464
stability of, 464–465
strong, 463
Promoters, 507. *See also* Expression vectors; Prokaryotic promoters
with baculovirus, 524
strengths of (table), 507
Proofreading activity, with polymerase chain reactions, 302
Proteases, 470
Protective clothing, for biosafety, 118–119
Protective gloves, for biosafety, 119
Protein A
problems with, 396
reactivity of, 386
secondary antibodies and, 384
Proteinases, in DNA extraction, 173
Protein digestion, problems with, 485–486
Protein expression systems
selecting, 465–475
troubleshooting, 480–486
working with, 475–480
Protein G
reactivity of, 386
secondary antibodies and, 384
Protein quantitation assays, 109–110
Protein resolution, with composite gels, 347–348
Proteins, 374–375
absorbance data and concentration for, 108–109
antibodies against, 378
databases for identifying, 367
elution from gels, 357, 358–359
for eukaryotic expression, 493–496, 501–502

in eukaryotic expression troubleshooting, 518
expressing glycosylated, 528
expressing more than one, 529
expressing secreted, 527–528
functionality of, 470
molecular weights of, 363–368
in native PAGE, 348–349
post-translational modification of, 469
in protein expression systems, 468–470
quantifying stained, 361–362
in RNA preparations, 202
solubility of, 481–483
stability of radiolabeled, 157–158
stains for (table), 360
stripping and reprobing and, 388–389
structural changes to, 470
toxicity of, 469–470
washing glassware for, 137
Western blotting of, 363
Protein samples, preparing for electrophoresis, 353–355
Protein size, in gene cxpression, 467
Protozoa, safe handling of, 126–127, 128
Pseudogenes, in polymerase chain reactions, 313
Pulsed field electrophoresis, preparing genomic digests for, 245–247
Purchasing radioisotopes, 147–148
Purification
after eukaryotic expression, 530
with fusion systems, 472
of polynucleotides, 284–285
of probes, 413
via centrifugation, 55–57
Purity. *See also* High purity samples
of eukaryotically expressed proteins, 502
of nucleotides, 269–270, 272–273
of oligonucleotides, 279
of reagents, 39–40, 343–345
of Western blotting antibodies, 383–384, 385
PVDF (polyvinyl difluoride) membranes
in hybridization experiments, 414, 416
with Western blotting, 380
Quality control, with Western blotting, 379
Quality control assays, for restriction endonucleases, 234–235
Quality control data, for restriction enzymes, 233–235
Quantitating ability
membranes for, 417
of storage phosphor imagers, 443–445
of Western blotting, 377–378
Quantitating nucleotides, 273–275
Quantitating oligonucleotides, 279–280
Quantitating polynucleotides, 284–285, 285–286
with polymerase chain reactions, 298–299
Quantitating radioactivity, 155
Quantitating stains, 361–362
Quantity needs, 18
Quartz filters, for spectrophotometers, 99
Quaternary ammonium compounds, as disinfectants, 132

Radiation dose
from *Bremsstrahlung*, 152
minimizing, 162–164
"Radiation equivalent man" (REM), 159
Radiation exposure, 159–164
measuring, 159
minimizing, 162–163, 164
monitoring, 159–161
Radiation safety officers (RSOs), 144, 163, 164
in monitoring radiation exposure, 161
radioactive shipments and, 150–152
radioactive waste disposal and, 158
Radioactive concentration, 164–165
molarity and, 154
of radioisotopes, 145–146
shelf life and, 149
in storing radioisotopes, 157
Radioactive detection method, with Western blotting, 375–376, 379
Radioactive labeling
for autoradiography film, 438–440
in hybridization experiments, 405–406, 409–413
stability of, 411

Radioactive materials. *See also* Radioisotopes
certification for handling of, 143–144
licensing for handling of, 143–144
safety with, 133, 142–165, 166
shelf life of, 148–149, 149–150
storing, 156–159
ten fundamental rules for handling, 142–143
Radioactive shipments, 150–153
arrival of, 150–151
problems with, 151–153
wipe test of, 151–152
Radioactive waste, disposal of, 158–159
Radioactive work areas
monitoring dosage in, 161
organizing, 164
Radioactivity. *See also* Specific activity
quantitating, 155
shelf life of, 148–149, 149–150
units of, 145–146*n*
Radioisotopes
autoradiography film and, 438–440
decay of, 148–150, 155, 156–159
describing radioactivity of, 145–146
designing experiments with, 153–155
handling, 159–165
handling shipments of, 150–153
label location of, 146
licenses to handle, 143–144
maximizing lifetime of, 156–157
minimizing radioactive dose from, 159–165
ordering, 147–148
physical properties of common (table), 166
selecting, 144–147
shelf lives of, 149–150
shielding for, 163
signal strengths of, 406
in solvents, 146–147
stability of, 157–158
storing, 156–159
Radiolytic decomposition, 148–150
causes of, 156
compensating for, 150
measuring, 155
Radionuclides. *See* Radioactive materials; Radioisotopes
Radius, of centrifuge rotors, 58–59
Ramp times, for polymerase chain reactions, 310
Rat, eukaryotic expression with, 504
Rate-zonal centrifugation, 55–56
rotors for, 57
Reaction time, for complex digests, 242
Reaction vessels, for polymerase chain reactions, 311
Reaction volume, for complex digests, 241
"Ready" indicator, of pH meters, 85
Reagents, 39–47
contaminated, 39
DEPC treatment and, 214
eukaryotic expression of, 493, 494, 495–496, 501–502
grades of, 39
judging purity of, 39–40
new from manufacturer, 39
from opened containers, 39
preparation of, 40
prepared by others, 40
purity of, 343–345, 362
refrigeration of, 41–42
secondary, 384–387
storage of, 39, 40–41, 41–42
water for, 42–47
Real time PCR detection strategy, 316
REBASE database, 227
RecA-assisted restriction endonuclease (RARE), with Achilles' heel cleavage, 253–255
Recognition sites, in genomic digests, 248
Recombinant baculovirus system, 522–524
preparing, 524
Recombinant DNA methods, with restriction enzymes, 226
Recombinant gene expression, 492–493. *See also* Eukaryotic expression; Gene expression
Records, of radioactive waste disposal, 158
Reducing agents, with simple digests, 239
Reference date, for radioisotopes, 148, 149–150

Reference electrodes
with pH meters, 77–78
in pH meters, 80
Refillable electrodes
in pH meters, 87, 91–92, 93
storage of, 91–92
Refrigerated centrifuges, troubleshooting, 66
Refrigerators, storing reagents in, 41–42
Refrigerator shelves, storing reagents on, 42
Regeneration, of oligo(dT)-cellulose, 211–212
Regulating eukaryotic expression, 509–510
Reimbursement, for company liability, 28
Relationships, with sales representatives, 16–17
Reliability
of balances and scales, 54–55
of data, 6–7
of products, 14
of reagents, 39–40
of suppliers, 13
Remeta, David, 267
Replication, data reliability and, 6
Reporting, of research results, 8–9
Reprobing
in hybridization, 432–433, 433–434
in Western blotting, 379, 380, 388–389
Reproducibility
of DNA purification methods, 169
in electrophoresis, 341–343
improving in electrophoresis, 353
with polymerase chain reactions, 298
of storage phosphor images, 443
Reproduction, of research results, 8
Research
defending, 8–9
motivation for doing, 8, 9
reporting of, 8–9
sales representatives in, 15–16
successful, 5–9
Research planning, 3
reporting results and, 8
Research style, 2
Resins, in RNA purification, 210
Resolution
of autoradiography film, 438
improving in electrophoresis, 353
of labels, 405–406
of pH meters, 85
of spectrophotometers, 97–98
of storage phosphor imagers, 441–442, 446
Resources, in project planning, 2–3
Response time, of pH meters, 93
Restriction endonucleases
properties of, 232–236
quality control assays for, 234–235
star activity of, 229–230
Restriction enzymes, 226–262
altering specificity of, 248–250
commercially available, 226–227
complex digests with, 239–244
costs of, 227–228
double digests with, 242–244
easily used, 229
genomic digests with, 244–255
methylation sensitivity of, 231
quality control data for, 233–235
restriction endonucleases as, 232–235
selecting, 229–231
shipments of, 259–260
simple digests with, 236–239
site preference of, 230
stability of, 236
titer assays for, 255–259
transformation failures with, 260–262
troubleshooting, 255–262
Restriction length fragment polymorphism (RFLP) analysis, 226
Restriction sites, creating rare or unique, 247–255
Reuse, of oligo(dT)-cellulose, 211–212
Reverse dot blots, in RNA purification, 200
Reverse osmosis, water purification via, 43–44
Review, data gathering and, 5
Ribogreen dye, quantitating dilute RNA via, 219
Ribonuclease protection assays, in RNA purification, 201, 203
Ribonucleotides
nomenclature of, 269
quantitating solutions of, 273–275
Riis, Peter, 373
Ring badges, monitoring radiation exposure with, 161

RNA. *See also* Nucleotides;
Oligonucleotides;
Polynucleotides
centrifugation of, 63
as hybridization target, 402
solutions of polymers of, 287–288
storage of, 214–215, 219, 222
RNA purification, 198–222
integrity of RNA from, 202–203
lysis in, 215
maximizing yield from, 212–219
pauses during, 218
predicting yield from, 201–202
storing RNA from, 219
strategies for, 198–212
troubleshooting, 220–222
RNA sample preparation, for
polymerase chain reactions, 312
RNase-free techniques, 212–214
RNase inhibition, 215
RNases
DNA contamination with, 169
in DNA extraction, 173
RNA contamination with, 212–213
in RNA degradation, 214, 215, 217,
219
Robinson, Derek, 225
Room temperature storage, of
reagents, 40
Rotor identification codes, for
centrifuges, 62
Rotors, for centrifuges, 57, 58–61,
62–63, 64
RT-PCR technique, 314
in RNA purification, 200, 201, 202,
203
troubleshooting, 221–222, 321–322
RTPs, nomenclature of, 269

Saccharomyces cerevisiae
as biohazard, 115, 128
recognition sites in, 248
S-adenosylmethionine, in simple
digests, 239
Safety
with acrylamide, 334–336
with biological materials, 114–140
electrical, 336–337
with radioactive materials, 133,
142–165, 166
Safety equipment, for biosafety, 119
Safety glasses, 118–119
Sales representatives, 14–18
expectations of, 15–16
functions of, 15
leverage via, 17–18
motivations of, 16
for pH meters, 94
relating to, 16–17
Salmonella typhi, as biohazard, 114,
115
Salmonella typhimurium, recognition
sites in, 248
Salt pellets, in RNA purification,
205–206
Salts
as buffers, 33, 35
in DNA precipitation, 174–175, 185,
189
in hybridization buffers, 427,
428–429
for sequential double digests, 243
for Western blotting buffers, 382
Sample collection, minimizing RNA
degradation during, 214–215
Sample concentration
absorbance and, 104–105, 108–109
in spectrophotometry, 103
Sample disruption, minimizing RNA
degradation during, 215–218
Sample handling. *See also* DNA
samples; High purity samples;
Hygroscopic samples; Low ionic
strength samples; Magnetic
samples; Semisolid samples;
Viscous samples
as affecting balance accuracy, 55
with pH meters, 93–94
in Western blotting, 382–383
Sample location, as affecting balance
accuracy, 55
Sample matrix, for pH meters, 85–86
Sample preparation, for polymerase
chain reactions, 311–312
Sample volumes
of cuvettes, 100
for pH meters, 86
for spectrophotometers, 95
Scales, 51–55
calibrating, 55
Scaling up
of eukaryotic expression, 514–515,
527
of gene expression, 478–479

Scheduling
adhering to, 7
in project planning, 4
Schlieren, in acrylamide polymerization, 342
Science, motivation for doing, 9
Scientific literature, 5–6
Scientists, companies and, 12
Scientist Web site, 14
Scintillation counters, 155
Screens, for storage phosphor imagers, 444, 445–447
SDS-PAGE, 337, 338, 349
electrical power for, 350–353
native PAGE versus, 345–348
problems with, 480–481
standardized gels for, 363
Sealed containment rooms, in biosafety, 117
Secondary antibodies, 384
problems with, 395
Secondary decomposition, of radioisotopes, 156
Secondary reagents
problems with, 395
species specificity in, 385
in Western blotting, 384–387
Secondary structures, in gene expression, 467
Secreted proteins, expressing, 527–528
Sedimentation coefficient, 55–56
Self-decontamination, 132
Self-inoculation, accidental, 123
Self-monitoring, in radioactive work areas, 161
Semimicro balances, 51
Semisolid samples, measuring pH of, 90
Sensing electrodes, in pH meters, 77, 80
Sensitivity
of autoradiography film, 438
of direct versus indirect labeling, 410
of hybridization experiments, 403
with polymerase chain reactions, 299–300
of storage phosphor imagers, 441–442, 445
in Western blotting, 377
Sequence information, for eukaryotic expression, 499
Sequencing cells, constant power electrophoresis with, 352–353
Sequencing gels, constant power electrophoresis with, 352
Sequential double digests, 243–244
Serial numbers, for products, 25
Serum
as blocking agent, 381
in eukaryotic expression, 511–512
Service calls
for balances, 55
for centrifuges, 64–66
for pH meters, 94
for spectrophotometers, 106
Service engineers, for pH meters, 94
Sharps, proper disposal of, 120
Shatzman, Allan R., 491
Shearing, in DNA purification, 169
Shelf life, 22
of acrylamide, 336
of hybridization buffers, 431–432
of labeled probes, 411
maximizing purified DNA, 172
of membranes after crosslinking, 423–424
of nucleotides, 270–271
of oligonucleotides, 280
of radioisotopes, 149–150, 156–157
of reagents, 40
of restriction endonucleases, 232
of restriction enzymes, 228
Shelves, storing reagents on, 40, 42
Shielding, in minimizing radiation exposure, 163
Shigella flexneri, as biohazard, 115
Shipments, of restriction endonucleases, 232, 259–260. *See also* Radioactive shipments
Short term monitoring, in radioactive work areas, 161
Shower, for biosafety, 119
Side-by-side comparisons, 26
Signal duration
in hybridization experiments, 407
in Western blotting, 377
Signal sequences, in protein expression, 468–469
Signal smears
in electrophoresis, 356–357
in genomic digests, 244
nuclease contamination and, 169

in polymerase chain reactions, 317–318
troubleshooting, 222
Signal strength, of labels, 405–406
Silica resins, in DNA extraction, 176–178, 186, 190
Silver/silver chloride reference system for pH meters, 77
Simple digests, 236–239
modifying reaction conditions in, 237–239
Simple two-fold titration, troubleshooting with, 255–257
Simultaneous double digests, 242–243
Single-beam spectrophotometers, 95
Single-coated autoradiography film, 438
Single nucleotide polymorphisms (SNPs), polymerase chain reactions for finding, 313
Single-stranded DNA, A_{260} values for, 278
Single-stranded markers, in electrophoresis, 356, 364
Single-stranded nucleic acid polymers
nomenclature of, 281–282
solutions of, 288
Single-vector systems, for eukaryotic expression, 510
Site preference, of restriction enzymes, 230
Six problem-solving steps, 20–23
example of, 23–25
Skin keratin, electrophoresis band from, 368
Slides, preparation of, 121
Small companies, 12–13
Smeared signals
in electrophoresis, 356–357
in genomic digests, 244
nuclease contamination and, 169
in polymerase chain reactions, 317–318
troubleshooting, 222
Smith, Tiffany J., 197
Sodium phosphate buffers, in DNA purification, 179
Software
for BLAST searches, 328
for selecting primers, 327
for storage phosphor imagers, 444
Solubility, of proteins, 481–483
Solubility problems, in baculovirus experiment, 532
Solution nucleotides
purity of, 269–270
stability of, 270–271
Solutions
quantitating dilute RNA, 218–219
RNase-free, 213
Solvents
radioisotopes in, 146–147
water as, 44
Sonication, in DNA extraction, 173
Southern blotting, 244, 449–453
Species specificity, in secondary reagents, 385
Specific activity, of radioisotopes, 145–146, 153–154. *See also* Radioactivity
Specificity, with polymerase chain reactions, 297
Spectinomycin, in plasmid purification, 182
Spectral bandwidth resolution, of spectrophotometers, 97–98
Spectrophotometers, 94–110
accuracy of, 96–97, 101–103, 103–104, 104–105, 105–106
calibrating, 96–97, 98–100
computers with, 95
cuvettes for, 100
light sources for, 95–96
limitations of, 102–103
maintenance of, 101, 107
operating, 107–110
resolution of, 97–98
selecting, 94–98
service calls for, 106
types of, 95–96
wavelength range of, 96
Spectrophotometry
quantitating dilute RNA via, 218–219
quantitating nucleotide solutions via, 273, 275
troubleshooting RNA, 221
Spectroscopy, quantitating nucleotide solutions via, 273, 275
Spending limits, 18
Spills
in biosafety, 123
in centrifuges, 66
Spin columns, in RNA purification, 211

Spinner flasks, in eukaryotic expression, 514
Spinning vacuum chambers, concentrating radioactive solutions via, 164–165
Spodoptera frugiperda, in eukaryotic expression, 503, 525
Spore-forming filamentous fungi, safe handling of, 127–128
Spun column chromatography through gel filtration resins, in DNA purification, 178–179, 184–185
Spyro Ruby stain, 359
Stability. *See also* Heat stability
  of cell proteins, 464–465
  of labeled probes, 411
  of nucleotides, 270–271, 275, 276
  of oligonucleotides, 280
  of radioisotopes, 157–158
  of restriction enzymes, 236
  of spectrophotometers, 99–100
Stable buffers, 37–38
Stable expression systems, for eukaryotic expression, 504–506
Stains
  high background, 362, 395–396
  for nucleic acids (table), 360
  for nucleic acid transfer, 419–420
  problems with, 395
  for proteins (table), 360
  selecting for electrophoresis, 357–363
Standard deviation, calculating, 75–76
Standards
  in experiments, 21
  NIST and, 83–84
  for protein quantitation assays, 109–110
  for spectrophotometers, 98–100
Standards and guidelines, for pipette calibration, 70–71
Staphylococcus aureus
  as biohazard, 115–116, 131
  preparing for pulsed field electrophoresis, 245–246
Star activity, of restriction endonucleases, 229–230
Start codons, in gene expression, 466
Statistics, data reliability and, 6
Sterilization, 116–117, 124–125. *See also* Autoclaves
  by media room, 136
  of nitrocellulose and nylon membranes, 417–418
  of plastic materials, 135
Stevens, Jane, 49, 77
Stirred tank bioreactors, in eukaryotic expression, 515
Stock solutions, buffers from, 36–37
Storage
  of acrylamide, 336
  of buffers, 37–38
  of hybridization buffers, 431–432
  of labeled probes, 411
  of membranes after crosslinking, 423–424
  of microbial strains, 125–126
  minimizing RNA degradation during, 214–215
  of nucleic acid polymers, 287–288
  of nucleotides, 270–271
  of oligonucleotides, 280
  of pH meters, 91–92
  of pipettes, 68
  of purified DNA, 172
  of purified RNA, 219, 222
  of radioactive materials, 156–159
  of reagents, 39, 40–41, 41–42
  of restriction endonucleases, 232
  of Western blots, 383
  of Western blotting antibodies, 384
Storage conditions, as source of problems, 21–22, 40–42
Storage phosphor imagers
  dynamic range of, 442–443
  erasure of, 447
  in nucleic acid hybridization, 441–448
  operation of, 441
  problems with, 447–448
  quantitative capabilities of, 443–445
  resolution of, 441–442
  screens for, 445–447
  sensitivity of, 441–442
  speed of, 441–442
Straight percent gels, 345–346
Stray light, as affecting spectrophotometer accuracy, 97, 99
Streaking, 122
Strength, of hybridization membranes, 414. *See also* Signal strength

Streptavidin
amplification and, 387–388
problems with, 395, 396
in Western blotting, 381–382, 386–387
Streptomyces achromogenes, restriction enzymes from, 227
Stripping
in hybridization, 432–433
in Western blotting, 379, 380, 383, 388–389
Strong acids, in buffering, 36
Strong bases, in buffering, 36
Structure, of polynucleotides, 283
Subcloning, in eukaryotic expression, 500
Suboptimal growth conditions, in baculovirus experiment, 530–531
Substrates, in complex digests, 239–241
Successful projects, 4
Successful research, 5–9
Sulfur radioisotopes, 157–158
autoradiography film and, 438, 439
shielding for, 163
signal strength of, 406
Supercoiled DNA, centrifugation of, 63–64
Suppliers, 11–29
communicating needs to, 18–19
contacting, 26–29
expectations of, 28
ordering custom products from, 18–19
problems with, 19–29
products from, 13–14
reliability of, 13
sales representatives of, 14–18
in solving problems, 25–26
working with, 12–14
Supplies, for biosafety, 119
Surgical instruments, autoclaving of, 134
Suspensions, proper handling of microbial, 122
Swinging bucket rotor, 58, 59
SYBR Gold dye, 420
quantitating dilute RNA via, 219
SYBR Green dye, 420
quantitating dilute RNA via, 219
Synaptic complex formation, restriction endonucleases and, 254
Synthetic polynucleotides, 281–288
Tags
in eukaryotic expression, 515–517
with fusion systems, 472–474, 474–475
Tap water, 42
organic compounds in, 45–46
Taq DNA polymerase
in DNA purification, 170–171
in polymerase chain reactions, 292
primers with, 313
*Taq*I methylase, in genomic digests, 250–251
Target pH, buffering toward, 33–34
Targets
for eukaryotic expression, 493–495, 501–502
of hybridization experiments, 402
Task planning, 2
TBE (Tris, borate, EDTA) buffer, in DNA purification, 170
Technical problems, in project planning, 3
Technical support, for pH meters, 94
TEMED potency, in acrylamide polymerization, 342–343
Temperature
in acrylamide polymerization, 343
as affecting balance accuracy, 52–53, 54–55
for autoradiography film detection, 440–441
for baking membranes, 422
in DNA precipitation, 174
in gene expression, 478
for hybridization, 424–425
nucleotides and, 275
pH meters and, 84–85, 86
during pipette testing, 73, 75
in plasmid purification, 180–181
for polymerase chain reactions, 302–303, 309–310
of restriction enzyme shipments, 259–260
storage phosphor imagers and, 446
in storing purified DNA, 172
in storing radioisotopes, 156–157
Temperature compensation, for pH meters, 84–85
Template contamination, minimizing in polymerase chain reactions, 306–308
Template modification, in polymerase chain reactions, 312

Templates, troubleshooting PCR, 315
TenBroeck homogenizer, cell disruption via, 216
Test liquids, in pipette testing, 73
Test points, in pipette testing, 74
Test volumes, in pipette testing, 74
Therapeutic proteins, eukaryotic expression of, 493, 494, 496, 501–502
Thermal degradation, of nucleotides, 275, 276
Thermocycling
  nucleotide stability and, 275, 276
  in polymerase chain reactions, 309–310
The Scientist Web site, 14
Thickness, of hybridization membranes, 414–415
30-mer oligonucleotide, labeling in hybridization experiments, 404
Tilt, as affecting balance accuracy, 53–54
Time
  autoclaving, 135, 136
  for autoradiography film detection, 440
  for complex digests, 242
  in DNA precipitation, 174
  in DNA purification, 168
  for hybridization, 426–427
  for media culture requests, 137
  in minimizing radiation exposure, 162
  for signal detection, 406
  for stains, 361
  for storage phosphor imager detection, 441–442
Tip immersion, of pipettes, 69
Tissue
  accidental self-inoculation with, 123
  isolating DNA from, 172–184
  total RNA yield from, 201–202, 203–206, 207–209
Titer assays, troubleshooting with, 255–259
Toluene, radioisotopes in, 147
Top-loader balances, 51
Total RNA, purification of, 198–201, 203–206, 207–209, 212
Toxicity, of proteins, 469–470. *See also* Neurotoxin
Transfection, in eukaryotic expression, 513
Transfer buffers, for nucleic acid transfer, 418–419
Transfer membranes, with Western blotting, 379–380
Transfer vectors, selecting, 524–525
Transformation failures, with restriction enzymes, 260–262
Transient expression systems, in eukaryotic expression, 502–503
Transilluminators, crosslinking on, 422
Translation, with prokaryotic promoters, 464
Translation termination, problems with, 483–484
Trichoplusia ni, in eukaryotic expression, 503, 525–526
Trill, John J., 491
Triple helix formation, with Achilles' heel cleavage, 253
Triple helix resins, in plasmid purification, 183
Tritium
  autoradiography film and, 438–439
  as radioactive waste, 158
  shielding for, 163
  signal strength of, 406
  storage phosphor imagers for, 446
Troubleshooting. *See* Problems
Troutman, Trevor, 49, 51
Trypanosoma, as biohazard, 128
Tubes, for centrifuges, 61
Tungsten lamps, in spectrophotometry, 107, 108
2-D gels, 365–366
  pH gradients and, 366–368
2-kilobase DNA fragment, labeling in hybridization experiments, 405
Two-phase extraction systems, for DNA and RNA precipitation, 176
Two-step mRNA (poly(A) RNA) purification, 209
Type A License of Broad Scope, for handling radioisotopes, 143
Tyre, Thomas, 11, 267

Ultracentrifuges, 57
Ultramicrobalances, 51
Ultraviolet-visible (UV/Vis) spectrophotometry. *See* Spectrophotometers
Unexpected results, planning for, 3

Unit definition
for restriction enzymes, 233
for storage phosphor imagers, 444
Upper management, directing complaints to, 28–29
UV (ultraviolet) crosslinking, of nucleic acids, 422
UV lamps
germicidal, 116
maintaining, 107
in spectrophotometers, 103

Vacuum baking, of membranes, 422
Variables, controlling, 7–8
Vented flammables cabinets, storing reagents in, 40
Vertical rotors, for centrifuges, 63, 65
Vibration, in centrifuges, 66
Viral lytic systems, for eukaryotic expression, 503–504
Viral production problems, in baculovirus experiment, 531
Viruses, safe handling of, 126–127
Virus purification, 171
Viscous samples, measuring pH of, 90
Volatile nuclides, hoods for, 163
Volny, William R., Jr., 141
Volume range
of eukaryotic expression, 514–515
of pipettes, 67–68, 71–72
Volumetric flasks
in eukaryotic expression, 514
refrigerating reagents in, 41
Volumetric titration, troubleshooting with, 257

Walking centrifuge, 66
Walsh, Paul R., 225
Wash buffers, for Western blotting, 382
Washing
in hybridization, 434–435
problems with, 395
in Western blotting, 382–383
Washing efficiency, in hybridization, 435
Wash solutions, in hybridization, 434–435
Waste, disposal of radioactive, 158–159
Waste decontamination, 139–140
Water, 42–47
grades of, 42–44
leaks and, 46–47
microbial contamination of, 45–46
organic compounds in, 45–46
pH of, 44
as solvent, 44
Water purification, via reverse osmosis, 43–44
Water purification systems, leaks in, 46–47
Water purity, in acrylamide polymerization, 344
Wave Bioreactor, in eukaryotic expression, 514–515
Wavelength accuracy, of spectrophotometers, 96, 97, 99, 105
Wavelength range, of spectrophotometers, 96
Wavelength reproducibility, of spectrophotometers, 99
Weak acids, as buffers, 33
Weak emitters, autoradiography film and, 438–439
Web sites
Biosci, 13
for polymerase chain reactions, 328–329
Weighing, quantitating nucleotide solutions via, 274, 287
Western blotting, 374–397
amplification in, 387–388
antibodies in, 378
blocking in, 380–382
detection strategies with, 375–380
gene expression and, 477
molecular weight and, 365
primary antibody in, 383–384
proteins and, 374–375
reprobing in, 379, 380, 388–389
secondary reagents in, 384–387
setting up new methods for, 396–397
stained proteins and, 363
stripping in, 379, 380, 383, 388–389
troubleshooting, 389–396
washing in, 382–383
Western blotting troubleshooting logic tree, 390–392
Wet membranes
in hybridization, 432
in nucleic acid transfer, 420–421
Wick junctions, in pH meters, 93
Wipe test, for radioactive shipments, 151–152

World Health Organization (WHO), 115

Xenon lamps, in spectrophotometry, 107
X rays
autoradiography film and, 439–440
in *Bremsstrahlung*, 152

Yeast
disruption of, 217
eukaryotic expression with, 505–506
minimizing degradation of RNA from, 215, 217
total RNA isolation from, 208–209
Yelling, 29
Yields
from fusion systems, 471–472
maximizing eukaryotic expression protein, 529–530

Z factor, of pipettes, 75, 76
"Zippering up," hybridization times and, 426
Zonal centrifugation, 55–56
Zwitterionic detergents, for native PAGE, 355
Zymolase, cell disruption via, 217